T0350994

DESIGN OF MULTIPHASE REACTORS

DESIGN OF MULTIPHASE REACTORS

VISHWAS GOVIND PANGARKAR

Formerly Professor of Chemical Engineering and Head of Chemical
Engineering Department
Institute of Chemical Technology
Mumbai, India

Published by John Wiley & Sons, Inc., Hoboken, New Jersey
Published simultaneously in Canada

For general information on our other products and services or for technical support, please contact our
Customer Care Department within the United States at (800) 762-2974, outside the United States at
(317) 572-3993 or fax (317) 572-4002.

Wiley also publishes its books in a variety of electronic formats. Some content that appears in print may
not be available in electronic formats. For more information about Wiley products, visit our web site at
www.wiley.com.

Library of Congress Cataloging-in-Publication Data:

Pangarkar, Vishwas G.
  Design of multiphase reactors / Vishwas G. Pangarkar.
    pages   cm
  Includes bibliographical references and index.
  ISBN 978-1-118-80756-9 (hardback)
1. Chemical reactors.   I. Title.
  TP157.P256 2014
  660′.2832–dc23
                                        2014020554

Cover image courtesy of Vishwas Govind Pangarkar

Printed in the United States of America

10   9   8   7   6   5   4   3   2   1

Prajá-pataye Swaáhá. Prajá-pataye Idam Na Mama.

This ancient Sanskrit Mantra can be explained as follows:
"I offer the spiritual and material resources used to produce
this work to Prajapaty (the Cosmic creator of Life).
Oh Creator, this is not mine, but Thine."

CONTENTS

FOREWORD

Multiphase reactors are widely used in the chemical industry. The design and scale-up of such reactors is always a difficult task and is not adequately covered in traditional chemical reaction engineering books. The book by Pangarkar is a welcome addition to this field and brings a new perspective of combining the theory with practice.

The book opens with examples of industrial applications and addresses many issues associated with the success of industrial multiphase processes, such as catalyst selection, selectivity, environmental issues. It proceeds then to address the key problem of design and scale-up. The transport–kinetic interaction is vital to understand the design of these reactors. The author illustrates this clearly for various cases of both gas–liquid and gas–liquid–solid systems. Next, reactor selection is discussed in detail including the need for efficient heat management culminating with guidelines for selection of reactors. This is vital in industrial practice as a wrong type of reactor can lead to inefficient or poor process. Fluid dynamic aspects and scale-up based on similarity principles are examined next. Stirred tank reactor design is then examined through detailed analyses of both conventional multiphase chemical reactions and cell culture technology. A second case study deals with venturi loop reactor that is widely used for high-pressure hydrogenations in the fine chemicals sector. The last case study deals with sparged reactors that are of great relevance in gas to liquid fuels in the current context. These case studies are provided in a clear manner with appropriate worked examples and show how the theory can be applied to practice. Overall the book will be valuable for industrial practitioners and will help them to design these reactors in a fundamentally oriented way.

From a teaching point of view, many schools do not offer advanced courses in multiphase reactor design. Further, existing courses also do not cover it adequately. Lack of appropriate textbook may be one reason why this material that is so vital to industrial practice is not effectively covered. By including a number of appropriate case studies, Pangarkar has remedied this and fulfilled an important gap in the current teaching of Chemical Reactor Design. This book could therefore be used for special topic course or in a second course in Chemical Reaction Engineering. I presume that such a course is currently taught in Institute of Chemical technology, Mumbai, formerly known as UDCT. Both Pangarkar and I were fortunate to do our doctoral work in this field in this prestigious school and I have contributed to this field with my earlier coauthored book (with R. V. Chaudhari) *Three Phase Catalytic Reactor*, which has been quite popular with industrial practitioners and academicians. I believe that this book by Pangarkar will be equally popular and I endorse and recommend it to my colleagues in academia and industry.

P. A. RAMACHANDRAN

Professor, Department of Energy, Environmental and Chemical Engineering, Washington University in St. Louis, MO, USA

PREFACE

A practicing chemical engineer invariably comes across one or the other type of a multiphase reactor. The preponderance of multiphase reactors in the chemical process industry has given rise to widespread research directed toward understanding their behavior particularly with a view to develop reliable design and scale-up procedures. The literature is replete with information on various types of multiphase reactors. Additional information is being regularly generated on various aspects of multiphase reactors. The efforts in this direction are using increasingly sophisticated tools like Laser Doppler Anemometry/Velocimetry, computational fluid dynamics, etc. It will be long before the last word is written on design of multiphase reactors. Although this fresh information is welcome, the time and money required for it is disproportionately high, and hence there is a need for simpler, yet theoretically sound, methods to be applied.

Engineering is the bridge that spans the distance between Art and Science. Engineering can convert a seemingly intractable problem into a technically approachable one. Danckwerts (1961) pointed out that the tendency to go more scientific calls for some caution and that "we ought to produce more powerful teaching methods for developing both insight and the qualitative analysis of the problem." The fine balance between Science and Engineering in any text must be maintained since as another doyen of chemical engineering, Thomas Sherwood (1961) noted "if perspective is lost through enthusiasm for scientific and mathematical analyses, an engineer will be less effective in industry." The profound statements of these all-time great chemical engineers are extremely relevant to the present state-of-the art methods for design of multiphase reactors. Astarita (1997) while agreeing with Danckwerts and Sherwood argued that the enthusiasm for computational analyses must be separated from simple creative arguments. This book is an attempt to practice what Albert Einstein suggested **Everything should be made as simple as possible, but not simpler**.

A number of excellent texts dealing with design of multiphase reactors have been published (Shah 1979; Ramachandran and Chaudhari 1983; Doraiswamy and Sharma 1984; Trambouze et al. 1988; Westerterp et al. 1988; Harriott 2003). This book attempts to provide process design procedures for a variety of industrially important multiphase reactors. The basis of the procedures developed is that whereas the intrinsic kinetics of any multiphase reaction do not vary with the type of the reactor used and its scale, the transport parameters (in particular the gas–liquid/solid–liquid mass transfer coefficients) depend on both the type of reactor and its size. The intrinsic kinetics can be determined on small scale under appropriate conditions, but the transport properties need to be specific to the type and size of the reactor. This book therefore focuses on the development of credible correlations for predicting the mass transfer coefficients. It relies extensively on the findings of my research group at the University Department of Chemical Technology (UDCT), Mumbai, in developing these simple procedures. The tradition of industrial consulting at UDCT, an institute established by the very desire of the chemical industry, enabled me to understand the industrial world and its problems. Industrial consulting is a rich source of valuable research tips. It leads to quality research of industrial relevance combined with academic punch. This consulting experience over the last 40 years starting with Late Mr. Chandrakant Khagram (Evergreen Pvt. Ltd., Mumbai) provided me with a perspective that no classroom learning could have substituted. The concepts of "turbulence similarity," "relative suspension," and "relative dispersion" used in Chapters 6, 7A, and 7B have their roots in this industrial interaction. Theory without practice is generally considered "dry." However, experience convinced me that practice without theory can be disastrous.

This book should be of special interest to process design engineers in the chemical, fine chemicals, and allied industries. Chapter 7A uses the concepts of relative suspension/dispersion mentioned earlier for a simple spreadsheet-based design procedure for the highly complex stirred reactor. Chapter 8 is the first comprehensive chemical engineering–based treatment of the venturi loop reactor. This reactor has no serious competition in the fine chemicals industry. Unfortunately, most of the information pertaining to it is proprietary. The spreadsheet-based design provided in Chapter 8 should be of special interest to the fine chemicals sector. Chapter 7B is also probably a similar, first of its type treatment of stirred reactors for cell culture technology, a frontier area in healthcare. As explained in Chapter 3, specific reactor types are best suited to specific applications in the chemical industry. The treatment of various reactors has been arranged into chapters, more or less, in a self-contained manner. There is, however, some inevitable repetition among chapters, which I hope, would only serve to reinforce understanding in the respective context. The commonality in terms of basic design features has, however, not been ignored as evinced, for instance, in Chapters 7A and 7B.

The chemical and allied industry is continuously evolving. Newer molecules/ processes are being developed. In most cases, the time span between discovery and commercial exploitation tends to be very short. Therefore, the process designer has to either quickly do a reliable design of a suitable multiphase reactor or use an available one. Either way, the simplified spreadsheet-based design procedures for the stirred and venturi loop reactor should find favor with the process design engineering fraternity.

The author is grateful to his students: Aditi Deshpande, Bhushan Toley, Biswa Das, Dhananjay Bhatkhande, Keshavan Niranjan, Manoj Kandakure, Niteen Deshmukh, Prasad Pangarkar, Prashant Mandare, Rajendra Prasad, Randheer Yadav, Sameer Bhongade, Sanjay Kamble, Satish Bhutada, Sri Sivakumar, Venkatesh Shirure, Yogesh Borole and colleague Professor Sudhir Sawant for help in literature details, checking of the spreadsheets and overall comments on the flow of information and readability of the book. Professor Sawant's support on a personal level at crucial stages is also gratefully acknowledged. Special thanks are due to Vishwanath Dalvi and Arun Upadhyaya for regressions of published data. The author gratefully acknowledges help from Mr. Manoj Modi (Reliance Industries Ltd., Mumbai, India) during the initial stages of the project on venturi loop reactor. The diligence of Rahul Bagul and timely corrections by Ajay Urgunde in the artwork is highly appreciated.

Finally, my family, who walked the path and endured my long working hours over the past 7 years, deserve all thanks.

The author sincerely hopes that the information given in this text will make the life of the process design engineer easier. The reductionist approach adopted should appeal to students who wish to unravel the complexities of chemical process equipments through simple arguments.

The workbook files are available at the Wiley Book Support site (http://booksupport. wiley.com).

VISHWAS GOVIND PANGARKAR
Nashik, India

REFERENCES

Astarita G. (1997) Dimensional analyses, scaling and orders of magnitude. Chem. Eng. Sci., 52:4681–4698.

Danckwerts PV. (1961) Review of BSL. Endeavour, XX(801):232–235.

Doraiswamy LK, Sharma MM. (1984) Heterogeneous reactions: analysis, examples, and reactor design, Vol. 2: Fluid-fluid-solid reactions, John Wiley Interscience, New York, USA.

Harriott P. (2003) Chemical reactor design. Marcell-Dekker, New York, USA.

Ramachandran PA, Chaudhari RV. (1983) Three phase catalytic reactors. Gordon & Breach Science Publishers Inc., New York, USA.

Shah YT. (1979) Gas-liquid-solid reactor design, McGraw-Hill International Book Company, New York, USA.

Sherwood TK. (1961) Review of BSL. Chem. Eng. Sci., 15(9):332–333.

Trambouze P, Euzen J-P. (2004) Chemical reactor design and operation. Editions Technip, Paris, France.

Westerterp KR, van Swaaij WPM, Beenackers AACM. (1988) Chemical reactor design and operation. 2nd ed. John Wiley & Sons Inc., New York, USA.

1

EVOLUTION OF THE CHEMICAL INDUSTRY AND IMPORTANCE OF MULTIPHASE REACTORS

1.1 EVOLUTION OF CHEMICAL PROCESS INDUSTRIES

Multiphase reactors have been at the cutting edge of technology development in the chemical industry. This premier status of multiphase reactors can be best appreciated in the context of the evolution of the chemical industry itself. It is therefore appropriate to discuss specific aspects relating to the growth and progress of the chemical process industries. This introduction starts with the evolution of the modern chemical industry and discusses the importance of green and sustainable methods and the inevitability of catalysts and multiphase catalytic reactors for carrying out highly efficient catalytic reactions.

The chemical process industries took a long and arduous road of development from modest beginnings through processes such as brewing and distillation; manufacture of soap, sugar, and paper, etc. in small-scale units. Most of the development was based on serendipity and empiricism rather than application of sound chemical engineering principles. In view of the poorly defined methodology, the advancement was slow. The seventeenth and eighteenth centuries witnessed practically no scientific progress that could bring about significant improvements in chemical engineering principles required for rational design.

In the early 1920s, a need was felt to have a unified approach for different disciplines of chemical engineering, and thus, the concept of unit operations (Walker et al. 1923) was introduced. Subsequently, the concept of unit processes (Groggins 1958), which allowed treatment of individual reaction types on a unified basis, was

Design of Multiphase Reactors, First Edition. Vishwas Govind Pangarkar.
© 2015 John Wiley & Sons, Inc. Published 2015 by John Wiley & Sons, Inc.

added. For example, hydrogenations, esterifications, nitrations, etc. were organized on the basis of related thermodynamics, kinetics, and, to a lesser extent, the hardware for each type of process. Groggins showed that unit operations and unit processes are intimately connected through the governing chemical engineering principles. Indeed, from this point onward, chemical engineering has been mathematically defined as ChE = Unit operations + Unit processes.

Extensive research efforts, particularly in the Western world that began in the second half of the twentieth century, laid the foundations of the modern chemical industry. The information generated is, however, still not sufficient for many objectives. For example, *a priori* design procedures for majority of the process equipment are still lacking. Indeed, in many chemical engineering design problems, we find that experience must supplement pure theory. This is an indication that chemical engineering is still largely an art rather than science, where we can evaluate the parameters exactly. For comparison, in electrical engineering, we can precisely calculate the drop in voltage, given the resistance of a conductor and the current it carries. It would be difficult to do the same for the pressure drop in a simple two-phase pipe flow. We need to resort to empirical/semiempirical approaches in a majority of the cases because of lack of sufficient knowledge of the phenomena involved. This example should serve as an indicator of the difference in chemical engineering and other basic engineering disciplines.

In the initial stages, the chemical process industries were restricted to inorganic chemicals (sulfuric acid, nitric acid), sugar, paper, fertilizers, etc. Most hydrocarbons were derived from coal.

From the mid-1950s onward, petroleum crude took over from coal as the principal supplier of hydrocarbons. The major impetus to the organic chemicals sector came from the availability of inexpensive petroleum crude in large quantities. Products derived from petroleum crude had capability to undergo a variety of complex reactions to yield different products that the evolving society required. Chemical engineering became a much more complex profession than in the 1920s. Some basic changes were occurring but were not obvious. The refining and petrochemical industries started producing specialized products with well-defined functions/ properties. Products such as high-octane gasoline and specific lubricants were considered as commodity products notwithstanding their careful formulation that gave the specific desired end result. These formulations also underwent changes brought about by various considerations such as environmental impact. An example is that of replacement of tetraethyl lead in gasoline by more benign antiknock compounds. The chemical process industries were slowly shifting from the commodity/bulk chemicals to specialty/functional products (Cussler and Moggridge 2001). This paradigm shift called for more specialists than generalist chemical engineers. According to a 2004 survey (Jenck et al. 2004), "the global chemical industry represents a significant part of world trade and economic activity with 10 million employees and a combined turnover of some USD 1600 billion excluding pharmaceuticals, and at USD 2200 billion including pharmaceuticals, representing 4–5% of world income. It contributes 9% of world trade whilst emitting only 4% of global carbon dioxide." Evidently, the carbon dioxide emissions of the chemical process industries are an insignificant fraction of the total global carbon dioxide emissions. In spite of these highly revealing statistics, most chemical majors are curtailing greenhouse gas emissions as a part of overall sustainability measures (McCoy 2008). Indeed, processes for utilizing this liability (CO_2, generated by other sectors) for value-added products are receiving increasing research attention (Section 1.3.1.1). A recent review by Muller et al. (2014) discusses the thermodynamic feasibility of potential reactions for converting CO_2 to value-added chemicals. This review points out the severe thermodynamic limitations imposed by the low energy level of CO_2. The following conclusion have been derived: Thermodynamically, favorable routes for producing useful chemicals require (i) high-energy reactants such as epoxide that overcome the low energy level of CO_2, (ii) *in situ* hydrogenation of the intermediate, or (iii) formation of at least two water molecules per mole of CO_2. Such efforts are indicative of an industry that is conscious of its societal responsibility despite the fact that it bears a very small burden of the CO_2 generated.

The chemical process industries have been at the receiving end of the regulatory authorities not because of their greenhouse gas emissions but due to their toxic emissions. Both of these prompted a drive for sustainable processing. The sustainability aspect needs to be dealt with in some detail. The rapid expansion of the petroleum refining and petrochemicals industry through a *laissez-aller* approach resulted in unbridled consumption of vital resources with simultaneous generation of hazardous waste. The development of the industry was disorganized with little attention being paid to the damage caused to the ecosystem. Over the past five to six decades, release of toxic wastes in all forms (solid/liquid/gaseous) has caused serious damage to the ecosystem. Rachel Carson's 1962 book *Silent Spring* was the first recorded warning of the catastrophic nature of the rapid, unsavory expansion. Rachel Carson argued that man is not above nature but is an integral part of it and hence must ensure peaceful coexistence with all species involved. The chemical and allied industries were the main culprits in the eyes of the society that suffered episodes such as the "Love Canal" and "Bhopal" tragedies. As a result of the severe criticism, the chemical industry is now closely looking at safety, health, and environment (SHE) issues while developing a new process or designing a new plant. However, the SHE aspects as practiced are related to decisions that are essentially of short-term nature to avoid contingencies typified by "firefighting" situations. In the 1980s, a more mature approach "sustainable development" was advocated as a long-term objective. *Our Common Future* published by the World Commission on Environment and Development defined sustainable development as "Development that meets the needs of the present without compromising the ability of future generations to meet their own needs." This definition abhors senseless consumption and waste creation. A complementary definition of sustainable development given in 1991 in *Caring for the Earth: A Strategy for Sustainable Living* by IUCN, UNEP, and WWF was "Improving the quality of human life while living within the carrying capacity of the supporting eco-system." The 2002 World Summit on Sustainable Development expanded the previous definitions, identifying "three overarching objectives of Sustainable Development as (1) eradicating poverty, (2) protecting natural resources, and (3) changing unsustainable production and consumption patterns." Sustainability has been at the core of human philosophy as evinced by the following excerpt from ancient Vedic literature:

"Whatever I dig from thee, O Earth, may that have quick growth again.
O Purifier; may we not injure thy vitals or thy heart (Atharva Veda)"

This quote should be a clear message to all engaged in various industries and in particular the chemical process industries. Sustainability was the theme of 21st International Symposium on Chemical Reaction Engineering (LaMarca et al. 2010).

1.2 SUSTAINABLE AND GREEN PROCESSING REQUIREMENTS IN THE MODERN CHEMICAL INDUSTRY

The reader may wonder about the need for all this philosophy and its relation to the design and scale-up of multiphase reactors. The origin, advent, and ubiquity of multiphase reactors lie in the quest for curtailing wastage of resources through highly selective alternatives to less selective processes practiced earlier. The statement of Sir John Cornforth (cf. Adzima and Bowman 2012) sums up the final goal of chemical synthesis on large scale:

The ideal chemical process is that which a one-armed operator can perform by pouring the reactants into a bath tub and collecting pure product from the drain hole.

Chemical engineers and chemists always envied the highly selective synthesis of complex molecules in nature. The specificity of enzymatic reactions was also known for a long time. The brewing industry that performed a simple sugar to alcohol reaction was the first known use of enzymes to provide a product of great joy to many. The chance discovery of penicillin triggered the quest for antibiotics to treat various diseases. The subsequent specific varieties of antibiotics were refinements brought about by the application of life sciences/synthetic chemistry. The specificity of enzymatic reactions created enormous interest, particularly for complex molecules, which could only be made with great difficulty and low selectivity by synthetic organic chemistry. This second paradigm shift arose out of the industry's aim of making complex products from inexpensive raw materials at selectivity levels achievable only in nature. The advantages of biotechnology brought it to the forefront. Consequently, chemical majors such as DuPont, DSM, etc. made major investments in life sciences divisions. In the recent past, there is a growing interest in biological therapeutics and stem cell therapy (subject of Chapter 7B).

Selectivity is important across the broad spectrum of the chemical process industry. Therefore, there is a continuous quest for more selective syntheses. The most prominent approach adopted by the team of chemical engineers and chemists is to develop highly selective catalysts (Section 1.3). The ultimate aim would be complete selectivity for conversion of a given reactant to the desired product with minimum energy requirement and without any hazard.

This approach called "green chemistry" was first enunciated by Anastas and Warner (1998) in their book *Green Chemistry: Theory and Practice*. The American Chemical Society's Green Chemistry Institute has identified the following 12 distinguishing features of green chemistry:

1. **Prevention**

   It is better to prevent waste generation than to treat or clean up waste after it has been created. The earlier messy Bechamp reductions using Fe–HCl and generating a highly acidic waste sludge are replaced by clean catalytic processes using suitably designed multiphase reactors (Section 1.3 and Chapters 7A and 8).

2. **Atom Economy**

   Synthetic methods should be designed to maximize the incorporation of all materials used in the process into the final product. Special types of processes that yield high selectivity are being introduced.

3. **Less Hazardous Chemical Syntheses**

   Wherever practicable, synthetic methods should be designed to use and generate substances that possess little or no toxicity to human health and the environment.

4. **Designing Safer Chemicals**

   Chemical products should be designed to perform their desired function while minimizing their deleterious effects. Product design has emerged as a new discipline.

5. **Safer Solvents and Auxiliaries**

   The use of auxiliary substances (such as solvents and separation agents) should be made unnecessary wherever possible. When unavoidable, the auxiliary substances should be safe to handle. Water-based systems are desirable because of the absence of toxicity and low (practically no) cost.

6. **Design for Energy Efficiency**

   Energy requirements of chemical processes should be an important aspect in view of their environmental and economic impacts. Minimization of energy requirement should be a perpetual exercise. If possible, synthetic methods should be conducted at ambient temperature and pressure. Section 1.3.1.1 discusses some recent trends for achieving this objective. All worked reactor design examples in this book (Sections 7A.10, 8.13, and 9.5) include estimation of energy requirement to compare the energy efficiency of different reactors.

7. **Use of Renewable Feedstocks**

   A raw material or feedstock should be renewable rather than depleting, wherever technically and economically practicable. Energy from biomass is a very important contemporary research activity. First-generation fuel (ethanol) technologies were based on plant sugars and starch. In second-generation biofuels, lignocellulosic species, considered to be much less expensive, were contemplated, while the third-generation biofuels comprise fuel components derived from algae. There are conflicting reports/conclusions for each type. Hammerschlag's (2006) analysis of literature data from 1990 on ethanol's return on investment indicated that corn-based ethanol can substantially reduce petroleum crude oil consumption when used to replace gasoline. This study further showed that cellulosic ethanol is still better in terms of reducing oil consumption. A later report (Pala 2010) also supported the energy efficiency of the biomass (grass) to ethanol route. However, in terms of cultivation, corn is advantageous because it has multiple uses against limited uses for grass. The idiom "There is no free lunch in science or nature" applies and, indeed, dictates the economics. In a rare paper, the proponents and opponents of bioethanol joined to analyze the discrepancies in their individual assessments (Hall et al. 2011). The conclusion was that the practicality of a biomass to fuel (ethanol, in this case) must be ascertained on the basis of energy return on investment (EROI) with the best data available. In another recent critical analysis, Bahnholzer and Jones (2013) have shown that biomass to chemicals route is ridden with several problems. The shelving of the USD 300 million worth biomass to ethanol project of BP has been cited as a commercially significant example. Estimates of Bahnholzer and Jones from the data of Salehi et al. (2013) indicate that even the gas to liquid (GTL) conversion is inefficient since it consumes 1.8 MJ energy from natural gas for producing 1 MJ in terms of the liquid fuel. However, GTL is relatively attractive because it converts a low-mass and low-energy density raw material to a higher density type, thereby improving the quality of the fuel. GTL has substantial commercial logic when there is a specific demand for liquid fuel, particularly when the raw material is "stranded gas." Further, liquid fuel is easier to transport, and therefore, GTL has some justification. An example of such positive upgradation cited by Bahnholzer and Jones (2013) is conversion of fossil fuels to electricity. This conversion also has an inherent inefficiency. However, the final energy product is far superior to the fossil fuel employed as the raw material. The latter is used mainly in utility and to an extent directly in transport. Electricity, on the other hand, has multiple applications besides being more easily transported as compared to a fossil fuel. In this regard, Bahnholzer and Jones suggested that EROI may not be a vital factor in evaluating an improved energy delivery service as compared to simple conversions such as biomass to fuel. Overall, for conversion of low-mass and low-energy density substances such as biomass to low-value chemicals, this approach has very little justification. The obvious reason is consumption of excessively high energy (very highly carbon negative) than what is available from the product of the process. The arguments made by Bahnholzer and Jones are compelling and cast a serious doubt on the biomass to chemical (ethanol)/algal routes. The situation could, however, be different for biomass-based specialty (nutraceuticals)/functional chemicals where the value addition and economic incentive is greater. The use of valuable multipurpose electricity for low-value products has also been heavily criticized by Bahnholzer and Jones (2013). They argued that "It simply is impractical to make fuels (and also low value chemicals) from electricity because the energy losses during the conversion are simply too large."

8. **Reduce Derivatives**

   Unnecessary derivatization (use of blocking groups, protection–deprotection, and temporary modification of physical–chemical processes) should be minimized or avoided if possible, because such steps require additional reagents and can generate waste. This also implies reduction of multistep syntheses to much lesser, preferably a single step. Examples of this strategy pertaining to selective aqueous-phase photocatalytic oxidations of hydrocarbons are discussed in Section 1.3.1.1 (Izumi et al. 1980; Yoshida et al. 1997; Gonzalez et al. 1999; Du et al. 2006).

9. **Catalysis**

   Catalytic processes are superior to processes that use stoichiometric reagents (Bechamp reduction cited earlier). There is a continuous improvement in catalysts. Many acid-catalyzed reactions (esterifications, hydrations, alkylations) have been replaced by benign ion exchange resins in the H^+ form (Wasker and Pangarkar 1992, 1993). The newer catalysts are not only more selective but also more active in enhancing reaction rates. In some cases, for instance, replacement of mineral acids by ion exchangers allows the use of low-cost material of construction and facile separation/recycle of the catalyst while simultaneously eliminating ultimate neutralization of the product by a base.

10. **Design for Degradation**

    Chemical products should be designed so that at the end of their function, they break down into innocuous products and do not persist in the environment.

11. **Real-Time Analysis for Pollution Prevention**

    Analytical methodologies need to be further developed to allow for real-time, in-process monitoring and control to avoid formation of hazardous substances.

12. **Inherently Safer Chemistry for Accident Prevention**

    Substances used in a chemical process should be chosen to minimize the potential for accidents, including release of toxic substances, explosions, and fires. Examples of such safe chemistries and the corresponding multiphase reactors available are discussed later in Sections 1.3.1.1 and 1.6. The possibility of a runaway reaction can be eliminated through the use of microstructured multiphase reactors, which afford order of magnitude of higher heat transfer coefficients as compared to the conventional reactor/heat exchanger combination. Section 1.3.1.1 describes "direct" introduction of a specific functional groups as an alternative to their "indirect" introduction.

Source: American Chemical Society's Green Chemistry Institute (cf. Ritter 2008).

In the foregoing, only points that are relevant to multiphase reactions/reactors are dealt. Points 1, 2, 6, 9, and 12 are synonymous, implying (directly or indirectly) high selectivity of conversion of a given reactant to the desired product. The limit of selective processes is the acid–base (electron transfer) reaction. This reaction goes not only to completion but is also instantaneous (Section 1.7). The same cannot be envisaged even of combustion reactions. These have a highly favorable free energy change, yet these reactions are never complete. If they were, we would not have the problem of presence of carbon monoxide in, for example, auto exhausts. A highly selective process also implies no or very little wastage of the reactant through formation of undesired products. When these undesired products have no utility, they have to be discarded. Further, when these species are toxic and are disposed without any treatment, serious ecological problems arise. Risking repetition, a highly selective process is also the least polluting or requires least treatment prior to disposal of the undesired by product.

The earlier discussion clearly points out the direction in which the modern chemical process industry is moving. Katzer (2010) has articulated the case for sustainable research, development, and demonstration (RD&D). As argued by Katzer, sustainable development needs to be an organized effort, combining skills and expertise of various disciplines. An excellent example is that of solar energy. Meaningful development in economically viable solar electricity generation will need combined efforts of material scientists, architects, and government agencies so that individual households and commercial premises can have sustainable supply that will always be carbon negative. Solar energy to electricity is the ultimate upgradation of energy in the sense of Bahnholzer and Jones. Even considering the energy used in making the solar energy conversion/storage systems, this approach should be a winner. There is an urgent need for increasing the conversion efficiency and reduction of capital costs through innovation in this field (Johnson 2013; Jacoby 2014).

There are several instances of innovative ideas originating from nonscientists. These will again need to be taken up and developed for the benefit of the society. Katzer has given examples of "success and failures" across a broad spectrum of scientific activities. In the field of chemical processing, the overriding considerations that must be employed in future plans involve sustainable processing, including green chemistry and engineering. An ideal reaction would have to be highly selective, nonhazardous, nonpolluting, and self-sufficient in terms of energy requirement. This is certainly a tall order. Current research activity both in terms of the process (Wiederkehr 1988; Choudary et al. 2000; Kolb et al. 2001; Chaudhari and Mills 2004; Jenck et al. 2004; Chen et al. 2005; Andrews et al. 2009, 2010; Buchholz 2010; Adzima and Bowman 2012) and hardware, in terms of multiphase reactors (Dudukovic et al. 2002; Stitt 2002; Boger et al. 2004), is directed at achieving at least some of the aforementioned objectives. What needs to be done at a much higher level can be illustrated by the excellent example of nitrogenous fertilizers based on ammonia. The latter is obtained industrially by the Haber–Bosch process. Andrews (1977) has shown that these fertilizers are thermodynamically extremely inefficient in providing nitrogen to the plants. Ammonia is made by an energy-intensive process that abstracts precious hydrogen from carbon–hydrogen feedstock. Eventually, prior to the final step of uptake of the nitrogen by the plants, this expensive hydrogen is almost totally wasted through oxidation to water by nitrate-forming bacteria. In turn, the plants have to spend more photosynthesis-derived energy to reduce the nitrate back to ammonia. Plant biochemistry does not allow swift absorption of large quantities of ammonia. The most efficient route would be through fixing of nitrogen as nitrates. However, commercially viable processes for this route do not exist (Andrews 1977). As a result, a very large amount of hydrocarbon feedstock is being consumed for the production of urea through ammonia by the Haber–Bosch process. There is an urgent need for INVENTING alternate sustainable methods for this very important activity that provides food to the ever-growing world population. As indicated earlier, the other area that requires immediate attention is conversion of solar energy to electricity. This field has witnessed innovation, but if the world is to meet its increasing energy needs, innovation may not be sufficient. There is a clear need for INVENTION in this field so that a quantum jump in efficiency with corresponding lower costs and longer life can be achieved.

1.3 CATALYSIS

A report of the American Chemical Society (1996) indicates that more than 90% of industrial chemicals are produced by catalytic processes. At the heart of a highly selective process is the catalyst. The term "catalyst" was first introduced by J. J. Berzelius in 1835. A catalyst is a material that promotes a given reaction without undergoing any change in itself. The catalyst only enhances the rate of the reaction without affecting any of the related thermodynamic phenomena. As compared to the noncatalytic path, the catalytic path offers a low-energy barrier route for the reactant to undergo the specified reaction. There are nearly 30,000 chemicals produced worldwide and production of most of the processes involves catalysis in at least some of the steps (Weissermel and Arpe 2003). In the beginning, the fine chemicals sector, in particular, used routes employing stoichiometric reagents (chromic acid/permanganate for oxidation, borohydrides, Bechamp reaction for reduction, etc.). These processes not only raised safety and health concerns but also created a waste stream containing hazardous inorganic salts. Current specifications on all chemicals for human consumption do not allow use of any heavy metal intermediates or even heavy metals as materials of construction in a process. Therefore, the use of the earlier routes has been completely phased out. These have been replaced by benign, selective, and eco-friendly multiphase catalytic routes. However, in the specialty sector involving small volume production of high-value products, some processes may still use less benign methods.

The Haber–Bosch catalytic process for production of ammonia is perhaps an invention that had the most dramatic impact on the human race (Ritter 2008). The inexpensive iron-based catalyst for ammonia synthesis, which replaced the original, more expensive osmium and uranium catalysts, made it possible to produce ammonia in a substantially effective manner. The objective here was not improvement in selectivity but higher reaction rates for rapid approach to the equilibrium conversion at the specified temperature and pressure. Higher rates meant lower catalyst volume and smaller high-pressure reactors. The iron catalyst was improved by addition of several promoters such as alkali metals. In contrast to this simple single reaction case of ammonia synthesis, most organic reactions are complex with multiple pathways.

Armor (2011) has given a brief but excellent account of the history of modern catalysis. The following major stages/product lines in which catalysts play an important role have been identified: (i) basic chemical industry consisting of petroleum refining and petrochemicals. Houdry's fluid catalytic cracking catalyst was developed around 1930. This was the harbinger of the burgeoning petrochemical industry that fed many other sectors such as polymers, pharmaceuticals, agrochemicals, and other specialty chemicals. Parallel to the petrochemicals from petroleum crude, pre-World War II Germany had developed the coal to chemicals route. This was later abandoned in the Western world. However, the same flourished in crude oil-starved South Africa, (ii) transportation fuels sector that later became an integral part of modern refineries. Development of catalysts for naphtha reforming, alkylation, isomerization, etc. was the first phase in this stage. With increasing demands on cleaner fuels, hydro-desulfurization catalysts were developed. Later with increasing crude prices, catalytic hydrocracking was added in the "bottom of the barrel" approach to crude utilization; (iii) polymers obtained from catalytic processes appeared on the scene with introduction of nylon by DuPont in the 1930s. A large variety of polymers with uses ranging from industrial to household to apparel were added to this list; (iv) beginning with 1950, a host of specialty and fine chemicals were made by catalytic routes. These included active pharmaceutical ingredients, agrochemicals, synthetic dyes, surface coatings (paints), fragrance and flavor chemicals, etc.; (v) automobile emission control catalysts were introduced in the later half of the 1970s as a sequel to the increasing concerns about NO_x, diesel soot emissions, etc.; and (vi) increase in petroleum crude prices brought into focus biodiesel obtained from catalytic transesterification of vegetable oils. Currently, better catalysts are also being developed for coal gasification and Fischer–Tropsch synthesis to yield a variety of products from coal. The latter is considered as a substitute for petroleum crude (Levenspiel 2005). Hydrogenation is an important process addressed in this book. Chen et al. (2005) have reviewed the literature on hydrogenation catalysts for fine and intermediate chemicals. With increasing demands on high selectivity combined with clean processes, the quest for appropriate catalysts intensified. Very few reactions such as the nitro to amino in the presence of abundant hydrogen supply and a good hydrogenation catalyst have high selectivity, such as those desired in industrial processes. An example is the extensively studied catalytic epoxidation of ethylene to ethylene oxide. This process is used worldwide on a very large scale. The silver-based catalyst with several additives can achieve selectivity very close to ~86% predicted by one of the mechanisms proposed. Yet, the quest for a better catalyst combination for this reaction continues. A catalyst that gives even 1% higher selectivity can reap rich dividends. This reaction is a classical example of application of catalysis science. With plant scales in the range of several hundred thousand tonnes per year, the stakes are very high. The margin for error is very small.

Fischer–Tropsch synthesis is another industrially important case where the quest for a catalyst with higher rate as well as selectivity continues. This synthesis is exothermic, and catalysts with higher activity (higher rates) will impose a burden on the heat exchanging capacity of the multiphase reactor used. Development of better catalysts must be accompanied by multiphase reactors that can cater to the higher exotherm associated with faster rates. Section 3.4.1.4 discusses the various available reactor options.

Catalysis is generally classified into two types depending on the physical nature of the catalyst employed: (1) heterogeneous, in which the catalyst is immiscible with the reaction medium and is present as a separate phase, and (2) homogeneous, in which the catalyst is soluble in the reaction medium. Multiphase reactors are used in both categories.

1.3.1   Heterogeneous Catalysis

Heterogeneous catalysis is the most preponderant type in industrial applications. Some of the common features of heterogeneous catalysis are (i) relatively severe temperature and pressure conditions, (ii) applicable for both batch and continuous modes of operation, (iii) relatively long catalyst life if poisons are eliminated, (iv) facile separation of the catalyst after completion of the reaction, etc. Therefore, almost all important areas, such as processing of petroleum crude (e.g., catalytic reforming, catalytic cracking, hydrocracking, hydro-desulfurization, etc.) and other processes for bulk chemicals such as ammonia and Fischer–Tropsch synthesis, manufacture of sulfuric acid, etc. use heterogeneous catalysts. This long list extends to fuels (for transportation, energy, etc.) and generation of building blocks (for further consumption by pharmaceutical and fine chemicals industries) from petroleum crude oil. For more details, the reader is referred to Cybulski et al. (2001), Weissermel and Arpe (2003), and Moulijn et al. (2013).

1.3.1.1   Selective Photocatalysis: A Paradigm Shift in Synthetic Chemistry
Photocatalysis has been widely investigated for degradation of refractory organic compounds as an advanced oxidation process (Bhatkhande et al. 2002, 2003, 2004; Kamble et al. 2003, 2006; Pujara et al. 2007). The major advantage of this approach is complete mineralization (complete reduction of chemical oxygen demand, COD) of refractory pollutants at ambient conditions. Semiconductor materials have been used as the photocatalyst. In majority of these investigations, Degussa P25, a 70:30 mixture of anatase and rutile forms of TiO_2, has been used. A typical photocatalytic mineralization reaction is described by the following stoichiometry:

$$C_\alpha H_\beta X_\delta + \left(X + \frac{\beta - \delta}{4} \right) O_2 \xrightarrow{h\nu} \alpha(CO_2) + \left(\frac{\beta - \delta}{2} \right) H_2O + \delta H^+ + \delta X^+$$

The generally accepted mechanism for photocatalytic transformations in aqueous media is the attack of OH^- on the organic moiety. Bhatkhande et al. (2002) have discussed the various mechanisms proposed for photocatalytic pathways. In the case of aromatic compounds, it has been shown that hydroxy aromatic compounds are formed through the mediation of OH^-. It has also been shown that a maximum of three hydroxyl groups can be attached after which the compound becomes highly unstable and decomposes to CO_2 and water. This is evident because no aliphatic compounds are formed. This mechanism can be used to obtain di- and trihydroxy compounds (Brezova et al. 1991; Centi and Misono 1998; Ye and Lu 2013). Other hydroxylated compounds such as *o* (salicylic acid) and *p* (-hydroxy benzoic acids) are also potential candidates. Salicylic acid is the precursor for aspirin, the more than century-old wonder drug manufactured on a very large scale. The conventional process for salicylic acid starting with phenol is relatively time-consuming and generates large quantities of waste streams. Photocatalytic hydroxylation of benzoic acid can be a neat and nonpolluting alternative.

Unfortunately, it is not easy to control the extent of oxidation of the substrate for selective formation of a partially oxidized species or addition of specific number of hydroxyl groups at the desired position in the substrate. Notwithstanding this draw-back, there has been considerable research interest in selective photocatalytic oxidation. Du et al. (2006) have shown that with a proper combination of the incident photon wavelength and photocatalyst, it is possible, for instance, to obtain almost complete selectivity for oxidation of cyclohexane to cyclohexanone. It may be noted that cyclohexanone is manufactured in large quantities as an intermediate for ε-caprolactam. A sizeable quantity of cyclohexanone is obtained by oxidation of cyclohexane, a process that has very bad memories (Flixborough, United Kingdom). The major problem in this conventional route is formation of cyclohexanol, which must be separated and catalytically dehydrogenated to improve the cyclohexanone yield. This latter part of the process consumes substantial quantities of energy. Therefore, there has been a quest for a selective, safe, and hazard-free process for oxidation of cyclohexane to cyclohexanone. The investigation of Du et al. (2006) shows great promise to replace the conventional oxidation route.

There is a surfeit of attractive applications of selective photocatalysis that can be game changers. An example is the elegant innovative approach suggested by Wang et al. (2005) that involves the use of inexpensive ethane and a liability greenhouse gas, CO_2, to obtain value-added products. The proposed reaction is:

$$CO_2 + C_2H_6 \xrightarrow{h\nu} \text{Hydrocarbon oxygenates} \begin{cases} C_2H_5CHO \\ C_2H_5OH \\ CH_3CHO \end{cases}$$

The use of an inexpensive feedstock such as natural gas for producing value-added products is a very interesting concept, particularly in view of large-scale availability of lower alkanes from shale gas in the United States and other new natural gas finds. The current route to formaldehyde through methanol involves relatively high capital cost in production of syngas and its conversion to methanol followed by oxidation to formaldehyde. Direct conversion of alkanes to aldehydes can allow significant reduction in both capital cost and energy consumption. Applications of photocatalysis in converting CO_2 to value-added products such as formaldehyde, formic acid, methanol, etc. is another highly attractive concept that has been exhaustively surveyed by Usubharatana et al. (2006).

Palmisano et al. (2007) have reviewed pertinent literature up to 2007. Fagnoni et al. (2007) have presented an excellent review of the literature on photocatalysis for the formation of the C–C bond. Table 1.1 summarizes some recent investigations besides those covered by Palmisano et al.

TABLE 1.1   Some Important Investigations on Selective Photocatalytic Oxidation

Substrate	Main products	Photocatalyst employed	References
Ethane	Acetaldehyde, ethanol, formaldehyde, CO_2	MoO_3/SiO_2, $ZnO-TiO_2/SiO_2$	Wada et al. (1991); Kobayashi (2001)
CO_2	Formaldehyde, formic acid, methanol, and trace amount of methane	Various photocatalysts reviewed	Usubharatana et al. (2006)
Styrene	Styrene oxide, phenyl acetaldehyde, benzaldehyde	TS-1	Laha and Kumar (2001)
Ethane	Acetaldehyde, ethanol, formaldehyde, CO_2	MoO_3/SiO_2, $ZnO-TiO_2/SiO_2$	Wada et al. (1991); Kobayashi (2001)
Benzene	Maleic acid, hydroquinone, resorcinol, catechol, phenol, CO_2 Very poor selectivity to non-CO_2 products	TiO_2/mica	Shimizu et al. (2002)
Ethane+CO_2	Propanal, ethanol, acetaldehyde	$ZnO-TiO_2/SiO_2$	Wang et al. (2005)
Ethyl benzene	Acetophenone	Degussa P25	Gonzalez et al. (1999)
Cyclohexane	Cyclohexanone (partially cyclohexanol). Total selectivity to cyclohexanone possible through careful control of operating conditions	V_2O_5/Al_2O_3; TiO_2	Giannotti et al. (1983); Mu et al. (1989); Gonzalez et al. (1999); Teramura et al. (2004); Du et al. (2006); Almeida et al. (2010)
Toluene	Benzaldehyde; benzyl alcohol	Degussa P25	Gonzalez et al. (1999)
Methyl cyclohexane	1-, 2-, 3-, and 4-methyl cyclohexanol, 2- and 3-methyl cyclohexanone.	Degussa P25	Gonzalez et al. (1999)
Benzyl alcohol	Benzaldehyde	Degussa P25/Cu (II)	Spasiano et al. (2013)
Nitrobenzene	Amino benzenes and ketones	Ti (IV) without precious metals	Imamura et al. (2013)
Ethanol	Acetaldehyde	V_2O_5 on TiO_2 on commercial ZnS-based phosphor	Ciambelli et al. (2011); Sannino et al. (2013)
Phenol	Catechol, hydroquinone	Specially prepared anatase TiO_2 nanocrystals	Ye and Lu (2013)
Amino alcohols	1,3-Oxazimes	$Ru(bpy)_3Cl_2$	Mathis et al. (2013)
Hydroamination of alkynes	Amines, enamines, and imines	Supported gold nanoparticles	Zhao et al. (2013)

It is evident from this summary that there is an increasing interest in heterogeneous photocatalysis as a green and safe substitute to conventional chemical transformations.

Bhatkhande et al. (2002) have surveyed the literature pertaining to the substitution of OH˙ on the aromatic ring. In most cases, the attack occurs at a position favored by the existing functional group. However, exceptions have been observed depending on various factors such as the source and strength of radiation, catalyst used, etc. For instance, nitrobenzene should favor formation of m-nitrophenol. Contrary to this meta-directing effect in nitrobenzene, Bhatkhande et al. (2003) observed large amounts of p-nitrophenol (almost twice as compared to m-nitrophenol) when concentrated solar radiation was used. On the other hand, with UV radiation ($\lambda_{max} = 253$ and 365 nm), no intermediates were observed, implying that this type of radiation was strong enough to completely degrade the substrate without noticeable formation of intermediates. Evidently, tuning of the oxidation level and selectivity for a desired isomer is possible through a combination of the related parameters (Du et al. 2006).

Photocatalytic reactions occur on the surface of the photocatalyst. Evidently, all participants in the reaction, the oxidant, and substrate should be present on the catalyst surface. Supply of the oxidant is obviously dictated by the process parameters that generate it. These are the type of photocatalyst and the strength and intensity of the incident photons. The substrate has to be supplied to the catalyst from the bulk liquid phase, and therefore, mass transfer factors play an important role. Finally, the affinity of the substrate with the surface of the photocatalyst decides its surface concentration. Therefore, factors that affect adsorption of the substrate on the photocatalyst play a crucial role. These are (Bhatkhande et al. 2002) (i) the nature of the photocatalyst surface, (ii) the substrate itself, (ii) the isoelectric pH of the photocatalyst, (iv) pH of the solution being treated, and (v) presence of anions. Degussa P25 has been used as the model photocatalyst in most investigations. It has been generally found that adsorption is maximum in the vicinity of the zero point charge pH (zpc). Further, ionized substances (organic acids at high pH or organic bases at low pH) suffer from poor adsorption. Therefore, when such substrates are involved, the solution pH must be such that it is near the zpc of the catalyst. Bhatkhande et al. (2002, 2003) and Kamble et al. (2003, 2006) found that anions present in the reaction mixture generally had a detrimental effect on the attack of OH˙. In particular, typical hydroxyl scavengers (Cl^-, HCO_3^-, $CO_3^=$) exhibited strong negative effect on the extent of adsorption. This effect can be advantageously applied for slowing the attack of OH˙ and regulate the selectivity for the intermediate in the photocatalytic attack on the substrate. This is in addition to the various other methods suggested for enhancing the selectivity (Du et al. 2006). In conclusion, it can be argued that it is possible to fine-tune selectivity for a given product by employing conditions that (i) regulate the adsorption of the substrate and desorption of the product desired, (ii) reduce the strength (λ) and dosage (intensity) of the photons, etc. as discussed by Du et al. (2006). Evidently, selective photocatalysis is a highly attractive route that can bring about paradigm changes in the current processes for certain chemical syntheses.

The discussion on this topic will not be complete without a brief discussion on different types of photocatalytic reactors that can be used. A photocatalytic reaction is essentially a heterogeneous reaction involving the substrate, an electron acceptor (oxygen from air in most cases), a solid photocatalyst, and the incident photons. Thus, all parameters that affect the performance of a three-phase (gas–liquid–solid) reactor need consideration while choosing the type of reactor to be used. An additional and probably the most important parameter that governs the rate of a photocatalytic reaction is the photon intensity distribution in the reactor (Pareek 2005; Pareek et al. 2008; Motegh et al. 2013). This additional variable needs to be included in the modeling and design of a photocatalytic reactor. Basically, two types of photocatalytic reactors have been employed: (1) fixed film of photocatalyst (Usubharatana et al. 2006) and (2) suspended photocatalyst. A comparison reveals the advantages and drawbacks of each type. The fixed bed type has the advantage that it does not require separation of the photocatalyst. However, this advantage is more than offset by problems concerning inadequate mass transfer from the liquid to the photocatalyst, lack of photon penetration beyond the top layer of the photocatalyst, and consequent nonutilization of major fraction of the photocatalyst. The suspended or slurry-type photoreactor provides excellent mass transfer and utilization of the photocatalyst. The requirement of photocatalyst is also relatively very low (<1 wt%; typically 0.1 wt%). Kamble et al. (2004) and Pujara et al. (2007) have described slurry reactors for both solar and artificial UV radiation.

Selective photocatalysis requires electrical energy to generate photons. The arguments of Bahnholzer and Jones (2013) against use of electricity for producing commodity chemicals have been discussed earlier. Notwithstanding, such clear exposition of this folly, examples of the same abound. An example is the use of ultrasound for intensification of reactions such as transesterification of vegetable oils, enzymatic hydrolysis of lignocellulose, etc. Efficiency of conversion of primary power to ultrasonic power is relatively very low (maximum 30–40%), and the actual utilization of ultrasound energy for reaction is unlikely to be far different than the above value. Contrary to this, for many examples in Table 1.1, the EROI should be attractive. Natural gas to aldehydes (instead of methanol route), benzoic acid to salicylic/p-hydroxy benzoic acid (instead of phenol-based Kolbe–Schmitt transformation), and cyclohexane to cyclohexanone (instead of conventional oxidation) are some examples for which a simple comparison of existing processes with selective photocatalytic oxidation route should reveal the immense benefits in terms of EROI.

Conclusion: The earlier discussion clearly shows the significant advantages of selective photocatalysis. It is evident that this route requires milder conditions, far less capital investment while yielding high selectivity under proper conditions. Thus, it can bring about radical changes in certain chemical syntheses. It must be emphasized that considerable efforts will be needed to develop the concepts discussed earlier to the level of mature technologies.

1.3.2   Homogeneous Catalysis

Homogeneous catalysis is of a more recent origin as compared to heterogeneous catalysis. Although industrial applications of homogeneous catalysis are much less in comparison to heterogeneous catalysis, it has some significant advantages. Some general features of homogeneous catalysts are as follows:

1. **Advantages**: mild reaction conditions, high activity and selectivity, better mechanistic understanding, etc.
2. **Drawbacks**: difficult and expensive separation of the soluble catalyst from the product.

The advantages of homogeneous catalysis outweigh the drawbacks, and therefore, the interest in homogeneous catalysis is growing. In view of this, some basic information on homogeneous catalysis is presented in the following.

Most of the processes involving homogeneous catalysis (e.g., carbonylation, hydroformylation, oxidation, telomerization, copolymerization, metathesis, etc.) utilize inexpensive feedstocks available from processing of petroleum crude and produce important bulk and fine chemicals for polymer, pharmaceutical, paints, and fertilizer industries (Parshall 1980, 1988a; Masters 1981; Cornils and Herrmann 1996, 1998; Beller and Bolm 1998; Choi et al. 2003; Weissermel and Arpe 2003; Gusevskaya et al. 2014). As mentioned earlier, the separation and recycle of the catalyst is relatively complex. This is achieved through precipitation of the catalyst by addition of nonpolar solvents, high vacuum distillation, extraction of products into a second phase, etc. In spite of this complicated recovery and recycle, about 25% of the industrial catalytic reactions involve homogenous catalysis. The most attractive feature of homogeneous catalysis is better mechanistic understanding of its micro "processes" (catalytic cycles), which allows the possibility of influencing steric and electronic properties of these molecularly defined catalysts. Therefore, homogeneous catalysis is highly promising for innovative chemistry for specialty materials that are otherwise difficult to produce by conventional processes. This is an overriding and clear advantage over heterogeneous catalysis. Understanding of heterogeneous catalysis is still poor despite the advent of highly sophisticated instrumental techniques (Section 1.5). Indeed, heterogeneous catalysis is termed as an alchemist's "black art" because of this poor mechanistic understanding (Schlogl 1993). This is clearly illustrated by a comparison of examples of hydroformylation on one hand and Fischer–Tropsch reaction on the other. While both represent catalytic carbon monoxide chemistry, the molecular structure of the homogeneous catalyst is precisely known to be trigonal-bipyramidal, $d8$-Rh-I, whereas for the Fischer–Tropsch reaction occurring on solid surfaces, no clear molecular level mechanism is known (Herrmann 1982). The progress of homogeneous catalysis is particularly significant and rapid in the area of fine chemicals. The main objective is to achieve high atom efficiency or E factors. Thus, the goal of "green chemistry" as a synonym to environmentally benign chemicals and processes, including sustainable development, is more reliably performed through homogeneous catalysis rather than heterogeneous catalysis (Cornils and Hermann 2003).

The performance of homogeneous catalysts depends not only on the physical conditions of the reaction but also on the type of the metal, ligands, promoters, and cocatalysts. Selectivity is one of the most important issues in a majority of the cases. Generally, transition metal complexes are used as catalysts because of their stability in varying oxidation states. Systematic catalyst characterization with vastly improved techniques has allowed significant progress in the field of coordination chemistry. Consequently, this high level of fundamental understanding has resulted in significantly higher selectivities for homogeneous catalytic processes. Further, it has also allowed use of milder conditions than before. In addition, for an important class of homogeneous-catalyzed reaction such as hydroformylation, the complex problem of catalyst–product separation has been effectively tackled with the development of various techniques reviewed by Cole-Hamilton (2003). A list of some important industrial applications of homogeneous catalytic processes is presented in Table 1.2.

The major defining characteristics for a homogeneous catalyst are (i) activity that is quantified by "turnover number (TON) or frequency" and (ii) selectivity.

TON indicates the number of product molecules that are produced per mole of the catalyst. Turnover frequency is simply the TON expressed in the rate form (i.e., TON per unit time). The difference between activities of homogeneous and heterogeneous catalysts is generally not larger than an order of magnitude when a given reaction can use either of them as a catalyst (Bhaduri and Mukesh 2000). As discussed earlier, in view of the increasing pressure on premium raw materials, there is growing need of catalytic technologies particularly for conversion of inexpensive feedstock to chemicals. These technologies are aimed at providing new routes for fine chemicals, pharmaceuticals and specialties, waste minimization, and conservation of energy. Higher selectivities can be realized with homogeneous catalysis as a result of its better mechanistic understanding. Combined with the use of less expensive feedstocks, such routes have some elements of sustainable processing. Among others, the major industrial products of homogeneous catalysis are products of hydroformylation reaction. Indeed, hydroformylation is acting as a model for other homogeneous catalytic reactions and is being viewed as a convenient way to produce fine chemicals from laboratory scale to industrial scale.

1.4 PARAMETERS CONCERNING CATALYST EFFECTIVENESS IN INDUSTRIAL OPERATIONS

Selectivity and activity of a catalyst has a profound influence on the economics of a commercial process. Selectivity of a reaction can be of different types. These types are explained in Figure 1.1 using the examples of reactions of vinyl acetate monomer.

TABLE 1.2   Industrial Applications of Homogeneous Catalysis

Process	Catalyst	References
Oxidation of ethylene to acetaldehyde	$PdCl_2/CuCl_2$	Jira (1969, 2009)
Oxidation of p-xylene to terephthalic acid/ester	Co/Mn salts + Br⁻	Partenheimer (1995)
Polymerization of ethylene to HDPE/LDPE	Ni complex	Lutz (1986)
Hydrocyanation of butadiene to adipic acid	Ni complex	Ludecke (1976)
Asymmetric hydrogenation of acetamido cinamic acid (3-methoxy-4-acetoxy derivative) (l-dopa process)	Rh (diene) (solvent)]⁺/ DIPAMP	Knowles (1983)
Hydroformylation of propene to butyraldehyde	$NaCo(CO)_4$ $HCo(CO)_3PBu_3$ $HRh(CO)(PPh_3)_3$ Rh/TPPTS	BASF AG (1977a, b) Johnson (1985) Anon (1977) Cornils and Kuntz (1995)
Hydroformylation of higher olefins to oxo alcohols	$HCo(CO)_3PBu_3$	Greene and Meeker (1967)
Hydroformylation of diacetoxybutene to 1-methyl-4-acetoxy butanal (vitamin A intermediate)	$HRh(CO)(PPh_3)_3$ Rh catalyst	Fitton and Moffet (1978) Pommer and Nuerrenbach (1975)
Carbonylation of methanol to acetic acid	Rh/iodide $Co_2(CO)_8$ Ir/iodide	Roth et al. (1971) Hohenshutz et al. (1966) Watson (1988)
Carbonylation of methyl acetate to acetic anhydride	Rh/MeI Rh/MeI	Coover and Hart (1982) Agreda et al. (1992)
Carbonylation of ethylene to propionic acid	$Ni(OCOC_2H_5)_2$	Hohenschutz et al. (1973)
Carbonylation of acetylene to acrylic acid	Ni salts or carbonyls	Blumenberg (1984)
Carbonylation of benzyl chloride to phenyl acetic acid	$Co_2(CO)_8$	Casssar (1985); Parshall and Nugent (1988b)
Carbonylation of 1-(4-isobutylphenyl) ethanol to ibuprofen	$PdCl_2(PPh_3)_2/HCl$	Elango (1990)
Oxidative carbonylation of methanol to dimethyl carbonate	$PdCl_2$–$CuCl_2$	Ugo et al. (1980)
Hydroformylation of ethylene oxide to 2-hydroxy propanal	$Co_2(CO)_8$	Powell et al. (1999)

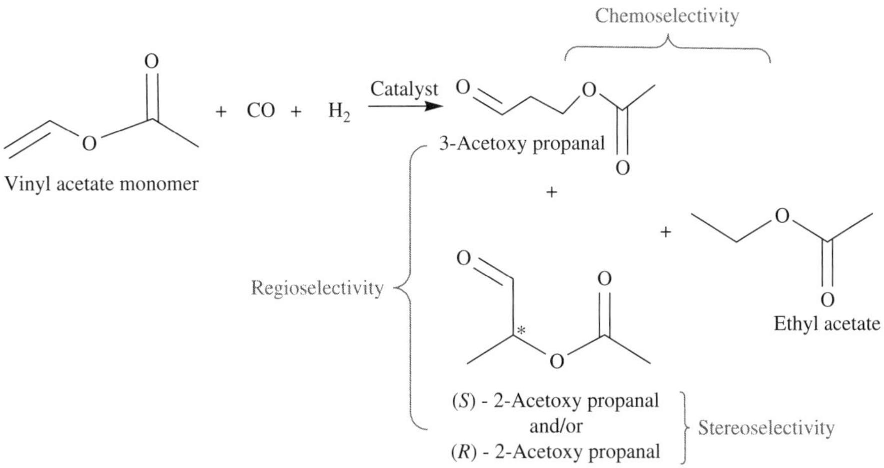

FIGURE 1.1   Examples of various types of selectivity.

1.4.1   Chemoselectivity

This refers to the selectivity based on major chemical transformation and side reaction. For example, hydroformylation of vinyl acetate monomer with synthesis gas can yield two species: (1) aldehydes, acetoxy propanals, and (2) ester, ethyl acetate, as shown in Figure 1.1. In this case, acetoxy propanals are the desired hydroformylation products, whereas ethyl acetate is formed by an undesired hydrogenation of vinyl acetate monomer. The chemoselectivity would therefore be decided by the relative amounts of the acetoxy propanals and ethyl acetate formed.

1.4.2   Regioselectivity

This refers to the selectivity based on the chemical isomers formed due to presence of two or more possibilities capable of bond making or breaking. Again, considering the earlier example of hydroformylation of vinyl acetate monomer, two isomers of acetoxy propanal are possible: (1) 3-acetoxy propanal and (2) 2-acetoxy propanal as shown in Figure 1.1. The 3-acetoxy propanal is obtained by formation of an aldehyde group at the terminal carbon of the vinyl acetate monomer double bond. On the contrary, in the case of 2-acetoxy propanal, the aldehyde group is formed at the internal carbon of the vinyl acetate monomer double bond. The regioselectivity would therefore be decided by the relative amounts of the 3- and 2-acetoxy propanals formed during the reaction.

1.4.3   Stereoselectivity

If one or more of the products has an element of chirality present in them, possibility of formation of products having different chiral structures emerges. The elements of chirality could be attributed to chiral carbon or absence of a plane of symmetry. In the example being discussed, vinyl acetate monomer is a prochiral species for hydroformylation. During hydroformylation, the 2-acetoxy propanal formed has a chiral carbon atom at position 2, indicated by * in Figure 1.1. Therefore, two stereo-isomers, *R* and *S*, 2-acetoxy propanal, are possible. The stereoselectivity would consequently be decided by the relative amounts of the *R*- and *S*-2-acetoxy propanals formed during the asymmetric hydroformylation reaction.

1.5   IMPORTANCE OF ADVANCED INSTRUMENTAL TECHNIQUES IN UNDERSTANDING CATALYTIC PHENOMENA

Advances in instrumental techniques have given a major boost to understanding of catalysis science. Several powerful tools such as high-resolution electron spectroscopy, X-ray photoelectron spectroscopy (XPS), low energy electron diffraction (LEED), secondary ion mass spectroscopy (SIMS), photoelectron spectroscopy, low energy ion scattering (LEIS), etc. have been of great use in understanding the catalyst surface structure (Thune and Niemantsverdriet 2009). An excellent contemporary example of use of these techniques is the work of Gerhard Ertl, which fetched him the Nobel Prize for Chemistry for 2007 (http://nobelprize.org/nobel_prizes/chemistry/laureates/2007/press.html). The citation applauded the groundbreaking efforts "for his studies of chemical processes on solid surfaces." The combination of random selection and evolutionary strategies with high-throughput approaches of combinatorial chemistry has also played a major role in the discovery of new heterogeneous catalysts. Multielement combinations appear to dominate catalyst evolution, particularly in the development of heterogeneous catalysts. The active catalytic species are generally impregnated on a suitable support. In most cases, the support is a material with a well-developed pore structure and high surface area. A major exception to this rule is the famous support for the catalyst used in oxidation of ethylene to ethylene oxide. This support is the antithesis of the general support theory. It has a surface area of ~1 m²/g as against several hundred square meters for other supports. The logic is obvious; the ethylene oxide formed is highly reactive due to the epoxide ring. It must be released quickly from the oxidation site to prevent its further oxidation to carbon dioxide and water. If the catalyst has extensive pores, the product ethylene oxide trapped in these pores is readily oxidized. This logic should also be applied to hydrogen peroxide based epoxidation of allyl chloride to epichlorohydrin using TS-1 catalyst. The latter with an epoxide ring tends to participate in side reactions if not released rapidly from the catalytic site. Some of these side products are very difficult to separate from the main product, epichlorohydrin. The latter has stringent purity requirement as a raw material for epoxy compounds.

The catalyst preparation process is highly specific to generate a catalyst for a specific reaction and also with specific selectivity. A large body of literature describes such procedures/recipes. Control of the nature (crystal type, size) of the active species on the atomic level is considered to be an important factor. Thus, atoms of a given metal can exhibit different behavior since these have different properties (energy levels) at different locations in the crystallite. The multielement catalysts widely used in current practice are much more complicated since atoms of the different metals that serve specific purposes must be at specific locations in a given cluster of the multi-metal structure. In the case of zeolites, the shape and size of pores allow shape selectivity because the active sites in the pores of the zeolites are tailored for the reactant of choice. In view of this, many reactions that involve isomers/reactants of different sizes use specific zeolites to keep undesired reactants out of the scene of the reaction.

1.6   ROLE OF NANOTECHNOLOGY IN CATALYSIS

As mentioned earlier, in the majority of heterogeneous-catalyzed reactions, a catalyst with high surface area is preferred because it presents a very large number of active sites. In addition, the nature of the catalyst and its distribution on the support are equally important. There has been a continuing effort to increase the catalyst surface area and simultaneously eliminate other drawbacks of conventional catalysts such as loss through leaching, sintering, etc. Nanotechnology is gaining increasing importance as a means of achieving the aforementioned goals. Sophisticated instrumental techniques (scanning tunneling microscopy, scanning probe microscopy, atomic force microscopy) have literally opened the gates for the "nano world" (Gerber and Lang 2006). For more details, the reader is advised to refer to Kung (2001) and Filipponi (2007).

1.7   CLICK CHEMISTRY

Conventional chemical synthesis usually employs a combination of several chemical reactions to obtain the desired end product. The overall process is tedious, results in a number of undesired by-products, etc. as discussed in Section 1.2. Adzima and Bowman (2012) have given an excellent and exhaustive review of a new paradigm termed "click chemistry" by Kolb et al. (2001). Sharpless and his group have done pioneering work on this new technique that uses a modular approach consisting of facile and dependable chemical transformations that yield relatively very high atom efficiencies and require only simple separations. These near-perfect reactions occur on a "click" as if they are "spring loaded" (Kolb et al. 2001). Kolb et al. (2001), Kolb and Sharpless (2003), and Adzima and Bowman (2012) list the following broad criteria that define a click reaction:

1. The feedstock comprises of readily available reactants.
2. It obeys the reaction stoichiometry.
3. The quantity of the final product is in precise stoichiometric proportion.
4. It is completely selective, implying no side reactions/by-products. This includes stereospecificity but not essentially enantioselectivity.
5. The main click reaction product is stable being in its lowest free energy state. It compares with the final state of a released spring (Kolb et al. 2001).
6. Product isolation is either not necessary or facile.

7. It occurs under relatively mild/ambient conditions.

8. Akin to the carbonylation reactions in nature, water is the favored solvent.

9. The reaction occurs rapidly although not necessarily as spontaneous as the release of a spring.

10. A highly favorable explicit thermodynamic driving force (>80 kJ/mol) is the key to the speedy completion of the reaction leading to a specific single product.

Click chemistry is likely to be a very attractive complement to combinatorial methods that are widely used in drug discovery and related process development. The latter is very arduous, involving a large number of complex individual chemical reactions to build the desired molecule. These individual reactions lack selectivity, and as a result, (i) the yield is low and (ii) isolation/purification poses serious problems. Indeed, in most cases, the complex separation processes contribute a significant portion to the overall cost. The major advantage of click chemistry is its potential for evolving novel structures that could be different than already identified pharmacophores (Kolb and Sharpless 2003).

Some of the important click reactions are (Kolb and Sharpless 2003; Adzima and Bowman 2012) azide-alkyne; azide-sulfonyl cyanide; benzyne-azide cycloaddition and hetero Diels–Alder reactions; nucleophilic ring opening of strained molecules such as epoxides, aziridines, etc.; non-aldol-type reactions; and additions to carbon–carbon multiple bonds through oxidation-type reactions such as epoxidation, hydroxylation, etc. For more details, the reader is referred to a recent review by Adzima and Bowman (2012).

1.8   ROLE OF MULTIPHASE REACTORS

It has been mentioned earlier that more than 90% of industrial chemicals are produced by catalytic processes. Most of these processes involve multiphase systems. Even those processes that use "homogeneous catalysts" such as hydroformylation, carbonylation, oxidation, etc. involve reactants originally present in two phases (Table 1.2). Three-phase reactions involving gas–liquid–solid (catalyst) contacting, such as hydrogenation, alkylation (solid catalyst), hydro-desulfurization, Fischer–Tropsch synthesis, etc. comprise the most widely used class. Fixed-bed, sparged, and stirred contactors were the workhorses for carrying out these reactions. For the last two, there was relatively poor understanding of the hydrodynamics involved, which reflected in lack of reliable procedures for predicting their performance/transport characteristics. It is needless to say that the hardware in terms of the multiphase contacting device is just as important as the catalyst system. Indeed, this hardware is an integral part of a reliable technology. For instance, new highly active catalysts for the Fischer–Tropsch process will require corresponding reactors that can cater to the increased amount of heat released. Catalytic reactions that involve very high pressures (>5 MPa) also pose serious problems for conventional (e.g., stirred) reactors, and hence, suitable alternatives need to be evolved. Sustained research and development efforts on various aspects of multiphase contacting devices have allowed better understanding of the underlying principles. These in turn have helped in developing relatively better design procedures. In addition, new types of multiphase contactors such as the venturi loop reactor and microstructured devices have also been introduced.

In summary, the progress of the modern chemical industry relies heavily on catalysis for high selectivity/green chemistry and sustainable technology. However, the effectiveness of a catalyst system is pivotally dependent on the availability of appropriate multiphase contacting devices, which are commonly referred to as multiphase reactors. Chapter 3 gives a detailed discussion on the types of multiphase reactors available and the rationale for their selection for a specific reaction system.

REFERENCES

Adzima BJ, Bowman CN. (2012) The emerging role of click reactions in chemical and biological engineering. AICHEJ, 58:2952–2965.

Agreda VH, Pond DM, Zoeller JR. (1992) From coal to acetic anhydride. Chemtech, 22(3):172–181.

Almeida AR, Carneiro JT, Moulijn JA, Mul G. (2010) Improved performance of TiO_2 in the selective photocatalytic oxidation of cyclohexane by increasing the rate of desorption through surface silylation. J. Cat., 273:116–124.

American Chemical Society. (1996) Report: Technology Vision 2020: the chemical industry. Report, December 1996. Available at http://www.chemicalvision2020.org/pdfs/chem_vision.pdf. Accessed on July 29, 2014.

Anastas PT, Warner JC. (1998) Green chemistry: theory and practice. Oxford University Press, Oxford, England and New York, USA.

Andrews SPS. (1977) Modern processes for the production of ammonia, nitric acid and ammonium nitrate. In: Thompson R, editor. The modern inorganic chemicals industry: the proceedings of a symposium organised by the Inorganic Chemicals Group of the Industrial Division of the Chemical Society, London, UK, March 31–April 1, 1977 (Vol. 31). Royal Society of Chemistry, London, UK, p 201–231.

Andrews I, Cui J, DaSilva J, Dudin L, Dunn P, Hayler J, Hinkley J, Kapteijn B, Lorenz K, Mathew S, Rammeloo T, Wang L, Wells A, White T, Zhang W. (2009) Green chemistry highlights: green chemistry articles of interest to the pharmaceutical industry. Org. Proc. Res. Dev., 14:19–29.

Andrews I, Cui J, DaSilva J, Dudin L, Dunn P, Hayler J, Hinkley J, Kapteijn B, Lorenz K, Mathew S, Rammeloo T, Wang L, Wells A, White T, Zhang W. (2010) Green chemistry highlights: green chemistry articles of interest to the pharmaceutical industry. Org. Proc. Res. Dev., 14:770–780.

Anon. (1977) Low-pressure oxo process yields a better product mix. Chem. Eng., 84(26):110–115.

Armor JN. (2011) A history of industrial catalysis. Cat. Today, 163:3–9.

Bahnholzer WF, Jones ME. (2013) Chemical engineers must focus on practical solutions. AICHEJ, 59:2708–2720.

BASF AG. (1977a) n-Butyraldehyde/n-butanol. Hydrocarb. Process., November:135.

BASF AG. (1977b) Higher oxo alcohols. Hydrocarb. Process., November:172.

Beller M, Bolm C, editors. (1998) Transition metals for organic synthesis: building blocks and fine chemicals. Vols. 1 and 2. VCH, Weinheim, Germany.

Bhaduri S, Mukesh D. (2000) Homogeneous catalysis: mechanisms and industrial application. John Wiley & Sons, New York, USA.

Bhatkhande DS, Pangarkar VG, Beenackers AACM. (2002) Photocatalytic degradation for environmental applications: a review. J. Chem. Technol. Biotechnol., 77:102–116.

Bhatkhande DS, Pangarkar VG, Beenackers AACM. (2003) Photocatalytic degradation of nitrobenzene using titanium dioxide and concentrated solar radiation: chemical effects and scale up. Water Res., 37:1223–1230.

Bhatkhande DS, Kamble SP, Sawant SB, Pangarkar VG. (2004) Photocatalytic and photochemical degradation of nitrobenzene using artificial UV radiation. Chem. Eng. J., 102(3):283–290.

Blumenberg B. (1994) Verfahrensentwicklung heute. Nachrichten aus Chemie, Technik und Laboratorium, 42:480–485.

Boger T, Heibel AK, Sorensen CM. (2004) Monolithic catalysts for the chemical industry. Ind. Eng. Chem. Res., 43:4602–4611.

Brezova V, Ceppan M, Brandsteterova E, Breza M, Lapcik L. (1991) Photocatalytic hydroxylation of benzoic acid in aqueous titanium dioxide suspension. J. Photobiol. Photochem., 59:385–391.

Buchholz S. (2010) Future manufacturing approaches in the chemical and pharmaceutical industry. Chem. Eng. Process. (Process Intensification), 49:993–995.

Carson R. (1962) Silent spring. Houghton Mifflin Co., New York, USA.

Cassar L. (1985) Process and product innovation in the chemical industry. Chem. Ind., 67:256–262.

Centi G, Misono M. (1998) New possibilities and opportunities for basic and applied research on selective oxidation by solid catalysts: an overview. Cat. Today, 41:287–296.

Chaudhari RV, Mills PL. (2004) Multiphase catalysis and reaction engineering for emerging pharmaceutical processes. Chem. Eng. Sci., 59:5337–5344.

Chen B, Dingerdissen U, Krauter JGE, Lansink Rotgerink HGJK, Mobus K, Ostgard DJ, Panster P, Riermeier TH, Seebald S, Tacke T, Trauthwein H. (2005) New developments in hydrogenation catalysis particularly in synthesis of fine and intermediate chemicals. App. Cat. A: Gen., 280:17–46.

Choi K-M, Akita T, Mizugaki T, Ebitani K, Kaneda K. (2003) Highly selective oxidation of allylic alcohols catalyzed by monodispersed 8-shell Pd nanoclusters in the presence of molecular oxygen. New J. Chem., 27:324–328.

Choudary BM, Kantam ML, Santhi PL. (2000) New and ecofriendly options for the production of specialty and fine chemicals. Cat. Today, 57:17–32.

Ciambelli P, Sannino D, Palma P, Vaiano V, Mazzei RS. (2011) Intensification of gas-phase photoxidative dehydrogenation of ethanol to acetaldehyde by using phosphors as light carriers. Photochem. Photobiol. Sci., 10:414.

Cole-Hamilton D. (2003) Homogeneous catalysis—new approaches to catalyst separation, recovery, and recycling. Science, 299 (5613):1702–1706.

Coover HW Jr., Hart RC. (1982) Turn in the road: chemicals from coal. Chem. Eng. Prog., 78(4):72–75.

Cornils B, Herrmann WA, editors. (1996) Applied homogeneous catalysis with organometallic compounds. Vols. 1 and 2. VCH, Weinheim, Germany.

Cornils B, Herrmann WA, editors. (1998) Aqueous-phase organometallic catalysis. VCH, Weinheim, Germany.

Cornils B, Herrmann WA. (2003) Concepts in homogeneous catalysis: the industrial view. J. Cat., 216:23–31.

Cornils B, Kuntz E. (1995) Introducing TPPTS and related ligands for industrial biphasic processes. J. Organomet. Chem., 502:177–186.

Cussler EL, Moggridge GD. (2001) Chemical product design. Cambridge University Press, Cambridge, UK.

Cybulski A, Moulijn JA, Sharma MM, Sheldon RA. (2001) Fine chemicals manufacture: technology and engineering. Elsevier Science B.V., Amsterdam, the Netherlands.

Du P, Moulijn JA, Mul G. (2006) Selective photo(catalytic)-oxidation of cyclohexane: effect of wavelength and TiO_2 structure on product yields. J. Cat., 238:342–352.

Dudukovic MP, Larachi F, Mills PL. (2002) Multiphase catalytic reactors: a perspective on current knowledge and future trends. Cat. Revs., 44(1):123–246.

Elango V, Murphy MA, Mott GN, Zey EG, Smith BL, Moss GL. (1990) Method for producing ibuprofen. Europe Patent 400892.

Fagnoni M, Dondi D, Ravelli D, Albini A. (2007) Photocatalysis for the formation of the C–C bond. Chem. Rev., 107:2725–2756.

Filipponi L. (2007) Applications of nanotechnology: catalysis. FP6 Nanocap Report, European Commission under the Sixth framework program. Available at http://www.nanocap.eu/Flex/Site/Downloaddfc2.pdf. Accessed on July 29, 2014.

Fitton P, Moffet H. (1978) Preparation of esters of hydroxy tiglic aldehyde. U.S. Patent 4124619.

Gerber C, Lang HP. (2006) How the doors to the nano world were opened. Nat. Nanotechnol., 1:3–5.

Giannotti C, Le Greneur S, Watts O. (1983) Photo-oxidation of alkanes by metal oxide semiconductors. Tetrahedron Lett., 24:5071–5072.

Gonzalez MA, Howell SG, Sikdar SK. (1999) Photocatalytic selective oxidation of hydrocarbons in the aqueous phase. J. Cat., 183:159–162.

Greene CR, Meeker RE. (1967) Hydroformylation of higher olefins to oxo alcohols, Shell Oil Co., DE-AS 1212953.

Groggins PH. (1958) Unit processes in organic synthesis. McGraw-Hill Kogakusha, Tokyo, Japan.

Gusevskaya EV, Jiménez-Pinto J, Börner A. (2014) Hydroformylation in the realm of scents. Chem. Cat. Chem., 6(2):363–662.

Hall CA, Dale BE, Pimentel D. (2011) Seeking to understand the reasons for different energy return on investment (EROI) estimates for biofuels. Sustainability, 3:2413–2432.

Hammerschlag R. (2006) Ethanol's energy return on investment: a survey of the literature 1990-Present. Environ. Sci. Technol., 40:1744–1750.

Herrmann WA. (1982) Organometallic aspects of the Fischer–Tropsch synthesis. Angew. Chem., Int. Ed. Engl., 21:117–130.

Hohenshultz H, von Kutepow N, Himmele W. (1966) Acetic acid process. Hydrocarb. Process., 45:141–144.

Hohenschutz, H, Franz D, Bulow H, Dinkhauser G. (1973) Method for manufacturing propionic acid (BASF AG), DE 2133349.

Imamura K, Yoshikawa T, Hashimoto K, Kominami H. (2013) Stoichiometric production of amino benzenes and ketones by photocatalytic reduction of nitro benzenes in secondary alcoholic suspension of titanium (IV) oxide under metal-free conditions. Appl. Cat. B: Environ., 134–135:193–197.

Izumi I, Dunn WW, Wilbourn KO, Fan F-RF, Bard AJ. (1980) Heterogeneous photocatalytic oxidation of hydrocarbons on platinized titanium dioxide powders. J. Phys. Chem., 84:3207–3210.

Jacoby M. (2014) Tapping solar power with perovskites. Chem. Eng. News, 98(8):10–15.

Jenck JF, Agterberg F, Droescher MJ. (2004) Products and processes for a sustainable chemical industry: a review of achievements and prospects. Green Chem., 6:544–566.

Jira R. (1969) Manufacture of acetaldehyde directly from ethylene. In Ethylene and its industrial derivatives. In Miller SA, editor. Ernest Benn Ltd., London, UK. p 650–659.

Jira R. (2009) Acetaldehyde from ethylene—a retrospective on the discovery of the Wacker process. Angew. Chem. Int. Ed., 48:9034–9037.

Johnson J. (2013) A new race for solar. Chem. Eng. News, 91(50):9–12.

Johnson TH. (1985) Hydroformylation process. U.S. Patent 4584411.

Kamble SP, Sawant SB, Pangarkar VG. (2003) Batch and continuous photocatalytic degradation of benzenesulfonic acid using concentrated solar radiation. Ind. Eng. Chem. Res., 42:6705–6713.

Kamble SP, Sawant SB, Pangarkar VG. (2004) Novel solar based photocatalytic reactor for degradation of refractory pollutants. AICHEJ, 50:1647–1650.

Kamble SP, Sawant SB, Pangarkar VG. (2006) Photocatalytic degradation of *m*-dinitrobenzene by UV illuminated TiO_2 in slurry photoreactor. J. Chem. Technol. Biotechnol., 81:365–373.

Katzer JR. (2010) Sustainable research, development, and demonstration (RD&D). Ind. Eng. Chem. Res., 49:10154–10158.

Knowles WA. (1983) Asymmetric hydrogenation. Acc. Chem. Res., 16:106–112.

Kobayashi T. (2001) Selective oxidation of light alkanes to aldehydes over silica catalysts supporting mononuclear active sites—acrolein formation from ethane. Cat. Today, 71(1–2):69–76.

Kolb HC, Sharpless KB. (2003) The growing impact of click chemistry on drug discovery. Drug Disc. Today, 8(24):1128–1137.

Kolb HC, Finn MG, Sharpless KB. (2001) Click chemistry: diverse chemical function from a few good reactions. Angew. Chem. Int. Ed., 40:2004–2021.

Kung HH. (2001) Nanotechnology for heterogeneous catalysis: active sites and beyond. Available at http://www.nacatsoc.org/18nam/Orals/104-Kung-Nanotechnology%20for%20 Heterogeneous%20Catalysis.pdf. Accessed on July 29, 2014.

Laha SC, Kumar R. (2001) Selective epoxidation of styrene to styrene oxide over TS-1 using urea–hydrogen peroxide as oxidizing agent. J. Cat., 204:64–70.

Levenspiel O. (2005) What will come after petroleum? Ind. Eng. Chem. Res., 44:5073–5078.

LaMarca C, Sundaresan S, Vlachos D. (2010) Preface: 21st International Symposium on Chemical Reaction Engineering (ISCRE 21). June, 13–16, 2010. Philadelphia, PA, USA.

Ludecke VD. (1976) Encyclopedia of chemical processing and design. In: McKetta JJ, Cunningham WA, editors. Marcel & Decker, New York, USA. p 146.

Lutz EF. (1986) Shell higher olefins process. J. Chem. Educ., 63:202–203.

Masters C. (1981) Homogeneous transition-metal catalysis. Chapman & Hall, London, UK.

Mathis CL, Gist BM, Frederickson CK, Midkiff KM, Marvin CC. (2013) Visible light photo oxidative cyclisation of amino alcohols to 1,3-oxazines. Tetrahedron Letts., 54:2101–2104.

McCoy M. (2008) Converging pathways: chemical companies and environmentalists edge closer together in the pursuit of sustainability. Chem. Eng. News, 86, Aug. 18, 2008.

Moulijn JA, Makkee M, van Diepen AE. (2013) Chemical process technology. 2nd ed. John Wiley & Sons, Chichester, UK.

Motegh M, van Ommen JH, Appel PW, Mudde RF, Kreutzer MT. (2013) Bubbles scatter light, yet that doesn't hurt the performance of bubbly slurry photocatalytic reactors. Chem. Eng. Sci., 100:506–514.

Mu W, Herrmann J-M, Pichat P. (1989) Room temperature photocatalytic oxidation of liquid cyclohexane into cyclohexanone over neat and modified TiO_2. Cat. Lett., 3(1):73–84.

Muller K, Mokrushina L, Arlt W. (2014) Thermodynamic constraints for the utilization of CO_2. Chem. Ing. Technol., 86(4):1–8.

Pala C. (2010) Study finds using food grain to make ethanol is energy-inefficient: processing alfalfa to make ethanol is much more energy-efficient than using corn grain. Environ. Sci. Technol., 44:3648–3653.

Palmisano G, Augugliaro V, Pagliarob M, Palmisano L. (2007) Photocatalysis: a promising route for 21st century organic chemistry. Chem. Commun., 33:3425–3437.

Pareek V. (2005) Light intensity distribution in a dual-lamp photoreactor. Int. J. Chem. Reactor Eng., 3 (Article A56):1–11.

Pareek V, Chong S, Tadé M, Adesina AA. (2008) Light intensity distribution in heterogeneous photocatalytic reactors. Asia-Pacific J. Chem. Eng., 3(2):171–201.

Parshall GW. (1980) Homogeneous catalysis, Wiley-Interscience, New York, USA.

Parshall GW, Nugent WA. (1988a) Making pharmaceuticals via homogeneous catalysis-I. Chemtech, 18(3):18–190.

Parshall GW, Nugent WA. (1988b) Functional chemicals via homogenous catalysis. II. Chemtech, 18(5):314–320.

Partenheimer W. (1995) Methodology and scope of metal/bromide auto-oxidation of hydrocarbons. Cat. Today, 23:69–158.

Pommer H, Nuerrenbach A. (1975) Industrial synthesis of terpene compounds. Pure. Appl. Chem., 43:527–551.

Powell JB, Slaugh LH, Mullin SB, Thomason TB, Weider PR. (1999) Cobalt catalyzed process for preparing 1,3-propanediol from ethylene oxide. Shell Oil Company. U.S. Patent 5981808.

Pujara K, Kamble SP, Pangarkar VG. (2007) Photocatalytic degradation of phenol-4 sulfonic acid using artificial UV radiation. Ind. Eng. Chem. Res., 46:4257–4264.

Ritter SK. (2008, Aug 18) The Haber–Bosch reaction: an early chemical impact on sustainability. Chem. Eng. News, 86:44–45.

Roth JF, Craddock JH, Hershmann A, Paulik FE. (1971, Jan) Low pressure process for acetic acid via carbonylation of methanol. Chemtech, 1:600–605.

Salehi E, Nel W, Save S. (2013, Jan) Viability of GTL for the North American gas market. Hydrocarb. Process., 92(1)41–48.

Sannino D, Vaiano V, Ciambelli P. (2013) Innovative structured VOx/TiO$_2$ photocatalysts supported on phosphors for the selective photocatalytic oxidation of ethanol to acetaldehyde. Cat. Today, 205:159–167.

Schlogl R. (1993) Heterogeneous catalysis—still magic or already science? Angew. Chem. Int. Ed. Engl., 32:381–383.

Shimizu K-I, Kaneko T, Fujishima T, Kodama T, Yoshida H, Kitayama Y. (2002) Selective oxidation of liquid hydrocarbons over photo irradiated TiO$_2$ pillared clays. Appl. Cat. A: General, 225:185–191.

Spasiano D, Rodriguez LDPP, Olleros JC, Malato S, Marottaa R, Andreozzi R. (2013) TiO$_2$/Cu(II) photocatalytic production of benzaldehyde from benzyl alcohol in solar pilot plant reactor. Appl. Cat. B: Environ., 136–137:56–63.

Stitt EH. (2002) Alternative multiphase reactors for fine chemicals: a world beyond stirred tanks? Chem. Eng. J., 90:47–60.

Teramura K, Tanaka T, Kani M, Hosokawa T, Funabiki T. (2004) Selective photo-oxidation of neat cyclohexane in the liquid phase over V$_2$O$_5$/Al$_2$O$_3$. J. Mol. Cat. A: Chem., 208:299–305.

Thune PC, Niemantsverdriet JWH. (2009) Surface science models of industrial catalysts. Surf. Sci., 603:1756–1762.

Ugo R, Tessel R, Mauri MM, Rebaro P. (1980). Synthesis of dimethyl carbonate from methanol, carbon monoxide, and oxygen catalyzed by copper compounds. Ind. Eng. Chem. Prod. Res. Dev., 19(3):396–403.

Usubharatana P, McMartin D, Veawab A, Tontiwachwuthikul P. (2006) Photocatalytic process for CO$_2$ emission reduction from industrial flue gas streams. Ind. Eng. Chem. Res., 45:2558–2568.

Wada K, Yoshida K, Watanabe Y, Suzuki T. (1991) Selective conversion of ethane into acetaldehyde by photo-catalytic oxidation with oxygen over a supported molybdenum catalyst. Appl. Cat., 74:L1–L4.

Walker WH, Lewis WK, McAdams WH. (1923) Principles of chemical engineering. McGraw-Hill, New York, USA.

Wang X-T, Zhong S-H, Xiao X-F. (2005) Photo-catalysis of ethane and carbon dioxide to produce hydrocarbon oxygenates over ZnO–TiO$_2$/SiO$_2$ catalyst. J. Mol. Cat. A: Chem., 229(1–2):87–93.

Watson DJ. (1998) The Cativa™ process for the production of acetic acid. In: Herkes FE, editor. Catalysis of organic reactions. Vol. 70. CRC Press, Boca Raton, FL, USA. p 369–380.

Wasker LB, Pangarkar VG. (1992) Catalysis with ion exchange resins: hydration of citronellol. Ind. Chem. Eng., 34(2–3):99–103.

Wasker LB, Pangarkar VG. (1993) Ion exchange catalyzed cyclisation of citronellal to isopulegol. Ind. Chem. Eng., 35(3):122–128.

Wiederkehr H. (1988) Examples of improvements in the fine chemicals industry. Chem. Eng. Sci., 43(8):1783–1791.

Ye H, Lu S. (2013, July 15) Photocatalytic selective oxidation of phenol in suspensions of titanium dioxide with exposed {0 0 1} facets. Appl. Surf. Sci., 277:94–99. http://dx.doi.org/ 10.1016/j.apsusc.2013.04.008. Accessed on August 6, 2014.

Yoshida H, Tanaka T, Yamamoto M, Yoshida T, Funabiki T, Yoshida S. (1996) Photo oxidation of propene by O_2 over silica and Mg-loaded silica. Chem. Commun., 18:2125–2126.

Yoshida H, Tanaka T, Yamamoto M, Yoshida T, Funabiki T, Yoshida S. (1997) Epoxidation of propene by gaseous oxygen over silica and Mg-loaded silica under photo irradiation. J. Cat., 171:351–357.

Zhao J, Zheng Z, Bottle S, Chou A, Sarina S, Zhu H. (2013) Highly efficient and selective photocatalytic hydroamination of alkynes by supported gold nano particles using visible light at ambient temperature. Chem. Commun., 49:2676–2678.

2

MULTIPHASE REACTORS: THE DESIGN AND SCALE-UP PROBLEM

2.1 INTRODUCTION

The overall process occurring in a multiphase reactor can be divided into a number of steps occurring in series. Figures 2.1 and 2.2 show typical concentration profiles for a reaction involving two (gas–liquid) and three (gas–liquid–solid) phases, respectively. The gaseous solute dissolves in the liquid phase and can react directly with the liquid-phase reactant as in Figure 2.1 or with it as an adsorbed species on a solid catalyst (Fig. 2.2). The various resistances required to be overcome in transferring a gaseous solute to the liquid-phase/solid catalyst surface where the reaction takes place are shown using the film model as the basis. This book is mainly devoted to the chemical, petrochemical, and fine chemicals industries. Majority of the reactions encountered in these industries, for example, hydrogenation and oxidation, involve sparingly soluble gases. For these reactions, the gas film resistance can be neglected since the gas-phase reactant has a very low solubility in the liquid phase. The notable exceptions are chlorination and sulfonation with gaseous SO_3. In view of this, prediction of the mass transfer coefficients is restricted to the gas–liquid ($k_L a$) and solid–liquid mass transfer (K_{SL}) steps in terms of general nature of concentration profiles for the gaseous solute. Concentration profiles for specific cases for the different regimes of mass transfer accompanied by a chemical reaction are covered under the relevant cases later in this chapter.

Since the two diffusional processes and the subsequent reaction occur in series, the step that is the slowest controls the overall rate. As the first step in the design

Design of Multiphase Reactors, First Edition. Vishwas Govind Pangarkar.
© 2015 John Wiley & Sons, Inc. Published 2015 by John Wiley & Sons, Inc.

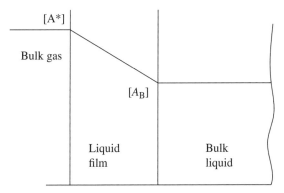

FIGURE 2.1 Typical concentration profiles for a gas–liquid reaction. (Reproduced with permission from Doraiswamy and Sharma (1984). © John Wiley & Sons Inc.)

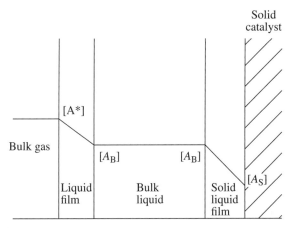

FIGURE 2.2 Typical concentration profiles for solid-catalyzed gas–liquid reaction. (Reproduced with permission from Doraiswamy and Sharma (1984). © John Wiley & Sons Inc.)

procedure, the rate-governing step(s) among those shown in Figures 2.1 and 2.2 needs to be identified. In this respect, two important aspects must be considered: (1) intrinsic kinetics and its invariance with the type/scale of multiphase reactor used and (2) rates of transport processes and their dependence on type/scale of the multiphase reactor chosen.

2.2 THE SCALE-UP CONUNDRUM

The process development starts at the laboratory level. For catalytic reactions, the first step obviously involves screening of various catalysts. Data on selectivity, activity, and life of the catalyst are acquired at this stage, typically approximately on gram level. Commercially available completely automated high-throughput screening techniques, which can simultaneously test a large number of catalysts, are used for this purpose to save time and labor costs (Archibald et al. 2002). Most benchtop experiments are conducted with relatively pure reactants. These studies are followed by testing with the feed composition that is expected in commercial operation. Issues regarding poisoning of the catalyst by species present in the commercial feed composition are also addressed at this stage. Most of the work up to this stage is carried out in glass kettles if the pressure is atmospheric or in laboratory autoclaves for higher pressures. It is important that the laboratory equipment should at least have similar geometry with the prototype that is likely to be used as discussed later in Section 5.2.1. The round bottom flask (RBF) fitted with a Teflon paddle that is popular with chemists should be avoided in this stage. The hydrodynamics (flow pattern) in this RBF is completely different than the fully baffled cylindrical variety used in industrial practice. Clearly, the hydrodynamic dependent transport properties are vastly different in these two equipments. After having identified a catalyst with high selectivity, activity, and good life, the process of scaling up of the bench-scale data starts. Most research and development departments have a multipurpose pilot plant that is used to ascertain the viability of the process on the kilogram scale. There is hardly any process that has not gone through the pilot plant stage. A notable exception is the first atomic bombs that were used in World War II. The reasons were clear (i) the relatively very small time frame available and (ii) the element of secrecy and surprise.

The reactor is the focus of this book and its scale-up is our foremost consideration. Because of the "art" involved in chemical engineering, an intermediate-scale study is required to gain the confidence for investing on the large scale. There is no "high-throughput screening" alternative for this case. The major question that arises is, what is this "intermediate scale" at which a pilot study needs to be carried out? The basic reactor choice has to be made on the basis of the thermodynamic and reaction engineering aspects of the reaction involved as discussed later in Section 3.2. Most process licensors have experience with running commercial plants set up on the technology supplied earlier. Besides this, they also have pilot plants for continuous improvements. For a given process, a number of critical equipments like reactors, separation trains, etc. have to be scaled up. In such cases, data on operating commercial plants always get precedence over theoretical estimates. An example in this author's experience was the packed height of a pump-around section in a fractionator. The mass transfer equipment supplier's correlation predicted a height that was at variance with the process licenser's data sheet by a factor of 6! The mass transfer equipment supplier insisted that they had a well-tested correlation, whereas the process licenser could show a working plant under practically the same conditions with their specified height providing the desired result. In this case, the choice was clear: "the cake had been made and it tasted good!" Hence, no alternate recipe (correlation for calculating the height) could challenge the successfully operating commercial unit.

In the absence of reasonable pilot-scale data, process designers tend to be conservative and substantial factors of safety are provided. At this stage, the dangers of overdesign need to be emphasized. As an example, consider a continuous flow reactor. If a substantially high residence time is considered to be "conservative," it may lead to side reactions and loss of selectivity. Similarly, a highly conservative design of a mass transfer equipment, such as a packed distillation column, is counter-productive. The low liquid velocities resulting from a substantially larger than necessary column diameter result in high height equivalent to a theoretical plate or low separation efficiency. To compensate for this loss of efficiency, a correspond-ingly larger packed height has to be provided. In such a case, the overall capital cost of the unit increases (cost ~ α [total volume]^n with $0.6 < n < 0.8$). On the other hand, if there is no such factor of safety in the height, the desired product quality is not achieved. Therefore, it must be borne in mind that "conservative design" in the chemical industry must be used with great caution!

Scale-up of modular equipment is easiest. Consider the case of a shell-and-tube-type reactor for epoxidation of ethylene. In commercial designs, 0.05 m diameter tubes are packed with the catalyst and surrounded by the heat transfer fluid on the shell side. It is known that the heat transfer resistance on the tube side (packed bed) is the controlling factor. Estimation of the heat transfer coefficient for individual tube requires the effective thermal conductivity of the packed bed. Considerable information is available on this aspect (e.g., Pangarkar et al. 2010), and for a given catalyst type and size, the effective thermal conductivity of tubular packed beds of catalysts is fairly well established. Therefore, temperature variation inside the bed along the packed height can be predicted with reasonable confidence given the reaction kinetics and heat of reaction. The industrial version of this type of the epoxidation reactor can be scaled up relatively easily by increasing the shell diameter and the number of tubes. Contrast this situation with that of an alternate fluidized bed reactor, which may appear attractive because of its far better heat transfer characteristics. This is a two-phase (gas-fluidized solid) system. It is very risky to extrapolate experimental results of a small 0.1 m diameter fluidized bed to a large (>3–4 m diameter) fluidized bed that may be required in the industrial version. This is due to the fact that the behavior of the dense and dilute phase at different scales is extremely difficult to pre-dict. This is clearly reflected in the small scale-up factors for fluidized bed in Table 2.1 (Vogel 2005). The situation for a stirred reactor is practically the same since it is not a modular equipment. Further, the hydrodynamics of a stirred reactor are not under-stood completely. If a single large-diameter stirred reactor has to be designed, most operators would prefer to go through a pilot stage owing to this complexity. However, it will be shown in Chapter 7A that using the concepts of "relative dispersion" and "relative suspension" uncertainty in scale-up can be significantly reduced.

A high scale-up factor in Table 2.1 reflects the ease in scaling up of the reactor. It can be seen that heterogeneous/multiphase reactors have substantially smaller scale-up factors than those for homogeneous systems. This is due to the complex nature of scale-dependent diffusional processes (Section 2.4).

The advent of microstructured devices should be viewed in terms of the greater confidence in going from small scale to the commercial scale as indicated in Table 2.1 (Jahnisch et al. 2004; Kashid and Lioubov 2009). Microstructured devices are modular in nature, and therefore, simple stacking up yields the desired high capacity. The true potential of microstructured devices as reactors will be realized when

TABLE 2.1 Scale-up Factors for Different Multiphase Reactors[a]

Reactor type	Scale-up factor
Shell and tube (packed)	>10,000[b]
Shell and tube (homogeneous reaction)	>10,000[b]
Stirred tank (homogeneous reaction)	>10,000[b,c]
Stirred tank (heterogeneous; two/three phase)	500[d]
Two-/three-phase sparged reactors	<1000[b,e]
Fluidized beds (two/three phase)	50–100[b]
Venturi loop reactor	500[f]; ~200[g]
Microstructured devices as reactors	No scale-up, only scaling out
Stirred bioreactor	10[h]

[a] Adapted from Vogel (2005) with permission from Wiley-VCH Verlag GmbH & Co. KGaA
[b] Vogel (2005).
[c] Not strictly applicable when the reaction is very fast and micromixing plays an important role.
[d] Chapter 7A.
[e] Chapter 10.
[f] Author's experience.
[g] Based on constant power dissipation per unit total volume in the ejector (Dierendonck et al. 1988, 1998).
[h] Votruba and Sobotka (1992), Junker (2004), Garcia-Ochoa and Gomez (2009).

problems with nonuniformity of distribution in multichannel devices that are used in commercial applications are resolved (Rebrov et al. 2010). The relatively very high heat transfer rates obtained in these devices is a significant advantage and should spur further research for development of the requisite hardware. It will not be long before these devices replace many conventional designs.

2.3 INTRINSIC KINETICS: INVARIANCE WITH RESPECT TO TYPE/ SIZE OF MULTIPHASE REACTOR

Kinetics are readily determined in the laboratory under carefully controlled conditions such that the effects of diffusional resistances are entirely eliminated. The intrinsic kinetics so determined yield the rate constant, the orders of the reaction with respect to the reactants, catalyst, and the activation energy. These are unique to a given gas–liquid or gas–liquid–solid catalyst system and do not vary with the type or size of the multiphase reactor. This matter is briefly discussed later in Section 2.5.

2.4 TRANSPORT PROCESSES: DEPENDENCE ON TYPE/SIZE OF MULTIPHASE REACTOR

In direct contrast to intrinsic kinetics, the transport processes (mass/heat transfer coefficient) depend on the type of multiphase reactor, its size, and operating parameters. Thus, one can have an order or two of magnitude changes in the gas–liquid mass transfer coefficient, $k_L a$, when shifting over from packed columns to stirred reactors and finally to the venturi loop reactor. As depicted in Figure 2.2, the mass transfer and reaction processes occur in series. When the rate of gas–liquid or solid–liquid mass transfer increases relative to the intrinsic reaction rate due to a change in the type of multiphase reactor, the controlling resistance may shift from mass transfer-controlled to surface reaction (on the catalyst surface)-controlled regime.

2.5 PREDICTION OF THE RATE-CONTROLLING STEP IN THE INDUSTRIAL REACTOR

The most important step in design of a commercial multiphase reactor is identification of the rate-controlling step on the large scale. Further, it requires a rate expression for this rate-controlling step in terms of known parameters.

Since the intrinsic kinetics are unaffected by scale-up, only the rates of the diffusional processes need to be predicted for the larger scale. The intrinsic kinetic rate is then compared with the rates of these diffusional processes at the specific industrial scale. This comparison allows identification of the controlling step in the large-scale reactor. A relevant expression for the rate-controlling step is then used to estimate the size of the reactor. The scale-up problem thus concerns prediction of the mass transfer coefficients in the industrial version. The mass transfer coefficients are dependent not only on the operating parameters (e.g., mode and extent of energy input) but also on the type of multiphase reactor and its size. There can be no better example of the striking effect of size and operating parameters than three-phase sparged reactors. The hydrodynamic regime in a laboratory scale (<0.1 m diameter) is totally different than a large-diameter industrial column (Fig. 10.1). Predictably, the transport characteristics of the two columns are drastically different as discussed in Chapters 6 and 10. Therefore, process design of multiphase reactors should focus on prediction of the mass transfer rates for desired scale and operating parameters. The prediction method should be simple to use. In addition, if this method is scale independent and also has a sound theoretical basis, it imparts a significantly higher level of confidence. This book deals with different multiphase reactors on individual basis. The focus is on the main problem concerning prediction of mass transfer coefficients. Attempts will be made to provide (as far as possible) scale- and geometry-independent correlations for the mass transfer coefficients. In addition, the endeavor will be to present procedures that use readily available information or require minimum measurements.

2.6 LABORATORY METHODS FOR DISCERNING INTRINSIC KINETICS OF MULTIPHASE REACTIONS

2.6.1 Two-Phase (Gas–Liquid) Reaction

This type of reaction is relatively simple for laboratory investigation. A number of inexpensive model contactors are available for these investigations. These include stirred cell and laminar jet apparatus (Danckwerts 1970). The gradient-less contactor proposed by Levenspiel and Godfrey (1974) is another useful model contactor, particularly when there is a need to agitate the gas phase for eliminating any gas side resistance that is likely to be present.

The model contactors mentioned earlier have a well-defined effective interfacial area, and hence, the true mass transfer rate per unit area can be determined. The theory of mass transfer accompanied by a chemical reaction has been discussed in detail in various texts (Astarita 1967; Danckwerts 1970; Ramachandran and Chaudhari 1983; Doraiswamy and Sharma 1984). Four different regimes of mass transfer accompanied by a chemical reaction are broadly identified. The regimes and the corresponding equations for the specific rate of mass transfer (flux) are summarized in Table 2.2. The classification is relative to the rate of mass transfer in the absence of a chemical reaction as explained briefly in the following. For detailed discussions on these regimes, the aforementioned texts can be referred. These regimes can shift when the model contactor is changed. The availability of the stirred cell ($k_L \sim 10^{-5}$ m/s) and laminar jet apparatus/gradient-less contactor ($k_L \sim 10^{-4}$ to 10^{-3} m/s) allows variation of the mass transfer rates by at least an order of magnitude such that the regime can be shifted by a simple change of the model contactor.

We consider a general type of reaction:

$$A + \psi B \rightarrow products$$

Here, ψ is the stoichiometric factor. We consider the general case where the orders of the reactions are p and q with respect to the gaseous and liquid-phase reactant, respectively. For the sake of simplicity, the film model is chosen for the discussion that follows.

2.6.1.1 Regime A: Extremely Slow Reaction That Occurs Only in the Bulk Liquid Phase

Referring to Figure 2.3a, for this situation, the rate of the reaction is so slow that no reaction occurs in the liquid film. Therefore, $A^* \approx A_B$. Evidently, the reaction occurs exclusively in the bulk liquid phase.

This situation can be explained through the concept of characteristic times for the two steps of diffusion and reaction (Astarita 1967). We define the characteristic diffusion time τ_D and reaction time, τ_R, respectively, for the diffusion across the liquid film and the reaction between the dissolved gaseous species and the liquid-phase reactant. The reciprocals of these characteristic times signify the respective rate coefficients. Thus,

$$\tau_D \propto \left(\frac{1}{k_L a} \right) \tag{2.1}$$

and

$$\tau_R \propto \left(\frac{1}{k_{p-q}[A^*]^p[B_B]^q} \right) \tag{2.2}$$

TABLE 2.2 Regimes of Mass Transfer Accompanied by a Chemical Reaction, Corresponding Conditions, and Rate Expressions for General (p, Q)th-order Reaction[a]

Regime	Conditions to be satisfied	Rate expression	Comment
A	$Ha \ll 1;\ k_L\underline{a} \gg \varepsilon_L k_2 B_B$	$R_{VA} = \varepsilon_L \times k_{pq}(A^*)^p(B_B)^q$ (mol/m^{3}s)	R_{SA} varies with $[A^*]$ and $[B_B]$ but is independent of hydrodynamics (speed of agitation of the stirrer/jet velocity/length)
B	$Ha \ll 1;\ k_L\underline{a} \ll \varepsilon_L k_2 B_B$	$R_{VA} = k_L\underline{a}(A^*)$(mol/m^{3}s^{-1}) ; $R_{SA} = k_L(A^*)$ (mol/m^{2}s)	R_{VA} depends on hydrodynamics and is always first order with respect to $[A^*]$ but independent of $[B_B]$
C	$Ha > 3$ when $\left(\dfrac{B_B}{\psi A^*}\right) \gg Ha$	$R_{SA} = A^*\left(\sqrt{\dfrac{2}{p+1}D_A k_{p-q}[B_B]^q[A^*]^{p-1}}\right)$ (mol/m^{2}s)	R_{SA} is independent of hydrodynamics but varies with $[A^*]$ and $[B_B]$
D	$Ha \gg 1;\ \left(\dfrac{B_B}{\psi A^*}\right) \ll Ha$	$R_{SA} = k_L\left(\dfrac{D_B B_B}{D_A \psi}\right)$ (mol/m^{2}s)	R_{SA} is (i) dependent on hydrodynamics; (ii) independent of $[A^*]$ but (iii) first order with respect to $[B_B]$

[a] Adapted from Doraiswamy and Sharma (1984) with permission from John Wiley & Sons Inc. © John Wiley & Sons Inc.

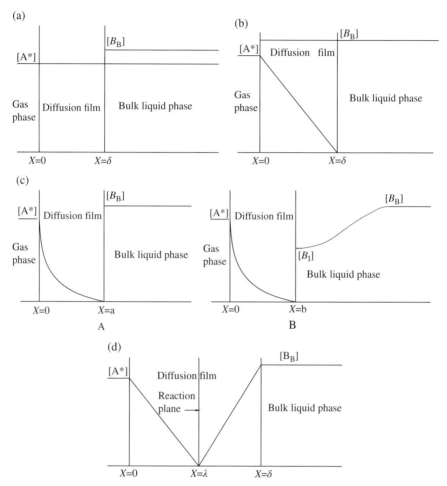

FIGURE 2.3   (a) Concentration profiles for mass transfer accompanied by a slow reaction occurring in the bulk liquid phase. (b) Concentration profiles for mass transfer accompanied by a diffusion-controlled slow reaction. (c) Concentration profiles for a fast gas–liquid reaction occurring in the liquid film. (A) Fast pseudo qth-order reaction with respect to liquid-phase reactant B (B) General $(p - q)$th-order reaction. (d) Concentration profiles for an instantaneous reaction occurring in the liquid film. (Reproduced with permission from Doraiswamy and Sharma (1984). © John Wiley & Sons Inc.)

When the rate of the reaction in the liquid film is very low ($\tau_D \ll \tau_R$), it is implied that the gaseous reactant diffuses across the diffusion film so rapidly that it does not have a chance to take part in the reaction with reactant B. However, once the gas has diffused to the bulk liquid phase, it has a very large time available for reaction with reactant B. For the (p, q) th-order reaction, the volumetric rate of reaction in the bulk liquid is given by

$$R_{VA} = \varepsilon_L \times k_{p-q}(A^*)^p \times (B_B)^q \quad (mol/m^3 s) \tag{2.3}$$

In this case, since the diffusion factor no longer masks the kinetics of the reaction ($\tau_D \ll \tau_R$), experimentally measured values of R_{VB} can be used in conjunction with known values of A^* and B_B to obtain the intrinsic reaction rate constant k_{p-q}. The values of p and q can be obtained by the usual technique of varying one variable at a time (constant B_B at different values of A^*, and vice versa). However, it has to be ensured that $\tau_D \ll \tau_R$.

The aforementioned criterion based on diffusion and reaction times is qualitative. To establish the regime more precisely, the Hatta number is popularly used. For the aforementioned general (p, q)th-order reaction, the Hatta number is given by

$$\text{Ha} = \frac{\left(\sqrt{\dfrac{2}{p+1} D_A k_{p-q} [B_B]^q [A^*]^{p-1}} \right)}{k_L} \tag{2.4}$$

For the sake of simplicity, we take $p=q=1$. This type of second-order reaction is a commonly encountered situation (Bawane and Sawant 2003, 2004). In this case, for regime A:

$$\text{Ha} = \frac{\sqrt{D_A k_2 [B_B]}}{k_L} \ll 1 \tag{2.5}$$

Further, the rate of diffusion across the liquid film is much greater than the rate of the reaction in the bulk liquid. For a second-order reaction, this is signified by

$$k_L \underline{a} \gg \varepsilon_L k_2 B_B \tag{2.6}$$

2.6.1.2 Regime B: Diffusion-Controlled Slow Reaction

In this case also, the reaction is sufficiently slow such that it does not occur in the diffusion film $\left(\text{Ha} = \left(\sqrt{D_A k_2 [B_B]} \right) / k_L \ll 1 \right)$. Thus, only physical diffusion occurs in the liquid film. The concentration profiles for this regime are shown in Figure 2.3b.

However, in contrast to regime A, the homogeneous reaction is significantly rapid as compared to the diffusion process in the film ($\varepsilon_L k_2 B_B \gg k_L \underline{a}$). This ensures almost complete consumption of the amount of A supplied to the bulk liquid phase. Effectively, this implies $A_B \sim 0$. In terms of Hatta number:

Ha $\ll 1$ and also

$$\varepsilon_L k_2 B_B \gg k_L \underline{a} \tag{2.7}$$

The volumetric rate of mass transfer in this case is

$$R_{VA} = k_L a (A^*) (mol/m^3 s) \tag{2.8}$$

The specific rate per unit effective interfacial area or flux of A is given by

$$R_{SA} = k_L (A^*) (mol/m^2 s) \tag{2.9}$$

Equations 2.8 and 2.9 are typical of a physical process in which the rate is always first order with respect to the diffusing species.

In the case of the laminar jet apparatus, the contact time of the jet can be precisely calculated from the jet velocity and its length. This in turn allows precise calculation of k_L using the penetration model:

$$k_L = 2 \left(\sqrt{\frac{D_A}{\pi \times t}} \right) \tag{2.10}$$

Here t is the contact time. The specific interfacial area of the jet is obtained from the jet dimensions $\left(4 \times \pi d_j L / \pi d_j^2 L \right)$. In the case of the stirred cell, such precise estimation of k_L is not possible, and therefore, experimental measurements of k_L using physical absorption in fresh liquid ($A_B \approx 0$) need to be carried out at different speeds of agitation for the gas–liquid system of interest.

2.6.1.3 Regime C: Fast Reaction Occurring in the Diffusion Film
There are situations where both τ_D and τ_R are low and comparable. Under these conditions, both reaction and diffusion occur simultaneously in the diffusion film. This regime is characterized by Hatta number > 3. Concentration profiles for this regime are shown in Figure 2.3c.

For this case, the specific rates of mass transfer accompanied by a general $(p\text{-}q)$ th-order reaction are given by

$$R_{SA} = A^* \left(\sqrt{\frac{2}{p+1} D_A k_{p-q} [B_1]^q [A^*]^{p-1}} \right) (mol/m^2 s) \tag{2.11}$$

When $\left(\dfrac{B_B}{\psi A^*} \right) \gg$ Ha [Fig. 2.3c(A)], the rate of diffusion of B from the bulk liquid is much higher than its consumption in the liquid film. In this case, $B_B^* \sim B_B$ and the corresponding specific rate of mass transfer is given by

$$R_{SA} = A^* \left(\sqrt{\frac{2}{p+1} D_A k_{p-q} [B_B]^q [A^*]^{p-1}} \right) (mol/m^2 s) \tag{2.12}$$

2.6.1.4 *Regime D: Instantaneous Reaction*   In the limit when the value of τ_R is extremely low, it implies an instantaneous reaction. An example of this type of reaction is a strong-acid–strong-base reaction represented by the equation

$$HCl_{(G)} + NaOH_{(L)} \rightarrow NaCl + H_2O \qquad (2.13)$$

Although the overall reaction is given by Equation 2.13, the actual reaction is a simple proton transfer type:

$$H^+ + OH^- \rightarrow H_2O$$

This charge transfer reaction is instantaneous. Examples of this type are absorption of hydrogen chloride/SO_2 in alkaline solutions. This situation may be fraught with a limitation on the supply of B into the diffusion film from the bulk. The latter occurs when

$$\left(\frac{B_B}{\psi A^*} \right) << Ha \qquad (2.14)$$

For such instantaneous reactions, the reactants A and B cannot coexist. They are instantaneously consumed in a very short zone called the reaction plane at a distance λ from the gas–liquid interface decided by the relative rates of diffusion of the reacting species as depicted in Figure 2.3d.

Under these conditions, the specific rate of mass transfer is given by

$$R_{SA} \sim k_L \left(\frac{D_B B_B}{\psi D_A} \right) \; (mol/m^2s) \qquad (2.15)$$

Table 2.2 summarizes the condition(s) to be satisfied and the corresponding expression for the volumetric (R_{VA}) and specific (R_{SA}) rates of mass transfer accompanied by a chemical reaction for the four different regimes discussed earlier.

Figure 2.4 gives a guide for discerning the regime from experimental data from a stirred cell. For the case of the laminar jet apparatus, a similar guide can be constructed from the details available in Doraiswamy and Sharma (1984).

2.6.2 Three-Phase (Gas–Liquid–Solid) Reactions with Solid Phase Acting as Catalyst

The majority of the applications of multiphase reactors in three-phase reactions involve a solid phase as the catalyst. An example of this class of reactions is catalytic hydrogenation that has now completely replaced the highly polluting reductions with iron-HCl. The concentration profiles for this type of systems have been shown in Figure 2.2. From an industrial point of view, the main focus should be on achieving the intrinsic rate afforded by a given catalyst. This is discussed later for the case of Fischer–Tropsch synthesis (Section 3.4.1.5), wherein new types of reactors are being proposed to accommodate new generation of highly active catalysts.

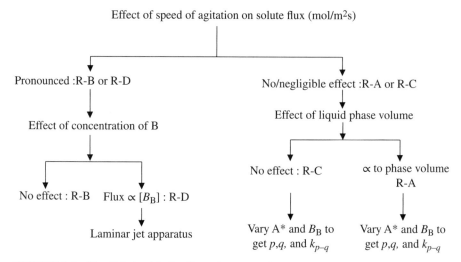

FIGURE 2.4 Simplified guide for discerning the regime of mass transfer accompanied by a chemical reaction from effects of major variables in experiments in a stirred cell. (Adapted from Doraiswamy and Sharma (1984) with permission from John Wiley & Sons Inc. © John Wiley & Sons Inc.)

At equilibrium (steady state), the rates of the various steps depicted in Figure 2.2 should be equal. For a simple first-order reaction, Equation 2.16 gives the rates of the three steps in series:

$$R_{VA} = k_L \underline{a}([A^*] - [A_B]) = \left(K_{SL} \left(\frac{A_P}{V} \right) \right)([A_B] - [A_S]) = k_R[A_S] \qquad (2.16)$$

The surface reaction rate constant may have a finite order with respect to the catalyst loading. It will be reflected through the total catalyst surface area, A_P, K_{SL} is based on catalyst surface area per unit volume of the reaction mixture, A_P/V. This factor is also related to the catalyst loading and surface area of the catalyst. For spherical catalysts, A_P is given by

$$A_P = \left(\frac{6w}{\rho_P \times d_P} \right) \qquad (2.17)$$

The overall rate constant k_R is then given by

$$k_R = k'_R(A_P) \qquad (2.18)$$

Equation 2.16 can be rearranged to yield

$$\left(\frac{[A^*]}{R_{VA}} \right) = \left(\frac{1}{k_L \underline{a}} \right) + \left(\frac{1}{K_{SL} \left(\dfrac{A_P}{V} \right)} \right) + \left(\frac{1}{k_R} \right) \qquad (2.19)$$

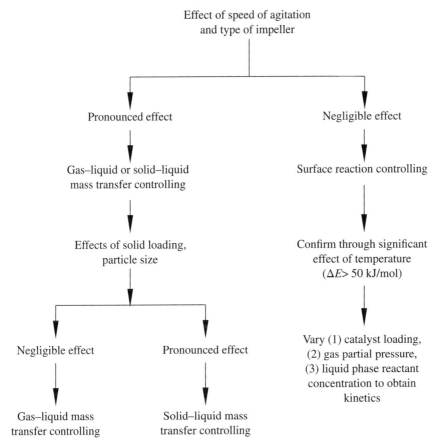

FIGURE 2.5 Simplified guide for elucidating rate-controlling step/kinetics of solid-catalyzed gas–liquid reaction (nonporous catalyst). (Adapted from Doraiswamy and Sharma (1984) with permission from John Wiley & Sons Inc. © John Wiley & Sons Inc.)

The relative values of $(1/k_L \underline{a})$, $(1/K_{SL}(A_p/V))$, and $(1/k_R)$ indicate the importance of the individual steps in series. It is evident that the intrinsic kinetics of the surface reaction can be determined only after the resistances of the gas–liquid and solid–liquid mass transfer steps are eliminated. For this purpose, generally an agitated reactor is used. For reactions involving relatively high-pressure conditions, an autoclave with the desired pressure, temperature rating, and provision for fitting standard impellers is the preferred laboratory equipment. Figure 2.5 shows a simplified guide for elucidating the controlling step and finally the kinetics of the surface reaction.

A very simple method of ascertaining the importance of diffusional processes $(k_L a \text{ and } K_{SL})$ is to study the effect of temperature on the overall rate of consumption of the gas-phase solute. An Arrhenius plot ($\ln R_{VA}$ vs. $1/T$) gives the activation energy, ΔE (slope $= \Delta E/R$), of the process consuming the gas-phase reactant. When ΔE is less than 20 kJ/mol, it is inferred that one or both of the two diffusional steps ($k_L a$

and/or K_{SL}) are posing a significant resistance. To eliminate the same and obtain intrinsic kinetics, it is then necessary to increase the rate(s) of the mass transfer steps by employing higher speeds of agitation. On the other hand, if the activation energy is greater than 50 kJ/mol, further experiments should be aimed at obtaining the kinetics through variation of the gas-phase partial pressure, liquid-phase reactant concentration, and catalyst loading as shown in Figure 2.5. In the case of a porous catalyst, it is desirable to eliminate (or estimate) the resistance due to internal diffusion by varying the catalyst particle diameter. Detailed discussion on this matter is available in several texts (Ramachandran and Chaudhari 1983; Levenspiel 1999).

As mentioned in Section 2.2, the main issue relating to scale-up of a multiphase reactor is the values of the transport coefficients, $k_L a$ and K_{SL}. Once these values are determined, comparison of the parameters in the parentheses of the right-hand side of Equation 2.19 allows the rate-controlling step for a given set of operating conditions in a given type of multiphase reactor. The worked examples in Chapters 7A, 7B, 8, and 9 illustrate this procedure.

NOMENCLATURE

A: Gaseous solute

$[A^*]$: Equilibrium concentration of at the solute at the interface (mol/m^3)

$[A]$:   Concentration of dissolved solute (mol/m^3)

$[A_B]$: Bulk liquid concentration of dissolved solute (mol/m^3)

A_p: Total surface area of the catalyst (m^2)

$[A_S]$: Concentration of dissolved solute on the catalyst surface (mol/m^3)

B: Liquid-phase reactant

$[B]$:   Concentration of liquid-phase reactant (mol/m^3)

$[B_B]$: Bulk liquid concentration of liquid-phase reactant (mol/m^3)

$[B_S]$: Concentration of the liquid-phase reactant on the catalyst surface (mol/m^3)

D_A: Molecular diffusivity of dissolve gaseous species A (m^2/s)

D_B: Molecular diffusivity of liquid-phase reactant B (m^2/s)

D_I: Molecular diffusivity of species I (m^2/s)

d_p: Diameter of the catalyst/solid phase (m)

Ha: Hatta number $\left(\sqrt{\dfrac{2}{p+1} D_A k_{p-q} [B_B]^q [A^*]^{p-1}} \right) \Big/ k_L$ (—)

K_{SL}: Particle–liquid mass transfer coefficient (m/s)

k_2:   Second-order reaction rate constant (m^3/mol s)

k_L: True gas–liquid mass transfer coefficient (m/s)

$k_L a$: Volumetric gas–liquid mass transfer coefficient (1/s)

k_{p-q}: Rate constant for general (p, q)th-order reaction (mol/m^3)$^{1-p-q}$/s^{-1}

k_R: Overall surface reaction rate constant given by Equation 2.18 (—)

k'_R: Surface reaction rate constant given by Equation 2.18 (1/kg catalyst s) or $(m^3/m^2 s)$.

p: Order of reaction with respect to the dissolved gaseous species (—)

q: Order of reaction with respect to the liquid-phase reactant (—)

R_{SA}: Rate of mass transfer per unit area or flux of A $(mol/m^2 s)$

R_V: Volumetric rate of mass transfer $(mol/m^3 s)$

R_{VB}: Volumetric rate of reaction in the bulk liquid $(mol/m^3 s)$

t: Contact time in Equation 2.10 (s)

V: Volume of the reaction mixture (m^3)

w: Total weight of the catalyst used (kg)

X: Distance in the liquid film (m)

Greek Letters

δ: Thickness of the diffusion film (m)

ε_L: Liquid volume fraction in the gas–liquid dispersion (—)

τ_D: Characteristic diffusion time (s)

τ_R: Characteristic reaction time (s)

ψ: Stoichiometric factor for reaction between dissolved species A and liquid-phase reactant B (—)

REFERENCES

Archibald B, Brummer O, Devenney M, Gorer S, Jandeleit B, Uno T, Weinberg WH, Weskamp T. (2002) Combinatorial aspects of material science. Chapter 32. In: Nicolaou KC, Hanko R, Hartwig J, editors. Handbook of combinatorial chemistry. Wiley-VCH, Weinheim, Germany. p 885–990.

Astarita G. (1967) Mass transfer with chemical reaction. Elsevier, Amsterdam, the Netherlands.

Bawane SP, Sawant SB. (2003) Kinetics of liquid-phase catalytic hydrogenation of benzophenone to benzhydrol. Org. Proc. Res. Dev., 7:769–773.

Bawane SP, Sawant SB. (2004) Liquid phase catalytic hydrogenation of p-chlorobenzophenone to p-chlorobenzhydrol over a 5% Pd/C catalyst. Chem. Eng. Technol., 27:914–920.

Danckwerts PV. (1970) Gas–liquid reactions. McGraw-Hill, New York, USA.

van Dierendonck LL, Meindersma GW, Leuteritz GM. (1988) Scale-up of G-L reactions made simple with loop reactors. In: Proceedings of Sixth European Conference on Mixing, Pavia, Italy, May 24–26. p 287–295.

van Dierendonck LL, Zahradnik J, Linek V. (1998) Loop venturi reactor—a feasible alternative to stirred tank reactors? Ind. Eng. Chem. Res., 37:734–738.

Doraiswamy LK, Sharma MM. (1984) Heterogeneous reactions: analysis, examples, and reactor design. Vol. 2: Fluid–Fluid–Solid Reactions. John Wiley Interscience, New York, USA.

Garcia-Ochoa F, Gomez E. (2009) Bioreactor scale up and oxygen transfer rates in microbial systems: an overview. Biotechnol. Adv., 27:153–176.

Jahnisch K, Hessel V, Lowe H, Baerns M. (2004) Chemistry in micro structured reactors. Angew. Chem. Int. Ed., 43:406–446.

Junker BH. (2004) Scale-up methodologies for escherichia coli and yeast fermentation processes. J. Biosci. Bioeng., 97:347–364.

Kashid MN, Lioubov K-M. (2009) Micro structured reactors for multiphase reactions: state of the art. Ind. Eng. Chem. Res., 48:6465–6485.

Levenspiel O. (1999) Chemical reaction engineering. John Wiley & Sons, New York, USA.

Levenspiel O, Godfrey JH. (1974) A gradient less contactor for experimental study of inter-phase mass transfer with/without reaction. Chem. Eng. Sci., 29:1723–1730.

Pangarkar K, Schildhauer TJ, van Ommen JR, Nijenhuis J, Moulijn JA, Kapteijn F. (2010) Heat transfer in structured packings with co-current downflow of gas and liquid. Chem. Eng. Sci., 65:420–426.

Ramachandran PA, Chaudhari RV. (1983) Three phase catalytic reactors. Gordon & Breach Science Publishers Inc., New York, USA.

Rebrov EV, Schouten JC, deCroon MHJM. (2010) Single-phase fluid flow distribution and heat transfer. Chem. Eng. Sci., 66:1374–1393.

Vogel GH. (2005) Process development: from the initial idea to the chemical production plant. Wiley-VCH Verlag GmbH & Co. KGaA, Weinheim, Germany.

Votruba J, Sobotka M. (1992) Physiological similarity and bioreactor scale-up. Folia Microbiol, 37(5):331–345.

3

MULTIPHASE REACTORS: TYPES AND CRITERIA FOR SELECTION FOR A GIVEN APPLICATION

3.1 INTRODUCTION TO SIMPLIFIED DESIGN PHILOSOPHY

Multiphase reactions may involve gas–liquid, gas–liquid–solid (solid as catalyst or reactant), liquid–liquid, liquid–liquid–solid reactions, etc. The reactions may vary from very slow to very fast, endothermic to highly exothermic. Based on the reaction characteristics, different types of multiphase reactors are used in industrial practice. A number of texts dealing with design of multiphase reactors are available (Satterfield 1970; Shah 1979; Ramachandran and Chaudhari 1983; Westerterp et al. 1988; Deckwer 1992). Considerable information on theoretical, hydrodynamics, and mass transfer aspects of different multiphase reactors has become available since the publication of the above texts. This recent information is likely to allow rational, simple and yet reliable designs of many industrially important multiphase reactors. In this book, different types of multiphase reactors falling under two categories—(1) gas–liquid and (2) gas–liquid–solid—are considered. The basic aim is to provide user-friendly, simple, and reasonably accurate design procedure for each multiphase reactor.

Multiphase systems of the above type have at least one free interface, that is, liquid film/drops/bubbles. The behavior of such a discrete dispersed phase is very complex. The films/drops/bubbles can break/coalesce repeatedly when undergoing turbulent in motion both time and space. Several factors such as the flow pattern and turbulent shear stress, the nature (coalescing/no coalescing), etc. affect the size of the drops/bubbles. The transport of mass from/to the surrounding liquid is affected by the turbulence

Design of Multiphase Reactors, First Edition. Vishwas Govind Pangarkar.
© 2015 John Wiley & Sons, Inc. Published 2015 by John Wiley & Sons, Inc.

parameters (Bennett and Myers 1962; Hinze 1975). Turbulence parameters in multiphase systems are being extensively investigated using laser Doppler anemometry and other techniques. Computational fluid dynamics is also being extensively used to model and understand the behavior of multiphase systems (Ranade 2002; Ranganathan and Savithri 2010). Inasmuch as these studies are essential for a scientific basis for design of multiphase reactors, in many cases, a much simpler yet reasonably accurate design can be performed using simple correlations available in the literature. If these simple correlations indeed have a reasonably consistent scientific basis, then the charge of *blind* use of empiricism leveled against Engineering analysis (Bird et al. 1958) can be negated. Astarita and Ottino (1995) have discussed the essential difference between Engineering and Engineering Science. They have further coined the term Engineering Art. To quote Astarita and Ottino (1995), "The **art** of engineering science is the art of seeing which steps should be taken to erase from real life the (hopefully) irrelevant details which make it so complex, to look at the essential aspects of it, or, in other words, to decide what is the appropriate meta-problem to look at: to conceive the appropriate chain of idealization."

It is precisely this approach that is adopted in this book. The use of blind empiricism so heavily criticized by our peers is resorted to only when unavoidable. This book considers the different multiphase reactors separately while retaining the basic objective mentioned earlier.

3.2 CLASSIFICATION OF MULTIPHASE REACTORS

Multiphase reactors can be broadly classified in two categories:

1. Gas dispersed in liquid
2. Liquid dispersed in gas

Heat and mass transport characteristics depend mainly on the flow rate of the dispersed phase.

3.3 CRITERIA FOR REACTOR SELECTION

The criteria that may be applied for selecting a multiphase reactor for a given application can be summarized as follows:

1. Kinetics *vis-à-vis* mass transfer rates
2. Flow patterns of the various phases
3. Ability to remove/add heat
4. Ability to handle solids
5. Operating conditions (pressure/temperature)
6. Material of construction

3.3.1 Kinetics *vis-à-vis* Mass Transfer Rates

As explained in Chapter 2, the kinetics of a given reaction is invariant with respect to the type and size of the reactor used. On the other hand, the rates of heat and mass transfer can vary by at least an order of magnitude depending on the type of reactor used. Therefore, reactions, which have inherently fast kinetics, can become mass transfer limited in a multiphase reactor with poor mass transfer characteristics. On the other hand, for a reactor that affords relatively high mass transfer rates, it may be kinetic limited. Thus, a comparison of the mass transfer rate with the intrinsic reaction rate is required to reveal the relative importance of kinetics and mass transfer steps (discern "irrelevant details" as Astarita and Ottino imply). It is possible that this rate-controlling step is a mass transport process. In this event, it will be desirable to improve its rate such that the intrinsic kinetic rate that yields the maximum productivity can be achieved. On the other hand, if the overall process is kinetic controlled, any improvement in the productivity of the reactor can be achieved by, for example, a more active catalyst/higher catalyst loading or changes in operating temperature/pressure.

Most catalytic hydrogenation reactions are fast and the rate of supply of hydrogen to the catalyst surface is the limiting factor in reactor productivity. For noble metal catalysts (Pd/Pt on porous support), it is reported that these have a tendency to form complexes with the liquid-phase organic reactant, particularly when the catalyst is starved of hydrogen. Such leaching can cause up to 40% catalyst loss (Greenwood 1986). Packed/trickle bed reactors that afford relatively low mass transfer coefficients are thus ruled out for these reactions. Gas-dispersed multiphase reactors (sparged/stirred/venturi loop reactors) offer relatively high mass transfer rates and therefore may be selected based on which one of these fulfills the other criteria listed earlier. A nitro to amino hydrogenation can be considered as a typical example. These reactions, depending on the economics, may use catalysts ranging from nickel to noble metals. The catalyst loadings vary from as low as less than 0.1 wt % to several percent by weight. A stirred reactor with cooling through the jacket may be satisfactory for pressures up to approximately 2 MPa. For higher heat duties, an external pump-around heat exchanger is preferred over internal coils since the coils can cause severe dampening of turbulence (Chapter 7A). For more complex compounds such as hydrogenation of *m*-nitro acetophenone to *m*-amino acetophenone, much higher pressure (7 MPa) is required (Nishimura 2001). For such reactions, the stirred reactors pose serious sealing problems and a venturi loop reactor (Chapter 8) is the best choice. There are no rotary parts in this latter reactor. Consequently, operating safety and ease of maintenance make it attractive. Further, this reactor provides much higher mass transfer rates. At such high mass transfer rates, the catalyst is completely saturated with hydrogen, thereby preventing leaching of the metal mentioned earlier. The external heat exchanger provides flexibility in meeting the heat transfer load.

Reactions that require elimination of trace quantities of certain compounds (hydro-desulfurization reactions in refineries, deep hydrogenation, or other intrinsically slow reactions) are normally kinetic controlled. The trickle bed reactor in the cocurrent downflow mode is a popular multiphase reactor for such applications. The overall rate being kinetically controlled, it is futile to waste energy in creating turbulence for achieving higher mass transfer rates. On the contrary, the packed catalyst bed that provides a much higher catalyst loading can effectively treat the trace-level impurities. This type of reactor yields higher productivity since the rate is proportional to the surface area of the packed catalyst.

3.3.2   Flow Patterns of the Various Phases

The order of reaction with respect to a reactant dictates the effect of flow pattern on the overall reaction rate in a multiphase reactor. For reactions having zero-order kinetics, the flow pattern is inconsequential. Similarly, for instance, hydrogenation reactions using pure H_2 (under conditions such that the liquid-phase reactant has negligible vapor pressure) mixing in the gas phase does not affect the rate. However, when the gas-phase reactant is derived as a mixture with an inert gas (O_2 from air), gas-phase back mixing is detrimental. Stirred multiphase reactors are generally operated at a speed slightly higher than that for complete gas dispersion (N_{CD}). In such a case, the gas phase may be assumed to be back mixed. For the liquid-phase reactant, most of the gas-dispersed multiphase reactors exhibit significant back mixing, and generally, the liquid phase is also assumed to be well mixed. Only packed columns below the loading point and trickle bed reactors exhibit liquid- and gas-phase behavior closer to plug flow.

The solid phase (catalyst/reactant) shows different behavior in different multiphase reactors. In sparged reactors, there is an exponential decay of the solid concentration along the vertical axis. For stirred multiphase reactors operated at the critical speed for just suspension of the solid, N_S, there is a substantial variation in axial solid concentration. The speed required for achieving uniform solid concentration N_{CS} and the corresponding power input are relatively very high (Nienow 1969, 2000; Shaw 1992). Hence, most stirred multiphase reactors operate at N_S rather than N_{CS}. For venturi loop reactors, Bhutada and Pangarkar (1989) have shown that above a certain power input at which the three-phase jet reaches the reactor bottom, the solid concentration is uniform both axially and radially. In this respect, venturi loop reactor is a definitely better option (Chapter 8).

For trickle beds or monoliths, the catalyst is fixed in space; and hence, a uniform concentration is available along the reactor axis. The problem of nonuniform catalyst wetting, however, needs to be addressed to achieve the desired performance.

3.3.3   Ability to Remove/Add Heat

Reactions that are highly exothermic or endothermic need withdrawal or addition of the heat of reaction. Most reactions, which fall under the purview of this book, are exothermic. Efficient heat transfer is an important criterion in deciding the type of multiphase reactor to be selected. Indeed, many reactions such as chlorination of hydrocarbons are highly exothermic to the extent that the heat of reaction is the controlling factor in reactor design. In the case of chlorination, for relatively small scale of production, the rate of heat release can be regulated by resorting to semibatch operation with additions of chlorine at a controlled rate to the material (e.g., paraffin wax) that is being chlorinated.

TABLE 3.1 Approximate Heats of Hydrogenations of Representative Unsaturated organic Compounds[a]

Reaction	ΔH_R (kJ/mole)
Double bond to single bond	117
Triple bond to double bond	155
$RC \equiv N \rightarrow RCH_2NH_2$	120
Benzene to cyclohexane	208
Nitrobenzene to aniline (ϕNO_2 to ϕNH_2)	493

[a]Reproduced from Nishimura (2001) with permission from John Wiley & Sons Inc., New York, USA. © 2001.

Table 3.1 reproduced from Nishimura (2001) gives the heats of reactions for various types of hydrogenations. It is evident that the nitro to amino conversion has the highest heat of reaction. Significantly, this reaction is also an important part of many processes for pharmaceutical/specialty products.

Similarly, oxidation reactions are also highly exothermic. If the reaction selectivity is affected by temperature, temperature control must be given due importance. Most gas-dispersed multiphase reactors yield fairly high heat transfer coefficients (~400 W/m²°C). However, as the reactor is scaled up, the heat transfer area may not be sufficient and additional heat transfer area must be provided. As an example, heat transfer area per unit volume for kettle-type reactors is given by (jacket heat transfer area/volume) $= (4 \times \pi TH/\pi T^2 H)$ or $(4/T)$. Most large stirred reactors face this problem (Sections 7A.10 and 9.5). A cooling jacket that is sufficient for smaller capacities (smaller diameter) is no longer able to provide the required rate of heat transfer. Many designers opt for providing the additional area through cooling coils located inside the stirred tank reactor. Such internal cooling coils have a profound negative effect on the performance of the stirred tank reactor (Section 7A.8). An alternative to the internal cooling coil is an external heat exchanger loop similar to that in venturi loop reactor. This loop achieves the desired objective of effective temperature control without significantly disturbing the hydrodynamics in the stirred tank. It has the advantage of flexibility in choosing the external heat exchanger area. However, additional power is required, although not as high as compared to the case of internal cooling coils (Chapter 7A). The capital cost of the external pump-around circuit is, on the other hand, substantially higher as compared to that of a mere internal coil located inside the reactor. This is in accordance with the principle of "no free lunch."

If we now compare a venturi loop reactor with this stirred reactor/external heat exchanger combination, the superiority of the former is immediately apparent. In the stirred reactor, power is required for agitation as well as the pump in the external heat exchanger loop. In the venturi loop reactor, the pump in the external loop is the sole consumer of power. This pump performs the dual functions of providing the required flow across the heat exchanger and gas induction. It also yields high rates of both gas–liquid and solid–liquid mass transfer. The built-in flexibility of the external heat exchanger makes the venturi loop reactor attractive, particularly if the same reactor is to be used for different product campaigns involving substantially varying heat duties. Selection of a multiphase reactor for a complex gas–liquid reaction based on the heat removal criterion can be best illustrated by the classical example of absorption of NO_x to yield nitric acid.

The NO_x absorption reaction system is extremely complex consisting of more than 40 possible reactions (Hupen and Kenig 2005). These include several gas- and liquid-phase reactions.

Gas-phase reactions:

$$2NO + O_2 \rightarrow 2NO_2 \qquad (G1)$$
$$2NO_2 \leftrightarrow N_2O_4 \qquad (G2)$$
$$NO + NO_2 \leftrightarrow N_2O_3 \qquad (G3)$$
$$NO + NO_2 + H_2O \leftrightarrow 2HNO_2 \qquad (G4)$$
$$3NO_2 + H_2O \leftrightarrow 2HNO_3 + NO \qquad (G5)$$

Liquid-phase reactions:

$$2NO_2 + H_2O \rightarrow HNO_2 + HNO_3 \qquad (L1)$$
$$N_2O_3 + H_2O \rightarrow 2HNO_2 \qquad (L2)$$
$$N_2O_4 + H_2O \rightarrow HNO_2 + HNO_3 \qquad (L3)$$
$$3HNO_2 \rightarrow HNO_3 + H_2O + 2NO \qquad (L4)$$

The important points to be noted are: (i) the homogeneous gas-phase oxidation of $NO \rightarrow NO_2$ is the slowest reaction and hence the controlling step in the overall absorption process and extent of formation of HNO_3 (Kenig and Seferlis 2009). (ii) The rates of reactions G1 and G2 increase with decrease in temperature. G1 is essentially irreversible at temperatures below $350\,^{\circ}C$ and its rate increases with the cube of the pressure. (iii) The solubility of NO_2 in water is higher than that of NO by an order of magnitude (Hupen and Kenig 2005). (iv) Absorption of NO_2 is exothermic and heat must be removed to maintain low temperatures. (v) As expected, the rate of absorption increases with increase in pressure. However, this is valid only for moderate pressures. At higher pressures, reaction G1 and equilibria of other gas-phase reactions assume greater importance. (vi) For a fixed composition of the gas phase with respect to NO_x, the maximum attainable concentration of HNO_3 increases with decrease in temperature.

Industrial nitric acid processes are basically of three types: (1) ambient, (2) medium, and (3) high pressure. For the 1 ambient pressure variety, a low pressure drop contactor like packed tower is desired. The other two types can be handled by a relatively high pressure drop contactor like plate column. It may be noted here that the NO_x absorption train constitutes a major portion of the nitric acid plant cost since the material of construction (SS-304 ELC) is relatively expensive. Table 3.2 compares packed and plate columns for NO_x absorption.

The above comparison indicates that packed columns are suitable for the ambient pressure version, whereas plate columns would be the optimum choice for the medium- and high-pressure versions.

TABLE 3.2   Comparison of Packed and Plate Columns for Absorption of NO$_x$ Gases

Criterion	Packed columns	Plate columns
Gas-phase pressure drop	Low 15–30 mm WC[a] per theoretical stage	Relatively high 60–120 mm WC[a] per theoretical stage
Heat transfer	Very poor/requires external pump-around heat exchanger	Very good with cooling coils on the tray floor
NO oxidation	Modern high voidage packings provide ease of homogeneous gas-phase NO oxidation with simultaneous NO$_2$ absorption	The intertray spacing above the dispersion is used for NO oxidation but NO$_2$ absorption takes place only on the next tray above
Total cost	Although basic tower cost is lower, the heat exchanger loops make significant additional contribution to the total cost. Complicated flow sheet	Tower cost is higher but no external heat exchangers required. Overall cost is practically the same as packed tower

[a]Water column.

3.3.4   Ability to Handle Solids

Reactions involving solids or reactions that generate solids must necessarily be carried out in a multiphase reactor that has no clogging problems. For instance, a packed gas–liquid reactor cannot be used for carbonation of lime as both the reactant Ca(OH)$_2$ and product CaCO$_3$ are insoluble and would plug the column rapidly. Most gas-dispersed multiphase reactors can handle solids or solid-generating systems. Tray-type contactors are an exception. Holes in sieve trays tend to plug. Similarly, solid deposits on tray floor in the bubble cap type can build up to overflow weir heights leading to flooding. Special reactor designs such as the horizontal sparged reactor have been reported for this special case (Joshi and Sharma 1976). The pressure at which the gas phase is available is an important factor as mentioned in Section 3.3.3. For reactions at high pressures involving a solid catalyst, the gas phase is compressed to the reactor pressure. The static head of the liquid, which the gas needs to overcome, is negligible as compared to the absolute pressure in the reactor. Therefore, a gas-dispersed reactor such as sparged/stirred tank reactor can be proposed. However, in cases where the available gas-phase pressure is low and solids are involved, a low gas-phase pressure drop is mandated. In such a case, a liquid-dispersed contactor affording low pressure drop may be chosen. An important example is that of absorption of SiF$_4$ from off-gases of a super phosphate-manufacturing "Den." The acidulation of phosphate rock releases fluorine in the rock as SiF$_4$. SiF$_4$ reacts instantaneously with water (Kohl and Riesenfeld 1985):

$$3SiF_4 + 2H_2O \rightarrow 2H_2SiF_6 + SiO_2$$

The "Den" works at a small negative pressure (of the order of 200–250 mm water column) produced by a fan. The equilibrium vapor pressure of SiF$_4$ over aqueous fluosilicic acid solutions is very low, and acid of strength ~32 wt % can be produced with a feed gas (exhaust from the "Den") containing as low as 0.3 vol. % SiF_4. The major problem is the formation of a very fine silica precipitate that rules out packed towers. Sparged reactors that can handle solids cannot be recommended because of the high hydrostatic head that must be overcome. Therefore, spray towers or venturi absorbers are recommended. Both of these suffer from extensive gas-phase axial mixing. The efficiency of spray towers decreases rapidly due to (i) coalescence of drops (reduction in effective gas-liquid interfacial area) as the spray travels away from the nozzle and (ii) coverage of the spray by the silica produced in the reaction that introduces a relatively high resistance to solute diffusion. However, despite these drawbacks the major advantage of very low gas-phase pressure drop favors the use of spray towers. Short spray towers in cascade are preferred to curtail loss of mass transfer efficiency due to (i) and (ii) listed earlier. The construction of bank of such towers made out of high-density poly(ethylene) (or made out of cement in some old plants) is relatively simple and cost-effective. High pressure drop venturi absorber, particularly in the ejector mode, may also be used, but the overall power requirement is higher than that for spray towers.

3.3.5   Operating Conditions (Pressure/Temperature)

Reactions that require much higher pressure (>2 MPa) than in the previous sections combined with high-temperature and inflammable reactants pose serious operational hazards. Multiphase reactors that have a rotary part requiring sealing need careful maintenance. In such cases, extreme caution regarding leaks must be exercised. The stirred (Chapter 7A) and gas-inducing reactors (Chapter 9), which require diligent maintenance of the sealing arrangement for the agitator, pose a serious safety risk. Therefore, for such duties, these reactors are generally not the first choice. On the other hand, sparged reactors (bubble columns) and venturi loop reactors do not have a moving part in the reactor proper and hence are safer. However, when an expensive gas such as pure hydrogen is involved in the reaction, it should either be consumed completely or recycled efficiently. Hydrogen recycle compressors are expensive and add to the capital cost. Therefore, for such duties, the sparged reactor must necessarily operate at very low superficial hydrogen velocities (<0.1 cm/s) to consume the hydrogen completely. It is evident that such low gas velocities are detrimental to the mass/heat transfer performance and also may be insufficient for suspending the solid catalyst. The venturi loop reactor, on the other hand, can automatically recycle the unconverted hydrogen and is therefore preferred over the sparged reactor. In the case of venturi loop reactor, the maintenance is restricted to the pump located outside the reactor and is therefore much easier. Consequently, the downtime due to maintenance in this case is also much lower as compared to a stirred reactor.

3.3.6   Material of Construction

In the manufacture of fine chemicals, SS-304/SS-316/SS-316L is a commonly used material of constructions. For other duties, special materials such as SS-304 ELC, glass lined, fiber reinforced plastic (FRP), graphite, etc. may be required to counter the corrosive action of the fluids. For nonmetallic constructions, only some types of multiphase reactor are amenable to fabrication. The choice of the reactor is then restricted to only these types of multiphase reactors. Reactions involving halogens may require glass-lined or other corrosion-proof equipment since hydrogen halide produced in the reaction poses severe corrosion problems. As an example, it may be noted here that hydrogen chloride produced in chlorination is generally recovered as hydrochloric acid. This application involves a highly corrosive liquid and the absorption process is highly exothermic (Norman 1962). Unless the heat of solution is removed, the solution reaches the azeotropic composition of 20 wt % HCl at 108.5 °C. In this extreme case, to produce high strength (35 wt %) hydrochloric acid, special graphite falling-film block absorbers are used. Graphite has the advantages of high thermal conductivity and inertness to hydrochloric acid.

The aforementioned discussion was general in nature and also included conventional contactors such as tray and packed columns. In the case of three-phase (G–L–S) reactions, such conventional contactors are not used. The stirred reactor is the workhorse of the fine chemicals industry. The gas-inducing reactor can be considered as an alternative to stirred reactors when a pure gas is used. However, this reactor has several drawbacks (Chapter 9). In view of this, the venturi loop reactor has been widely used as a safe and energy-efficient alternative to the conventional stirred reactor. Table 3.3 summarizes the preceding discussion in the form of a multiphase reactor selection guide.

3.4   SOME EXAMPLES OF LARGE-SCALE APPLICATIONS OF MULTIPHASE REACTORS

3.4.1   Fischer–Tropsch Synthesis

The increasing pressure on petroleum crude has advanced the case of coal to oil processes. Levenspiel (2005) has analyzed the postpetroleum crude situation. He concluded that coal can indeed be advantageously used as a substitute for petroleum crude. Fischer–Tropsch synthesis for coal to hydrocarbons was first practiced in Germany. By 1938, these plants had a collective capacity of ~7 million tonnes per year (Anderson 1984) of various types of hydrocarbons from coal. Subsequently, major impetus to this route came from Sasol in South Africa. A large body of literature is available on Fischer–Tropsch synthesis. Dry (1982) has presented the basic aspects of Fischer–Tropsch synthesis as practiced in Sasol since 1955 when the first coal to hydrocarbon Fischer–Tropsch plant started production. Sasol added two more Fischer–Tropsch plants in the 1980s. South Africa had no oil reserves but has large coal reserves. Further, South Africa could not rely on imported crude oil. India also did not have oil reserves and could have adopted the South African approach. However, there was no economic isolation similar to South Africa. Besides the hydrocarbon products from coal-derived synthesis gas, Levenspiel (2005) lists a variety of commodity chemicals that can be made from synthesis gas. Indeed, two ammonia plants based on coal as feedstock were built in India in the 1960s. The coal-based

TABLE 3.3 Multiphase Reactor Selection Guide[a,b]

	Criteria	Trickle bed	Stirred tank reactor	Gas-inducing reactor	Venturi loop reactor	Sparged reactors	Microstructured[c]	Structured trickle beds[c]
1. Kinetics	Slow	✓	✓[a]	✓[e]	✓	✓	✓	✓
	Fast		[d]	[e,d]	✓	✓	✓	
2. Reaction Pressure	Low/moderate	✓	✓[d]	✓[e,d]	✓	✓	✓	✓
	High/very high	✓			✓[f]	✓	✓	✓
3. Heat of reaction	Low/moderate	✓	✓	✓[e]	✓	✓	✓	✓
	High/very high				✓√[f]	✓	✓	
4. Gas-phase reactant	Inexpensive (air)		✓			✓	✓	✓
	Expensive (H_2/pure O_2)		[g]	[e,d]	✓[f]	✓[g]	✓	✓

[a] Pressure: Low/moderate ≤ 2 MPa.
[b] Heat of Reaction: Low/moderate ≤ 150 kJ/mole.
[c] Relatively new technologies.
[d] Please refer to Section 7A.10/9.5.
[e] Reports indicate application on large scale not economical (also see worked example Section 9.5).
[f] Particularly advantageous for relatively very high pressure and highly exothermic reactions employing expensive/pure gas.
[g] Requires compressor for recycle of unreacted gas (expensive in case of H_2).

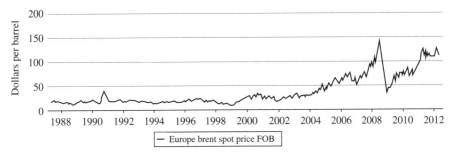

FIGURE 3.1 Variation of crude oil price. (Reproduced from U.S. Energy Information Administration Website. Independent Statistics & Analysis. http://www.eia.gov/dnav/pet/hist/ LeafHandler.ashx?n=PET&s=RBRTE&f=D, Accessed Sept. 2013.)

TABLE 3.4   U.S. Department of Energy Estimate of World Fossil Fuel Reserves[a,b]

Petroleum crude	Coal	Natural gas
6063×10^{15} kJ	$32,000 \times 10^{15}$ kJ	5400×10^{15} kJ

[a]Adapted from Levenspiel (2005) with permission from American Chemical Society. ©2005.
[b]Additional shale gas (456 trillion m^3 as per U.S. Energy Information Administration (www.eia.gov), 2011) probably not included.

routes were mostly sidelined because of the availability of relatively inexpensive petroleum crude and maturity of the refining technologies. International interest in the coal to oil route dwindled with the discovery of the massive oil fields in the Middle East. The period following the first Gulf war saw a sudden increase in oil prices and not surprisingly the additional capacity of Sasol. Petkov and Stratiev (2008) have analyzed the variation in crude oil prices along with specific crude consumption. Their results showed three stages of major increase in price: (1) 1940–1950, (2) 1970–1980, and (3) post-2000 up to 2008. Stage 2, which was the result of the Gulf war in the 1970s, showed the steepest increase followed by stage 3 between 2000 and 2008. After a short-lived peak in 2009, the prices fell in the later part of 2009 only to increase again over the next 3 years. Figure 3.1 shows the different stages of variation in international crude prices post-1988.

The interest in coal conversion processes, however, is increasing due to the order of magnitude higher price for the petroleum crude as compared to the pre-1970s prices.

Table 3.4 gives the estimates of the total world reserves of hydrocarbons and coal.

A relatively recent development in the supply of lower hydrocarbons is shale gas. Siirola (2014) has discussed the impact of shale gas on the chemical industry in detail. Existence and utility of shale gas was known almost 200 years ago. The incentive for shale gas production came from the high natural gas prices in 2000. However, technology for this unconventional natural gas was first demonstrated by Mitchell Energy, which established the first commercial shale gas production facility in the Fort Worth Basin in 1998 using slick water fracturing. Other breakthrough technologies resulting from concerted research efforts were directional (horizontal) drilling, hydraulic fracturing, and microseismic monitoring technologies. It is difficult to conclude on the production technologies due to unresolved issues like contamination of near-surface aquifers and environmental impact of chemicals used in fracturing. Nevertheless, the statistics give a clear indicator. From the modest beginning, the production has risen to 250 billion m^3 in 2013. This amounts to about 35% of the total natural gas produced in the United States. "Dry" shale gas (mostly CH_4) is mainly used as fuel substitute for coal, but the "wet" type ($+C_2$, $+C_3$) can be a good feedstock for the petrochemical industry, particularly where scale of production and long-term availability are sufficient to sustain production of olefins (through cracking) and aromatics (platforming). The latter will require considerable improvements in catalyst selectivity and activity over prolonged periods of operation.

The current (2013) high prices of oil and the fact that no major fields have been discovered make the coal (or inexpensive hydrocarbons) to oil alternative commercially attractive for countries that have far greater coal reserves than petroleum crude. This can also be judged from the fact that the number of research publications dealing with Fischer–Tropsch process in 2009 was approximately three times those in 1998 (Zhang et al. 2010).

The overall coal to oil conversion occurs in three stages: (1) mining of coal, (2) preparation of synthesis gas, and (3) the Fischer–Tropsch synthesis along with downstream processing. According to Dry (1990), the second and third stages contribute ~23 and 30% of the total capital investment, respectively.

The original aim of the Fischer–Tropsch synthesis was to produce transportation fuels. Coal to hydrocarbon conversion can give both types of fuels: gasoline and diesel. An ideal gasoline should consist of highly branched hydrocarbons in the desired boiling range. The standard for gasoline is isooctane, which is assigned a road octane number (RON) of 100. Indeed, to achieve this, n-paraffins are alkylated in the fuels section of a refinery. Diesel quality is rated in terms of cetane number. n-Hexadecane is assigned a cetane number of 100. Diesel engines have a better fuel efficiency than engines running on gasoline. On this count, diesel should be the preferred transportation fuel. Further, since diesel is a fully saturated mixture, it is expected to produce low CO and more water vapor. A variant of the conventional Fischer–Tropsch route combined with two subsequent process steps: (1) oligomerization of olefins on crystalline silica alumina catalysts of ZSM-5 type with intermediate pore sizes has been found to be particularly useful. The limitations caused by the pore size preclude formation of highly branched oligomers. Therefore, the process yields predominantly straight-chain species with only methyl branches at most. The product thus obtained has relatively high viscosity and a cetane number of ~50. (2) Controlled cracking/hydrocracking of the long-chain wax on Ni/Mo supported on an amorphous silica–alumina carrier can yield high (~65) cetane number. The diesel so obtained has practically no aromatics and further can more than meet the increasingly strict limits that are being imposed on sulfur and nitrogen compounds (Dry 2001). However, diesel engines have been castigated for soot pollution, and efforts have been made to address this problem to maintain the advantage of its high efficiency. Although, the coal to liquid (CTL) route has been in commercial operation over the last six decades, there is considerable scope for improvement. Glasser et al. (2012) have reviewed recent developments in Fischer–Tropsch synthesis that include kinetic description, reactors, effects of water and CO_2, and product upgrading. Modeling of the kinetics requires introduction of interaction between reaction and phase equilibria. In view of the similarity in chemical properties of the components, Raoult's law can be used to predict vapor–liquid equilibrium (Masuku et al. 2011). Besides coal as the raw material, the other major Fischer–Tropsch raw materials are natural gas/shale gas (gas to liquid, GTL) and biomass (biomass to liquid, BTL). A 1911 analysis by Milmo (2011) indicated that the GTL capacity is likely to rise to several million barrels equivalent of oil per day in 2030 as compared to about 0.3 million barrels of oil equivalent in 2011. This is in addition to the two million barrels per day from coal and a few million barrels per day from biomass (BTL). Equivalent data for biomass (similar to coal) are not available. Milmo (2011) has also presented a brief survey on the problems of emerging Fischer–Tropsch technologies for synthesis gas derived from biomass and municipal solid waste. A recent analysis, however, shows that the overall conversion in BTL is carbon negative (Bahnholzer and Jones 2013: Section 1.2 for recent contradictory analyses).

3.4.1.1 Catalysts

3.4.1.1 Catalysts   The two catalysts used in Fischer–Tropsch synthesis are (1) cobalt and (2) iron based. Dry (1990) has given an excellent discussion on the catalyst compositions that are suitable in terms of stability, activity, and cost. Zhang et al. (2010) have presented a comprehensive review on developments of novel catalysts for Fischer–Tropsch synthesis with emphasis on regulation of selectivity. Besides Co and Fe, other metals such as Rh, Ni, and Ru also display good activities. Virgin Ru yields the highest activity for hydrogenation of CO even at relatively low temperatures ($\sim 150\,^{\circ}C$). However, the high cost of Ru as compared to Fe or Co has precluded its industrial use. Between Fe and Co, the latter affords higher activity and selectivity for long-chain linear species. They are also more resistant to deactivation by water. Therefore, Co-based catalysts are preferred for production of diesel and waxes. Fe-based catalysts offer the following advantages over Co-based catalysts: (i) broader range of CO/H_2 ratio and operating temperatures without significant CH_4 formation, (ii) suitability for production of alkanes/alkenes and oxygenates, and (iii) suitability for higher CO_2 in the feed, which can be converted by the water gas reaction. However, Fe-based catalysts require careful preparation and addition of specific promoters (Dry 1990). Overall, Fe catalysts are good for syngas with low H_2/CO ratio (coal) but may not be suitable for H_2 rich syngas derived from natural gas. Catalyst deactivation is also a problem with these catalysts. In terms of catalyst cost, assigning a unit value to Fe-based catalyst, the costs of other catalysts are as follows: Ni, 250; Co, 1000; and Ru, 50,000 (Dry 2002). The review of Zhang et al. (2010) may be referred for more details. Both types (Co/Fe) of catalysts are poisoned by sulfur compounds, and hence, sulfur has to be removed to levels much less than $0.1\,mg/Nm^3$ of synthesis gas. Similarly, since the catalysts usually use alkaline promoters, any chlorine or acidic material must be avoided. The effectiveness of catalyst is closely linked to improved gasification and cleaning technologies for nonconventional raw materials such as biomass and municipal solid waste.

The major reactions in Fischer–Tropsch reactors are:

$$\text{Paraffins: } n\text{ CO} + (2n+1)\text{H}_2 \rightarrow \text{C}_n\text{H}_{2n+2} + n\text{ H}_2\text{O}$$

$$\text{Olefins: } n\text{ CO} + 2n\text{ H}_2 \rightarrow \text{C}_n\text{H}_{2n} + n\text{ H}_2\text{O}$$

For the iron-based catalyst, the water gas shift (WGS) reaction also occurs:

$$\text{CO} + \text{H}_2\text{O} \leftrightarrow \text{CO}_2 + \text{H}_2$$

WGS reaction is mildly exothermic (~41 kJ/mol, Bustamante et al. 2004).

The utilization of CO_2 varies based on the prevailing temperature. For low temperature operation, the water gas shift reaction is relatively slow. This results in a H_2/CO utilization factor of ~1.7. Higher temperatures result in a close approach to equilibrium affording higher utilization of CO_2 (Dry 2002). Further, if the synthesis gas composition is such that $(H_2/(2CO+3CO_2))$ is 1.05, Dry (2002) indicates that nearly all of CO, CO_2, and H_2 can be utilized for making Fischer–Tropsch hydrocarbons. CO_2-rich gases yield high CH_4/H_2 ratios. However, an optimized mixture of CO and CO_2 can yield improved hydrocarbon formation, particularly for an iron-based catalyst. Product upgrading can add value through wax cracking, isomerization, etc.

Most major improvements in performance of Fischer–Tropsch and other catalysts have been achieved through advanced imaging technologies. For example, different types of electron microscopy (Florea et al. 2013; Thomas et al. 2013), X-ray, and neutron powder diffraction (Rozita et al. 2013) facilitate characterization of catalyst particles and the included pores ranging in size from micron to a few nanometers. These advanced tools lend a capability to analyze down to the level of an atom. Further, the same facilitate introduction of efficient promoters and also distribution of smaller catalytic species over greater surface areas. The last feature implies a catalyst with high activity and surface area and should allow higher rates of reaction.

3.4.1.2 *Operating Temperature*

3.4.1.2 Operating Temperature Two modes, low (225–250°C) and high (300–350 °C) temperatures, are used in typical Fischer–Tropsch reactors (Dry 2002).

For maximum gasoline production, the high-temperature mode with an iron catalyst is used. This scheme produces approximately 40% straight run gasoline with a low octane number and about 20% C_3–C_4 alkenes. The alkenes are oligomerized to yield branched species having high octane number. C_5 and higher products need hydrogenation and isomerization to obtain higher octane numbers. The overall process is quite intricate. Therefore, production of gasoline is a less attractive option for Fischer–Tropsch synthesis. On the other hand, high-temperature operation coupled with iron catalyst can be conveniently used for producing feedstock for the petrochemical industry (Dry 2002). The high olefin content (85% in C_3 cut, 70% in C_5–C_{12} cut, and 60% in C_{13}–C_{18} cut) is useful for high-value products as compared to transportation fuels. Ethylene, for example, feeds poly(ethylene) and poly(vinyl chloride); propylene can be used for poly(propylene) and longer-chain olefins to alcohols via hydroformylation. This list shows the versatility of Fischer–Tropsch synthesis (Dry 2002). The low-temperature mode is more suited for obtaining diesel. As mentioned earlier, requirements for diesel are distinctly different from those for gasoline. The product obtained in the low-temperature mode comprises roughly 20% straight run diesel, which on hydrotreatment yields a cetane number of 75. Since a cetane number of 45 is sufficient for most markets, this premium straight run diesel can be used for improving low-quality diesel or directly as auto fuel in countries with stricter regulations. The rest of the product comprising of heavier hydrocarbons (~45% of the total) is subjected to relatively mild hydrocracking, resulting in high-quality diesel sans aromatics. The combined (straight run and hydrocracked) diesel is also a high-quality product due to a cetane number in excess of 45 (Dry 2002).

3.4.1.3 Reactors for Fischer–Tropsch synthesis Several authors (Dry 1990, 2001; Eilers et al. 1990; Fox 1990; Sie and Krishna 1999; Krishna and Sie 2000; De Deugd et al. 2003; Davis 2003; Guettel et al. 2008; Pangarkar et al. 2008, 2009; Woo et al. 2010; Deshmukh et al. 2010; Glasser et al. 2012) have presented excellent discussion on the conventional as well state-of-art reactor types for Fischer–Tropsch synthesis. The most important consideration in reactor design is adequate removal of the relatively high heat of reaction for conversion of the synthesis gas into hydrocarbons (55,000–60,000 kJ/kmol of synthesis gas reacted; Fox 1990). Almost all types of multiphase reactors have been used in this synthesis. In the conventional class, four types of reactors have been used. These are (i) shell- and tube-type reactors with catalyst packed in the tubes—this type is known as Arge reactor—(ii) slurry type (Fig. 3.2a and b, three-phase sparged reactors in the nomenclature adopted in this book), (iii) airlift (external downcomer type) slurry or three-phase sparged reactor (Fig. 3.2c), and (iv) fluidized bed reactors, both conventional and moving types (Synthol reactors; Fig. 3.2d and e, respectively).

The new reactor types that have been proposed are (i) structured-type monoliths and packed beds with structured packings (de Deugd et al. 2003; Pangarkar et al. 2008) and (ii) microstructured devices with catalyst coating (Deshmukh et al. 2010; Glasser et al. 2012).

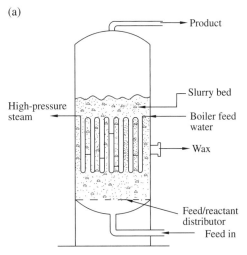

(a)

FIGURE 3.2 (a) Conventional slurry or three-phase sparged reactor. (Reproduced from Dry (2002) with permission from Elsevier. © 2002.)

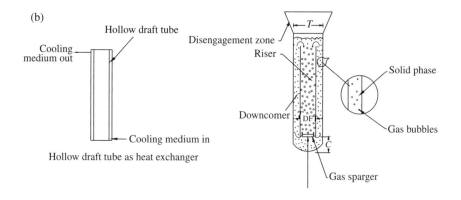

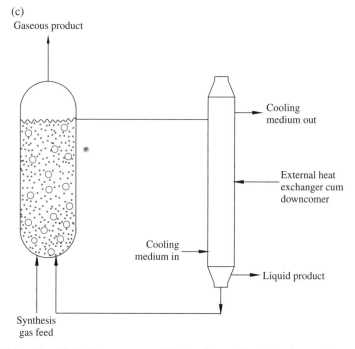

FIGURE 3.2 (Cont'd) (b) Slurry reactor or bubble column with a draft tube cum heat exchanger. (c) Airlift (external downcomer type) slurry or three-phase sparged reactor. (Reprinted from de Deugd et al. (2003) with kind permission from Springer Science + Business Media. © 2003.)

Reactors for High-Temperature Operation   Krishna and Sie (2000) have described the five variants of fixed-bed reactors. Sasol has been using four types of reactors ranging from fixed to two types of fluidized beds to the three-phase sparged reactors (Dry 1981). Dry (2002), however, indicated that the conventional fluidized bed (Synthol reactor) has replaced most moving fluidized beds in Sasol for the following reasons: (i) the capital cost of the Synthol reactor of equivalent capacity is 40% lower; (ii) there is better heat transfer; (iii) since there is no downcomer, the entire catalyst charge actively participates in the reaction; (iv) operation at higher pressures

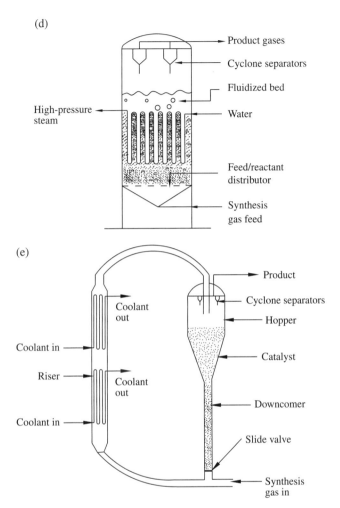

FIGURE 3.2 (Cont'd) (d) Conventional three-phase fluidized bed reactor. (Reproduced from Dry (2002) with permission from Elsevier. © 2002.) (e) Circulating fluidized bed reactor. (Reproduced from Dry (2002) with permission from Elsevier. © 2002.)

(~4 MPa) lowers coking that allows more "on time" operation for a given catalyst charge; and (v) the relatively low velocity in the Synthol reactor lowers abrasion. In moving fluidized bed, the ceramic downcomer requires frequent maintenance due to abrasion by the high-velocity iron carbide particles. Conventional fluidized bed does not have this problem and consequently yields longer "on time" between maintenance.

Reactors for Low-Temperature Operation   The preceding discussion in Section 3.3.1b shows that the low-temperature mode is used for obtaining higher carbon number hydrocarbons (diesel/waxes). Sasol started this mode with five shell- and tube-type fixed beds in 1955. These reactors operated at 2.7 MPa and 230 °C, producing 21,000 tonnes of higher hydrocarbons per year. Additional capacity was added in 1987. Two plants based on methane came up in 1992 and 1993 (Mossgas (Sasol) in South Africa and Bintuli (Shell) plant in Malaysia, respectively; Dry 2002). The last one employed a more active cobalt catalyst probably because of a higher H_2/CO ratio obtained from the predominantly CH_4 feed. These reactors had to use smaller tubes to accommodate the larger heat transfer area required due to a more vigorous reaction. Studies on three-phase slurry reactor were reported in the 1950s (Kölbel et al. 1956). The interest, however, remained dormant although Sasol R&D also confirmed the reproducibility of Kölbel et al.'s results (Dry 1981). The major problem was reliable separation system for the fragile catalyst. It must also be mentioned that extensive studies on two-/three-phase bubble columns that started in the 1970s and continue till date have given enhanced confidence and insight for the design procedure (Chapter 10). Once this filtration problem was overcome, successful pilot studies on 1 m scale heralded the arrival of the three-phase sparged reactor (Sasol–Oryx reactor) as the successor to the shell and tube fixed beds (Espinoza et al. 2000). This plant started operating at 100,000 tonnes per year scale in 1993. This single, three-phase slurry reactor-based plant has a capacity equal to the combined capacity of the five original fixed-bed-type reactors (Dry 2002; Woo et al. 2010). The major advantages are as follows: (i) low cost (40% of the Arge type according to Woo et al. (2010)) as compared to the heavy shell and tube Arge reactors; (ii) low pressure drop that entails low running cost (power for compression is proportional to $\Delta P \times$ gas flow rate); (iii) small catalyst particle size typical of slurry systems gives much higher surface area on weight basis as compared to larger pellets used in fixed beds; (iv) lower internal diffusion resistance; (v) far better bed to wall heat transfer ensures isothermal conditions, etc. Further improvement in the performance can be obtained by a simple modification of the conventional sparged reactor as discussed in Section 10.9. Both cobalt and iron catalysts used in this case are permanently poisoned by sulfur. Therefore, pretreatment consists of essentially complete removal of sulfur in the feed (Dry 2002).

3.4.1.4 Recent Developments in Reactors for Fischer–Tropsch Synthesis Davis (2003) has briefly discussed the evolution of design methodologies for reactors employed for highly exothermic reactions. Over the last eight decades, much has been learnt about dissipation of heat from various types of reactors. The order of increasing heat transfer coefficient is packed beds < fluidized beds < sparged reactors ~ stirred reactors. Guettel et al. (2008) have also discussed the various reactor options available and concluded that none of the currently available reactor designs is "ideal" for this highly exothermic synthesis. According to these authors, the "ideal" reactor should have the following attributes: (i) catalyst disposed in the form of a fixed bed in shell and tube configuration, (ii) efficient catalyst utilization through elimination of internal diffusion resistance, (iii) high gas–liquid mass transfer coefficients, and (iv) high bed to wall heat transfer coefficients. The heat and mass transfer problems will be further accentuated with the developments of catalysts with relatively higher activities than the currently used types (Pangarkar et al. 2008). Considering these problems, Davis (2003) commented that making a prediction of

the path that reactor development may take is "fraught with danger." Nonetheless, Davis did mention two important alternatives that were receiving attention beginning with the later part of the last century: (i) monoliths/structured reactors and (ii) micro-channel reactors. These two alternatives are briefly discussed in the following with reference to their applicability in Fischer–Tropsch synthesis.

Monolithic Reactors   de Deugd et al. (2003) have applied the reactor selection criteria of Krishna and Sie (1994) for suitability of monolithic reactors in Fischer–Tropsch synthesis. The three levels are (1) catalyst design; (2) feeding, dispersing, and removal of reactants and products along with heat and mass transfer factors; and (3) hydrodynamics. The first level of catalyst design must take into account the substantial difference in the diffusivities of hydrogen and carbon monoxide. Any resistance to mass transfer (internal or external) reduces the mass transfer rate for carbon monoxide more in comparison with that of hydrogen. In this event, the hydrogen-rich catalyst surface tends to promote formation of shorter products. Purely from a standpoint of curtailing formation of shorter products, diffusion limitations (external/internal) should be avoided. On the other hand, for larger chain length products, this issue has also to be considered in the light of the longer time spent by the products (e.g., α-olefins) of the reaction. Longer contact times of these reactive intermediates promote chain growth. However, after examining the available literature, de Deugd et al. (2003) concluded that internal diffusion limitation displays only a mild positive effect on chain growth. de Deugd et al. (2003) also pointed out that due to several factors, prominent among which are different solubilities and diffusion coefficients of carbon monoxide and hydrogen, the local synthesis gas consumption may not be strictly according to the stoichiometry of the reaction. This problem is enhanced at higher conversions. However, it can be overcome by staged feeding of the synthesis gas. At higher conversions, water, which is a product of the reaction, can be a significant diluent, besides being active as an inhibitor. Monolithic reactors in which the catalyst is coated as a thin coating can reduce the problem of internal diffusion without compromising on the active catalyst surface area. Further, the pressure drop of monolith structures is also substantially low. Among the different possible flow regimes, the slug flow regime has been suggested by de Deugd et al. (2003) for rapid mass transfer. Radial heat transfer so important for selectivity remains a problem. de Deugd et al. suggested recycle of the heavy products through an external heat exchanger loop as a solution to this problem. The energy requirement for this loop can be substantial, counteracting the low pressure drop advantage. In summary, the monolithic reactor has the advantage of a fixed bed (no filtration of the catalyst) and small particle size (no internal diffusion resistance). However, it still falls short of the required rate of heat transfer.

Multitubular Reactors with Structured Packings   Structured packings are widely used in high-vacuum distillation columns and in particular for difficult separations requiring a large number of stages (Niranjan et al. 1982). Their main advantages are high gas–liquid interfacial area and low pressure drop per theoretical stage. The former is a result of wetting and spreading of the liquid by capillary action, while the latter is attributed to the high free volume (>90%) of these packings. In the recent past, structured packings have also been used in reactive distillation (Siirola 1996). Pangarkar et al. (2008) have presented an exhaustive review on the application of various types of structured packings as supports for catalyst packed in small tubes similar to the Arge reactors used by Sasol. This review covered a wide variety of structured packings available. The wire mesh-type packings can be readily coated with the active catalyst by any of the conventional methods (e.g., dip coating). Literature data on the effective gas–liquid interfacial area, volumetric gas–liquid mass transfer coefficient, pressure drop, and some aspects of heat transfer were critically examined. In subsequent studies, Pangarkar et al. (2009; 2010) studied heat transfer aspects of various types of structured packings and packed beds of glass beads with cocurrent downflow of gas and liquid. Pangarkar et al. (2009) specifically studied the application of open–closed and closed cross (OCFS and OCCS, respectively) structured packings for Fischer–Tropsch synthesis. Glass bead-packed columns were also studied as representative of typical ceramic supports used in the Arge-type packed beds for comparison with the structured packings. Heat transfer measurements in the gas–liquid cocurrent downflow mode yielded the effective radial thermal conductivity and packing to tube wall heat transfer coefficients for the different packings. The measured values were used in 1D and 2D models along with the kinetic data of Yates and Satterfield (1991) and temperature dependence functions given by Maretto and Krishna (also see Section 10.8: Worked example for Fischer–Tropsch synthesis reactor) to obtain effects of various parameters on the conversion, space–time yield (STY), and selectivity. The results of the model calculations yielded the following conclusions: (i) the selectivity and space–time yield of the structured configuration for C_{5+} hydrocarbons are significantly better than the packed bed of glass beads. The difference between the performances widens with increasing catalyst activity, evidently because of the better heat transfer provided by the structured configuration. (ii) Doubling the tube diameter results in a small decrease in the STY for the structured configuration as compared to a significant decrease for the glass bead-packed bed. This is a very important conclusion which indicates that the capital cost of the reactor can be significantly lowered through the use of a multitubular structured packing but with larger tube diameters. Similar conclusions have been drawn by Hooshyar et al. (2012) from their investigation of structured packings in both slurry (packed bubble column) and fixed-bed mode (Arge type). Introduction of packings in a bubble column reduces axial mixing (Niranjan and Pangarkar 1984). Hooshyar et al. showed that this results in 20% higher STY due to reduction in axial mixing. For Arge-type reactors employing structured packings, increase in radial heat transfer and decrease in diffusional resistance were shown to yield 40% higher STY.

Microstructured Reactors   These types of devices have been the subject of investigation over the last two decades. The following advantages have been claimed for microchannel reactors: (i) order of magnitude higher heat transfer coefficients; (ii) lower fouling tendency; (iii) order of magnitude higher catalyst productivities; (iv) excellent economy, particularly for small-scale on-site operations like converting

BTL for "next-generation" fuels; (v) short "construct" times leading to shorter gestation periods; and (vi) minimum downtime if components/catalysts need to be replaced.

Several studies on the use of microstructured reactors for Fischer–Tropsch synthesis have been reported (Myrstad et al. 2009; Cao et al. 2009; Knochen et al. 2009; Deshmukh et al. 2010). Deshmukh et al.'s study spanned single channel to multiple channels and lengths from 0.04 to 0.62 m. It clearly demonstrated the advantages of microchannel devices specifically for the Fischer–Tropsch synthesis. The selectivity and carbon number distribution were found to be identical for single and multiple channels of different lengths under otherwise similar conditions. This conclusion is not surprising since microstructured systems need only stacking (scaling out) instead of scale-up. Their excellent heat transfer performance for the relatively exothermic Fischer–Tropsch synthesis was also apparent from the fact that the various configurations tested gave almost isothermal (±1 °C) operation. Facile regeneration of the catalyst fouled due to wax deposition could restore the original activity and selectivity. Excellent operational flexibility with respect to CO:H_2 ratio (1.5:2.0), temperature (210–230 °C), and pressure (1.8–2.6 MPa) allows optimization of the flow sheet to arrive at the best configuration in terms of operating and capital costs. In a recent study, LeViness et al. (2011) have given a quasiquantitative comparison of microchannel and conventional reactors for Fischer–Tropsch synthesis. Compact nature of microchannel devices allows utilization of natural gas reserves in remote places ("stranded" gas) that cannot be flared or reinjected due to economic reasons. Small capacity microstructured systems are the only alternative in this case since conventional Fischer–Tropsch reactors are economical only for relatively very large capacities (Tonkovich et al. 2009, 2011).

3.4.1.5 Prognosis for Fischer–Tropsch Synthesis Reactors   The aforementioned discussion clearly indicates the advantages of shell- and tube-type reactors with structured packings and microstructured reactors over the other types currently in practice. Between the two short-listed reactors, the microstructured reactors look more attractive both in terms of effectiveness and ultimate capital cost. Current (2013) prices for the latter may appear high in comparison with the other types. This must be attributed to the inclusion of developmental costs. Increasing number of applications of microstructured reactors is likely to attract more manufacturers. The resulting competition in the international market is expected to result in equipment prices at least an order of magnitude lower than the present. This is likely to happen in the next 8–10 years.

3.4.2 Oxidation of *p*-Xylene to Purified Terephthalic Acid for Poly(Ethylene Terephthalate)

Polyester made from terephthalic acid and monoethylene glycol is one of the largest products of the petrochemical industry. Polyester fiber has main outlets in wash-and-wear and drip-dry fabrics. These fabrics do not require ironing. A popular fabric is a 40:60 cotton–polyester blend that has similar advantages in addition to the feel and absorbency of cotton. The other major markets of this polymer are (i) tire cord, (ii) blow molded bottles, and (iii) film for food wrapping. The transparency and durability of polyester have lent it a leading edge over other polymers. The immense commercial potential of this polymer spurred research on development of its raw materials in quantities as large as the demand for the polyester itself. This is evident from the fact that the world production of polymer-grade purified terephthalic acid (abbreviated as PTA), a major ingredient of poly(ethylene terephthalate) (PET), increased by 68% from 1999 to 2010. Current world production capacity for PTA is 50 million tonnes per annum. Major introduction of capacity has occurred in the Asian region, with China contributing a major share. China's PTA production capacity increased by 11 million tonnes during this period. In September 2010, the Chinese PTA production capacity had reached 15.74 million tonnes/year. This amounted to 31% of world's production and 41% of Asian production of PTA (http://www.cxtextiles.com/html_news/YUZHIGUO-Web-Design-Studio-1.html).

Such large production capacities called for a very well-developed, mature technology, particularly for PTA since the second ingredient, ethylene glycol, can be obtained by hydration of ethylene oxide made by a relatively straightforward and established vapor-phase oxidation of ethylene on a silver catalyst. In the following, a brief history and evolution of the PTA technology is presented (Landau and Saffer 1968; McIntyre 2003).

The first recorded synthesis of PET by a reaction of PTA and ethylene glycol, popularly termed polyester, is attributed to Whinfield and Dickson (1941). The first commercially suitable process used was oxidation of p-xylene to terephthalic acid with dilute nitric acid under pressure (McIntyre 2003). Besides the pollution due to the oxidant used, the product obtained was colored or had a tendency to form colored impurities that necessitated conversion to di-methyl terephthalate by esterification with methanol and subsequent purification by recrystallization and distillation. As a result, the process was cumbersome even by the standards of the late 1940s. The polymer formed with monoethylene glycol, however, could be converted into a fiber having very good tensile properties (specific strength of 4.95 g/denier and extension at break of 11.8%). Subsequently, in 1943, Imperial Chemical Industries (ICI) was invited to further improve the process. ICI in turn revealed the ideas to Du Pont with whom there was an agreement for exchange of information. In the absence of a credible alternative, in the initial stages, all PET was made by transesterification of dimethyl terephthalate (DMT) with ethylene glycol. The nitric acid-based oxidation was replaced in 1953 by air oxidation of p-xylene to p-toluic acid, which was esterified with methanol to yield p-methyl toluate. This intermediate ester was then converted to DMT by first oxidation to monomethyl terephthalate and then esterification with methanol (Katzschmann 1966). Subsequently, an improved process that resulted in savings of operating and capital cost (which could be classified as process intensification) was developed. This innovation combined the two oxidations followed by another single esterification reactor for the two esterification steps (Katzschmann 1966). This process was variously known as Witten, Imhausen, Hercules, and Dynamit Nobel process. Even as the DMT-based manufacture of PET was commercialized with a capacity of 3 million tonnes per annum (Weissermel and Arpe 2003), the quest for direct air oxidation of *p*-xylene to terephthalic acid continued. Availability of *p*-xylene in large quantities was a major driving force for development of the direct oxidation route. This reaction is represented by:

$$CH_3C_6H_4CH_3 + 3O_2 \rightarrow HOOCC_6H_4COOH + 2H_2O \quad [\Delta H_R = -1360\,kJ/mol]$$

Landau and Saffer (1968) have described the development of this direct air oxidation route in detail. A previous attempt had given *p*-toluic acid in good yield but suffered from the strong resistance of the methyl group in *p*-toluic acid to further oxidation. Effectively therefore, the oxidation stopped at *p*-toluic acid. Landau and Saffer (1968) described the breakthrough in which manganese bromide in acetic acid at ~2.8 MPa and 200 °C could oxidize *p*-xylene to terephthalic acid at a yield of ~77%. According to these authors, "the critical combination of reaction ingredients required for high efficiency oxidation of *p*-xylene to terephthalic acid was a source of bromine, a metal catalyst preferably in the form of manganese or cobalt or mixtures thereof in acetic acid." Amoco bought the patent rights for the catalyst patented by Saffer and Barker (1958) and went on to develop the now famous Amoco-Mid-century (Amoco-MC) technology. Partenheimer (1995) has described and compared the various direct oxidation processes available commercially. The main advantages of the Amoco process listed by Partenheimer (1995) are as follows: (i) the first is the high activity as compared to other auto-oxidation catalysts. The catalyst activity of a Br–Mn–Co catalyst system can be 16 times higher (Sakota et al. 1968). This high activity has been attributed to the fact that the bromine atom can abstract hydrogen from the methyl group far more rapidly than the Co (III) species. This in turn increases the rate of initiation (presented in Section 3.4.2.1). Further, for commercially available supplies of the hydrocarbons, the induction time for the oxidation reaction is negligible. (ii) The second is the high selectivity over a wide temperature range (25–300 °C). The cobalt (alone) catalysts have an upper limit of 150 °C. In comparison, the Amoco-MC catalyst can be used at much higher temperatures without loss of yield of the aromatic acid. The amounts of carbon monoxide and dioxide in the vent are reduced by an order of magnitude that is indicative of selective oxidation to the desired product rather than complete oxidation. The process is therefore well suited for conversion of a variety of methyl-substituted benzenes to the corresponding carboxylic acids. (iii) Saffer and Barker (1958) suggested acetic acid as the solvent. It has a much better stability at the reaction conditions than other acid solvents. The carboxylic acids produced have a relatively low solubility in acetic acid (used as solvent) at ambient conditions. This allows facile separation of the product from the solvent-soluble catalyst and recycle of the same. (iv) The fourth is the order of magnitude lower catalyst requirement and smaller reactor volumes as compared to the Witten process (Brill 1960). In the Amoco-MC process, all organic compounds undergo oxidation. However, acetic acid is relatively inert, and oxidation of acetic acid is in fact lower than that of the methylbenzene, which is the main reactant. Claims of less than 10% loss of acetic acid per unit weight of terephthalic acid produced have been made (Nowicki and Lowry 1990). The major drawback of the Amoco-MC process is the highly corrosive nature of bromine–acid system that necessitates the use of titanium as the material of construction. The reactor cost can be minimized by using a titanium cladding on mild steel. However, in the case of a stirred reactor, the impeller assembly comprising of the shaft, the hub, and the impeller must be made out of titanium. Li and Li (2008) proposed $CoBr_2$–$MnBr_2$ as the bromine sources as an improvement. This catalyst is less corrosive as compared to the HBr containing catalyst. Further, theses investigators found that the catalyst system proposed operates at much lower temperature of 100 °C as compared to 175–225 °C in the conventional Amoco-MC process. Recovery of the catalyst is also claimed to be simpler as compared to the Co/Mn/Br catalyst containing NaBr. With an optimum Br/Co atomic ratio of three, the maximum yield of terephthalic acid was reported to be 93.5%, which is comparable to the conventional process.

Besides the Amoco-MC process, several other processes such as Eastman–Kodak, Teijin, Maruzen, Henkel I (based on phthalic anhydride), Henkel II (based on benzoic acid), Lummus (*p*-xylene ammoxidation route), and Toray (acetaldehyde/paraldehyde in acetic acid) have been developed (Raghavendrachar and Ramachandran 1992; Weissermel and Arpe 2003). Efforts on development of bromine free heterogeneous catalytic routes appear attractive. Futter et al. (2013) have presented an excellent contemporary review on application of heterogeneous catalysts for cumene to phenol, cyclohexane to adipic acid, and auto-oxidation of toluene, xylene, and 1,3,5-trimethylbenzene. This review has brought out the associated problems. These include leaching and fouling besides suitable framework or the support for the redox-active transition metal ions that must allow redox chemistry without compromising the structure of the catalyst. Another exhaustive review by Tomas et al. (2013) covers the recent trends in terephthalic acid synthesis, which include use of subcritical/supercritical water/ionic liquids as solvents, alternate catalyst systems, etc. Tomas et al. (2013) also reviewed the literature pertaining to process optimization and related developments. The aim of research in this field will obviously be to develop a benign, less corrosive, and environment-friendly but equally efficient process to replace the Amoco-MC process.

3.4.2.1 Mechanism of the Amoco-MC Process (Sheldon and Kochi 1981; Partenheimer 1995)

The discussion that follows is based on an exhaustive review by Partenheimer (1995) of metal-/bromide-promoted auto-oxidation of hydrocarbons. The homolytic oxidation process has three steps commonly encountered in free radical-assisted reactions. Considering toluene as the simplest methylbenzene and I as a radical initiating species, the steps involved have been listed as follows:

1. Initiation: $C_6H_5CH_3 + I \rightarrow C_6H_5CH_2^{\bullet} + IH$
2. Propagation: $C_6H_5CH_2^{\bullet} + O_2 \rightarrow C_6H_5CH_2O_2^{\bullet}$

   $C_6H_5CH_2O_2^{\bullet} + C_6H_5CH_3 \rightarrow C_6H_5CH_2OOH + C_6H_5CH_2^{\bullet}$
3. Termination: $C_6H_5CH_2^{\bullet} + C_6H_5CH_2O_2^{\bullet} \rightarrow C_6H_5CH_2OOCH_2C_6H_5$

   $C_6H_5CH_2O_2^{\bullet} + C_6H_5CH_2O_2^{\bullet} \rightarrow C_6H_5CH_2O_4CH_2C_6H_5$

At temperatures exceeding 50 °C, the primary free peroxide ($C_6H_5CH_2OOH$) produced in the propagation stage slowly dissociates according to:

$$C_6H_5CH_2OOH \rightarrow C_6H_5CH_2O^{\bullet} + OH^{\bullet}$$

The propagation step is the slowest and therefore rate controlling. In general, catalysts do not have any significant effect on the rate of the propagation (Roby and Kingsley 1996). Metals (cobalt/manganese) present in the homogeneous system act as strong modifiers for the above scheme in the following manner: cobalt (II) reacts rapidly with the primary peroxide radicals forming Co (III), which converts methylbenzenes to benzylic radicals ($C_6H_5CH_2^{\bullet}$). Thus, cobalt helps in the initiation step. Peracids originating from corresponding aldehydes rapidly react with cobalt species. There is an intricate synergy between the solvent and the catalyst. It is claimed that the dimeric form of cobalt (II) in acetic acid–water mixtures allows a nonradical, very fast, and selective formation of corresponding carboxylic acids from peracids. The solvent used in most MC type oxidations is a carboxylic acid (acetic acid in the case of terephthalic acid owing to its stability). For more details, the reader may refer to Partenheimer (1995). This paper provides detailed information on the mechanism of oxidation and types of catalysts, followed by discussion of oxidation of individual alkyl aromatics and other important organic species.

3.4.2.2 Operating Conditions and Safety Aspects

Typical operating conditions for the Amoco-MC process are as follows: pressure, 1.5–3.0 MPa, and temperature, 190–210 °C. Catalyst used comprises of cobalt acetate, manganese acetate, and bromine in the form of ammonium bromide or HBr. Acetic acid–water mixture (90:10) is used as a solvent. *p*-Xylene to solvent ratio is approximately 1:3, which works out to a *p*-xylene concentration range of 1.5–2.0 kmol/m³. Under these conditions, typical yield of terephthalic acid is ~90%. The catalyst system is highly active and therefore the commonly encountered induction period is practically absent. The product terephthalic acid has relatively low solubility in acetic acid at reaction conditions (1.8 wt % at 200 °C). Therefore, the crude terephthalic acid product is separated from the reaction mixture by conventional filtration (the solubility of terephthalic acid in acetic acid is a function of both its purity and temperature). Recently, Malpani et al. (2011) have determined Hildebrand solubility parameters for *p*-xylene oxidation products with a view to provide better understanding of the solubility of the various products. These authors have also provided the 3D Hansen solubility parameters. More work in this area is required to design a suitable solvent that may permit a relatively simpler and inexpensive method to obtain polymer-grade terephthalic acid. There are several claims of use of other solvents including water (Hronec and Ilavsky 1982).

Safety Aspects: Liquid-phase oxidations of hydrocarbons/aldehydes and similar flammable compounds have an inherent safety problem. Alexander (1990) has presented a comprehensive discussion on the causes of conflagration in air oxidation of hydrocarbons and possible methods of preventing the same. In a majority of the cases involving such oxidations, atmospheric air compressed to the desired pressure is used as the source of oxygen. As the air bubbles contact the process, liquid oxygen dissolves in it. Simultaneously, depending on the vapor pressure of the process liquid at the operating temperature, a certain amount of liquid is vaporized and incorporated in the bubbles. The operating conditions should be such that the oxygen-combustible vapor mixture is below the lower flammability limit or above the upper flammability limit. In the present case of p-xylene oxidation, the conversion of p-xylene is above 90%. Thus, the major component in the outlet gas mixture is acetic acid, which is used as the solvent. The flash point of acetic acid is 39 °C and it falls in class IC of combustible and flammable liquids. The operating temperature range for the Amoco-MC process is always above the flash point of acetic acid. Therefore, care must be taken to curtail the oxygen partial pressure in the exhaust gases to ensure safe operation. The corresponding oxygen content should be below 8–9 vol. %. Under actual operating conditions, a factor of safety may be applied that limits the oxygen content to 5 vol. % (Roby and Kingsley 1996; Mills and Chaudhari 1999). A conservative estimate of airflow rate can be made on the basis of (i) 90% conversion of p-xylene in the feed, (ii) expected consumption of oxygen in side reactions, (iii) evaporation of acetic acid corresponding to saturation of the exit gas at the reactor temperature, and (iv) a final oxygen volume percentage less than 5%. Arpentinier (2006) has stressed the importance of safety measures particularly during the start-up of the reactor when the uptake of oxygen is negligible but stripping of the solvent is inevitable. Hence, oxygen may build up to levels much higher than 7–8 vol. %. Similarly, precaution must also be exercised during shutdown of the reactor. Addition of an inert gas (nitrogen/steam) in the disengagement space of the reactor can avert a dangerous conflagration in such situations.

3.4.2.3 Regime of Mass Transfer Accompanied by a Chemical Reaction in the Amoco-MC Process

As mentioned earlier, in the presence of the MC catalyst, the induction period is absent. The reaction is typically autocatalytic, and as it progresses, the rate increases. For such a reaction, the regime of mass transfer accompanied by a chemical reaction may shift from an initially kinetic-controlled regime (Regime A in Table 2.2, due to low rate constant) to mass transfer-controlled regime (Regime B in Table 2.2), as the rate constant increases due to the autocatalytic nature of the reaction. For liquid-phase oxidation of cyclohexane, Suresh et al. (1988) have observed a similar shift. It is also important to note that the reaction mixture contains a large amount of suspended terephthalic acid that is likely to decrease the overall gas–liquid mass transfer coefficient. Thus, although hydrocarbon oxidations are known to exhibit zero-order kinetics with respect to dissolved oxygen (Sheldon and Kochi 1981) under typical Amoco-MC operating conditions, the system may suffer from substantial mass transfer limitation as pointed out by Suresh et al. (1988). The propagation reaction $C_6H_5CH_2^{\bullet} + O_2$ becomes rate controlling when the oxygen partial pressure is below 0.15 atm (Arpentinier 2006). For such low partial pressures of oxygen, the overall absorption process is controlled by mass transfer of oxygen from the gas to the liquid phase. Several investigators have studied the kinetics of this reaction. It is generally accepted that oxidation of p-toluic acid to 4-carboxy benzaldehyde (4-CBA) is the rate-controlling step in the consecutive reaction scheme mentioned earlier (Cincotti et al. 1999; Wang et al. 2005; Li et al. 2014). Wang et al. (2007) have indicated that the Hatta number is <0.3 and further the ratio of the homogeneous reaction rate to the gas–liquid mass transfer rate ($\varepsilon_L \times k_2 \times B_B / k_L \underline{a}$ for a second-order reaction) is 0.2–0.3 for a stirred tank reactor and 0.3–0.4 for a sparged reactor. This indicates that the operating regime corresponds partially to Regime A. Li et al. (2013) have also used the reported kinetic data to conclude that Hatta number is <1. These investigators, however, concluded that oxygen mass transfer is the controlling step (Regime B). They argued that low dissolved oxygen concentrations are responsible for the formation of the undesired products (particularly, 4-CBA). For large industrial stirred reactors, inhomogeniety in hydrodynamics can lead to regions having low oxygen concentration and consequently low overall selectivity and product purity (Hobbs 1972). In view of this, local oxygen mass transfer rates assume importance. Computational fluid dynamics, as used by Patwardhan et al. (2005), can identify such regions and corrective action can lead to improvements. Osada and Savage (2009) have suggested that instead of sparging the entire amount of oxygen required at a single location (in the discharge stream of the impeller in the case of a stirred reactor), feeding at multiple locations leads to remarkable increase in the yield of terephthalic acid. Evidently, a strategy employing feeding at multiple locations maintains consistent availability of oxygen at different locations and improves selectivity (Patwardhan et al. 2005; Wang et al. 2007). Local values of $k_L \underline{a}$ are not available. However, global values of $k_L \underline{a}$ for stirred tank, and sparged reactors can be predicted using fairly reliable correlations as discussed in Chapters 7A and 10, respectively.

3.4.2.4 *Reactor Selection for the Amoco-MC Process*   Majority of liquid-phase oxidations use relatively high pressure (>0.5 MPa). Oxidations are characterized by a substantial heat of reaction that must be removed to prevent side reactions. Therefore, the reactor selected should have good heat transfer characteristics. In the present case of *p*-xylene oxidation using the Amoco-MC process, the heat of reaction is generally removed by evaporation of the solvent. The solvent vapor is condensed and returned to the reactor. The enthalpy of condensation of the solvent is used to raise high-pressure steam for use elsewhere in the plant. The temperature–pressure condition is auto-related by the refluxing media.

A special feature of the direct oxidation process is the large amount of suspended crude terephthalic acid (25–35 wt % depending on the feed concentration of *p*-xylene, temperature/pressure/catalyst concentration, residence time in the reactor, etc.) that the reactor must handle. This solid phase must be kept in suspension. At the same time, uniformity in supply of oxygen also must be maintained. Therefore, the impeller speed should be above N_{SG} or N_{CD}, whichever is higher. Provision of multiple impellers matching with oxygen feed locations should be an ideal option to maintain uniformity in solid suspension/gas dispersion and gas–liquid mass transfer.

Considering the aforementioned requirement, both sparged and stirred reactors can be used. Both these reactors provide good mass transfer. Their relative merits are discussed in the following.

1. Sparged Reactors

These reactors (Fig. 3.2a) have been suggested for the oxidation of *p*-xylene to terephthalic acid (Carra and Santacesaria 1980; Arpentinier 2006; Wang et al. 2007).

The major advantage is simple construction with no moving parts that require maintenance. The power input for mass transfer/solid suspension in the case of the sparged reactor depends on the air velocity. For suspending large solid populations such as in the present case, a relatively high air velocity is required. Three major negative features arising out of the high air velocity required are apparent: (i) the compression power $(P_G = Q_G \times \Delta P_G) = V_G \times A_C \times \Delta P_G)$ increases linearly with air throughput (ΔP_G is approximately equal to the hydrostatic head). Hence, with larger air throughputs, there is a proportional increase in the compression cost. (ii) Higher air throughputs also imply correspondingly higher oxygen throughput. Since the amount of oxygen required (mainly for the major reaction) remains the same, the extra oxygen can lead to an unsafe situation (exit volume percentage >7–8%; please refer to Section 3.4.2.1: **Safety Aspects**). (iii) Assuming that the vent gas contains acetic acid at a level that corresponds to saturation at the reactor temperature (partial pressure = vapor pressure), higher air throughputs also increase the amount of solvent stripped. This results in a greater load on the downstream vent gas treatment system and thereby higher capital and operating costs for this system. In addition, Wang et al. (2007) indicate that due to a lower global value of $k_L a$, the volume of a sparged reactor is about 30–50% higher than that of a stirred reactor for equivalent duty. Hobbs et al. (1972) and Jacobi and Baerns (1983) argued that in a sparged reactor the oxygen partial pressure at the bottom is substantially high. As a result, in this zone, the system corresponds to Regime A. On the other hand, toward the top of the column where the oxygen partial pressure is reduced due to (i) consumption of oxygen in the reaction and (ii) decrease in static head, the system can correspond to Regime B. Wang et al. (2007) have also pointed out that in sparged reactors there is a significant variation in gas holdup in the radial direction. This gives rise to problems in scaled-up versions. The gas velocity in the area close to column axis varies from 4 to 10 times the value near the wall and increases with $(T)^{0.3–0.5}$ (Wang et al. 2007). The consequence of this is channeling of gas, which can lead to safety problems. Wang et al. have suggested the use of internals (resistance internals) to break up the dominant upward gas flow at the center. These internals also reduce the radial variation in gas velocity and hold up. As a natural consequence of energy dissipation at the internals, the global $k_L a$ increases. The lower the free area of these internals, the higher the energy dissipation and $k_L a$. The practical use of such internals in a system with relatively high solid loading is doubtful. In fact, the major advantage pertaining to ability to handle solids is likely to be lost when such internals are used. The earlier discussion clearly outlines the problems that will be faced when a conventional sparged reactor/bubble column is used. Based on the information available, most large Amoco-MC plants use stirred reactors rather than sparged reactors.

The concept of three-phase sparged reactor or bubble column with a draft tube (BCDT) can be advantageously applied in this oxidation process. Section 10.9 presents a detailed discussion of various aspects of BCDT. The BCDT (Fig. 3.2b) is a simple variation of the conventional slurry or three-phase sparged reactor. The major conclusions that can be drawn with respect to the present application are as follows: (i) the overall gas holdup in a BCDT is approximately the same as that in a conventional bubble column. Further, the gas holdup is independent of solid loading. (ii) There is a well-directed liquid circulation—upward in the draft tube (riser) and downward in the annulus (Koide et al. 1988; Kushalkar and Pangarkar 1994). The respective liquid circulation velocities ranges are relatively high: 0.4–0.7 m/s and 0.2–0.4 m/s for the riser and downcomer sections. As a result, homogeneous regime, which is marked by smaller bubbles and higher gas holdup, prevails in the riser. The recirculation of the gas via the downcomer improves gas utilization. (iii) The radial and axial solid concentration profiles are uniform (Muroyama et al. 1985; Kushalkar and Pangarkar 1994). This is a very important advantage over conventional bubble column in which the solid concentration decreases exponentially along the height. Similar advantage of a draft tube has been reported by Conway et al. (2002) for the liquid oxidation reactor discussed in Section 3.4.2.4 (3). In addition, due to the high circulation velocities, solid suspension in a BCDT can be achieved at relatively low power inputs as compared to conventional bubble column (Jones 1985; Goto et al. 1989). This implies that a lower sparging rate can be used with consequent higher conversion if a pure gas is used (see Sections 3.4.2.4 (3) and 3.4.2.2: **Safety Aspects**). (iv) When required, the draft tube can also be used as an internal heat exchanger to facilitate rapid removal of heat (Koide 1983; Pangarkar 2002). Mudde et al. (2002) and de Deugd et al. (2003) have claimed similar advantages with an external downcomer in their gas lift recycle reactor. In conclusion, a BCDT is a distinct improvement over the conventional bubble column. An existing bubble column reactor is readily amenable to a retrofit since incorporation of the draft tube does not entail any major changes.

2. Stirred Tank Reactor

In the stirred tank reactor for a given set of operating conditions (gas velocity and solid loading in particular) and system geometry, solid suspension can be achieved by suitable choice of the agitator, its location, and the power input (this matter is discussed in detail in Section 7A.6.1). Thus, relatively large solid loading can be handled at air velocities small enough to maintain the vent oxygen partial pressure dictated by safety consideration. For large-scale manufacture of terephthalic acid, the L/T ratio may be higher than unity. In such cases, multiple impellers coupled with multiple spargers for air may be needed. The influence of impeller type on gas dispersion and solid suspension has been discussed in detail in Chapters 7A and 7B. Upflow impellers have been shown to be more energy efficient. Correlations for N_{CD} and N_{SG} are presented in Chapter 7A. For most cases, the value of critical impeller speed for solid suspension (N_{SG}) is higher than impeller speed for complete dispersion of the gas phase (N_{CD}) except for some low-density particles such as polystyrene (Chapman et al. 1983). In view of this observation, the operating impeller speed is fixed at a value slightly higher than N_{CD} ($\sim 1.1 \times N_{CD}$). Under these conditions, both the liquid and gas phases can be assumed to be wholly mixed. The oxygen and organic species concentrations in the reactor therefore correspond to the vent gas and product (slurry) concentrations, respectively. A unified correlation presented in Section 7A.5 can be used for predicting $k_L a$ for the given set of operating conditions.

3. Praxair Liquid Oxidation Reactor

The importance of operational safety has been mentioned several times in the above discussion on oxidation of *p*-xylene. The upper limit of oxygen partial pressure

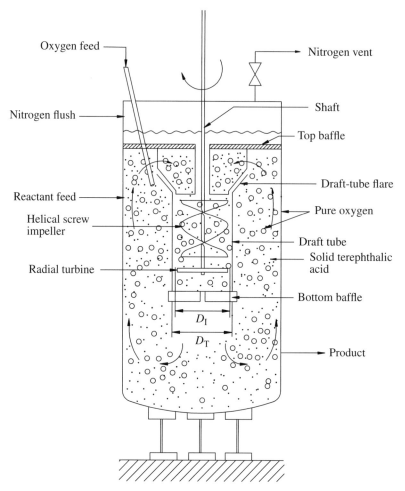

Oxygen feed

Nitrogen vent

Nitrogen flush

Shaft

Top baffle

Draft-tube flare

Reactant feed

Pure oxygen

Helical screw
impeller

Draft tube

Solid terephthalic
acid

Radial turbine

Bottom baffle

D_I

D_T

Product

FIGURE 3.3 Praxair liquid oxidation reactor. (Reproduced from Roby and Kingsley (1996) with permission from Praxair Technology Inc.)

severely restricts the productivity of the reactor (Roby and Kingsley 1996). Further, a large amount of nitrogen present in air-based oxidation increases the size of the vent scrubbing/solvent recovery system. The productivity of the reactor can be greatly improved by employing a gas-inducing-type agitator in conjunction with pure oxygen, provided the safety norms are not transgressed. A unique combination of hardware and controls developed by Praxair has been claimed to allow the use of pure oxygen or enriched air without sacrificing safety (Litz 1985). The Praxair liquid oxidation reactor has been suggested for oxidation of p-xylene to terephthalic acid (Roby and Kingsley 1996). The severity of the operating conditions is a clear incentive for such innovations (Moulijn et al. 2013). The main features of the liquid oxidation reactor are shown in Figure 3.3.

Conway et al. (2002) have studied the performance of a system similar to the Praxair liquid oxidation reactor. The equipment details were as follows: (i) a four-flight helical screw that yields a downward flow, (ii) a secondary six-flat-blade turbine that serves the purpose of shearing the gas entrained by the helical screw, and (iii) a draft tube of diameter equal to 0.33 of the main reactor diameter. Both the primary and secondary impellers are mounted on the same shaft inside the draft tube. A conical flared entry (bell shaped) at the top of the draft tube affords a smooth flow of the multiphase mixture into the draft tube. The helical screw creates a powerful swirling flow of the liquid in the downward direction. The turbine disperses the entrained gas and finally the flow is driven further in an axial direction by the cruciform baffles at the bottom section. The discharge from the draft tube impinges on the bottom of the reactor vessel and reverses direction to move upward in the annulus. This upward flow further reenters the draft tube directed by a horizontal baffle and under the suction created by the helical screw. The upper cruciform baffles create surface vortices that cause ingestion of the gas above a certain critical speed of agitation. Recirculation of entrained gas improves its utilization. Conway et al. made a systematic study of hydrodynamics of the aforementioned system. The power input (maximum up to 3 W/kg) and solid loading (0–25 wt %) were varied over a wide range. The conclusions derived were as follows: (i) starting with a low solid loading, the critical speed for gas ingestion initially increased and then decreased to a value approximately the same as that in the absence of solids. (ii) The effect of solid loading on the gas holdup was similar to that on the critical speed in (i). (iii) The critical speed for solid suspension, N_{SG}, depended on the liquid level above the draft tube and also its clearance from the bottom of the reactor. A correlation similar to that of Zwietering (1958) was satisfactory for predicting, N_{SG}. (iv) The gas–liquid mass transfer coefficient $k_L a$ was found to be proportional to $(\varepsilon_M)^{0.82}$. This dependence is better than that for a conventional stirred reactor (Section 7A.6) probably because of the well-directed circulation afforded by the draft tube. (v) The solid concentration was nonuniform at low solid loadings. However, at higher solid loadings and $N > N_{SG}$, the difference between the annulus and draft tube solid concentrations was much less. This observation is significant inasmuch as the highest solid loading employed is relevant to the Amoco-MC process. The main advantage of the Praxair liquid oxidation reactor is probably derived from the well-directed mean flow produced through the incorporation of the draft tube (Conway et al. 2002). The importance of mean flows in stirred reactors is discussed in Sections 6.3.1.2, 7A.5.3, and 7A.7.1. Roby and Kingsley (1996) have reported an oxygen utilization efficiency of 99% for the Praxair liquid oxidation reactor. Oxygen concentration in the gas space is monitored and any transgression of the safety norms is claimed to be effectively nullified by a control strategy that shuts down oxygen supply with concomitant increase in the inert (nitrogen) gas supply.

The major advantages of the liquid oxidation reactor are (Roby and Kingsley 1996) as follows: (i) the use of pure oxygen implies a partial pressure of oxygen 20 times higher as compared to the limit of 5 vol. % in air prescribed by safety considerations. Since the solubility of a gas is proportional to its partial pressure, this implies a corresponding 20-fold increase in its dissolved concentration. In the case of a mass transfer-controlled regime (Regime B), since the rate of absorption is first order with respect to the dissolved concentration of the solute gas, the use of pure oxygen yields correspondingly 20 times higher rate of absorption. (ii) Since the volume of the gas fed to the liquid oxidation reactor is proportionately lower than that in the case of a conventional air-based system, there is a concomitant decrease in the compression cost. (iii) By the same argument, the volume of the vent gas is also lower than that for a conventional system. Correspondingly, the capital and operating costs of the vent scrubbing system are lower. (iv) Because of the high oxygen concentration, the reactor can be operated at lower temperatures that suppress the formation of 4-CBA. This compound and other impurities that have one carboxylic acid group act as chain terminators and hence must be eliminated in the polymer grade of terephthalic acid (Tashiro et al. 2001). The specifications for polymer-grade terephthalic acid require that 4-CBA and p-toluic acid should be less than 25 and 150 ppm, respectively (Hashmi and Al-Luhaidan 2006). The latter has a higher solubility in water and hence does not crystallize with terephthalic acid. 4-CBA unfortunately cocrystallizes with terephthalic acid and needs to be eliminated separately. In the conventional stirred tank reactor, 4-CBA concentrations as high as 3000–4000 ppm are obtained. In current practice, 4-CBA content is reduced through a rather expensive hydrogenation. This purification section of PTA plant contributes 50% of the total capital cost and also increases the operating cost substantially (Nextant 2005). A lower 4-CBA content in the crude terephthalic acid allows substantial cost reduction.

It should be pointed out that the Praxair liquid oxidation reactor is a variant of gas-inducing reactors discussed in Chapter 9. Several investigators (Forrester et al. 1998; Conway et al. 2002) have shown that conventional gas-inducing reactors are not suitable for large-scale applications. However, the well-directed mean flow in the Praxair liquid oxidation reactor allows facile solid suspension as compared to conventional gas-inducing systems (Sections 9.4.4 and 9.5). The manifold theoretical advantages mentioned are obvious and cannot be ignored. Indeed, a process intensification idea can evolve with proper application of the concept. An important point that needs to be considered in developing this concept to the level of a mature, safe, and reliable technology is the propensity of titanium to combustion, particularly at high temperatures and high oxygen content in the gas phase. It may also be noted that there is very little published information on commercial utilization of this reactor.

4. Venturi Loop Reactor

Duveen (1998) has suggested the use of a venturi loop reactor for oxidations with pure oxygen in a manner exactly analogous to the Praxair liquid oxidation reactor. The operation in a dead-end mode has been claimed to produce practically no vent gas. It must be noted here that "burning" of acetic acid and the consequent products cannot be avoided. CO_2 and other products formed must be purged. To this extent, the operation is likely to be similar to the Praxair liquid oxidation reactor.

As will be seen in Chapter 8, the venturi loop reactor yields order of magnitude higher values of gas–liquid mass transfer coefficient in comparison with a stirred reactor or gas-inducing system. This is particularly advantageous in the mass transfer-controlled regime (Section 3.4.2.3) that may prevail at high conversions of p-xylene. However, this needs verification in view of the high solid loadings. The advantages associated with use of pure oxygen discussed in Section 3.4.2.4 (3) are applicable here as well. The most prominent pertains to the absence of a rotary part as compared to the different versions of the stirred reactor including the Praxair liquid oxidation reactor. These require careful maintenance of the sealing arrangement for the shaft. Other advantages of the venturi loop reactor are discussed in Chapter 8 along with the requisite information on the design of this alternative. The major consideration in developing this application to the level of a mature technology is operational safety, since pure oxygen is used (Section 3.4.2.2: **Safety Aspects**).

5. Spray Column Reactor

In a recent article, Li et al. (2013) have shown that a spray column has the potential to curtail formation of 4-CBA in the oxidation process to levels required for PTA and thus eliminate the expensive hydrogenation step. The premise for this innovation is the fact that a small droplet such as in a fine spray affords relatively very small diffusional path for the oxygen. Assuming a representative k_L value of 6×10^{-5} m/s and diffusivity of 5×10^{-9} m²/s, the film model yields a diffusion film thickness, $\delta = (D_M/k_L) = 1 \times 10^{-4}$ m or 100 µm. The typical droplet radius afforded by the spray nozzle in Li et al.'s study was 25 µm. The diffusion path (from drop surface to its center) is therefore much shorter than the diffusion film thickness, implying that the gas–liquid mass transfer resistance is relatively very low. Li et al. used the kinetic data from various literature investigations and estimated the characteristic diffusion and reaction times, τ_D and τ_R (Eqs. 2.1 and 2.2, respectively). They found that τ_D was one to two orders of magnitude lower than τ_R. These calculations reflect the fact that the relatively very small spray droplet is saturated with oxygen at the prevailing partial pressure of oxygen. The study was carried out in a titanium spray reactor using a mixture of oxygen and CO_2 (oxygen concentration limited to 50%). The total amount of oxygen fed was such that the excess or unconverted oxygen (after complete oxidation of p-xylene to terephthalic acid) in the reactor amounted to 3 mol. %. CO_2 was used instead of nitrogen (please refer to Section 3.4.2.4 (3)) because of its higher heat capacity and better inhibition of possible combustion of the solvent and titanium. This arrangement maintained the vapor phase outside the flammability region. The following conclusions were derived:

1. Concentrations of 4-CBA, p-toluic acid, 4-bromomethyl benzoic acid, and p-tolualdehyde in the product were, respectively, <25 ppm and negligible as compared to >1000 ppm of 4-CBA and 450–750 ppm of p-toluic acid, 4-bromomethyl benzoic acid, and p-tolualdehyde in a stirred autoclave.

2. Optimum conditions for best product quality were found to be 200 °C and 1.5 MPa. The lower temperature also provides a stable operation with heat of the reaction being dissipated rapidly through evaporation of acetic acid. Temperature overshoot is undesirable because of possibility of rapid combustion of the solvent and titanium.

3. The high purity product does not require further purification such as recrystallization followed by hydrogenation of 4-CBA. Lower solvent loss and a smaller exhaust gas treatment unit are major additional benefits.

4. Although Li et al. do not explicitly mention it, the spray tower has no moving parts. This is a significant advantage over the stirred reactor since the impeller and sealing arrangement in the latter require careful maintenance.

The foregoing discussion shows the superiority of the spray tower for oxidation of *p*-xylene to terephthalic acid. Li et al. (2013) suggest that this concept can be extended to replace existing stirred reactors for other similar gas–liquid mass transfer limited reactions.

Drawbacks: It is well known that the spray generated by a nozzle rapidly degrades due to coalescence, loss of momentum of the droplets, etc. In general, a spray is effective over not more than 1 m from the source of generation. In addition, precipitation of PTA formed is likely to increase the resistance to diffusion of oxygen in the droplet (Section 3.3.4). This implies that for large-scale applications, short spray columns with a parallel bank of sprays located to avoid overlap will be required. The problem of back mixing of the gas phase is similar to that in a stirred reactor operated above N_{CD}.

NOMENCLATURE

A_C: Tower cross-sectional area (m²)
D: Impeller diameter (m)
H: Tower height (m)
$k_L a$:   Volumetric gas–liquid mass transfer coefficient (1/s)
N_{CD}:   Critical speed of agitation for dispersion of gas (rev/s)
N_{CS}:   Critical speed of agitation for complete suspension of solids (rev/s)
N_S: Critical speed of agitation for just suspension of solids in solid–liquid system (rev/s)
N_{SG}:   Critical speed of agitation for just suspension of solids in gas–liquid–solid system (rev/s)
P_G: Power input (W)
Q_G: Volumetric flow rate of gas (m³/s)
T: Tower diameter (m)
V_G: Superficial gas velocity (m/s)
X: Solid loading (wt %)

Greek letters

ΔH_R: Heat of reaction (J/mol or kJ/mol)
ΔP_G: Gas-phase pressure drop (Pa or MPa)

Superscript

•:   Free radical

REFERENCES

Alexander JM. (1990) Gas phase ignition in liquid phase oxidation processes: a recipe for disaster. Trans Inst. Chem. Eng. (UK), B68:17–23.

Anderson RB. (1984) The Fischer–Tropsch synthesis. Academic Press, New York, USA.

Andrews SPS. (1977) Modern processes for the production of ammonia, nitric acid and ammonium nitrite. In: Thompson R, editor. The modern inorganic chemicals industry. The proceedings of a symposium organized by the Inorganic chemicals group of the industrial division of the Chemical Society, London, March 31 to April 1, 1977. Soc. Chem. Ind., London, UK.

Arpentinier P. (2006) Oxidation processes in liquid phase with oxygen. Section 11.1.2. Vol. II: Synthesis of intermediates for the petrochemical industry. In: Amadei C, editor in chief. Encyclopaedia of hydrocarbons, Istituto della Enciclopedia Italiana Fondata da Giovanni Treccani S. p. A., Rome, Italy. http://www.treccani.it/export/sites/default/Portale/sito/altre_aree/Tecnologia_e_Scienze_applicate/enciclopedia/inglese/inglese_vol_2/615-686_ING3.pdf. Accessed on August 6, 2014.

Astarita G, Ottino JM. (1995) 35 years of BSL. Ind. Eng. Chem. Res., 34:3177–3184.

Bahnholzer WF, Jones ME. (2013) Chemical engineers must focus on practical solutions. AICHEJ, 59:2708–2720.

Bennett CO, Myers JE. (1962) Momentum, heat and mass transfer. McGraw Hill, New York, USA.

Bhutada SR, Pangarkar VG. (1989) Solid suspension and mixing in liquid jet loop reactors. Chem. Eng. Sci., 44:2384–2387.

Bird RB, Stewart WE, Lightfoot EN. (2007). Transport phenomena. John Wiley & Sons., Inc, New York, USA.

Brill WF. (1960) Terephthalic acid by single-stage oxidation. Ind. Eng. Chem., 52:837–845.

Bustamante F, Enick RM, Cugini AV, Killmeyer RP, Howard BH, Rothenberger KS, Ciocco MV, Morreale BD, Chattopadhyay S, Shi S. (2004) High-temperature kinetics of the homogeneous reverse water–gas shift reaction. AICHEJ, 50:1028–1041.

Cao C, Hu J, Li S, Wilcox W, Wang Y. (2009) Intensified Fischer–Tropsch synthesis process with microchannel catalytic reactors. Cat. Today, 140:149–156.

Carra S, Santacesaria E. (1980) Engineering aspects of gas-liquid catalytic reactions Cat. Rev. Sci. Eng., 22(1):75–140.

Chapman CM, Nienow AW, Cooke M, Middleton JC. (1983) Particle-gas-liquid mixing in stirred vessels. Part III: Three-phase mixing. Chem. Eng. Res. Des., 61(3):167–181.

Cincotti A, Orrù A, Cao G. (1999) Kinetics and related engineering aspects of catalytic liquid-phase oxidation of p-xylene to terephthalic acid. Cat. Today, 52:331–347.

Conway K, Kyle A, Rielly CD. (2002) Gas–liquid–solid operation of a vortex-ingesting stirred tank reactor. Chem. Eng. Res. Des., 80:839–845.

Davis BH. (2003) Fischer–Tropsch synthesis: overview of reactor development and future potentialities. Prep. Pap. Am. Chem. Soc., Div. Fuel Chem., 48(2):787–790.

Deckwer W-D. (1992) Bubble column Reactors (English translation). John Wiley & Sons, Chichester, UK.

de Deugd RM, Kapetijn F, Moulijn JA. (2003) Trends in Fischer–Tropsch reactor technology—opportunities for structured reactors. Top Cat, 26(1–4):29–39.

Deshmukh SR, Tonkovich AL, Jarosch KT, Schrader L, Fitzgerald SP, Kilanowski DR, Lerou JJ, Mazanec TJ. (2010) Scale-up of microchannel reactors for Fischer-Tropsch synthesis. Ind. Eng. Chem. Res., 49:10883–10888.

Dry ME. (1981) The Fischer–Tropsch Synthesis. In: Anderson JR, Boudart M, editors. Catalysis—science and technology. Vol. 1. Springer, Berlin, Germany. Chapter 4, p 159–255.

Dry ME. (1982) Catalytic aspects of industrial Fischer–Tropsch synthesis. J. Mol. Cat. 17:133–144.

Dry ME. (1990) The Fischer–Tropsch process—Commercial aspects, Cat. Today, 6:183–206.

Dry ME. (2001) High quality diesel via the Fischer–Tropsch process—a review. J. Chem. Technol. Biotechnol., 77:43–50.

Dry ME. (2002) The Fischer–Tropsch process: 1950–2000. Cat. Today, 71:227–241.

Duveen RF. (1998, Oct) High-performance gas–liquid reaction technology. In: Proceedings of the Symposium New Frontiers in Catalytic Reactor Design, The Royal Society of Chemistry, Applied Catalysis Group, Billingham, UK..

Eilers J, Posthuma SA, Sie ST. (1990) The Shell middle distillate process. Cat. Letters, 7:253–270.

Espinoza RL, Du Toit E, Santamaria J, Menendez M, Coronas J, Irusta S. (2000). Use of membranes in Fischer–Tropsch reactors. Studies in Surface Science and Catalysis 130A (International Congress on Catalysis, 2000, Pt. A), Elsevier BV, the Netherlands. p 389–394.

Florea I, Liu Y, Ersen O, Meny C, Pham-Huu C. (2013) Microstructural analysis and energy-filtered tem imaging to investigate the structure–activity relationship in Fischer–Tropsch catalysts. ChemCatChem, 5:2610–2620.

Forrester SE, Rielly CD, Carpenter KJ. (1998) Gas inducing impeller design and performance characteristics. Chem. Eng. Sci., 53:603–615.

Fox JM III. (1990) Fischer–Tropsch reactor selection. Cat. Lett., 7:281–292.

Futter C, Prasetyo E, Schunk SA. (2013) Liquid phase oxidation and the use of heterogeneous catalysts—a critical overview. Chem. Ing. Technol. 85(4):1–18.

Glasser D, Hildebrandt D, Liu X, Lu X, Masuku CM. (2012) Recent advances in understanding the Fischer–Tropsch synthesis (FTS) reaction. Curr. Opin. Chem. Eng., 1(3):296–302.

Goto S, Matsumoto Y, Gaspillo P. (1989) Mass transfer and reaction in bubble column slurry reactor with draft tube. Chem. Eng. Commun., B5:181–191.

Greenwood TS (1986) Loop reactors for catalytic hydrogenations. Chem. Ind., 3:94–98.

Guettel R, Kunz U, Turek T. (2008) Reactors for Fischer-Tropsch synthesis. Chem. Eng. Technol., 31(5):746–754.

Hashmi SMA, Al-Luhaidan S. (2006) Process for preparing purified terephthalic acid. EP 1671938 A1.

Hinze JO. (1975) Turbulence. 2nd ed. McGraw-Hill, Inc, New York, USA.

Hobbs CC, Drew EH, Van't Hof HA, Mesich FG, Onore MJ. (1972) Mass-transfer rate-limitation effects in liquid-phase oxidation. Ind. Eng. Chem. Prod. Res. Dev., 11:220–225.

Hooshyar N, Vervloet D, Kapteijn F, Hamersma PJ, Mudde RF, van Ommen JR. (2012) Intensifying the Fischer–Tropsch Synthesis by reactor structuring—a model study. Chem. Eng. J., 207–208:865–870.

Hronec M, Ilavsky J. (1982) Oxidation of polyalkylaromatic hydrocarbons. 12. Technological aspects of p-xylene oxidation to terephthalic acid in water. Ind. Eng. Chem. Prod. Res. Dev., 21:455–460.

Hupen B, Kenig EY. (2005) Rigorous modeling of NOx absorption in tray and packed columns. Chem. Eng. Sci., 60:6462–6471.

Jacobi R, Baerns M. (1983) The effect of oxygen transfer limitation at the gas–liquid interphase. Kinetics and product distribution of the p-xylene oxidation. Erdöl Kohle Erdgas Petrochemie, 36:322–326.

Jones AG. (1985) Liquid circulation in a draft tube bubble column. Chem. Eng. Sci., 40:449–462.

Joshi JB, Sharma MM. (1976) Mass transfer characteristics of horizontal sparged contactors. Trans. Inst. Chem. Eng. (UK), 54(1):42–53.

Katzschmann E. (1966) Oxidation of alkyl aromatic compounds. Chem. Ing. Tech., 38:1–10.

Kenig E, Seferlis P. (2009) Modeling reactive absorption. Chem. Eng. Prog., 105(1):65–73.

Knochen J, Guttel R, Knobloch C, Turek T. (2010) Fischer–Tropsch synthesis in milli-structured fixed-bed reactors: experimental study and scale-up considerations. Chem. Eng. Process., 49:958–964.

Kohl AL, Riesenfeld FC. (1985) Gas purification, 4th ed. Gulf Pub. Co., Houston, TX, USA.

Koide K, Sato H, Iwamoto S. (1983) Gas holdup and volumetric liquid-phase mass transfer coefficient in bubble column with draught tube and with gas dispersion into tube. J. Chem. Eng. Jpn., 16:413–419.

Koide K, Kimura M, Nitta H, Kawabata H. (1988) Liquid circulation in bubble column with draught tube. J. Chem. Eng. Jpn., 21:393–399.

Kölbel H, Ackermann P, Engelhardt F. (1956) Nue Entwicklungen zur Kohlenwasserstoff-Synthese. Erdöl und Kohle, 9:303–307.

Krishna R, Sie ST. (1994) Strategies for Multiphase Reactor Selection. Chem. Eng. Sci., 49:4029–4065.

Krishna R, Sie ST. (2000) Design and scale-up of the Fischer–Tropsch bubble column slurry reactor. Fuel Proc. Technol., 64:73–105.

Kushalkar KB, Pangarkar VG. (1994) Particle-liquid mass transfer in a bubble column with a draft tube. Chem. Eng. Sci., 49:139–144.

Landau R, Saffer A. (1968) Development of the M-C process. Chem. Eng. Prog., 64 (10):20–26.

Levenspiel O. (2005) What will come after petroleum? Ind. Eng. Chem. Res., 44:5073–5078.

LeViness S, Tonkovich AL, Jarosch K, Fitzgerald S, Yang B, McDaniel J. (2011) Improved Fischer–Tropsch economics enabled by microchannel technology. Velocys© 2009. http://www.velocys.com/press/wp/wp110224_microchannel_FT_white_paper_24Feb11.pdf. Accessed on August 6, 2014.

Li K-S, Li S-W. (2008) $CoBr_2$–$MnBr_2$ containing catalysts for catalytic oxidation of p-xylene to terephthalic acid. App. Cat. A: General, 340:271–277.

Li M, Niu F, Zuo X, Metelski PD, Busch DH, Subramaniam B. (2013) A spray reactor concept for catalytic oxidation of p-xylene to produce high-purity terephthalic acid. Chem. Eng. Sci., 104:93–102.

Litz LM. (1985, Nov) A novel gas–liquid stirred reactor. Chem. Eng. Prog., 81(11):36–39.

Malpani V, Ganeshpure PA, Munshi P. (2011) Determination of solubility parameters for the p-xylene oxidation products. Ind. Eng. Chem. Res., 50:2467–2472.

Maretto C, Krishna R. (1999) Modeling of a bubble column slurry reactor for Fischer–Tropsch synthesis. Cat. Today, 52:279–289.

Masuku CM, Hildebrandt D, Glasser D. (2011) The role of vapour–liquid equilibrium in Fischer–Tropsch product distribution. Chem. Eng. Sci., 66:6254–6263.

McIntyre JE. (2003) The historical development of polyesters. In: Scheirs J, Long TE, editors. Modern polyesters: chemistry and technology of polyesters and co-polyesters. Vol. 3. John Wiley & Sons Ltd. Chichester, UK. p 1–28.

Mills PL, Chaudhari RV. (1999) Reaction engineering of emerging oxidation processes. Cat. Today, 48:17–29.

Milmo S. (2011, May 9) Go faster FT catalysts. Chem. Ind., (9):17–19.

Moulijn JA, Makkee M, van Diepen AE. (2013) Chemical process technology. 2nd ed. John Wiley & Sons, Chichester, UK.

Mudde RF, Ghotage AV, van den Akker HEA, Grievink J. (2002) Design of a gas lift loop reactor for Fischer–Tropsch synthesis. ISCRE 17, 17th International Symposium on Chemical Reaction Engineering, Session A: Novel Reactors and Processes, Hong Kong, China.

Muroyama K, Mitani Y, Yasunishi A. (1985) Hydrodynamic characteristics and gas–liquid mass transfer in a draft tube slurry reactor. Chem. Eng. Commun., 34:87–98.

Myrstad R, Eri S, Pfeifer P, Rytter E, Holmen A. (2009) Fischer–Tropsch synthesis in a micro structured reactor. Cat. Today, 147S:S301–S304.

Nexant Inc. (2005) Squeezing profitability from the PTA/PET value chain: impact of the latest technology advances. http://www.chemsystems.com/reports/search/docs/prospectus/mc-polyester-pros.pdf. Accessed on Feb. 2014.

Nienow AW. (1969) Dissolution mass transfer in turbine agitated baffled vessels. Can. J. Chem. Eng., 47:248–256.

Nienow AW. (2000) The suspension of solid particles. Chapter 16. In: Harnby N, Edwards MF, Nienow AW, editors. Mixing in the process industries. 2nd ed. Butterworth-Heinemann, Oxford, UK. p 364–393.

Niranjan K, Pangarkar VG. (1984) Gas holdup and mixing characteristics of packed bubble columns. Chem. Eng. J., 29:101–111.

Niranjan K, Sawant SB, Joshi JB, Pangarkar VG. (1982) Counter current absorption using wire gauze packings. Chem. Eng. Sci., 37:367–374.

Nishimura S. (2001) Handbook of heterogeneous catalytic hydrogenation for organic synthesis. John Wiley & Sons Inc., New York, USA.

Norman WS. (1962) Absorption, distillation and cooling towers. Longmans, Green & Co. Ltd., London, UK.

Nowicki NR, Lowry JD. (1990, Jan 9) Inventors; Amoco Corp. Assignee. Continuous two-stage oxidation of aromatic hydrocarbons to aromatic carboxylic acids in an aqueous system. U.S. Patent 4892970.

Osada M, Savage PE. (2009) Terephthalic acid synthesis at higher concentrations in high temperature liquid water. 2. Eliminating undesired by-products. AICHEJ, 55(6): 1530–1537.

Pangarkar VG. (2002) Use of draft tube for effective heat removal in stirred and sparged reactors. Unpublished work.

Pangarkar K, Schildhauer TJ, van Ommen JR, Nijenhuis J, Kapteijn F, Moulijn JA. (2008) Structured packings for multiphase catalytic reactors. Ind. Eng. Chem. Res., 47:3720–3751.

Pangarkar K, Schildhauer TJ, van Ommen JR, Nijenhuis J, Moulijn JA, Kapteijn F. (2009) Experimental and numerical comparison of structured packings with a randomly packed bed reactor for Fischer–Tropsch synthesis. Cat. Today, 147 (Suppl.):S2–S9.

Pangarkar K, Schildhauer TJ, van Ommen JR, Nijenhuis J, Moulijn JA, Kapteijn F. (2010) Heat transfer in structured packings with co-current downflow of gas and liquid. Chem. Eng. Sci., 65:420–426.

Partenheimer W. (1995) Methodology and scope of metal/bromide autoxidation of hydrocarbons. Cat. Today, 23:69–158.

Patwardhan AW, Joshi JB, Fotedar S, Mathew T. (2005) Optimization of gas–liquid reactor using computational fluid dynamics. Chem. Eng. Sci., 60:3081–3089.

Petkov K, Stratiev D. (2008) Long-term prognosis of crude oil price variation. Petroleum Coal. 50(3):62–66.

Raghavendrachar P, Ramachandran S. (1992) Liquid phase oxidation of *p*-xylene. Ind. Eng. Chem. Res., 31:453–462.

Ramachandran PA, Chaudhari RV. (1983) Three phase catalytic reactors. Gordon and Breach Science Publishers Inc., New York, USA.

Ranade VV. (2002) Computational flow modeling for chemical reactor engineering. Academic Press, San Diego, CA, USA.

Ranganathan P, Savithri S. (2010) Computational flow modeling of multiphase mechanically agitated reactors. In: Ho HW, editor. Computational Fluid Dynamics. ISBN: 978-953-7619-59-6, InTech. Available at: http://www.intechopen.com/books/computational-fluid-dynamics/computational-flow-modeling-of-multiphase-mechanically-agitated-reactors. p 307–334. Accessed Aug. 2013.

Rozita Y, Brydson R, Comyn TP, Scott AJ, Hammond C, Brown A, Chauruka S, Hassanpour A, Young NP, Kirkland AI, Sawada H, Smith RI. (2013) A study of commercial nanoparticulate g-Al$_2$O$_3$ catalyst supports. ChemCatChem, 5:2695–2706.

Saffer A, Barker RS. (1958, May 6) Inventors; Mid-Century Corp. assignee. Aromatic polycarboxylic acids. U.S. Patent 2,833,816.

Sakota K, Kamiya Y, Ohta N. (1968) The autoxidation of toluene catalyzed with cobalt monobromide in acetic acid. Bull. Chem. Soc. Jpn., 41:641–648.

Satterfield CN. (1970) Mass transfer in heterogeneous catalysis, M.I.T. Press, Cambridge, MA, USA.

Shah YT. (1979) Gas–liquid–solid reactor design. McGraw-Hill International Book Company, New York, USA.

Shaw JA. (1992) Succeed at solid suspension. Chem. Eng. Prog., 88(5):34–41.

Sheldon RA, Kochi JK. (1981) Metal-catalyzed oxidations of organic compounds. Academic Press, New York, USA.

Sie ST, Krishna R. (1999) Fundamentals and selection of advanced Fischer–Tropsch reactors. App. Cat. A: General, 186:55–70.

Siirola JJ. (1996) Industrial applications of chemical process synthesis. In: Anderson JL, editor. Advances in chemical engineering, Process Synthesis. Vol. 23. Academic Press, New York, USA. p 1–31.

Siirola JJ. (2014) The impact of shale gas in the chemical industry. AICHEJ, 60:810–819.

Suresh AK, Sridhar T, Potter OE. (1988) Autocatalytic oxidation of cyclohexane modelling reaction kinetics. AICHEJ, 34:69–80.

Tashiro Y, Iwahama T, Sakaguchi S, Ishii Y. (2001) A new strategy for the preparation of terephthalic acid by the aerobic oxidation of *p*-xylene using *N*-hydroxyphthalimide as a catalyst. Adv. Synth. Cat., 343:220–225.

Thomas JM, Ducati C, Leary R, Midgley PA. (2013) Some turning points in the chemical electron microscopic study of heterogeneous catalysts. ChemCatChem, 5:2560–2579.

Tomas RAF, Bordado JCM, Gomes JFP. (2013) *p*-Xylene oxidation to terephthalic acid: a literature review oriented toward process optimization and development. Chem. Rev., 113:7421–7469.

Roby AK, Kingsley JP. (1996) Oxide safely with pure oxygen. Chemtech, 26(2):39–47.

Tonkovich AL, Mazanec T, Jarosch K, Fitzgerald S, Yang B, Taha R, Kilanowski D, Lerou JJ, McDaniel J, Atkinson D, Dritz T. (2009) Gas-to-liquids conversion of associated gas enabled by microchannel technology. Velocys © 2009. http://www.velocys.com/press/wp/wp091504_%20associated_gas_white_paper_may09.pdf. Accessed on August 6, 2014.

Tonkovich AL, Jarosch K, Fitzgerald S, Yang B, Kilanowski D, McDaniel J, Dritz T. (2011) Micro channel gas-to-liquids for monetizing associated and stranded gas reserves. Oxford Catalyst Group, ©2011 http://www.velocys.com/press/wp/wp110206_microchannel_GTL_White_Paper_060211%5B1%5D.pdf; http://www.oxfordcatalysts.com/press/wp/wp130618_microchannel_GTL_White_Paper_061813.pdf. Accessed on August 6, 2014.

Wang Q, Li, X, Wang L, Cheng Y, Xie G. (2005) Kinetics of p-xylene liquid-phase catalytic oxidation to terephthalic acid. Ind. Eng. Chem. Res., 44:261–266.

Wang L, Cheng Y, Wang Q, Li X. (2007) Progress in the research and development of *p*-xylene liquid phase oxidation process. Frontiers Chem Eng China, 1(3):317–326.

Weissermel K, Arpe HJ. (2003) Industrial organic chemistry. 4th completely revised ed. Verlag Chemie, Weinheim, Germany.

Westerterp KR, van Swaaij WPM, Beenackers AACM (1988) Chemical reactor design and operation. 2nd ed. John Wiley & Sons Inc., New York, USA.

Whinfield JR, Dickson JT. (1941) Improvements relating to the formation of highly polymeric substances. Great Britain Patent 578079, 1946 June 14.

Woo KJ, Kang S-H, Kim S-M, Bae JW, Jun KW. (2010) Performance of a slurry bubble column reactor for Fischer–Tropsch synthesis: determination of optimum condition. Fuel Proc. Technol., 91:434–439.

Yates IC, Satterfield CN. (1991) Intrinsic kinetics of the Fischer–Tropsch synthesis on a cobalt catalyst. Energy Fuels, 5:168–173.

Zhang Q, Kang J, Wang Y. (2010) Development of Novel catalysts for Fischer–Tropsch synthesis: tuning the product selectivity. ChemCatChem, 2(9):1030–1058.

4

TURBULENCE: FUNDAMENTALS AND RELEVANCE TO MULTIPHASE REACTORS

4.1 INTRODUCTION

All multiphase reactors require efficient contacting of the various phases (gas/liquid/solid) involved. For this purpose, expenditure of energy is required. The energy expended generally manifests itself as turbulence. All gas-dispersed multiphase reactors, as well as some liquid-dispersed multiphase reactors, operate at a high level of turbulence. The turbulence created by the energy input is the cause of the high transport rates. Turbulence in single phase (homogeneous gas or homogeneous liquid) has been the focus of study of physicists and engineers since Reynolds first noted turbulence in pipe flow in 1883. Significant progress has been made in both the understanding of turbulence and its measurement since then. Powerful computational aids have also played an important role in both solution of complex problems and data collection. Sreenivasan (1999) has presented an extensive and critical analysis of the literature available on fluid turbulence and has concluded that despite the Herculean efforts of scientists all over the world, the knowledge about turbulence is still in its "infancy." Computational fluid dynamics coupled with powerful computers is being used to study flow structures, but again, as Sreenivasan (1999) points out, "computing is not understanding." Turbulence manifests when a certain critical Reynolds number is reached. For flow in pipes, the first appearance of turbulence begins above $Re_D > 2100$. In the case of boundary layer, flow turbulence manifests at $Re_L > 5 \times 10^5$. These lower limits are not necessarily fixed. For example, in the case of pipe flow, if the fluid enters the pipe

Design of Multiphase Reactors, First Edition. Vishwas Govind Pangarkar.
© 2015 John Wiley & Sons, Inc. Published 2015 by John Wiley & Sons, Inc.

smoothly without any disturbance, laminar flow can persist even at relatively higher values of Re_D than 2100. On the other hand, small disturbances can cause premature manifestation of turbulence. Schlichting (1955) gives details of the perturbation analysis for determination of the critical Reynolds number for onset of turbulence. Turbulence can have both negative and positive results. In the case of transport vehicles ranging from a simple automobile to the most sophisticated aircrafts, all attempts are made to suppress turbulent drag. A sleek, streamlined shape such as an aerofoil that affords lower resistance to flow reduces the drag on aircrafts, ships, automobiles, and consequently their fuel consumption. A wrong design apparent from the higher fuel consumption is an expression of the undesirable effect of turbulence. On the other hand, in a washing machine, the turbulent motion is directed at producing positive results by forcing the detergent solution through the clothes. This loosens the dirt adsorbed/absorbed by the cloth and eventually removes it from the same. The efficiency of a given washing machine depends on the extent of contact of the detergent solution with the soiled clothes. In the case of chemical process equipments, the effort is aimed at enhancing the positive effect. Turbulence is employed to enhance contact between various phases and thereby improve the rate of the transport processes. Physicists have labored for several decades to understand turbulence and develop a credible theory of turbulence. In the context of this book, the main objective is to predict the effect of turbulence and harness it to improve performance of chemical process equipments. Performance indicators in this case are the transport properties.

4.2   FLUID TURBULENCE

Turbulence in single phase has been extensively dealt with in various texts (Schlichting 1955; Bennett and Myers 1962; Brodkey 1967). Although this type is the simplest among the different varieties, it is still not fully understood. The situation becomes much more complicated when dealing with multiphase systems. Multiphase systems comprising of free interfaces are far more complex than a homogeneous system comprising of a single phase. The complexity arises because of a deformable interface. The particle phase (gas bubbles/liquid drops) can deform/break or coalesce to yield a size distribution in time and space. The deformation/break up or coalescence itself is a result of the chaotic motions that represent turbulence besides the phase volume fractions. The last parameter is a variable in process equipments and can vary from relatively very low to as high as 30–40 vol%. The high phase volume example is the most difficult. In summary, turbulence in multiphase systems is a conundrum that continues to challenge both theoretical and experimental approaches. In the following, basic definitions of turbulence parameters relevant to the subject of discussion are presented. The focus will be on prediction of the mass transfer parameters.

Turbulence in any system is characterized by chaotic motion of the fluid/particles in time and space. The fluid velocity vector, u_x, at any time is the sum of $\overline{u_x}$, the time average velocity, and u'_x, the instantaneous fluctuating component:

$$u_x = \overline{u_x} + u'_x$$

$$(4.1)$$

The instantaneous fluctuating component (u'_x) can be negative or positive. However, the time average of u'_x, $\overline{u'_x} = 0$. The amplitude of (u'_x) is conventionally expressed as $\sqrt{\overline{u'^2_x}}$ and is always positive. $\sqrt{\overline{u'^2_x}}$ is considered as a measure of the energy associated with the fluctuating component or the intensity of turbulence in the corresponding direction, in this case the x direction. The intensity of turbulence is also termed as the root-mean-square value of the fluctuating velocity in that direction, $i{:}u'_i = \sqrt{\overline{u'^2_i}}$. In this definition, the turbulence intensity has the units of velocity. u'_i reflects the importance of the fluctuating components relative to the mean flow. It will be seen later that the mean flow has an important bearing on bulk phenomena such as gas dispersion or solid suspension (Wu et al. 2001a, b) in stirred reactors, gas-inducing reactors, and venturi loop reactors. On the other hand, gas–liquid and particle–liquid mass transfer coefficients are decided by the intensity of turbulence as determined by u'_i. This definition is used later in Chapters 6, 7A, 7B, and 10 for characterizing turbulence for a given set of conditions in stirred and sparged reactors. Turbulence intensity is also expressed particularly in parallel flow as the percentage or fraction of the mean velocity, $I = \sqrt{\overline{u'^2_x}}\,/\overline{u}_x$. In this latter form, the turbulence intensity is dimensionless. A three-dimensional turbulence structure will have, apart from u'_x, u'_y and u'_z with similar interpretations of the intensity of turbulence in the respective directions.

Turbulence intensity is a major factor governing transport processes. It will be shown later that simple, measurable parameters or parameters that can be readily calculated are useful as a substitute for the somewhat esoteric turbulence intensity. Such a substitute can then be used to obtain correlations for mass transfer coefficient in multiphase reactors.

Most complex phenomena are made tractable using cases that are far simpler than the real phenomena. Examples of this type of approach abound in many areas of science/engineering starting with the ideal gas law for predicting fluid properties. This ideal case is then considered as the basis, and the deviations from this ideal case (fugacity/activity coefficient) represent the behavior of the real system. In the case of turbulence, two classes are defined.

4.2.1   Homogeneous Turbulence

In this class, the fluctuations in the velocity are truly random such that averaged values of turbulence parameters are independent of the position in space. Thus, the fluctuations do not vary with the axis of translation (Brodkey 1967). This type of turbulence can be generated with a well-defined equipment geometry. However, it can only be sustained over relatively short distances. An important parameter that can be used to represent turbulence is the root-mean-square velocity defined earlier. For homogeneous turbulence, the u'_i values need not be equal but they should not vary over the field under consideration.

4.2.2 Isotropic Turbulence

For this type of turbulence, the three fluctuating components are identical:

$$\sqrt{\overline{u'^2_x}} = \sqrt{\overline{u'^2_y}} = \sqrt{\overline{u'^2_z}}$$

For completely random fluctuations in space, there is no correlation between the different directional fluctuations, which can be written as

$$\overline{u'_x u'_y} = 0$$

for one pair and similarly for other pairs. As defined earlier, isotropic turbulence is always homogeneous. Isotropic turbulence is more amenable to mathematical analysis. Free turbulence not bounded by a solid boundary can often be approximated by isotropic turbulence. Isotropic turbulence can also be generated artificially.

4.2.3 Eddy Size Distribution and Effect of Eddy Size on Transport Rates

Although the primary energy input in most multiphase reactors results in anisotropic turbulence, the ultimate dissipation of the input energy by viscous dissipation through the smallest eddies yields isotropic turbulence. This matter will be discussed in some detail under Kolmogorov's theory.

A multiphase reactor has a wide distribution of eddy sizes having different energies. This can be illustrated using the example of a stirred reactor:

1. Primary eddies: These have low frequency and hence long life. They can contain 20–30% of the kinetic energy of the overall turbulence. Their length scale is of the order of the major length scale of the equipment used (e.g., impeller diameter/pipe diameter).

2. Secondary eddies: These "energy-containing" eddies contain a major part of the turbulent kinetic energy. Their length scale is a characteristic value called the "Prandtl mixing length." The latter is defined as the travel length over which an eddy maintains its identity.

3. Smallest eddies: These eddies are small enough for the viscous forces to predominate. They continuously receive energy along the energy cascade (primary → secondary → smallest eddies). These eddies more often than not show isotropy, although isotropy may not prevail with increasing length scale (smallest → secondary → primary).

The effect of eddy length scale on the rate of mass transfer depends on the relative size of the eddy and that of the bubble/drop/particle (d_p) that is involved in the transport process:

1. For eddy size $>> d_p$, the eddy simply engulfs the particulate phase and carries it along it flight path. Very little mass transfer takes place in this case.

2. When the eddy size is (i) smaller than d_p (ii) but not too small as to have very low energy content, its interaction with the particle is significant. Such interaction can lead to higher rates of mass transfer provided the eddy frequency is also high. For multiphase systems, these eddies are also responsible for breakage of drops/bubbles (Section 7A.5.1 and Fig. 7A.11).

3. For eddy size $<<d_p$, such eddies have very low kinetic energy and hence cannot penetrate the hydrodynamic boundary layer. Consequently, such eddies do not play an important role in mass transfer.

The earlier discussion relates to some basic parameters in turbulence. The turbulence intensity, u', is the most important parameter for the purpose of this book. u' can vary significantly in time and space in a given multiphase reactor. Further, since u' is the main parameter affecting heat and mass transfer, its variations in time and space cause corresponding variations in the rates of heat and mass transfer. For process design, the overall intensity of turbulence is more important rather than these variations. It will be shown later that, indeed, an overall u' can be obtained, though indirectly, through simple considerations.

NOMENCLATURE

D:　　Diameter of pipe (m)

d_p:　　Particle size (m)

I:　　Turbulence intensity defined as $\dfrac{\sqrt{\overline{u_x'^2}}}{u_x}$ (—)

L:　　Distance in the direction of flow (m)

Re_D:　　Reynolds number based on pipe diameter $\left(\dfrac{Du\rho}{\mu}\right)$ (—)

Re_L:　　Reynolds number based on the length along the flat plate $\left(\dfrac{Lu\rho}{\mu}\right)$ (—)

u':　　Overall intensity of turbulence (m/s)

u_x:　　Fluid velocity vector in the x direction (m/s)

u_x':　　Instantaneous fluctuating component in the x direction (m/s)

u_y':　　Instantaneous fluctuating component in the y direction (m/s)

u_z':　　Instantaneous fluctuating component in the z direction (m/s)

$\overline{u_x'}$:　　Time average of u_x' in the x direction (m/s)

$\sqrt{\overline{u_x'^2}}$:　　Intensity of turbulence in the x direction (m/s)

REFERENCES

Bennett CO, Myers JE. (1962) Momentum, heat and mass transfer. McGraw Hill Book Co., New York, USA.

Brodkey RS. (1967) The phenomena of fluid motions. Addison-Wesley Pub. Co., Reading, MA, USA.

Schlichting H. (1955) Boundary-layer theory. McGraw-Hill Book Co., New York, USA.

Sreenivasan KR. (1999) Fluid turbulence. Rev. Mod. Phys., 71:S383–S395.

Wu J, Zhu Y, Pullum L. (2001a) Impeller geometry effect on velocity and solids suspension. Trans. IChemE, 79(Part A):989–997.

Wu J, Zhu Y, Pullum L. (2001b) The effect of impeller pumping and fluid rheology on solids suspension in a stirred vessel. Can. J. Chem. Eng., 79:177–186.

5

PRINCIPLES OF SIMILARITY AND THEIR APPLICATION FOR SCALE-UP OF MULTIPHASE REACTORS

5.1 INTRODUCTION TO PRINCIPLES OF SIMILARITY AND A HISTORIC PERSPECTIVE

Bench-scale laboratory experiments are inadequate to give meaningful information for scale-up to the prototype, industrial situation. Therefore, scale-up of many chemical processes is preceded by pilot plant testing. As Baekeland (1916) suggested, the philosophy behind pilot plant testing is, "Commit your blunders on a small scale and make your profits on a large scale." In the first *Handbook of Chemical Engineering*, Davis (1901) clearly spelt out the difference between bench-scale and pilot plant testing by the statement: "A small scale experiment made in the laboratory will not be of much use in guiding to the erection of large scale works, but there is no doubt that an experiment based on a few kilograms will nearly give all the data required…" Since pilot plant testing itself involves significant capital and time, there is a need to have a rational basis for the design of the pilot plant and the corresponding experiments. The objective of these experiments is to obtain reasonably reliable information for designing the prototype in as short a time as possible. In view of this, chemical engineers have strived to provide rational guidelines for pilot plant testing. In many cases, statistical approaches are used to limit the number of experiments at the pilot plant level with the sole view of reducing the associated time and costs. Thus, statistical experimental design and analysis has been incorporated as a course in many chemical engineering programs, particularly at the postgraduate level. Montgomery (2003) has given an excellent discussion on the statistical methodology.

Design of Multiphase Reactors, First Edition. Vishwas Govind Pangarkar.
© 2015 John Wiley & Sons, Inc. Published 2015 by John Wiley & Sons, Inc.

The process development work starts at bench scale, typically less than a 100 g of the product. If the results are encouraging, a small plant is used to test the reliability/reproducibility of the results obtained on the bench scale. These tests may be conducted on kilogram scale of the product. Such pilot plant testing has another purpose. The product so obtained may be used for strategic marketing purposes. If the pilot plant contains individual equipments such as reactor, distillation columns, etc. these individual process equipments need to be scaled up using different rationale. Scale-up of multiphase reactors is a complex process unlike the scale-up, for instance, of an oil pumping system from a low-rate and low-distance pumping to a high-rate and long-distance pumping system. In the latter, the scale-up involves calculation of pumping power. Most standard texts give Fanning's equation and the single-phase flow friction factor, which can be used to calculate single-phase pressure drop. This pumping scale-up is easy to the extent that a scientific calculator can do the job effectively. The problem with multiphase reactor scale-up revolves around the understanding of the flow structure in which there is at least one deformable phase in the system. This free interface phase has a chaotic motion in both time and space domains, which can be significantly different in the small-scale and large-scale equipments. This chapter reviews the principles of similarity as commonly used for fixed interfaces (pipe flow) with a view to bring out the inadequacy in the classical definitions of similarity. An attempt is then made to resolve the inadequacies through two new concepts of "hydrodynamic similarity" and "turbulence similarity."

A large number of texts discuss the concept of similarity as applied to scale-up. The earliest unified approach to the application of concept of similarity in chemical process equipments was propounded by Johnstone and Thring (1957). The concept of dimensional analysis based similarity has become a part of all standard chemical engineering texts. Astarita (1985) has presented a lucid exposition of the subject matter. A recent book covering various aspects of dimensional analysis and scale-up in chemical engineering by Zlokarnik (2002) is recommended for more detailed information.

The principle of similarity was first enunciated by Newton for systems composed of solids in motion. It is concerned with the relations between physical systems of different sizes. For chemical engineers, the added dimension of chemically reacting systems is of great importance.

5.2   STATES OF SIMILARITY OF RELEVANCE TO CHEMICAL PROCESS EQUIPMENTS

The following different states of similarity are important:

1. Geometric similarity
2. Mechanical similarity
3. Thermal similarity
4. Chemical similarity

For bioreactors, an additional similarity, physiological similarity, is important.

5.2.1   Geometric Similarity

Two bodies are geometrically similar when for every point in one body, there exists a corresponding point in the other. Thus, if X, Y, and Z are the coordinates of a given point, then the relation that defines geometric similarity between two bodies denoted by 1 and 2 is

$$\frac{X_1}{X_2} = \frac{Y_1}{Y_2} = \frac{Z_1}{Z_2} = L \qquad (5.1)$$

This type of similarity should be maintained from the beginning with laboratory-scale and bench-scale equipment up to the commercial scale. The example of round bottom flasks with a Teflon paddle stirrer used by most chemists for preliminary laboratory tests has been briefly mentioned in Section 2.2. This type of small-scale reactor has several drawbacks. The absence of baffles creates a swirling motion and the inevitable vortex. In such a vortex, the tangential velocity component is predominant. There is practically no axial/radial mixing. Therefore, for example, if such a system is used for carrying out a solid-catalyzed gas–liquid reaction, the contact between the catalyst and the reacting phases is poor. Gas–liquid/solid–liquid mass transfer is also poor because of very poor dispersion of the gas/suspension of solids in the liquid. The reactor type used on the large scale is invariably a cylindrical vessel with a dish end and is fully baffled (four baffles of width equal to 10% of the vessel diameter spaced at 90°). In view of this, to maintain geometric similarity, the laboratory experiments should also be carried out in a similar vessel. Such resin kettles with flanged lids/covers having provision for temperature indication, sampling, etc. are available in glass. These can be equipped with a stuffing box and practically all types of impellers made in glass except a perfect propeller. Glass baffles are also available to break any vortex that may form. In such a system, the laboratory equipment serves the desired purpose (gas–liquid mixing, solid suspension, etc.) and also maintains geometric similarity with the large-scale equipment that is to be proposed. Many chemical process equipments have multiple elements as internals and multiple geometric ratios. Some important examples are the stirred tank reactor and packed towers for gas–liquid and liquid–liquid contacting. In the case of the stirred tank reactor, different impellers are available. Although most impellers have well-defined dimensions with respect to the diameter of the impeller (blade width, height, thickness, etc.), there are other ratios that are equally important. The ratio of the impeller diameter to the vessel diameter and the ratio of the clearance of the impeller from the bottom to the vessel diameter are two important parameters that have an impact on the overall performance of the stirred vessel. These multiple elements (e.g., impeller, baffles) and their dimensional relationship to the vessel diameter should ideally be the same in the laboratory and prototype equipment to have ideal geometric similarity. Further, if the large-scale vessel is likely to have a height to diameter ratio greater than 1 (i.e., 2 or 3), then the laboratory reactor to be used must also have similar height to diameter ratio besides the multiple element and spatial geometric similarity mentioned earlier. These are some of the minimum requirements if meaningful data are expected from a good laboratory set-up. Geometric similarity is in-built in devices that are structured. Examples are microstructured devices, monoliths, etc. In such devices, scale-up is through simple increase in numbers without disturbing the basic channel dimensions or any other related parameters such as the plate thickness.

5.2.2 Mechanical Similarity

Mechanical similarity introduces the major element of motion. It basically deals with systems in motion. It can be divided into two major subclasses.

5.2.2.1 Dynamic Similarity

Geometrically similar moving systems are dynamically similar when the ratios of various forces affecting the motion are equal:

$$\frac{F_1'}{F_1} = \frac{F_2'}{F_2} = \cdots = \frac{F_N'}{F_N} = \text{constant} \tag{5.2}$$

Thus, the parallelograms or polygons of forces for corresponding particles should be geometrically similar. A corollary of this definition is that the ratios of different forces in the same system shall be constant:

$$\frac{F_1'}{F_2'} = \frac{F_1}{F_2}, \frac{F_1'}{F_N'} = \frac{F_1}{F_N} = \text{constant} \tag{5.3}$$

Applying the well-established technique of dimensional analysis to all the variables affecting the process yields the ratios of various relevant forces, termed dimensionless groups. It must be mentioned that the magnitude of these dimensionless numbers is based on global values of the respective variables. Table 5.1 lists some dimensionless groups that are pertinent to the case of multiphase reactors. A detailed list of dimensionless groups used in various disciplines is given by Fulford and Catchpole (1968).

One major prerequisite of this approach is that the physics of the process must be clearly understood (Zlokarnik 2002). Most multiphase reactors operate in the highly turbulent regime. As discussed in Chapter 4, the physics of single-phase turbulence (a situation far simpler than multiphase flow) itself has not been properly understood. The understanding of turbulence in multiphase flows and its consequent effects on heat and mass transfer is, to say the least, very poorly comprehended. It is probably this reason that *unique* correlations based on dynamic similarity (or dimensionless groups) cannot be derived for mass transfer in multiphase reactors. Another drawback of this approach is that it uses global values of the forces in Equation 5.3. For example, Reynolds number is a ratio of parameters representing inertial and viscous forces. It is well known that these forces can vary greatly in space in a given multiphase reactor as is evident from Kolmogorov's theory on decay of turbulence from a large scale to the final stage in which viscous dissipation is most important. Considering the case of a stirred reactor, the inertial energy is highest in the stream discharged from the impeller. The same decays as the fluid travels away from the

TABLE 5.1 Some Dimensionless Groups Relevant to Multiphase Systems[a]

Name	Notation used	Formula	Significance	Field of use	Comment
Euler number	Eu	$Eu = \left(\dfrac{\Delta P}{\rho u_C^2}\right) ; \dfrac{gS}{\left(Q_G/D_{RT}^2\right)^2}$	Ratio of pressure force and inertial force	Decrease in pressure due to increase in velocity	Rate of gas induction (Chapter 9)
Eötvös number	Eo, N_{Eo}	$\left(\dfrac{\Delta \rho g l^2}{\sigma_L}\right)$	Ratio of buoyancy force and surface tension force	Moving drops/bubbles in a liquid	Characterizes shape of the particulate phase
Froude number	Fr, N_{Fr}	$\left(\dfrac{N^2 D}{g}\right)\left(\dfrac{N^2 D^2}{S \times g}\right) ; \left(\dfrac{V}{\sqrt{gl}}\right)$	Ratio of inertial force and gravitational force. For liquids undergoing rotational motion: ratio of centripetal force and gravitational force	Formation of surface vortices in stirred vessels, condition for gas induction in flotation cells/gas-inducing reactors	Prediction of surface aeration/critical speed for gas induction (Chapter 9)
Hatta number	Ha	$\left(\dfrac{\sqrt{2/(p+1)D_A k_{p-q}[B_B]^q[A*]^{p-1}}}{k_L}\right)$	Ratio of rate of mass transfer in the presence of a chemical reaction and that in the absence of a chemical reaction	Fluid–fluid, fluid–fluid–solid (catalyst) reactions	Indicates the relative importance of reaction and diffusion for a given hydrodynamic situation in multiphase system
Morton number	Mo, N_{Mo}	$\left(\dfrac{g\mu_C^4 \Delta \rho}{\rho_C^2 \sigma^3}\right) ; \left(\dfrac{g\mu_C^4}{\rho_C^2 \sigma^3}\right)$	Ratio of viscous force and interfacial force	Characterization of shapes of bubbles/drops in motion (in conjunction with Eötvös number)	Prediction of mass ratio of secondary and primary fluid in a liquid jet ejector (Chapter 8)
Nusselt number	Nu, N_{Nu}	$\left(\dfrac{hl}{k_C}\right)$	Ratio of heat transfer by convection and that by conduction	Convective heat transfer	Indicates the relative rates of convective and conductive heat transfer

(Continued)

TABLE 5.1 (Continued)

Name	Notation used	Formula	Significance	Field of use	Comment
Peclet number	Pe, N_{Pe}	$\left(\dfrac{lu_C}{D_{Ax}}\right)$	Ratio of advective and diffusive transport	Used to characterize axial mixing in flow systems	Indicative of flow pattern
Power number	N_P, N_{PG}	$\left(\dfrac{P}{\rho N^3 D^5}\right)$	Ratio of drag force on impeller and inertial force	Estimation of power drawn by an impeller	Requires modification when a gas phase is introduced
Prandtl number	Pr, N_{Pr}	$\left(\dfrac{C_p \mu}{k_C}\right)$	Ratio of momentum and thermal diffusivity	Convective heat transfer	Ratio of thickness of hydrodynamic and thermal boundary layers
Reynolds number	Re, N_{Re}, etc.	$\left(\dfrac{lu_C \rho}{\mu}\right)$	Ratio of inertial and viscous force	Single-/multiphase fluid flow	u_C and l need to be defined appropriately in multiphase systems
Schmidt number	Sc, N_{Sc}	$\left(\dfrac{\mu}{\rho D_M}\right)$	Ratio of kinematic viscosity and molecular diffusivity	Convective mass transfer	Ratio of thickness of hydrodynamic and mass transfer boundary layers
Sherwood number	Sh, N_{Sh}	$\left(\dfrac{k_M l}{D_M}\right)$	Ratio of mass transfer by convection and that by molecular diffusion	Fluid–fluid, fluid–solid mass transfer	Indicates the relative rates of convective and diffusive mass transfer
Weber number	We, N_{We}	$\left(\dfrac{N^2 D^3 \rho_L}{\sigma}\right)$; $\left(\dfrac{\rho V_G^2 l}{\sigma}\right)$	Ratio of inertial and surface tension force	Fluid–fluid systems involving breakup/coalescence of bubbles/drops (defined here for stirred/three-phase sparged reactors)	Useful in determining condition for bubble/drop breakup

[a]Adapted from Fulford and Catchpole (1968).

reactor and is least at the farthest point from the impeller. At the latter location, viscous forces predominate. Thus, the Reynolds number based on fluid stream velocity varies from a very high value in the vicinity of the impeller to the lowest at the wall. Considering an overall value of Re based on, for example, the tip speed of the impeller, $ND^2\rho/\mu$ will be a gross simplification of the true phenomena. In a rigorous treatment, the representative Reynolds number should be a value obtained by triple integration of the local Reynolds number over the entire reactor space using the velocity that prevails at the various locations. This is clearly not practicable. Therefore, gross assumptions such as the aforementioned Reynolds number based on tip speed are made to empirically correlate experimental data. Most such correlations available are restricted to the limited parameters studied and do not apply to other system configurations. Shaikh and Al-Dahhan's (2010) study, discussed in Sections "Hydrodynamic Regime Similarity," "Turbulence Similarity," and "Empirical Correlations Using Conventional Dimensionless Groups" clearly brings out the inadequacy of dimensionless group-based methods for scaling up multiphase systems. This matter will be discussed in more detail in Chapter 6.

5.2.2.2 Kinematic Similarity   Geometrically similar moving systems are kinematically similar when corresponding particles trace out geometrically similar paths in corresponding intervals of time. Kinematic similarity introduces the additional dimension (besides space) of time since it refers to intervals of time. Part of the hypothesis proposed in the work of Shaikh and Al-Dahhan (2010) on hydrodynamic similarity in bubble columns arises out of the conditions imposed by kinematic similarity. These investigators have shown that similarity of axial velocity profiles is an essential part of the overall similarity, which is akin to the earlier definition of kinematic similarity.

Hydrodynamic Regime Similarity   In multiphase reactors, the hydrodynamic regime may vary depending on the equipment geometry and operating conditions. For instance, in two- and three-phase sparged reactors (bubble columns), three distinct regimes are identified based on tower diameter and gas velocity (Fig. 10.1). Similarly, in two-/three-phase stirred vessels, different hydrodynamic regimes have been identified (Nienow et al. 2000). Inasmuch as these regimes have different flow behaviors and hence different transport characteristics, the model and prototype must operate in the same hydrodynamic regime. Thus, hydrodynamic regime similarity forms an integral part of kinematic similarity in multiphase systems. Hydrodynamic regime similarity is based on global parameters that are indirectly related to the local parameters as shown by several investigators (Nikhade and Pangarkar 2007; Shaikh and Al-Dahhan 2010). It must be mentioned here that mere hydrodynamic similarity is not sufficient to yield similar transport properties in multiphase systems. In the case of gas holdup in bubble columns, Shaikh and Al-Dahhan (2010) have shown that hydrodynamic similarity is a *necessary* but *not sufficient* condition. They have shown that mere hydrodynamic similarity on the global level does not yield similar mixing or turbulence structure.

Turbulence Similarity   For multiphase reactors, kinematic similarity is the most important among the mechanical similarities since it implies that the corresponding particles trace out similar paths in the model and prototype. However, it still falls short of the requirement because it does not specify the level of turbulence experienced by the particles as they trace out similar paths. As discussed in Chapter 4, the level and scale of turbulence are the most important parameters that determine rates of heat and mass transfer. Thus, to complete the definition of similarity in multiphase systems, we define a new state of similarity, turbulence similarity. Nikhade and Pangarkar (2007) have presented an exhaustive discussion on this concept. Kinematically similar systems (which also exhibit hydrodynamic regime similarity) exhibit turbulence similarity when the particles in their path experience the same turbulence intensity, u'. This similarity state is related to the microscopic phenomena. However, a global value of u' may be an adequate representation as shown by Nikhade and Pangarkar (2007). This definition in terms of u' includes both global and microscopic hydrodynamics. Consequently, existence of turbulence similarity between two systems results in similar transport behavior. Shaikh and Al-Dahhan's (2010) investigation corroborate the hypothesis of corresponding hydrodynamic states proposed by Nikhade and Pangarkar (2007). The former state that besides global hydrodynamic, local hydrodynamic similarity is essential. This implies that particles while traveling similar paths (hydrodynamic regime similarity) also must experience similar local hydrodynamics (i.e., turbulence). The implications of this investigation will be discussed further in Section 6.3.1.1.

5.2.3   Thermal Similarity

Geometrically similar systems are thermally similar when the corresponding temperature differences bear a constant ratio to one another and when the systems if moving are also kinematically similar. For multiphase reactors, an additional requirement will be that of hydrodynamic regime and turbulence similarity since the turbulence structure decides the rate of heat transfer.

In many industrially important situations, it is impossible to maintain geometric, mechanical (kinematic/hydrodynamic and turbulence similarities), and thermal similarities simultaneously. Consider a stirred tank reactor with heat exchange only through a jacket on its external surface. The jacket heat transfer area to vessel volume ratio is proportional to $(1/T)$. Evidently, with scale-up, this ratio decreases, and it is difficult to maintain the same heat transfer area per unit volume as in the small-scale unit. Additional heat transfer area is required to cater to the extra heat load resulting from increase in reactor volume. This area can be provided in the form of a coil inside the reactor or an external heat exchanger circuit. In both cases, the flow patterns are significantly different than the model contactor used in bench-scale studies and kinematic similarity is violated. This is the classic dilemma of a chemical engineer; it is impossible to preserve the different types of similarities simultaneously.

5.2.4   Chemical Similarity

This is an obvious requirement to be satisfied by the model and prototype reactor. Since the rate of a reaction in a multiphase reactor is dependent on mass and heat transfer rates, hydrodynamic regime and turbulence similarity (with underlying kinematic similarity) are essential preconditions. Thus, the new definition of chemical similarity relevant to multiphase reactors is as follows: geometrically and thermally similar systems, which also exhibit hydrodynamic regime and turbulence similarity, are chemically similar when the corresponding concentration differences bear a constant ratio to one another.

The requirement for chemical similarity can be explicitly stated as follows: The model and prototype should exhibit equality of the following ratios:

1. Rate of product formation to the rate of bulk flow
2. Rate of product formation to the rate of convective mass/heat transfer

The multiple and simultaneous requirements imply that it is practically impossible to have chemical similarity between the model and prototype.

5.2.5 Physiological Similarity

Biological processes are distinctly different than chemical processes. The former are mediated by living organisms. Factors affecting the physiological state of organisms/ cells are (i) physical, (ii) chemical, and finally (iii) biological (Malek 1982). Physical factors are represented by measurable quantities such as pressure and temperature (Onken et al. 1984). The optimum temperature window for biological processes is relatively very small as compared to chemical processes. The same applies to chemical factors such as pH and redox potential of the broth. Further, in some cases such as mammalian cells, the cells are also shear sensitive. Other factors that relate to homogeneity of the broth are supply of nutrients and removal of toxins, which are clubbed under physiological homogeneity. Physical homogeneity allows a comparison of the required rates of mass and energy transport from the living cell to/from its immediate neighborhood. The latter are intimately related to mixing and mass transfer factors that are essentially part of chemical engineering. The overall issue under consideration is therefore a dynamic equilibrium between the biological factors that govern cell growth and the environment of the cell (von Bertalanffy 1953). Votruba and Sobotka (1992) have proposed the concept of physiological similarity for scale-up of bioreactors. According to this principle, the microenvironment of the living cell must be identical to reproduce the same physiological function in the model and prototype. In terms of rate processes, this implies that the rates of physical processes (mixing/heat/mass transfer) and physiological processes (governed by the microenvironment of the organism) should be comparable. From a chemical engineering perspective, it is evident that scale-up of a bioreactor is critically related to mixing and transport processes. This matter is discussed in detail in Chapter 7B.

5.2.6 Similarity in Electrochemical Systems

Electrochemical reactions are typically heterogeneous reactions occurring on the surface of the respective electrodes. Analogous to other heterogeneous reactors, these reactors have (i) mass transfer controlled regime referred to as concentration overpotential/polarization and (ii) activation-controlled regime, which is similar to kinetic controlled regime in conventional chemical reactions. The concepts of geometric similarity and hydrodynamic similarity are both relevant in these reactors. Since the reaction occurs on the electrode surface, the geometries and area ratio of the electrodes must necessarily be the same in the model and prototype. Hydrodynamic regime similarity implies equal solid–liquid (electrolyte–electrode) mass transfer coefficients in the model and prototype. If the model experiments are performed under conditions such that the diffusional resistance has been eliminated and activation regime is controlling, then hydrodynamic regime similarity would ensure that the prototype is also operating under activation-controlled regime. In the case of flow reactors (plate and frame-type electrolyzers), the electrodes are flat plates of equal area, and geometric similarity can be ensured. However, in some industrial situations such as aluminum smelting, this may not be true. Further, the flat plate anodes used in these applications may be blanketed by the CO_2 that evolves causing substantial increase in electrical resistance (IR drop). This may alter the energy consumption in the large-scale version. In addition, problems related to heat transfer are also important since the heat transfer surface area to volume ratio in the large scale will be lower than the model. This application is a special case, which is not considered further in this text.

5.2.7 Similarity in Photocatalytic Reactors

Photocatalytic reactors are being investigated for both selective oxidation and photocatalytic degradation. This application has been discussed in Section 1.3.1.1, and a large number of industrially important reactions have been summarized in Table 1.1. In this case, besides similarity in transport (solid–liquid, in particular), similarity in radiation intensity in the small scale and prototype is essential. Photocatalytic reactor design is far more complex than chemical reactor design because of an additional important factor: distribution of the radiation in a highly complex three-phase system. Other parameters such as source of radiation, arrangement of the light source in the reactor system, and selection of materials for reactor can also affect the reactor performance.

NOMENCLATURE

$[A^*]$:	Equilibrium concentration of at the solute at the interface (mol/m^3)
$[B_B]$:	Bulk liquid concentration of liquid-phase reactant (mol/m^3)
C_p:	Specific heat of the species (J/kg K)
D_{Ax}:	Axial dispersion coefficient (m^2/s)
D_M, D_A:	Molecular diffusion coefficient (m^2/s)
D:	Impeller diameter (m)
D_{RT}:	Diameter of rotor in gas-inducing impeller (m)
Eo:	Eötvös number $\left(\dfrac{\Delta \rho g l^2}{\sigma_L} \right)$ (—)

Eu, N_{Eu}:   Euler number $\left(\dfrac{\Delta P}{\rho u_C^2}\right)$ or $\dfrac{gS}{(Q_G/D_{RT}^2)^2}$ (—)

F_i: Force of type i in model equipment (N)

F_i':   Force of type i in prototype equipment (N)

Fr: Froude number $\left(\dfrac{N^2 D}{g}\right)$; $\left(\dfrac{V_G}{\sqrt{gl}}\right)$ (—)

g: Gravitational acceleration constant (m/s²)

h: Heat transfer coefficient (W/m² K)

Ha: Hatta number $\left(\dfrac{\sqrt{(2/(p+1))D_A k_{p-q}[B_B]^q[A^*]^{p-1}}}{k_L}\right)$ (—)

k_C: Thermal conductivity (W/m K)

k_M: Mass transfer coefficient (m/s)

k_{p-q}:   Rate constant for general p, qth-order reaction [(mol/m³)$^{1-p-q}$/s]

L: Ratio of distances in respective coordinate directions (—)

l, l_C: Characteristic linear dimension (m)

Mo, N_{Mo}:   Morton number $\left(\dfrac{g\mu_F^4 \Delta\rho}{\rho_C^2 \sigma_L^3}\right)$ (—)

N: Speed of agitation of the impeller (1/s)

N_P: Power number $\left(\dfrac{P}{\rho_F N^3 D^5}\right)$ (—)

P: Power (W)

p: Order of reaction with respect to the dissolved gaseous species (—)

Pe, N_{Pe}:   Peclet number $\left(\dfrac{l u_C}{D_{Ax}}\right)$ (—)

Pr, N_{Pr}:   Prandtl number $\left(\dfrac{C_P \mu}{k_C}\right)$ (—)

q: Order of reaction with respect to the liquid-phase reactant (—)

Q_G: Gas flow rate (m³/s)

Re, N_{Re}:   Reynolds number $\left(\dfrac{l u_C \rho_F}{\mu_F}\right)$; $\left(\dfrac{ND^2 \rho_F}{\mu_F}\right)$ (—)

S: Submergence of gas-inducing impeller (m)

Sc, N_{Sc}:   Schmidt number $\left(\dfrac{\mu}{\rho D_M}\right)$ (—)

Sh, N_{Sh}:   Sherwood number $\left(\dfrac{k_M l}{D_M}\right)$ (—)

u':　　　　Turbulence intensity (m/s)

u_C:　　　　Characteristic fluid velocity (m/s)

V_G:　　　　Superficial gas velocity (m/s)

We, N_{We}:　Weber number $\left(\dfrac{N^2 D^3 \rho_L}{\sigma} \right)$ or $\left(\dfrac{\rho_L V_G^2 l}{\sigma_L} \right)$ (—)

X_i:　　　　Coordinate in the X direction for body I (m)

Y_i:　　　　Coordinate in the Y direction for body I (m)

Z_i:　　　　Coordinate in the Z direction for body I (m)

Greek Symbols

μ:　Viscosity (Pa s)

ρ:　Density (kg/m^3)

ν:　Kinematic viscosity (m^2/s)

σ:　Surface tension (N/m)

Subscripts

F:　Fluid phase

G:　Gas phase

L:　Liquid phase

REFERENCES

Astarita G. (1985) Scale up: overview, closing remarks, and cautions. In: Bisio A, Kabel RL, editors. Scale-up of chemical processes: conversion from laboratory scale tests to successful commercial size design. Wiley, New York, USA. p 678.

Baekeland LHJ. (1916) Synthetic phenol resins. Ind. Eng. Chem., 8(6):568–570.

Davis GE. (1901) A handbook of chemical engineering. Davis Bros., Manchester, England.

Fulford GD, Catchpole JP. (1968) Dimensionless groups. Ind. Eng. Chem., 60(3):71–78.

Johnstone CO, Thring MW. (1957) Pilot plants, models and scale-up methods in chemical engineering. McGraw Hill Book Company, New York, USA.

Malek I. (1982) Vectorial methodology of physiological state studies in continuous culture. In: Krumphanzl V, Sikyta B, Zanek Z, editors. Overproduction of microbial products. Academic Press, London, UK. p 702–717.

Montgomery DC. (2003) Design and analysis of experiments. 5th ed. John Wiley & Sons, Inc., New York, USA.

Nienow AW, Harnby N, Edwards F. (2000) Introduction to mixing problems. Chapter 1. In: Harnby N, Edwards F, Nienow AW, editors. Mixing in the process industries. Butterworth-Heinemann, Oxford, UK. p 1–23.

Nikhade BP, Pangarkar VG. (2007) A theorem of corresponding hydrodynamic states for estimation of transport properties: case study of mass transfer coefficient in stirred tank fitted with helical coil. Ind. Eng. Chem. Res., 46:3095–3100.

Onken U, Jostman T, Weiland P. (1984) Effects of hydrostatic pressure on microbial growth for fermentation process modeling. In: Proceedings of Third European Congress on Biotechnology, Verlag Chemie, Weinheim, Germany. p 481–495.

Shah YT (1979) Gas–liquid–solid reactor design. McGraw Hill International Book Company, New York, USA.

Shaikh A, Al-Dahhan M. (2010) A new methodology for hydrodynamic similarity in bubble columns. Can. J. Chem. Eng., 88:503–517.

von Bertalanffy L. (1953) Biophysic des Fliessgleichgewichts. In: Sammlung Vieweg 124, S. 15-80, Braunschweig, Germany. cf: Votruba and Sobotka (1992).

Votruba J, Sobotka M. (1992) Physiological similarity and bioreactor scale-up. Folia Microbiol., 37(5):331–345.

Zlokarnik M. (2002) Scale-up in chemical engineering. Wiley-VCH, Weinheim, Germany.

6

MASS TRANSFER IN MULTIPHASE REACTORS: SOME THEORETICAL CONSIDERATIONS

6.1 INTRODUCTION

Chapters 1, 4, and 5 emphasized the fact that the rate of mass transfer in multiphase reactors depends on the type and size of the equipment used. The reactors dealt with in this and subsequent chapters are of the type in which the gas phase is dispersed in a continuous liquid phase. The various phases taking part in the overall reaction sequence experience chaotic, turbulent motion in time and space. Under such conditions, mass transfer mainly occurs by a mechanism in which different eddies that come to the interface deliver/receive the solute during their lifetime at the interface and return back to the bulk phase. This unsteady-state mass transfer process has been exhaustively discussed in several texts (Astarita 1967; Danckwerts 1970). In the following, the various approaches to predict mass transfer coefficients in different multiphase reactors are discussed along with the advantages/drawbacks of each approach.

The understanding of mass transfer in multiphase reactors has closely followed the understanding of the underlying hydrodynamics. Therefore, in the ensuing discussion, a historical perspective of the evolution of approaches for predicting mass transfer coefficients in multiphase systems is presented. The theoretical background for the same is also included.

Design of Multiphase Reactors, First Edition. Vishwas Govind Pangarkar.
© 2015 John Wiley & Sons, Inc. Published 2015 by John Wiley & Sons, Inc.

6.2 PURELY EMPIRICAL CORRELATIONS USING OPERATING PARAMETERS AND PHYSICAL PROPERTIES

This approach was predominant in the early stages of studies on mass transfer in multiphase reactors. As an example, for the case of stirred reactors, several correlations of the following type were proposed:

$$k_L \text{ or } K_{SL} = f(V_G, T, N, D/T, C/T, H/T, \text{ etc. and Sc}) \tag{6.1}$$

Similarly, for bubble columns (two-phase sparged reactors), Sharma and Mashelkar (1968) proposed Equation 6.2 for the volumetric liquid-side mass transfer coefficient, $k_L \underline{a}$, based on measurements by the chemical method:

$$k_L \underline{a} = a(V_G)^b \tag{6.2}$$

A singular lack of understanding of the hydrodynamics in the different multiphase contacting devices was the main reason for resorting to such decidedly empirical and system-dependent correlations.

Evidently, the variables in the parenthesis of the right-hand side of correlations represented by Equation 6.1 or 6.2 depended on the type of multiphase reactor being considered. Thus, for bubble columns, only the superficial gas velocity was used if the column diameter exceeded 0.1 m. The latter criterion is required to satisfy hydrodynamic regime similarity (Section 5.2.2.1) when the laboratory data are to be used for large-scale industrial columns. The Schmidt number, Sc, was incorporated when the correlation was required to be used for wide variation in relevant system properties. For stirred reactors, the right-hand side of Equation 6.1 incorporated the system geometry through the type of impeller used, D/T, C/T, H/T, and the speed of agitation besides the Schmidt number. The effects of other important physical properties were subsequently introduced. Correlations of the type represented by Equation 6.1 or 6.2 had the advantage that they were simple to use. The functional dependence of the mass transfer coefficient (empirical constants in the right-hand side of Equations 6.1 and 6.2) was evaluated by regressing the experimental data. These empirical constants were distinctly specific for a given combination of system geometry and also the range of physical properties covered in the experimental investigations. Consequently, such correlations were not satisfactory for substantially different operating parameters and system geometries. The problem was particularly acute when empirical correlations were further extrapolated to substantially different hydrodynamic regimes. Therefore, in most cases, these empirical correlations obtained from small-scale laboratory data had very limited utility, particularly for large-scale industrial applications wherein these regimes could be totally different. There will be frequent reference to this lack of similarities in the rest of this chapter.

In majority of experimental investigations, overall volumetric mass transfer coefficients comprising the true mass transfer coefficient and the interfacial area are reported. To obtain the true mass transfer coefficient from such data, knowledge of the interfacial area is required. There is a wide variation in bubble/drop sizes in a gas–liquid or liquid–liquid dispersion. Therefore, it is not possible to define a unique bubble size. As a result, simultaneous experimental determination of the overall mass transfer coefficient and interfacial area is required to derive the true mass transfer coefficient. In the case of the solid–liquid mass transfer coefficient, K_{SL}, the solute can be chosen such that the particle size and shape vary in a relatively narrow range. Therefore, it is possible to have a relatively accurate estimate of the solid–liquid interfacial area. In view of this simplification in the discussion that follows, the case of solid–liquid mass transfer coefficient, K_{SL}, is used to illustrate methods employed in the literature for correlating mass transfer data. The hydrodynamics of suspended particle (solid–liquid) systems in turbulent motion is not significantly different for gas–liquid systems. The only difference lies in the varying bubble sizes as compared to constant solid size (when it is the catalyst) and the direction of the gravity force. Therefore, the underlying treatment is equally applicable for both gas–liquid and solid–liquid mass transfer. Indeed, exactly similar approaches for correlating these two mass transfer steps (e.g., Kolmogorov's theory of isotropic turbulence) have been used in the literature.

6.3 CORRELATIONS BASED ON MECHANICAL SIMILARITY

The concept of mechanical similarity has been discussed in Chapter 5. This can be divided into two major subclasses: (1) dynamic similarity and (2) kinematic similarity. The correlations based on these two subclasses are discussed separately.

6.3.1 Correlations Based on Dynamic Similarity

6.3.1.1 Empirical Correlations Using Conventional Dimensionless Groups   The literature abounds with correlations based on dynamic similarity, starting with Froessling (1938, 1940) and Ranz and Marshall (1952a, b). The equations proposed by these investigators were of the following form:

$$Sh_p = Const \ (Re)^{0.5} \ (Sc)^{0.33} \tag{6.3}$$

$$Sh_p = 2 + 0.6 \ (Re)^{0.5} \ (Sc)^{0.33} \tag{6.4}$$

Equations 6.3 and 6.4 assume that at equal Re and Sc the value of Sh_p should be the same irrespective of the type and size of the contacting device/reactor. However, the definitions of Re and Sh_p pose a formidable problem. The linear dimension term in Re and Sh_p and the velocity term in Re need to be defined with relevance to the type of system/reactor. Most of the investigators used simple definitions for the velocity term. For instance, in bubble columns, the superficial gas velocity was used, whereas for stirred vessels the tip speed ($N \times D$) was used. The discussion presented in Chapter 5 clearly indicates that these simplified definitions cannot form the basis of convective mass transfer in a highly turbulent multiphase reactor. The linear dimension to be used in Sh_p and Re is similarly elusive, particularly in the case of mass transfer from bubbles/drops. These bubbles/drops have a wide size distribution under varying operating conditions, and hence, it is difficult to assign a representative value for d_p. In the case of solid (catalyst)–liquid mass transfer, the catalyst particle size can be used as the linear dimension. The velocity term, however, poses a major problem in this case also. Simple correlations based on dimensionless groups such as in Equations 6.3 and 6.4 are satisfactory in very limited situations. An example is that of a relatively simple multiphase reactor such as a three-phase sparged reactor. When the data in the laboratory scale cover the same range of physical properties and operating conditions and further when the laboratory column has a diameter greater than at least 0.1 m, correlations based on these data may yield reasonably good estimates for the large-scale reactor. This result is primarily because of hydrodynamic regime similarity. The importance of the concept of hydrodynamic regime similarity can be further elaborated from some investigations on gas holdup in two-/three-phase sparged reactors. Macchi et al. (2001) in their study on three-phase fluidized beds observed that two systems having similar overall gas holdups can differ in flow patterns and mixing intensities. More than five dimensionless groups are required to adequately characterize the hydrodynamics. This is an indication that the dynamic similarity approach is unable to give unique correlations. Subsequently, Shaikh and Al-Dahhan (2010) also reached a similar conclusion based on an extensive study of fractional gas holdup and mixing intensity. These investigators measured the fractional gas holdup and mixing intensity using single-source γ-ray computed tomography (CT) and computer-automated radioactive particle tracking (CARPT), respectively. Mixing intensity was defined in terms of turbulent kinetic energy (TKE), which accounts for trace of a turbulence stress tensor. The experimental investigations covered both homogeneous bubble flow and churn turbulent (heterogeneous flow) regimes. In the former case, it was found that the radial holdup variation was insignificant. This also led to similar hydrodynamics on global and microscopic levels. The results for the churn turbulent regime, however, showed a totally different trend. It is well known that in the churn turbulent regime, there is substantial variation in gas holdup and liquid velocity in the radial direction. This leads to a substantial difference between global and microscopic hydrodynamics. Detailed examination of various sets of data having *similar* radial fractional gas holdup profiles showed that the measured axial velocity profiles were also *similar*. In an exactly analogous manner, *dissimilar radial* fractional gas *holdup* profiles were found to yield dissimilar axial liquid velocity profiles. Shaikh and Al-Dahhan tried to correlate their own data as well as the extensive data available in the literature using the following dimensionless groups based on the concept of dynamic similarity: Morton number, Mo; Eötvös number, Eo; Reynolds number, Re; density ratio, ρ_R; Froude number, Fr; Weber number, We; Galileo number, Ga; and aspect ratio (dispersion height to column diameter ratio), H/T. In addition, they also included a dimensionless group proposed by Fan et al. (1999). These individual dimensionless groups are based on global parameters. Shaikh and Al-Dahhan found that none of the combinations of the above dimensionless groups could correlate the data satisfactorily, although the data belonged to the same hydrodynamic regime. This conclusion is supported by the observation of Pangarkar et al. (2002), who showed that the use of dimensionless groups in correlating mass transfer data is fraught with uncertainty and a universal correlation satisfying all geometric and physical property combinations is impossible. Shaikh and Al-Dahhan concluded that mere similarity of hydrodynamic regime is not sufficient. The results showed that besides similarity in global hydrodynamics, similarity of local hydrodynamics is also important. The following hypothesis was proposed based on the data analyzed by them:

"Overall gas holdup and its time-averaged radial profile or cross-sectional distribution should be the same for two reactors to be dynamically similar."

The above hypothesis implicitly suggests that turbulence regime similarity needs to be satisfied. As discussed in Chapter 5, hydrodynamic regime similarity is an integral part of turbulence regime similarity. On the basis of their scrutiny of large sets of data, Shaikh and Al-Dahhan hypothesized that "hydrodynamic regime similarity is necessary but not sufficient." Similar liquid circulation and mixing intensities were obtained in different systems only when the radial gas holdup profiles were similar. Dimensionless group-based correlations using global parameters were found to be inadequate. The above observations are indirect statements of the requirement of turbulence regime similarity besides hydrodynamic regime similarity. The hypothesis has considerable similarity with the theorem of corresponding hydrodynamic states proposed by Nikhade and Pangarkar (2007), which is discussed later in Chapter 7A, Sections 7A.5.3, 7A.6, and 7A.7.2.

6.3.1.2 Approach Based on Kolmogorov's Theory   The first attempt at putting forth a credible theory of turbulence is attributed to Kolmogorov (1941). A simple description of this theory is as follows:

We consider the simple case of single-phase flow in a pipe. The finite viscosity of the fluid generates friction at the pipe wall. The resulting shear force is related to the fluid velocity past the wall and is the source of turbulence in the fluid. Thus, formation of large vortices or eddies is initiated at the pipe wall. The scale of such eddies is comparable to the major system dimension, the pipe diameter in the present case. The shear force decreases with distance from the wall. Kolmogorov argued that there is a succession of eddy sizes, starting with the primary eddies (scale ~ pipe diameter) and progressively decreasing in size to smallest eddies. Inasmuch as all physical transitions must be continuous, the large energy-containing eddies transfer their energy to smaller eddies in a continuous cascade. The principle of energy conservation implies that the energy transferred must finally manifest in a pertinent manner. In the energy cascade, the final receiving eddies are small in size such that only viscous forces decide the behavior of these eddies. Therefore, the energy received by these small eddies is dissipated through viscosity (viscous dissipation). An example of such viscous dissipation is the heating of a liquid observed when it is continuously recycled through a pump. The smallest eddies are universal, although those belonging to the large-scale range are specific to a given type of flow/class of flows with similar boundary conditions. In conjunction with pressure forces acting in the appropriate direction, the flow becomes isotropic. Kolmogorov argued that at sufficiently high Reynolds number, the motion of the smallest eddies is isotropic. Thus, although the system can be anisotropic on the scale of the primary eddy, isotropy prevails at the microscopic level. These small eddies have received the energy without any memory and therefore would not be affected by the primary source of generation of turbulence by the shear at the wall (Schlichting 1955; Hinze 1959; Brodkey 1967). Kolmogorov's theory states that "At sufficiently high Reynolds numbers there is a range of high wave numbers where the turbulence is statistically in equilibrium and is uniquely determined by the parameters, ε_M (power dissipated per unit mass) and ν (kinematic viscosity). This state is universal" (Hinze 1959). Further, it was stated that if the Reynolds number of the main stream is relatively large as compared to that required for the assumption of "universal equilibrium," then an inertial subrange exists for which dissipation due to viscosity is not important.

Using dimensional reasoning, Kolmogorov's length scale, η_K, and velocity, u_K, can be defined, respectively, as follows:

$$\eta_K = \left(\frac{\nu^3}{\varepsilon_M} \right)^{1/4} \tag{6.5}$$

$$u_K = (\nu_K \varepsilon_M)^{1/4} \tag{6.6}$$

Further, according to Kolmogorov, if distances and velocities are referred to these scales, a universal function β exists such that

$$\frac{u_{rms}}{u_K} = \beta \left(\frac{d}{\eta_K} \right) \tag{6.7}$$

Here u_{rms} is the root-mean-square (rms) relative velocity between two points in the fluid separated by a distance d. For very large Reynolds numbers of the main stream (much larger than the Re value required for assumption of universal equilibrium), Kolmogorov's theory proposes that the turbulence spectrum be divided into two subranges. The "inertial subrange" is that part of the spectrum in which viscous dissipation is unimportant and

$$u_{rms} = \acute{\kappa} (\varepsilon_M d)^{1/3} \quad \text{for} \quad \eta_K << d << L \tag{6.8}$$

For the second subrange, "viscous subrange," Kolmogorov suggested that

$$u_{rms} = k''(\nu\varepsilon_K)^{1/4} \quad \text{or} \quad k''d_p \left(\frac{\varepsilon_K}{\nu} \right)^{0.5} \quad \text{for} \quad d_p \ll \eta_K \ll L^2 \tag{6.9}$$

where L is the macroscale of the turbulence (i.e., column/impeller diameter). According to Kolmogorov, if a particle having size that is large in relation to η_K but much smaller in relation to the macroscopic system dimension (impeller diameter for stirred reactors) is freely suspended in a turbulent system, the hydrodynamics of the particle is solely determined by its size, d_p, and the rate of energy dissipation per unit mass, ε_M.

Based on the aforementioned analysis, different investigators assumed that the relative velocity in the neighborhood of a particulate phase depends on the length scale, η_K and $(v_K \varepsilon_M)$. Using the velocity term, u_K, derived from Kolmogorov's theory and the particle diameter as the linear dimension, a power dissipation Reynolds number Re_K for particle–fluid mass transfer was defined as follows:

$$Re_K = \left(\frac{\varepsilon_M d_P^4}{v^3} \right) \tag{6.10}$$

The Sherwood number, Sh_p, was correlated in the following variant of the Ranz and Marshall correlation (Eq. 6.4):

$$Sh_p = 2 + \text{constant} \, (Re_K)^a (Sc)^{b_1} \tag{6.11}$$

Equation 6.10 is a definition of the Reynolds number based on power dissipation using a velocity term derived from dimensional arguments. This prompted Middleman (1965) to comment that derivation of Equation 6.11 using Kolmogorov's theory is a sophisticated form of dimensional analysis. Even with this oversimplification, Equation 6.10 still needs to be modified for gas–liquid mass transfer for which d_B has a wide variation, and hence, an average d_B is difficult to define. In view of this, most investigators resorted to correlating the volumetric mass transfer coefficient, $k_L a$. The correlations proposed for stirred tank reactors were therefore of the form (Hickman 1988; Middleton 2000)

$$k_L a = a_2 \left(\frac{P}{V_L} \right)^{b_2} (\text{vvm})^{c_2} \tag{6.12}$$

or

$$k_L a = a_3 \left(\frac{P}{V_L} \right)^{b_3} (V_G)^{c_3} \tag{6.13}$$

Shinnar and Church (1960) were the first to suggest the use of this theory for correlating mass transfer data. Equations 6.12 and 6.13 contain empirical constants that must be obtained by regression of experimental data. It will be shown later that these empirical constants are not universal but are specific to the system geometry and scale employed for collecting the data. The main violating feature of Kolmogorov's theory when applied to process equipment such as multiphase reactors lies in the argument that the small-scale motions are the only determining factors and all types of transport processes can be completely described through ε_M and v. This postulate arises out of Kolmogorov's argument that the small-scale phenomenon can be understood independent of nature/origin of the large scale. Kolmogorov's theory developed on the above presupposition is appropriate for describing the decay of turbulence. However, it completely ignores the role that large scale or mean flows play in real situations such as dispersion of a gas or suspension of a solid phase in a liquid medium (Nienow et al. 1977). Kolmogorov (1962) made a small yet vital refinement to his previous hypothesis. According to this refinement, the various scaling exponents that characterize small-scale statistics are anomalous (Stolovitzky and Sreenivasan 1994; Sreenivasan 1999). Without going into the details of the anomaly, it can be acknowledged that they depend on the particular class of flows. Further, finite Reynolds number effects, large-scale anisotropies, etc. may be the cause of these delicate differences and the dependence of the scaling region on a finite power of Reynolds number and the large-scale forcing (Sreenivasan 1999).

Importance of Mean Flows   It is well known that mean flows (including the scale of largest eddies) developed by different reactors are different and depend on the hydrodynamic regime prevailing. Inasmuch as these mean flows in different multiphase reactors are different, their effect on performance of different multiphase reactors is also different. A sparged reactor that develops vertically upward flow at the center and downward flow at the wall has a totally different gas holdup structure than a stirred reactor. Further, for a given stirred reactor itself, the mean flow is dependent on certain system internals and variables. For example, mean flows in stirred reactors differ greatly in strengths (turbulence intensity and pumping capacity) and direction (axial, mixed, or radial) for different types of agitators and geometric proportions (C/T, D/T). The stirred reactor is therefore a very good case in point to illustrate the importance of mean flows in multiphase systems. A wide range of impellers is available (downflow, upflow, radial flow, and mixed flow) for two- and three-phase contacting. The direction of mean flow generated is different for these impellers, and therefore, their performance with regard to solid suspension and gas dispersion is substantially different (Chapman et al. 1983a, b; Nienow 2000). Specific discussions on the importance of mean flows and their relationship with hydrodynamic regimes are provided in Sections 7A.5, 7A.6, 7A.7, 7A.10, and 8.8. The large-scale flow of a propeller is truly axial. The downflow and upflow pitched blade turbines yield a mixed axial–radial flow that further depends on the impeller to vessel diameter ratio, and finally, the disc turbine exhibits a radial flow. The flow direction of a propeller or a downflow pitched blade turbine is opposite to the vertical buoyant motion of the gas bubbles rising from a gas sparger that is generally located below it. These two flows oppose each other. This can result in a tendency of the gas phase to collect or even stagnate in the region near the impeller. Due to this accumulation of the gas phase, the density of the mixture surrounding the impeller drops drastically. The overall consequence is a steep decrease in the power drawn by the impeller (Fig. 7A.7). At relatively high gas flow rates and low speeds of agitation, the pumping action or mean flow (dictated by the dimensions of the impeller besides its speed of agitation) is incapable of dispersing the gas being sparged. This is the classic case of flooding of the impeller. It is a direct consequence of the strength of the mean flow being insufficient, although its direction is favorable for driving the gas bubbles downward. Under this condition of flooding, the performance of the reactor is poor. Fluid turbulence is also a part of these mean flows. In the previous example, the turbulent fluctuating components are useful only to the extent of reducing the rise velocity of the gas bubbles due to the additional drag on the bubble created by the turbulence. On the other hand, the momentum of the downward mean liquid flow generated by the impeller is responsible for propelling the gas downward. The direction of the mean flow is also equally important. Contrary to the downflow variety, upflow impellers generate a mean flow that cooperates with the natural buoyancy of the gas bubbles, and akin to a cocurrent packed column, flooding does not occur. This cooperation also makes the upflow impellers exhibit better stability (Fig. 7A.7) and superior efficiency for gas dispersion and solid suspension (Section 7A.10). There are many other similar examples of multiphase systems in which the large-scale mean flows play an important role. When the strength (momentum) and direction of the mean flow are favorable, the combined effect causes dispersion of the gas in the region below the gas sparger. Kolmogorov's theory does not make a distinction between these situations in which the sparged gas is dispersed or not. Alternatively, it does not distinguish between different regimes (flooding, dispersion (N_{CD}), and recirculation (N_R) in Fig. 7A.8) and therefore cannot uniquely represent the hydrodynamics.

The importance of mean flows can be further buttressed by a comparison between stirred and three-phase sparged reactors in terms of solid suspension behavior/gas dispersion. A very large number of studies are available in the literature for predicting critical speeds for solid suspension and gas dispersion (Sections 7A.6 and 7A.7) for the wide variety of impellers available. On the other hand, although solid suspension (also gas dispersion) is as important in a three-phase sparged reactor (slurry bubble column), the attention it has got is insignificant. The explanation for this evident disregard lies in the fact that the mean flows of gas and liquid phases developed at the bottom in a sparged reactor are always upward in the central core. These flows are cooperative and, therefore, yield facile solid suspension. Solid suspension is achieved at superficial gas velocities as low as 0.02 m/s for typical particle sizes (~50 μm) and catalyst loading of relevance to three-phase catalytic reactions (Deckwer 1992). In reality, the operating gas velocities may be substantially higher than this lower limit, and suspension will be further assisted by the net liquid flow in the case of continuous operation. This observation explains the relative disregard for solid suspension in three-phase sparged reactors. As regards gas dispersion, there is no special effort required. The gas phase introduced at the bottom occupies the entire column height due to the natural upward flow generated by the buoyancy of the gas.

Other inadequacies of Kolmogorov's theory in the manner of its use by Shinnar and Church (1960) and Brian et al. (1969) to correlate solid–liquid mass transfer in stirred reactors have been dealt in some detail by Levins and Glastonbury (1972a). Stewart and Townsend (1951) have estimated the value of Re_K to be 1500 for the existence of the inertial subrange. Taylor (1935) estimated that for the region outside the impeller, the value of Re_K for conditions typical of stirred reactors is ~144. This is much smaller than that required by Stewart and Townsend's estimated value of 1500. Therefore, the validity of Equation 6.8 outside the impeller discharge stream is doubtful. Further, the use of d_p in Equation 6.10 for Re_K is not justified by

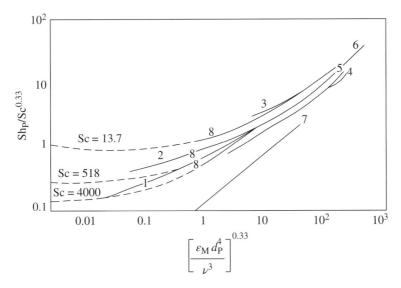

FIGURE 6.1 Comparison of K_{SL} data of different workers based on Kolmogorov's theory: (1) Harriott (1962a), Sc > 3600; (2) Harriott (1962a), Sc = 518; (3) Brian et al. (1969), Pr = 13.8; (4) Harriott (1962a), 1000 < Sc < 11,000; (5) Barker and Treybal (1960), 735 < Sc < 1328; (6) Wilhelm et al. (1941), Sc = 950; (7) Calderbank and Moo-Young (1961); and (8) Sano et al. (1974), 217 < Sc < 1410. (Reproduced with permission from Pangarkar et al. (2002). American Chemical Society. © 2002.)

Kolmogorov's theory itself. These correlations merely propose that for different systems (with same Sc) when Reynolds number based on Kolmogorov's theory is equal, the corresponding Sh_p is equal. This statement is too simplistic and ignores the importance of hydrodynamic and turbulence similarities (Sections "Hydrodynamic Regime Similarity" and "Turbulence Similarity" in Chapter 5, respectively). For instance, Jadhav and Pangarkar (1991a) have shown that particle–liquid mass transfer coefficients in small- and large-diameter bubble columns are significantly different inasmuch as the hydrodynamic regimes (and hence the turbulence structure) are different. As discussed earlier, this conclusion has also been substantiated through analysis of carefully measured fractional gas holdup, liquid circulation velocity, and turbulence parameters (Shaikh and Al-Dahhan 2010). Most investigators who used Kolmogorov's theory as the basis for correlating mass transfer data in stirred vessels advocated caution, citing several of its limitations (Middleman 1965; Brian et al. 1969, for instance). Extending the argument further at equal power inputs, Equations 6.10 through 6.13 predict equal Sh_p for different multiphase reactors ranging from three-phase sparged reactors, stirred reactors, to venturi loop reactors. This is evidently not true. Pangarkar et al. (2002) have discussed this matter in detail. Figure 6.1 from Pangarkar et al. (2002) shows the correlating lines for data from different investigators. It is evident that Kolmogorov's theory-based correlations yield widely varying predictions of K_{SL}. This is also evident from the waning of the enthusiasm of 1960s and early 1970s for using Kolmogorov's theory as a tool to correlate mass transfer in multiphase systems.

In conclusion, Kolmogorov's theory is a realistic description of decay of turbulence. However, its application in a sophisticated form of dimensional analysis (Middleman 1965) is stretching the theory beyond its aim. Such attempts and the resulting correlations cannot lead to a unique scheme for predicting transport properties.

A simplified approach that gives due importance to mean flows was hypothesized by Nikhade and Pangarkar (2007) and will be discussed later.

6.4   CORRELATIONS BASED ON HYDRODYNAMIC/TURBULENCE REGIME SIMILARITY

Major improvements in understanding particle–liquid mass transfer started with the development of the equation of motion of a particle in a turbulent fluid. Tchen (cf. Hinze 1959) derived the equation of motion for small spherical particles suspended in a turbulent flow. Various investigators used different forms of Tchen's original equation to understand particle–liquid relative motions and mass transfer process and obtain correlations for the relevant mass transfer coefficient.

6.4.1   The Slip Velocity Approach

This approach, which is a significant improvement over the dynamic similarity-based correlations, was first proposed by Calderbank and Moo-Young (1961) and subsequently developed, among others, by Harriott (1962a, b). Harriott used the mass transfer coefficient based on terminal settling velocity or velocity of a fluid flowing past a solid particle fixed in space as the base value. This base value was then modified by including relevant turbulence parameters. For particle Reynolds numbers relatively higher than unity, the Ranz and Marshal (1952b) correlation was taken as the basis:

$$Sh_P = 2 + 0.6(Re_T)^{0.5}(Sc)^{0.33} \qquad (6.14)$$

and for values of Reynolds numbers less than unity, Friedlander's (1957) correlation was considered to be a more accurate alternative:

$$Sh_P = 0.99(Re_T \, Sc)^{1/3} \quad \text{for} \quad Re_T < 1 \qquad (6.15)$$

According to Harriott, Equations 6.14 and 6.15 should yield the minimum value of the mass transfer coefficient, K_{SLT} (value of K_{SL} at $u = V_T$). Harriott's (1962a) study can be broadly divided into two parts: (i) the effect of physical properties, in particular the liquid viscosity, density, and solute diffusivity (components of Schmidt number) along with the particle diameter, was derived from the variation of K_{SLT} as calculated from Equation 6.14 or 6.15 for the respective parameters. (ii) The effect of power input (obtained from the impeller diameter, type, and speed of agitation) was derived from experimental measurements of mass transfer coefficients using different solutes. Interestingly, Harriott also examined the behavior of the mass transfer coefficient with the state of suspension of the solids and found that there was a distinct change in the plot of K_{SL} versus speed of agitation at the point when the particles were "completely" suspended. This important observation, as early as 1962, indicates that the state of "complete" suspension as defined by Harriott is exclusive in relation to the value of the mass transfer coefficient. We will be frequently referring to this state in this chapter. An exhaustive discussion on this unique state will be presented in Chapter 7A.

6.4.1.1 Effect of Physical Properties and Particle Size as Discerned from the Variation of K_{SLT}
Figure 6.2, reproduced from Harriott (1962a), shows the variation of K_{SLT} with d_p at different values of the density difference $(\rho_S - \rho_L)$.

The major conclusions of Harriott (1962a) for K_{SLT} are as follows:

1. The lack of dependence of K_{SLT} on particle size for particle size range of 100–1000 μm observed by various investigators was explained by Harriott (1962a) as follows: according to the Ranz and Marshal correlation (Eq. 6.14), the particle Sherwood number $(K_{SL} d_p / D_A)$ is proportional to $(Re_T)^{0.5}$. The terminal settling velocity, V_T, for particle sizes ranging between 100 and 1000 μm falling in still liquid is approximately a linear function of the particle diameter. Accordingly, $(Re_T)^{0.5} \propto d_p$. Therefore, the terms representing d_p on the two sides of the Ranz and Marshall correlation cancel out. Consequently, K_{SLT} is independent of d_p as observed by most investigators. Figure 6.2 shows that K_{SLT} initially decreases with increase in particle diameter starting in the range of $d_p \sim 10$–20 μm. With increasing particle size beyond $d_p = 200$ μm, K_{SLT} is almost independent of d_p up to $d_p = 2000$ μm. It should also be noted that majority of the previous investigations on K_{SL} in stirred vessels (Hixon and Baum 1941, 1942; Mack and Marriner 1949; Kneule 1956; Humphrey and Van Ness 1957; Mattern et al. 1957b; Barker and Treybal 1960; Calderbank and Jones 1961; Calderbank and Moo-Young 1961) employing particles in the 200–2000 μm did not reveal any dependence on the particle diameter (explained later). For liquids with relatively high viscosity $(\sim \mu_L = 20\,cP)$, the intermediate particle size range in which K_{SLT} is independent of d_p starts at $d_p \sim 600$ μm; and therefore, for such liquids, the threshold value of d_p is different.

2. For very large particles such that settling occurs in Newton's law region, V_T is proportional to $d_p^{0.5}$, and therefore, the Ranz and Marshal correlation predicts a mass transfer coefficient dependence: $K_{SLT} \propto d_p^{-0.25}$. For particles larger than 2000 μm, there is a small decrease in K_{SLT} as can be seen in Figure 6.2.

3. On the other hand, for very small particles $(Re_T \ll 1$ or $d_p \rightarrow 0)$, mass transfer can be depicted by diffusion in an infinite quiescent medium. For such a situation, the limit of $Sh_p = 2$ implicit in Ranz and Marshall's correlation suggests that with decreasing particle diameter, K_{SLT} must increase in a linearly reciprocal manner.

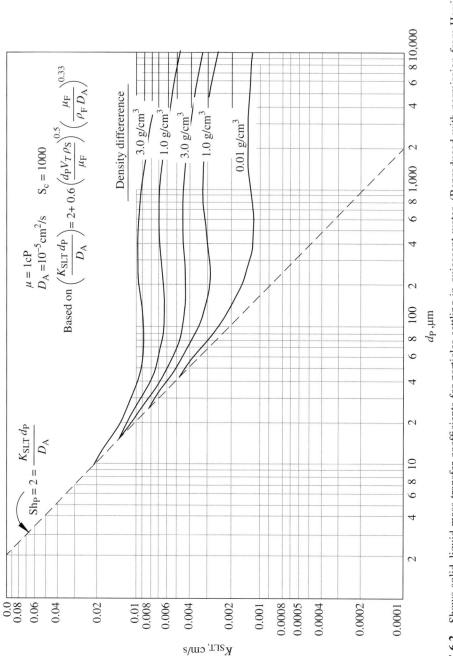

FIGURE 6.2 Shows solid–liquid mass transfer coefficients for particles settling in quiescent water. (Reproduced with permission from Harriott (1962a). American Institute of Chemical Engineers. © 1962.)

4. For relatively low viscosity liquids, K_{SLT} is proportional to $D_M^{0.8}$. This dependence decreases to 0.6 for high viscosity liquids. It must be noted that the diffusivity has a negative dependence on liquid viscosity, which is partially reflected in the exponent on D_M as given earlier. In addition, at low diffusivities, the transient mode contributions to the mass transfer process assume greater importance (Harriott 1962a). If the viscosity contribution to diffusivity is incorporated in variation of K_{SLT}, its true dependence on viscosity is still negative but varies with the particle size. Thus, $K_{SLT} \propto \mu_L^{-0.06}$ for small d_p and $K_{SLT} \propto \mu^{-0.22}$ for large d_p.

5. The dependence of K_{SLT} on the density difference between the solid and liquid phases can be described by $K_{SLT} \propto (\Delta \rho)^{0.3-0.4}$ when $\Delta \rho$ exceeds $\sim 400\,\text{kg/m}^3$. For highly soluble salts, the relatively high K_{SLT} was attributed to convection caused by rapid changes in the density of the liquid in the concentration boundary layer surrounding the particle. In such cases, the change in the particle size with time can also be substantial and is an additional complicating factor. Indeed, methods for measurement of mass transfer coefficient based on complete dissolution of the solid particle greatly suffer from this change in particle size from a finite size to nearly zero (Nienow 1969). The effect of particle diameter on K_{SL} is bound to cast serious doubts on the accuracy of such measurements in which d_p varies so widely.

6.4.1.2 Effect of Turbulence Parameters as Discerned from the Variation of K_{SL} *in a Stirred Tank*   Harriott's (1962a) exhaustive study of particle–liquid mass transfer in stirred tanks covered tank diameters ranging from 0.1 to 0.53 m. However, majority of the experiments were conducted in the 0.1 m vessel. The H/T used was unity in most cases with a small positive deviation for the smaller vessel. Tchen's (cf. Hinze 1959) equation of motion for a particle in a turbulent fluid has been widely used to explain the effect of turbulence on mass transfer coefficient.

The equation of motion derived by Tchen is as follows:

$$\frac{\pi}{6} d_P^3 \rho_P \frac{du_P}{dt} = 3\pi \mu d_P (u_F - u_P) + \frac{\pi}{6} d_P^3 \rho_F \frac{du_F}{dt} + \frac{\pi}{12} d_P^3 \rho_F \left(\frac{du_F}{dt} - \frac{du_P}{dt} \right)$$
$$+ \frac{3}{2} d_P^2 \left(\sqrt{\pi \rho_F \mu} \right) \int_0^t \frac{\{(du_F/dt) - (du_P/dt)\}}{\left(\sqrt{t - t'} \right)} dt' + F \tag{6.16}$$

In Equation 6.16, subscripts P and F denote the particle and fluid phase, respectively. The equation has been derived from a force balance on a particle in a turbulent fluid. It has the following components: the left-hand side represents the force necessary to accelerate the particle under a given set of conditions. On the right-hand side, the various forces acting on the particle are included. The resultant corresponds to the net force on the left-hand side. These forces in the order they appear in the right-hand side are (i) the drag force due to a finite fluid viscosity, (ii) the pressure gradient force that causes fluid acceleration, (iii) the added mass correction that accounts for the relative fluid acceleration required to accommodate a relative particle acceleration, (iv) the Basset term, and (v) net vertical force acting on the particle. Hinze (1959) has provided a detailed discussion of the assumptions made in deriving Tchen's equation of motion. To apply this equation for a highly turbulent system such as a conventional stirred tank, several investigators included various correction factors, particularly for the term due to viscous drag. This also included replacement of the viscous drag expression by a general drag force law so as to extend the range of applicability of the entire equation of motion. For more details on these extensions/improvements, the reader is referred to an excellent discussion by Maxey and Riley (1983). Equation 6.16 presents a formidable task, and at the time when attempts were made to solve it, the computational facilities were not as powerful as they are at present. However, even with the current powerful machines, some simplifications are necessary. Consequently, most investigators had to make simplifying assumptions for the afore-mentioned formidable equation to be tractable. It must be pointed out that although the following discussion pertains to solid phase, Tchen's equation is also valid for gas bubbles dispersed in a turbulent liquid (Hughmark 1974), albeit with the under-standing that the buoyant force is always vertically directed.

Harriott (1962a) used a simplified version of Tchen's equation of motion (Hinze 1959) for the estimation of fluid and particle fluctuating velocities. From these, average particle slip velocity in the turbulent liquid is determined from Equation 6.17:

$$u_{\text{Slip}} = \left(\sqrt{\overline{u_F'^2}} \right) - \left(\sqrt{\overline{u_P'^2}} \right) \tag{6.17}$$

For an arbitrary value of the Lagrangian timescale, $\tau_L = 0.001$ s and $\rho_p = 1240$ kg/m^3, Harriott estimated that the slip velocity for 50 and 300 μm particles should be approximately between 2 and 14% of the fluctuating velocity, respectively.

Harriott presented three steps for estimating the characteristic slip velocity. In the first step, the terminal settling velocity of a particle was used to calculate the Reynolds number. This yielded the minimum value of the mass transfer coefficient from Equation 6.14 or 6.15 as the case may be. The second step limited to small par-ticles was based on Equation 6.17. For much larger particles, however, Harriott (1962a) concluded that the turbulent parameters approach is not applicable and only approximate prediction of the limits of the slip velocity is possible. The minimum value of the mass transfer coefficient, K_{SLT}, corresponded to the particle terminal set-tling velocity in a quiescent liquid having the same physical properties as the actual liquid employed. The enhancement by the prevailing turbulence was defined relative to K_{SLT}, which served as the datum. The enhancement was calculated as explained later in this section. Subsequently, Nienow (1969) and Miller (1971) also tried this approach. The natural argument used was that the value of K_{SL} prevailing in a highly turbulent liquid will always be higher than this minimum value, K_{SLT}. Further, it was argued that this actual K_{SL} value can be represented as a multiple of K_{SLT}. This approach of correlating K_{SL} in a stirred tank was denoted by Harriott (1962) as the terminal settling velocity–slip velocity approach since it referred to enhancement over K_{SLT}.

Harriott rightly argued that the actual value of the particle–liquid mass transfer coefficient in a stirred reactor depends on parameters such as type of impeller, speed of agitation (power input), etc. It was found that with increasing impeller speed, there is a rapid increase in the mass transfer coefficient from a relatively low initial speed N to the speed N_S at which the particles are suspended. Once N_S is reached, the mass transfer coefficient becomes a relatively weak function of the impeller speed. For particle diameter >100 μm, Harriott (1962a) suggested that this dependence can be described by $K_{SL} \propto (d_p)^{0.5}$ for low-to-high viscosity liquids. For particles smaller than 100 μm, this exponent was further lower with a value of 0.3 for 15 μm particles. These apparently varying dependences can be explained as arising out of variation in the degree of suspension of particles of different sizes. For example, the 15 μm particles are suspended at relatively lower speeds of agitation as compared to those above 100 μm size. Obviously, the plateau in the K_{SL} versus N plot will be reached much earlier for the 15 μm particles, thereby resulting in a weaker dependence of K_{SL} on N (0.3 as compared to 0.5 for 100 μm particles). Harriott did not measure the power input directly but estimated the same based on the power number for the impeller available in the literature. Such estimation indicated that K_{SL} varied with the power input in a rather small range of $K_{SL} \propto (\varepsilon_M)^{0.1-0.16}$.

As indicated earlier, Harriott's derivation of the equation for particle fluctuating velocity was based on a simplified version of Equation 6.16. The major simplification was that the particle is at all times surrounded by fluid moving in the same direction. This implies that eddies comprising the mean flow are substantially large in size as compared to the particle and engulf it. This unidirectional motion is unlikely in any turbulent multiphase system. Harriott in the beginning pointed out that for commonly used operating conditions, the utility of this equation is limited to particle size of ~50 μm. He pointed out that the assumptions made in deriving the equation for particle fluctuating velocity limit its utility even if all relevant turbulent parameters are available. However, Harriott went on to argue that the velocity fluctuations arising from the smallest eddies (~50 μm) are likely to be relatively much lower than those created by larger eddies. Thus, it is reasonable to assume that the u_{Slip} calculated by Equation 6.17 is likely to hold for particles up to a few hundred micrometers. Harriott (1962a) further proposed that for particles having practically the same density as the fluid ($\Delta\rho = \rho_S - \rho_L = 0$), there is no slip between the two phases and hence the effective slip velocity is zero. For this case, mass transfer would occur mainly due to an unsteady-state mechanism caused predominantly by eddies penetrating the concentration boundary layer and renewing the liquid surface exposed to the particles. Harriott proposed a modified mechanism for unsteady-state mass transfer. According to this mechanism, only a part of the fluid surrounding the particle is renewed during the displacement of the particle from one spatial position to another. Thus, for eddies penetrating to a distance of H_P units from a plane surface every t seconds, the time-averaged mass transfer coefficient could be correlated to within 5% by Equation 6.18:

$$\frac{1}{K_{SL}} = \left(\frac{H_P}{D_A}\right) + \frac{1}{2}\overline{\sqrt{\frac{\pi t}{D_A}}} \qquad (6.18)$$

In Equation 6.18, H_p denotes the depth of penetration by an eddy in the neighboring mass transfer boundary layer. Harriott's analysis of his data indicated that both slip velocity and transient effects are important in determining the overall mass transfer process. Although Harriott could not provide actual slip velocity and unsteady-state mass transfer data, his conclusions based on effects of diffusivity and particle density on K_{SL} in stirred tanks are quite logical. In his conclusion, Harriott argued that the slip velocity theory is relatively easy to use for quantitative predictions as compared to the modified penetration theory.

Harriott's data on K_{SL} in stirred tanks were found to be 1.5–8 times higher as compared to those estimated from the steady-state correlations of Ranz and Marshall (1952b) and Friedlander (1957) using the particle Reynolds number based on V_T in Equations 6.14 and 6.15, respectively. In view of the complex procedure for calculating the true slip velocity of a particle suspended in stirred tank from Equation 6.16, Harriott (1962a) suggested a modified, simple procedure to estimate K_{SL}: (i) compute V_T and Re_T for the given particle size and density. For very light particles, $\Delta \rho = 300\,kg/m^3$ was suggested instead of the true density difference. (ii) Use Ranz and Marshall's (1952b) correlation (Eq. 6.14) to compute K_{SLT}. (iii) Estimate K_{SL}/K_{SLT} from the graph showing effect of d_p and diffusivity on K_{SL}/K_{SLT} (Figure 9 in Harriott's paper). This value is for situation with $\varepsilon_M^{1/3} = 0.29$ (W/kg)$^{1/3}$ and $D/T = 0.5$. Using K_{SLT} obtained in step 2 and the ratio K_{SL}/K_{SLT} in step 3, compute K_{SL}. (iv) Assuming a dependence given by $K_{SL} \propto N^{0.5}$ for $d_p > 100\,\mu m$, extrapolate K_{SL} value obtained in step 3 to the actual value of N.

There are several flaws in Harriott's above procedure. These are as follows: (i) it should be noted that K_{SLT} is based on steady-state conditions. On the other hand, actual K_{SL} in a stirred vessel is a global mass transfer coefficient decided by different degrees of turbulence that the particle experiences in space and time. (ii) $K_{SL} \propto N^{0.5}$ for fully suspended particles in the modified procedure must be used with caution. It is now well established that for speeds of agitation in the region slightly lower than N_S, K_{SL} is almost directly proportional to N. Above N_S, however, K_{SL} is approximately independent of N (Fig. 6.3).

The last suggestion ($K_{SL} \propto N^{0.5}$ for $d_p > 100\,\mu m$) of Harriott could again be due to incomplete suspension for some data sets rather than "complete" suspension. Indeed, Figure 6.3 does indicate that for the region A–B, the dependence of K_{SL} on N could be in the range of 0.5 observed by Harriott. (iii) As noted by Sykes and Gomezplata (1967), besides the effect of turbulence, the extent of entrainment of the particles in different eddies in the turbulent field has not been accounted. (iv) Different impeller–vessel configurations give different degree of particle suspension at equal power inputs (Nienow and Miles 1978). In view of this, the values of K_{SL}/K_{SLT} vary based on the impeller–vessel configuration. Considering this fact, K_{SL}/K_{SLT} plots of Harriott (1962a) for disc turbine cannot be extended to estimate K_{SL} for other types of impellers such as mixed/axial flow impellers along with different impeller–vessel geometries. The latter include the impeller location (C/T) and its relative size with respect to the vessel (D/T). However, Harriott correctly observed and explained the negligible effect of particle density under conditions such that the particles are suspended.

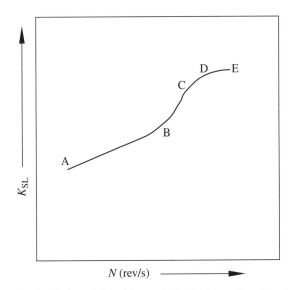

FIGURE 6.3   Typical variation of K_{SL} with speed of agitation under otherwise similar conditions. (Adapted from Nienow (1969) with permission from John Wiley & Sons. © 1969, Canadian Society for Chemical Engineering.)

Levins and Glastonbury (1972a, b, c), Kuboi et al. (1974a), and Lee (1981, 1984) refined Harriott's approach subsequently. The importance of turbulence parameters, such as intensity/scale of turbulence, in deciding the mass transfer coefficient was expanded in these studies. Evidently, application of this approach required more elaborate information on the relevant turbulence parameters. Some of the experimental investigations, particularly in stirred vessels, are discussed in the following. The early investigations used equipment with limited capabilities and accuracy. Contemporary techniques such as laser Doppler anemometry/velocimetry have far more sophistication and accuracy. Nonetheless, the initial studies deserve mention because many of them included simultaneous measurements of mass transfer.

6.4.1.3 Investigations Subsequent to Harriott (1962a)   Schwartzberg and Treybal (1968a, b) employed streak photography of tracer particles to measure rms values of both the fluid turbulent velocity u'_F and particle turbulent velocity u'_F. They did not measure rms value of relative velocity $\left(u'_F - u'_F\right)$, possibly due to limitations of the equipment used and the intricacies involved in measuring the same. These investigators made an attempt to solve Tchen's equation of motion for a particle in a turbulent fluid (Eq. 6.16) with the aim of estimating the relative particle–liquid slip velocity. A number of simplifying assumptions were made toward this end. These are as follows: (i) it was assumed that the relative acceleration caused by external forces at the solid–liquid interface was absent. This assumption amounted to neglecting the Basset term in Tchen's equation. (ii) The viscous drag term (first term on the right-hand side of Equation 6.16) was linearized by replacing the original term $C_D\left(u_F - u_P\right)$ with $C_D\left(u_F - u_P\right)_{rms}$. (iii) The complex flow pattern was described by a simple harmonic motion. It is well known that the drag coefficient increases with increase in the turbulence level. Schwartzberg and Treybal (1968b) employed a trial-and-error procedure to estimate the drag coefficients and Lagrangian scale of turbulence under turbulent conditions. They found that the agreement between the estimated and measured values was poor. This lack of agreement could be attributed to several reasons, prominent of which are (i) the number of simplifying assumptions made for the various terms in Tchen's equation to make its solution tractable and (ii) the inherent limitations of the first-generation technique for measurement of turbulence parameters.

Streak photography of tracer particles was also used by Levins and Glastonbury (1972b) to determine the turbulence parameters in a stirred vessel equipped with six blade pitched turbines. Two impellers with $D/T = 0.3$ and 0.4 were studied. The experimental results were analyzed to obtain the values of the fluctuating particle and fluid velocities. Naphthalene (density of $1018\,kg/m^3$ and average particle size of $110\,\mu m$) was used for measurements representing neutrally buoyant solid–fluid system. For the two metals (for iron, density: $7280\,kg/m^3$, and for aluminum, density: $2690\,kg/m^3$; particle diameter: 80–$90\,\mu m$), the ratio of the solid and fluid fluctuating velocities was found to be very close to unity. These streak photography data were employed to estimate the rms fluctuating velocity components for both neutrally buoyant and high-density particles. For this purpose, a variant of Tchen's equation of motion that neglected the effect of external gravity force was employed. In an extension of the aforementioned work, Levins and Glastonbury (1972c) employed Gaussian distribution to describe the turbulent eddy spectrum. Similar to their previous observation (Levins and Glastonbury 1972b), this latter investigation also showed very little difference in rms values of both the heavy (iron) and light (aluminum) particles and the fluid. This observation is open to debate since a higher density difference should manifest itself in relatively higher contributions from the inertial and gravitational components. Levins and Glastonbury (1972c) challenged the simple power per unit volume approach advocated by Calderbank and Moo-Young (1961) and Brian et al. (1969). The latter could not yield a satisfactory correlation for their data. Levins and Glastonbury (1972c) observed a perceptible effect of the D/T ratio on K_{SL}, which also conflicted with the power input approach. Thus, Equation 6.19 that included the D/T effect was proposed:

$$\left(\frac{K_{SL}d_P}{D_M}\right) = 2 + 0.47\left(\frac{d_P^{4/3}\varepsilon^{1/3}}{v}\right)\left(\frac{D}{T}\right)^{0.17}\left(\frac{v}{D_M}\right)^{0.36} \qquad (6.19)$$

Komasawa et al. (1974) and Kuboi et al. (1974a) used high-speed visual photography for determining the turbulence parameters and the particle–fluid slip velocity. The analysis presented by Kuboi et al. (1974b, c) was again based on a simplifying assumption that the viscous drag constituent in Tchen's equation can be replaced by the term $(\pi/8)d_P^2\rho_F C_D\left(\overline{|u_F'-u_P'|}\right)(u_{Fi}-u_{Pi})$. In addition to the above simplification, Kuboi et al. (1974a) neglected the Bassett term in Tchen's equation. The amplitude of the nth harmonic represented by the Fourier coefficient along with the drag coefficient in the highly turbulent field was thus determined using the aforementioned technique and assumptions. The drag coefficient affects the terminal settling velocity. The values of $\left(u'_{Fi} - u'_{Pi}\right)$ predicted from the modified form of Tchen's equation were then compared with the values measured using high-speed visual photography. In contrast to the findings of Levins and Glastonbury (1972b), Kuboi et al. (1974a) found that there was a substantial difference between the fluctuating velocities of the particle and fluid phases. Thus, for the high-density particles, the particle fluctuating velocities were ~30–75% lower than the corresponding fluid phase components. In an analogous manner, the neutrally buoyant solid–fluid systems also exhibited lower values for the solid phase in comparison to the fluid phase. Kuboi et al. (1974a) argued that the major differences between their measured parameters and those of Levins and Glastonbury (1972b) were due to the inaccuracies in the streak photography technique used by the latter. Kuboi et al. (1974a) proposed that the actual particle–fluid relative velocity is a resultant of the effective relative velocity of particles of similar size and having a vanishing density difference $[(\rho_P - \rho_F) \sim 0]$, effective velocity generated by inertial forces, and finally the settling velocity of the particle as affected by the extent of turbulence. It has been mentioned earlier that turbulence increases the drag coefficient. Consequently, the settling velocity in a turbulent field is lower than that in a still liquid, V_T.

In a significant departure from Harriott's (1962a) approach, Sage and Galloway (1964) modified the Ranz and Marshal correlation itself for the case when the turbulent intensity becomes large and its effect on mass transfer can no longer be neglected. The modifications proposed incorporated the effect of the turbulence intensity and were also later supported in principle by Kuboi et al. (1974b) and Hughmark (1969). Sage and Galloway proposed Equation 6.20:

$$Sh_P = 2 + A'' \operatorname{Re}_T{}^{0.5} Sc^{0.33} \tag{6.20}$$

Here, $A'' = 0.439 + 0.0513d_P + 0.234u'(u' + 0.05)\operatorname{Re}_T{}^{0.5}$, where u' is the prevailing turbulence intensity.

In many of the aforementioned investigations, Ranz and Marshal's and Friedlander's correlations were used as the basis. The basic reservations about these latter correlations are as follows: (i) they are derived for steady-state flows. On the other hand, a particle that is freely suspended in a stirred vessel exhibits chaotic motion both in time and space. The particle motion is affected by both the entraining mean liquid flow and the turbulent components (Nienow 1969; Kuboi et al. 1974b). (ii) Further, starting with the work of Harriott (1962a), it is now clearly established that the use of any form of V_T in defining the Reynolds number in a highly turbulent field is a gross approximation. (iii) Unlike the fixed exponents on the Reynolds and Schmidt numbers used earlier, these exponents vary with the magnitude of Re and Sc in the case of forced convection that is typical of transport in multiphase systems (Keey and Glen 1964).

Nienow (1969) studied the variation of K_{SL} as a function of the extent of suspension of the particles. The region A–B (Fig. 6.3) is marked by particles that are either motionless or exhibiting relatively insignificant motion on the bottom of the vessel.

As point B is passed, the extent of particles that are suspended increases up to point C. At and above point C, majority of the particles are suspended. Above point C, K_{SL} becomes a relatively weak function of the speed of agitation. Between C and D, the particles are suspended, but interplay of two different phenomena assumes importance. In the region D–E, particularly in the case of small-vessel diameters, this is a pseudo-constant K_{SL}. In this case, high values of N can cause surface aeration (Nienow 1969). It is known that introduction of a gas causes a reduction in the power number (Fig. 7A.6). However, since N is increasing along D–E, the reduction in power number is counteracted. The two opposing effects probably nullify each other with the result that K_{SL} remains unaffected by increase in N. For larger $(T>0.5\,\text{m})$ fully baffled vessels, the possibility of surface aeration is relatively remote and a truly constant K_{SL} is achieved. The dependence of K_{SL} on N varies from $K_{SL} \propto (N)^{1.6}$ (for region B–C) to $K_{SL} \propto (N)^0$ (for region D–E). Statistical analysis of all the data in the regions from B to D yielded the average dependence as $K_{SL} \propto (N)$. However, such an analysis has no logic since data for varying degrees of suspension or turbulence are clubbed together to obtain the average dependence.

Kneule (1956), Mattern et al. (1957), Kolar (1959), Nagata et al. (1960), Calderbank and Moo-Young (1961), Harriott (1962a), Madden and Nelson (1964), and Lal et al. (1988) also made similar observation about the role of particle suspension. According to Nienow (1969), the reason for the differences in the dependence of K_{SL} on N observed by different investigators is primarily due to differences in the operating ranges in their studies. These ranges varied from point A (particles settled/moving around the bottom of the vessel) to point E at which almost all the particles are suspended. The condition at point D represents minimum impeller speed for just suspension of the particles (N_S) originally introduced by Zwietering (1958) and later supported by the work of Nienow (1968). Further, according to Nienow (1969), at the just-suspended condition (N_S), the value of K_{SL} is independent of physical properties (ρ_p, d_p) and impeller–vessel geometry (impeller type, D/T, C/T). On the other hand, Nienow (1969) found that at equal power input, there is significant variation in the value of K_{SL} when different geometrical configurations $(D/T, C/T)$ and types of impeller are employed. This observation is an additional proof of the inadequacy of Kolmogorov's theory for predicting K_{SL} as discussed earlier in this chapter. Nienow (1969) tried to explain the discrepancies in Harriott's (1962a) data through the extent of particle suspension. He indicated that in the case of the larger particle data, the particles were most likely not suspended at the arbitrary power input used in step 3 of the procedure suggested by Harriott. Indeed, Nienow (1969) emphasized the importance of suspension and, as shown in Figure 6.3, concluded that there is insignificant variation in the value of K_{SL} beyond the just-suspended condition (point D in Fig. 6.3). Nienow agreed that the concept of slip velocity proposed by Harriott is applicable for estimating K_{SL}. However, he continued his advocacy of the importance of the just-suspended condition and argued that a modified slip velocity theory based on the impeller speed for just suspension is a better alternative to the arbitrary power input suggested by Harriott.

Nienow (1969) introduced the concept of an enhancement factor, E, which is a ratio of the actual value of K_{SL} in a stirred vessel at the just-suspended condition, K_{SLS}

to K_{SLT}. This enhancement factor $E = K_{SLS}/K_{SLT}$ is evidently greater than unity because of the effect of the imposed turbulence. The enhancement was found to be a function of the particle size, d_p. Nienow (1969) found that K_{SLS}/K_{SLT} is greater for relatively large and dense particles. This was ascribed to the higher level of turbulence in the fluid at the relatively higher speeds of agitation required to suspend such large and heavy particles. Nienow (1969) observed that for a fixed C/T ratio, the dependence of K_{SL} on N increases with increasing particle size and the particle–fluid density difference. The lesser effect observed by Harriott (1962a) is most likely because majority of his experiments used small particles ($<300\,\mu$m) and systems of low density difference ($<240\,$kg/m^3). The latter data showed that for a given d_p besides N, the system geometry (C/T, D/T) and physical properties (Schmidt number, density difference, etc.) also affect K_{SL}. Similar observations were made by Nienow and Miles (1978), Sicardi et al. (1979), Conti and Sicardi (1982), Ovsenik (1982), and Aussenac et al. (1982).

In an analysis of various literature data following his earlier work, Nienow (1975) compared the two main methods of predicting K_{SL}: (1) terminal settling velocity–slip velocity theory and (2) correlations based on Kolmogorov's power input criterion. This analysis did not find much difference in the dependence of K_{SL} on physical properties such as density difference, kinematic viscosity, and molecular diffusivity besides d_p for $d_p < 1500\,\mu$m. The discrepancies of various investigators' data for $d_p > 1500\,\mu$m must be due to different degrees of suspension arising out of different impeller–vessel configurations (type, D/T, C/T, etc.). This is evident because smaller particles are relatively readily suspended as compared to the larger particles. It will be seen later that the discrepancies in various data sets are most likely due to varying degrees of suspension (Hughmark 1974; Nienow 1969, 1975).

Hughmark (1974) argued that the terminal settling velocity of a particle in a turbulent fluid corresponds to the minimum speed of agitation for their suspension, N_S. The latter, according to Hughmark (1974), could be derived from Zwietering's (1958) definition. According to this argument, the characteristic relative velocity derived by Levins and Glastonbury (1972c) conforms to the terminal settling velocity of the particle at the prevailing level of turbulence. This is represented by $\bar{u}$. Hughmark (1974) estimated $\bar{u}$ from the following equation:

$$\bar{u} = \overline{\sqrt{V_T^2 + u_R^2}}\,.$$

Hughmark employed this $\bar{u}$ to derive a correlation for K_{SL}. Son and Hanratty (1967) and Hughmark (1971, 1974) correlated wall to fluid heat transfer in pipe flow based on the relatively simple and well-established boundary layer theory. In the case of pipe flow, momentum transfer is solely by skin friction because of the geometry involved. Nonetheless, this approach was extended to particle–fluid mass transfer in turbulent flow. The correlation proposed was of the following form:

$$K_{SLP} \propto \bar{u} \times \left(\overline{\sqrt{\frac{f}{2}}} \right) \times Sc^{-0.66} \tag{6.21}$$

The friction factor, f, is as usual a function of the Reynolds number. Hughmark (1974) suggested correlation of $K_{SL} \times (Sc^{0.66} / \bar{u})$ with the particle Reynolds number based on $\bar{u}$ and d_p, $d_p \bar{u} \rho_L / \mu_L$. Hughmark used a large number of data from the experimental investigations of Barker and Treybal (1960), Harriott (1962a), Keey and Glen (1966), Sykes and Gomezplata (1967), Brian et al. (1969), Miller (1971), and Levins and Glastonbury (1972b). Besides the impeller–vessel configuration (impeller diameters from 5 to 10 cm), these data also covered a wide range of physical properties. The systems used varied from solids dissolution in water, aqueous sucrose, ethanol, and methanol. The range of density difference covered was similarly relatively wide with $(\Delta \rho / \rho)$ ranging from 0.05 to 6.14. It should be noted that Brian et al.'s (1969) measurements also covered a system in which solid was lighter than the liquid $(-\Delta \rho)$. The particle size ranged from 80 μm in the case of dense metal particles, 250 μm for ion exchange resins, and up to a maximum of 0.4 cm for the benzoic acid particles.

For the impeller diameters covered, Hughmark (1974) employed the Lagrangian time scale obtained from the experimental data of Levins and Glastonbury (1972c) to estimate particle–turbulent fluid relative velocity, u_R. Then, u_R was used to estimate $\bar{u}$—the vectorial sum of the terminal settling velocity in a turbulent fluid and the particle–fluid velocity. Hughmark (1974) proposed the following correlation obtained by regression of these relatively very large sets of data:

$$\left(\frac{K_{SL}}{\bar{u}} \right) \times Sc^{0.66} = 3.6 \left(\frac{d_p \bar{u} \rho_L}{\mu} \right)^{-0.66} \tag{6.22}$$

Equation 6.22 gave satisfactory agreement for situations in which the particle–fluid velocity was higher than the terminal settling velocity. In cases where the reverse was true (incomplete suspension of the particles), the correlation predicted values of K_{SL} higher than the measured values. Further, the correlation could not give satisfactory agreement for particles smaller than 250 μm and also for relatively high viscosity liquids. Hughmark (1974), however, showed that Equation 6.22 could also adequately describe Calderbank's (1959) data for gas–liquid mass transfer coefficient for absorption of carbon dioxide in water and aqueous solutions of glycol and glycerol. Hughmark's correlation was limited to vessel diameters much smaller than those used in industrial practice. He noted that the absence of Lagrangian time scale data for larger vessels precludes the use of his correlation for the purpose of scale-up. Hughmark (1974) supported the observations of Nienow (1969, 1975) that particle suspension is an important factor in deciding the value of K_{SL}. The contradictions observed in the literature data were attributed to different degrees of suspension ranging from particles settled at the bottom of the vessel to those that are completely suspended. However, if the degree of suspension is identical, then the small-scale data can satisfactorily relate to the large scale. The origin of the concept of "relative suspension" for solid–liquid and "relative dispersion" for gas–liquid systems lies in this observation as well those of earlier investigators, in particular Nienow (1969, 1975). These concepts of "relative suspension" and "relative dispersion" will be used extensively later to yield reliable scale- and configuration-independent correlations for solid–liquid and gas–liquid mass transfer coefficients.

Hughmark (1980) subsequently regressed data for small particles (low Reynolds number) independently and identified two distinct regions in which separate correlations need to be used. The data of Harriott (1962a) and Barker and Treybal (1960) provided a range of Reynolds numbers (based on u_R as the characteristic velocity) from 10^{-6} to 100 for this regression. A plot of $K_{SL}/\bar{u}$ against Reynolds number exhibited the above referred two distinct regions. The separation of the correlating lines occurred between Reynolds number 0.1 and 1. The lower range Reynolds number data gave the following correlation that was found satisfactory:

$$\left(\frac{K_{SL}}{u_R}\right) \times Sc^{0.66} = 2.58 \left(\frac{d_p u_R}{v}\right)^{0.8} \tag{6.23}$$

Equation 6.23 used the particle–fluid relative velocity, u_R, obtained from Hughmark's (1974) study rather than $\bar{u}$. In view of the success of Equation 6.23 for low Reynolds number data, Hughmark (1980) attempted to verify the efficacy of the following velocity terms for the higher Reynolds number: (i) $\bar{u} = \sqrt{u_T^2 + u_R^2}$, (ii) u_R, and (iii) V_T. The three sets of regressions indicated that the use of $\bar{u}$ yielded the best fit for the larger Reynolds number data. The correlating lines showed a distinct transition point at a Reynolds number value of ~17. According to Garner et al. (1959), this is the maximum value of Reynolds number corresponding to an approximately symmetric flow. The high Reynolds number data were found to correlate well with the following expression:

$$\left(\frac{K_{SL}}{\bar{u}}\right) \times Sc^{0.75} \left(\frac{d_p \bar{u} \rho_L}{\mu_L}\right)^{0.5} = 3.04 \tag{6.24}$$

In view of the Schmidt number exponent ($Sc^{0.75}$) in Equation 6.24 being substantially different than those for the surface renewal ($Sc^{0.5}$) or boundary layer flow ($Sc^{0.66}$) models, Hughmark (1980) introduced a new argument that the mechanism of solid–liquid mass transfer in a stirred vessel is not the same as described by the two models mentioned earlier.

The other set of analysis of K_{SL} in stirred vessels was due to Lee. Lee (1981) argued that the overall particle–fluid relative velocity is the resultant of effects due to inertia and gravity. The inertial force precludes the motion of the solid particles precisely along the chaotic turbulent motion of the fluid. The result of this nonconformity is a fluctuating relative velocity. The density difference between the two phases introduces the gravity force. This effect causes a motion of the particle in a direction [depending on whether the particle is light (upward) or heavy (downward)] at its terminal velocity in the turbulent field. According to Lee, the relative velocity of the particles is the vector sum relative velocity due to inertial and gravity forces. Lee (1981) also used a modified form of Tchen's equation (Eq. 6.16). The modification introduced was in terms of the correction factor to the Stokes' drag term. This was further assumed by Lee to be a function of the rms value instead of the instantaneous relative velocity. In an additional simplification, Lee neglected the

Basset term in Tchen's original equation. Lee (1981) derived the values of the rms particle fluctuating velocity, u'_P, and fluid–particle relative turbulent velocity, u'_R, based on the assumption that the particle experiences Lagrangian fluid energy spectrum for particle size d_P up to 1000 μm. The experimental data of Snyder and Lumley (1971) and Kuboi et al. (1974a) exhibited a satisfactory agreement with Lee's (1981) estimated velocities. In particular, both the experimental and theoretical predictions mentioned earlier gave different values of the particle velocity and the velocity of the fluid surrounding it. This observation clearly indicates the existence of a "slip" between the particle and fluid phases, which is the basis of the slip velocity theories discussed earlier. The turbulent fluctuating particle velocity, u'_P, obtained from Lee's (1981) theoretical analysis showed fairly good agreement with the experimental values of Kuboi et al. (1974a). Further, the use of this estimated particle velocity u'_P in a conventional Sh=f(Re, Sc) correlation yielded values of K_{SL}, which agreed with the experimental data of Harriott (1962a) and Nienow (1969). Lee (1984) used a different approach than his previous study. In this latter work, Lee used a form of Tchen's equation in which the pressure gradient term was ignored. According to Maxey and Riley (1983), this assumption is not correct since the pressure gradient of the basic flow contributes to the net fluid force on the particle. Lee based the analysis on the assumption that Lagrangian motion prevails. He further employed a correction factor to the viscous drag term. From the resulting solution of the equation of motion, the rms relative velocity $(u'_R = u'_F - u'_P)$ was estimated. Lee (1984) argued that u'_R is the correct representative velocity that should be used in mass transfer correlations of the type Sh=f(Re, Sc). This concept was applied to the data of Harriott (1962a) for particle sizes <500 μm. It was found that the use of u'_R gave distinctly better agreement with the data. Analogous to the previous studies, Lee (1984) argued that for small particles such that $Re_P < 1$, the unsteady-state transport mode is dominant. The major change that Lee (1984) suggested over his previous analysis was in the type of mass transfer process that is relevant to the upstream and downstream halves of a spherical particle in a turbulent fluid. For the case of larger particles, the flow field is different on the upstream and downstream sides. Thus, the exponent on Schmidt number of 0.33 suggested in the Ranz and Marshall (1952b) correlation is valid for the developing boundary layer flow prevailing in the upstream region. On the other hand, for the downstream half where wake formation due to boundary layer separation is a distinct possibility, unsteady-state mass transfer may be important and the exponent on Schmidt number could increase to 0.5. Lee's (1984) argument implies that a variable Schmidt number exponent is necessary to describe mass transfer separately for the two regions. Therefore, the use of Equations 6.25 and 6.26 given by Friedlander (1957) and Whitaker (1972), respectively, was suggested to predict K_{SL}:

$$Sh_P = 0.99 \, Pe^{0.33} \quad \text{for} \quad Re_P < 1 \tag{6.25}$$

$$Sh_P = 2 + (0.4 \, Re_P^{0.5} + 0.06 \, Re_P^{0.667}) Sc^{0.4} \quad \text{for} \quad Re_P > 1 \tag{6.26}$$

The particle Reynolds number in Equations 6.25 and 6.26 is based on u'_R.

Fishwick et al. (2003) measured the particle–fluid slip velocity and K_{SL} simultaneously. Positron emission particle tracking (PEPT) technique was used to measure time-averaged slip velocities. Dissolution of salicylic acid impregnated on silica particles was used to determine the mass transfer coefficient. An upflow pitched blade turbine with the standard geometric configuration ($D/T = C/T = 0.33$) was used. The mass transfer results showed a dependence on the particle density that could not be ignored. The same was represented by $K_{SL} \propto \rho_P^{0.252}$ for two phase (without gas) and $K_{SL} \propto \rho_P^{0.255}$ in the presence of gas sparging. This agrees very well with the index of 0.25 on ρ_P observed by Blasinski and Pyc (1975). The PEPT measurements showed a very wide distribution of the slip velocity. The results showed that the slip velocity is highest at the tip of the impeller and decreases with increasing distance from the impeller. A major portion (~50%) of the vessel exhibited slip velocity lower than 20% of the maximum velocity measured. According to Fishwick et al. (2003), this problem will be accentuated in larger vessels where the region of maximum slip velocity will represent a much smaller portion of the vessel volume. Consequently, the rates of mass transfer will also be correspondingly lower in the zones of low turbulence. Fishwick et al. (2003) suggested the use of a draft tube to increase the circulation velocity as explained in Section 3.4.2.4 and/or multiple impellers to overcome this problem during scale-up. It will be seen later that a much simpler scale-up rule that gives reliable results can be devised.

6.4.1.4   Limitations of Previous Approaches in Predicting Particle–Liquid Mass Transfer Coefficient Based on Tchen's Equation of Motion as Applied to a Stirred Tank Reactor

1. In the foregoing, the reservations against the terminal velocity–slip velocity approach of Harriott (1962a) have been elaborated. This was, however, an important stage of theorizing the role of turbulence in mass transfer processes. It was a definite improvement on the dynamic similarity or power input-based empirical correlations. As pointed out by Nienow (1975), the estimation of the characteristic slip velocity involves a complicated process. Further, this approach cannot be used when the solid–fluid density difference approaches zero.

2. The particle–fluid slip velocity approach based on Tchen's equation is a refinement of the terminal settling velocity–slip velocity approach (e.g., Lee 1981, 1984). However, most investigators made simplifying assumptions that may not be tenable. The assumptions were inevitable since the original equation is far too complex. Even with these simplifications, determination of characteristic particle–fluid slip velocity (representative velocity) responsible for mass transfer still required numerical solution of the remaining equations. Further, it required information on turbulence characteristics over a wide range of conditions and is therefore tedious.

Overall, it may be concluded that the fluid mechanics-based approaches improved the understanding of particle–fluid mass transfer process. However, estimation of the mass transfer coefficient requires voluminous data. The procedures prescribed are complicated; and therefore, there is a need for an alternative predictive approach that is simple and reliable.

6.4.2 Approach Based on Analogy between Momentum and Mass Transfer

Osborne Reynolds first noted the similarity between momentum transfer and heat transfer in 1874 (Reynolds 1900). Reynolds stated that "in the transport of heat or momentum between a fluid and a solid surface, two mechanisms contributed to the transport process:

1. The natural internal diffusion of the fluid when at rest.
2. The eddy caused by visible motion which mix the fluid up and continuously bring fresh particles into contact with the surface."

The first mode was ascribed to the nature of the fluid (physical properties), whereas the second mode arose out of the fluid velocity past the solid surface. Thus, the first mode can be denoted as the molecular transport mode in a quiescent fluid, whereas the second mode pertains to systems dominated by strong convection. Industrial operations are carried out under conditions such that the second mode is dominant. Reynolds' propositions yielded equivalence between the drag coefficient (representing momentum transfer) and heat transfer. In the case of pipe flow, two distinct regions were visualized: (1) a turbulent core and (2) a laminar sublayer adjacent to the pipe wall. Reynolds' hypothesis was quantified subsequently, and equations relating to the drag coefficient and the heat transfer coefficient were developed (Bennett and Myers 1962). A major drawback of the equations based on Reynolds analogy was that for $Pr \neq 1$ the molecular contributions of thermal diffusivity and kinematic viscosity were disregarded. Eventually, such equations gave good agreements for systems with $Pr = 1$, such as gases, but deviated significantly when $Pr \neq 1$. In the turbulent core, transport is mainly by what was subsequently denoted as "eddy diffusivity," but in the laminar sublayer, the eddy coefficients are vanishingly small and, hence, offer major resistance to transport. In this laminar sublayer, the molecular mode cannot be ignored. Prandtl–Taylor analogy was aimed at removing this drawback by writing separate transport equations for the turbulent core and the laminar sublayer (Bennett and Myers 1962). Further improvement was made by von Karman (Bennett and Myers 1962). In this last approach, it was visualized that there is a buffer zone between the turbulent core and the laminar sublayer. In this case also, separate equations were written for the three regions. The major difference was that for the buffer zone contributions of both molecular and eddy transport modes were included. These analogies were later extended to mass transfer. In all these treatments, it was implied that the exchange of momentum, heat, and mass occurred across the same surface. As an example, for heat transfer through pipe walls or mass transfer from pipe walls to the fluid flowing inside a pipe, the analogies hold rigorously because the momentum is transferred by friction at the pipe wall and the same pipe wall is also the surface for heat/mass transfer. This type of drag wherein the fluid streamlines follow the surface is denoted as "skin friction." The situation is different for a bluff object (cylinder/sphere) present in the fluid stream. As mentioned earlier, in the frontal part of the object, there is a developing boundary layer with the fluid streamlines following the shape of the object. However, at sufficiently high Reynolds numbers, the streamlines diverge at the rear half and boundary layer separation occurs. In this latter case, most of the drag is caused by the pressure difference between the frontal and rear regions. Such a drag is called form drag. In separated flows, the drag coefficient is not only a function of the Reynolds number but also the shape of the object. The interface for heat, mass, and momentum transfer is no longer the same and analogies do not hold. When the analogies are applicable (only skin friction), they lead to a simple equation relating to the drag coefficient, heat transfer coefficient, and mass transfer coefficient. In a major development, Friend and Metzner (1958) showed that the Prandtl–Taylor analogy was valid for Schmidt numbers up to 3000. They used literature data for dissolution mass transfer from pipe wall to a fluid flowing inside the pipe (skin friction). Equation 6.27 representing the Prandtl–Taylor analogy was found to give satisfactory agreement with the experimental data:

$$\frac{K_{SL}}{u_F} = \frac{f/2}{1.20 + 11.8(Sc - 1)(Sc)^{-1/3}\sqrt{f/2}} \tag{6.27}$$

For $Sc \gg 1$, Equation 6.27 can be simplified as (Hughmark 1974)

$$\frac{K_{SL}}{u_{FB}} = D'(Sc)^{-2/3}\sqrt{\frac{f}{2}} \tag{6.28}$$

In the case of pipe flow for which Equation 6.28 has been derived, the relation between the bulk fluid velocity, u_{FB}, and the friction factor is given by (Bennett and Myers 1962)

$$u_{FB}\sqrt{\frac{f}{2}} = u^* \tag{6.29}$$

where u^* is the friction velocity related to the wall shear stress. Further, in the case of momentum transfer solely due to skin friction such as in pipe flow, it has been shown by Davies (1972) that the friction velocity is identical to the rms fluctuating velocity or turbulence intensity, u', normal to the wall. Therefore,

$$u_{FB}\sqrt{\frac{f}{2}} = u' \tag{6.30}$$

The equivalence of u^* and u' has been shown to hold for two- and three-phase fluidized beds (Hanratty et al. 1956; Handley et al. 1966). The case of fluidized beds is significantly different than pipe flow because of the existence of significant form drag. Thus, for a particle embedded in a turbulent fluid, Equation 6.29 needs to be adapted to this situation wherein significant form drag is also present:

$$u_{FB}\sqrt{\frac{f}{2}} = f(u')$$

(6.31)

Combining Equations 6.31 and 6.28, we obtain

$$Sh_P = f\left(\frac{d_p u' \rho}{\mu}, Sc\right)$$

(6.32)

The functional relationship in Equation 6.32 can be obtained by correlating experimental data with the observed values of K_{SL}. It has been mentioned earlier that the drag in turbulent flow is always higher and is decided by the extent of turbulence. The latter can be represented by the turbulence intensity, u'. The intensity of turbulence is dependent on the mode (type of impeller and geometric proportions, namely, C/T, D/T of the stirred reactors, gas sparging in bubble columns) and extent of power (impeller speed in stirred reactors, gas velocity in sparged reactors, etc.) provided. The important role of u' in deciding the mass transfer coefficient can be judged from the experimental investigations of Nikov and Delmas (1991, 1992). These investigators showed that the mass transfer coefficient is same for single particle–liquid and multiparticle–liquid turbulent flow systems as long as the particles experience the same turbulence intensity. Evaluation of u' is relatively straightforward as compared to the complex procedure and voluminous data required for estimating u_R, u_p, etc. For stirred reactors, u' can be estimated from the quotient of actual agitation speed to the agitation speed for just suspension of solids/complete dispersion of gas for correlating K_{SL} and $k_L a$, respectively (Jadhav and Pangarkar 1991b; Kushalkar and Pangarkar 1994, 1995; Dutta and Pangarkar 1994, 1996; Pangarkar et al. 2002; Yawalkar et al. 2002a, b) as shown later in Chapter 7A. For the case of three-phase sparged reactors, Jadhav and Pangarkar (1991a) evaluated u' using the energy balance approach. This u' was used as the characteristic velocity in equations of the type given by Equation 6.32 along with the particle diameter, d_p, to correlate K_{SL} (Chapter 10). This approach can be extended for correlating k_L in sparged reactors. However, in this case, there is a substantial spatial variation in d_B, the representative linear dimension to be used in place of d_p on the right-hand side of Equation 6.32. Therefore, it is not possible to define a unique linear dimension for use in Equation 6.32. It may be possible to use a representative value of d_B obtained from an existing literature correlation such as that of Parthasarathy and Ahmad (1991) and verify the applicability of Equation 6.32 in sparged reactors.

It is important to note that Equation 6.32 has a sound fundamental basis elaborated in the discussion preceding it and also later in Chapter 7A. The use of turbulence intensity, u', as the representative characteristic velocity ensures that turbulence similarity is incorporated in Equation 6.32. Hydrodynamic regime similarity, which is a part of kinematic similarity, is also given due importance in Equation 6.32 since a multiphase reactor will exhibit turbulence similarity only when the operating hydrodynamic regime is same. As an example, a bubble column operating under different hydrodynamic regimes (slug flow/homogeneous bubble flow/churn turbulent) will yield different values of u', implying that turbulence similarity is violated. Alternatively, for the same hydrodynamic regime and for identical system properties, u' will be same, and hence, turbulence similarity will be maintained. The work of Shaikh and Al-Dahhan (2010) referred to earlier in Chapter 5 and discussed at some length in this chapter indirectly supports the aforementioned argument.

The logic behind correlations based on u' is discussed in detail in Chapters 7A and 10 on stirred tank and three-phase sparged reactors, respectively. The most important aspect of these correlations as discussed in Chapter 7A is that they are scale independent even for a highly complex system such as a stirred reactor with a variety of impellers and impeller–vessel configurations. This strikingly important result is most likely due to the use of u' as the characteristic velocity in defining the particle Reynolds number. Equality of Reynolds number thus defined ensures turbulence similarity in systems that may differ widely. Further, the important aspect of hydrodynamic regime similarity that is a part of kinematic similarity is also given due importance in Equation 6.32. In a recent exhaustive study of hydrodynamics of various impellers for aerobic fermenters, Collignon et al. (2010) observed that similar velocity fields are realized when impellers rotate at their respective just-suspended speed. This is another important supporting observation for the role of the just-suspended condition in two-/three-phase stirred tank reactors.

NOMENCLATURE

a, a_1, a_2, a_3:	Empirical constants in various equations
b:	Constant in Equation 6.2
C:	Clearance of the impeller from the bottom (m)
D:	Diameter of the impeller (m)
D_M, D_A:	Diffusivity of solute (m^2/s)
d:	Linear dimension of particulate phase (m)
d_B:	Bubble diameter (m)
d_P:	Linear dimension of particle/bubble/drop (m)
Eo:	Eötvös number $\left(\dfrac{\Delta \rho g L^3}{\sigma} \right)$ (—)
F:	Net vertical (downward) force acting on the particle in Tchen's equation of motion in Equation 6.16 (N)
Fr:	Froude number $\left(\dfrac{V_G}{\sqrt{g d_B}} \right)$ (—)
f:	Friction factor (—)
Ga:	Galileo number $\left(\dfrac{g l^3}{v^2} \right)$ (—)
H:	Height of the liquid level in the tank reactor (m)

H_P: Penetration depth in the mass transfer boundary layer in Harriott's analysis (m)

k, k', k'':   Constants in Equations 6.8 and 6.9

k: True mass transfer coefficient (m/s)

k_L: True liquid-side mass transfer coefficient (m/s)

$k_L a$:   Volumetric liquid-side mass transfer coefficient (1/s)

K_{SL}:   Solid–liquid mass transfer coefficient (m/s)

K_{SLP}:   Pipe wall to fluid mass transfer coefficient (m/s)

K_{SLS}: K_{SL} at the just-suspended condition (m/s)

K_{SLT}: K_{SL} at the terminal settling velocity (m/s)

L: Macroscale of turbulence (m)

M: Mass of liquid/dispersion (kg)

Mo: Morton number $\left(\dfrac{g \mu_P^4}{\rho_P \sigma_P^3} \right)$ (—)

N: Speed of revolution of the impeller (rev/s)

N_S: Speed of agitation for just suspension (rev/s)

P: Power (W)

P_{Tot}:   Total power input (W)

Q_G: Gas flow rate (m³/s)

Re: Reynolds number $\left(\dfrac{l u_C \rho}{\mu} \right)$ (—)

Re_K: Reynolds number based on power dissipation criterion of Kolmogorov $\left(\dfrac{\varepsilon_M d_P^4}{v^3} \right)$ (—)

Re_P: Reynolds number based on particle diameter $\left(\dfrac{d_P u_C \rho}{\mu} \right)$ (—)

Re_T: Particle Reynolds number based on terminal settling velocity $\left(\dfrac{d_P V_T \rho_F}{\mu} \right)$ (—)

Sc: Schmidt number $\left(\dfrac{\mu_L}{\rho_L D_M} \right)$ (—)

Sh_P: Sherwood number based on particle/bubble/drop diameter $\left(\dfrac{K_{SL} d_P}{D_M} \right)$ (—)

T: Diameter of the stirred tank, sparged reactor (m)

t, t':   Time (s)

u':   rms velocity between two points in the fluid separated by a distance d in Equation 6.7 (m/s)

$\bar{u}$:   Vectorial sum of the terminal settling velocity in a turbulent fluid and the particle–fluid velocity (Hughmark 1974), $\left\{ \sqrt{V_T^2 + u_R^2} \right\}$ (m/s)

u^*:   Friction velocity (m/s)

u':   Turbulence intensity (m/s)

u_C: Characteristic fluid velocity (m/s)

u_F: Superficial fluid velocity in a pipe (m/s)

u'_{Fi}:   Instantaneous fluctuating fluid velocity (m/s)

u_K: Characteristic velocity based on Kolmogorov's theory (m/s)

u_P: Particle velocity (m/s)

u'_{Pi}:   Instantaneous fluctuating particle velocity (m/s)

u'_R: $(u'_R u'_F - u'_P)$ rms relative velocity (Lee 1984) (m/s)

u_R: Particle–fluid relative velocity (Hughmark 1974) (m/s)

u_{Slip}:   Slip velocity between fluid and particle phase defined by
$$\left(\overline{\sqrt{\overline{u_F^2}}}\right) - \left(\overline{\sqrt{\overline{u_P^2}}}\right) \text{ (m/s)}$$

V_T: Terminal settling velocity (m/s)

V_D: Volume of dispersion (m³)

V_G: Superficial gas velocity (m/s)

V_L: Total liquid volume in the tank reactor (m³)

v_K: Velocity based on Kolmogorov's power dissipation approach (m/s)

v_P: Time-averaged particle velocity (m/s)

vvm: Volume of gas supplied in vessel volume per minute (m³/(m³/min)).

We: Weber number $\left(\dfrac{\rho V_G^2 d_B}{\sigma}\right)$ (—)

Greek Letters

β: Constant in Equation 6.7

ρ: Density (kg/m³)

ν: Kinematic viscosity (m²/s)

η_K: Length scale of eddy dissipating energy by viscous dissipation (m)

μ_L: Liquid-phase viscosity (Pa s)

ε_M: Power input per unit mass (P/M) (W/kg)

μ_L: Viscosity of liquid (Pa s)

$\Delta\rho$:   Density difference ratio $\left(\dfrac{\rho_P - \rho_F}{\rho_F}\right)$ (—)

ρ_R: Density ratio $\left(\dfrac{\rho_P}{\rho_F}\right)$ (—)

Subscripts

B: Bubble, bulk phase

F: Fluid

G: Gas

L: Liquid

P: Particulate phase (bubbles/solid particles)

REFERENCES

Astarita G. (1967) Mass transfer with chemical reaction. Elsevier, Amsterdam, the Netherlands.

Aussenac D, Alran C, Couderc JP. (1982) Mass transfer between suspended solid particles and a liquid in a stirred vessel. In: Proceedings of Fourth European Conference on Mixing, Noordwijkerhout, the Netherlands. p 417–421.

Barker JJ, Treybal RE. (1960) Mass transfer coefficients for solids suspended in agitated liquids. AICHEJ, 6:289–295.

Bennett CO, Myers JE. (1962) Momentum, Heat and Mass transfer. McGraw Hill Book Company, Inc., New York, USA.

Blasinski H, Pyc KW. (1975) Mass transfer in chemically reacting solid–liquid systems subjected to agitation in baffles mixed tanks. Int. Chem. Eng., 15:73–79.

Brian PLT, Hales HB, Sherwood TK. (1969) Transport of heat and mass between liquids and spherical particles in as agitated tank. AICHEJ, 15:727–733.

Brodkey RS. (1967) The phenomena of fluid motions. Addison-Wesley Pub. Co., Reading, MA, USA.

Calderbank PH. (1959) Physical rate process in industrial fermentation. Part II: Mass transfer coefficients in gas liquid contacting with and without mechanical agitation. Trans. Inst. Chem. Eng. UK, 37:173–185.

Calderbank PH, Jones SRJ. (1961) Physical rate processes in industrial fermentation. Part III. Mass transfer from fluids to solid particles suspended in mixing vessels. Trans. Inst. Chem. Eng. UK, 39:363–368.

Calderbank PH, Moo-Young MB. (1961) The continuous phase heat and mass transfer properties of dispersion. Chem. Eng. Sci., 16:39–54.

Chapman CM, Nienow AW, Cooke M, Middleton JC. (1983a) Particle–gas–liquid mixing in stirred vessels. Part I: Particle–liquid motion. Chem. Eng. Res. Des., 61(2):71–81.

Chapman CM, Nienow AW, Cooke M, Middleton JC. (1983b) Particle–gas–liquid mixing in stirred vessels. Part III: Three phase mixing. Chem. Eng. Res. Des., 61(3):167–181.

Collignon M-L, Delafosse A, Crine M, Toye D. (2010) Axial impeller selection for anchorage dependent animal cell culture in stirred bioreactors: methodology based on the impeller comparison at just-suspended speed of rotation. Chem. Eng. Sci., 65:5929–5941.

Conti R, Sicardi S. (1982) Mass transfer from freely suspended particles in stirred tanks. Chem. Eng. Commun., 14(1–2):91–98.

Danckwerts PV. (1970) Gas–liquid reactions. McGraw Hill Book Company Inc., New York, USA.

Davies JT. (1972) Turbulence phenomena: an introduction to the eddy transfer of momentum, mass, and heat, particularly at interfaces. Academic Press, New York, USA.

Dutta NN, Pangarkar VG. (1994) Particle-liquid mass transfer in multi-impeller agitated contactors. Chem. Eng. Commun., 129:109–121.

Dutta NN, Pangarkar VG. (1996) Particle-liquid mass transfer in multi-impeller agitated three phase reactors. Chem. Eng. Commun., 146:65–84.

Fan L-S, Yang GQ, Lee DJ, Tsuchiya, K, Luo X. (1999) Some aspects of high-pressure phenomena of bubbles in liquids and liquid–solid suspensions. Chem. Eng. Sci., 54:4681–4689.

Fishwick RP, Winterbottom JM, Stitt EH. (2003) Explaining mass transfer observations in multiphase stirred reactors: particle–liquid slip velocity measurements using PEPT. Cat. Today, 79–80:195–202.

Friedlander SK. (1957) Mass and heat transfer to single spheres and cylinders at low Reynolds numbers. AICHEJ, 3:43–48.

Friend WL, Metzner AB. (1958) Turbulent heat transfer inside tubes and analogy among heat, mass and momentum transfer. AICHEJ, 4:393–402.

Froessling N. (1938) Uber die verdunstung fallender tropfen. Gerlands Beitrag zur Geophysik, 52:170–215. *cf*: Brodkey (1967).

Froessling N. (1940) Verdunstung, Wärmeübertragung und Geschwindigkeitverteilung bei Zweidimensionaler und Rotationssymmetrischen Laminarer Grenzschichtstromung, Lunds Universitets Arsskrift, N.F., Avd 2, Bd. 36:4–10. *cf*: Brodkey (1967).

Garner FH, Jenson VG, Keey R. (1959) Flow pattern around spheres and the Reynolds analogy. Trans. Inst. Chem. Eng., 37:191–197.

Handley D, Doraisamy A, Butcher KL, Franklin NLA. (1966) Study of fluid and particle mechanics in liquid fluidized beds. Trans. Inst. Chem. Eng., 44:T260–T273.

Hanratty TJ, Ratinen G, Wilhelm RH. (1956) Turbulent diffusion in particulately fluidized beds of particles. AICHEJ, 2:372–380.

Harriott P. (1962a) Mass transfer to particles. Part I. Suspended in agitated vessels. AICHEJ, 8:93–101.

Harriott P. (1962b) Random eddy modification of the penetration theory. Chem. Eng. Sci., 17:149–155.

Hickman AD. (1988) Gas–liquid oxygen transfer and scale-up: a novel experimental technique with results for mass transfer in aerated agitated vessels. In: Proceedings of Sixth European Conference on Mixing, Pavia, Italy. p 369–377.

Hinze JO. (1975) Turbulence. 2nd ed. McGraw-Hill Book Company Inc., New York, USA. p 460–471.

Hixson AW, Baum SJ. (1941) Mass transfer coefficients in liquid–solid agitated systems. Ind. Eng. Chem., 33:478–485.

Hixson AW, Baum SJ. (1942) Performance of propeller in liquid–solid systems. Ind. Eng. Chem., 34:120–125.

Hughmark GA. (1969) Mass transfer for suspended solid particles in agitated liquids. Chem. Eng. Sci., 24:291–297.

Hughmark GA. (1971) Heat and mass transfer for turbulent pipe flow. AICHEJ, 17:902–909.

Hughmark GA. (1974) Hydrodynamics and mass transfer for suspended solid particles in a turbulent liquid. AICHEJ, 20:202–204.

Hughmark GA. (1980) Power requirements and interfacial area in gas–liquid turbine agitated systems. Ind. Eng. Chem. Process Des. Dev., 19:638–641.

Humphrey DW, Van Ness HC. (1957) Mass transfer in a continuous-flow mixing vessel. AICHEJ, 3:283–286.

Jadhav SV, Pangarkar VG. (1991a) Particle–liquid mass transfer in three phase sparged reactors: scale up effects. Chem. Eng. Sci., 46:919–927.

Jadhav SV, Pangarkar VG. (1991b) Particle–liquid mass transfer in mechanically agitated contactors. Ind. Eng. Chem. Res., 30:2496–2503.

Keey RB, Glen JB. (1964) Mass transfer from solid spheres. Can. J. Chem. Eng., 42:227–235.

Keey RB, Glen JB. (1966) Mass transfer from fixed and freely suspended particles in an agitated vessel. AICHEJ, 12:401–403.

Kneule F. (1956) The testing of agitators by means of solubility determination. Chem. Ingr. Tech., 28:21–26.

Kolar V. (1959) Studies on mixing. Part V. Effect of mechanical mixing on the rate of mass transfer from granular solids in a liquid. Collect. Czech. Chem. Commun., 24:3309–3319.

Kolmogorov AN. (1941) The local structure of turbulence in incompressible viscous fluids at very large Reynolds numbers. Dokl. Akad. Nauk. SSSR, 30:299–303. Reprinted in Proc. R. Soc. Lond. A, 434:9–13.

Kolmogorov AN. (1962) A refinement of previous hypotheses concerning the local structure of turbulence in a viscous incompressible fluid at high Reynolds number. J. Fluid Mech., 13:82–85.

Komasawa I, Kuboi R, Otake T. (1974) Fluid and particle motion in turbulent dispersion-I. Measurement of turbulence of liquid by continual pursuit of tracer particle motion. Chem. Eng. Sci., 29:641–650.

Kuboi R, Komasawa I, Otake T, Iwasa M. (1974a) Fluid and particle motion in turbulent dispersion-II. Influence of turbulence of liquid on the motion of suspended particles. Chem. Eng. Sci., 29:651–657.

Kuboi R, Komasawa I, Otake T, Iwasa M. (1974c) Fluid and particle motion in turbulent dispersion-III. Particle-liquid hydrodynamics and mass transfer in turbulent dispersion. Chem. Eng. Sci., 29:659–668.

Kushalkar KB, Pangarkar VG. (1994) Particle-liquid mass transfer in mechanically agitated three phase reactors. Ind. Eng. Chem. Res., 33:1817–1820.

Kushalkar KB, Pangarkar VG. (1995) Particle-liquid mass transfer in mechanically agitated three phase reactors: power law fluids. Ind. Eng. Chem. Res., 34:2485–2492.

Lal P, Kumar S, Upadhyay SN, Upadhyay YD. (1988) Solid-liquid mass transfer in agitated Newtonian and non-Newtonian fluids. Ind. Eng. Chem. Res., 27:1246–1259.

Lee LW. (1981). The relative fluid–particle motion in agitated vessels. AICHE Symp. Ser., 77(208):162–169.

Lee LW. (1984) The heat and mass transfer of a small particle in turbulent flow. In: Veziroglu, TN, Bergles, AF, editors. Proceedings of Third Multi-Phase Flow and Heat Transfer Symposium-Workshop, Part B: Applications. Elsevier Science Publishers, Amsterdam, the Netherlands. p 605–611.

Levins DM, Glastonbury JR. (1972a) Application of Kolmogorov's theory to particle-liquid mass transfer in agitated vessels. Chem. Eng. Sci., 27:537–543.

Levins DM, Glastonbury JR. (1972b) Particle-liquid hydrodynamics and mass transfer in a stirred vessel: Part I. Particle–liquid motion. Trans. Inst. Chem. Eng. UK, 50:32–41.

Levins DM, Glastonbury JR. (1972c) Particle-liquid hydrodynamics and mass transfer in a stirred vessel: Part II. Mass transfer. Trans. Inst. Chem. Eng., 50:132–146.

Macchi A, Bi H, Grace JR, McKnight CA, Hackman L. (2001) Dimensional hydrodynamic similitude in three-phase fluidized beds. Chem. Eng. Sci., 56:6039–6045.

Mack DJ, Marriner RAA. (1949) Method of correlating agitator performance. Chem. Eng. Prog., 45(9):545–552.

Madden AG, Nelson DG. (1964) A novel technique for determining mass transfer coefficients in agitated solid–liquid systems. AICHEJ, 10:415–430.

Maxey MR, Riley JJ. (1983) Equation of motion for a small rigid sphere in a nonuniform flow. Phys. Fluids, 26:883–889.

Middleman S. (1965) Mass transfer from particles in agitated systems. Application of Kolmogoroff theory. AICHEJ, 4:750–761.

Middleton JC. (2000) Gas–liquid dispersion and mixing. In: Harnby N, Edwards MF, Nienow AW, editors. Mixing in the process industries. 2nd ed. Butterworth-Heinemann, Oxford, UK. p 322–363.

Miller DN. (1974) Scale-up of agitated vessels: gas–liquid mass transfer. AICHEJ, 20:445–453.

Nagata S, Yamaguchi I, Yabuta S, Harada M. (1960) Mass transfer in solid–liquid agitation systems. Soc. Chem. Eng. Jpn., 24:618–624.

Nienow AW. (1968) Suspension of solid particles in turbine agitated baffled vessels. Chem. Eng. Sci., 23:1453–1459.

Nienow AW. (1969) Dissolution mass transfer in a turbine agitated baffle vessel. Can. J. Chem. Eng., 47:248–258.

Nienow AW. (1975) Agitated vessel particle–liquid mass transfer: a comparison between theories and data. Chem. Eng. J., 9:153–160.

Nienow AW. (2000) The suspension of solid particles. In: Harnby N, Edwards MF, Nienow AW, editors. Mixing in the process industries. 2nd ed. Butterworth-Heinemann, Oxford, UK. p 364–393.

Nienow AW, Wisdom DJ, Middleton JC. (1977) The effect of scale and geometry on flooding, recirculation and power in gassed stirred vessels. In: Proceedings of Second European Conference on Mixing, Cambridge, UK. BHR Group, Cranfield, UK. p F1-1–F1-16.

Nikhade BP, Pangarkar VG. (2007) A theorem of corresponding hydrodynamic states for estimation of transport properties: case study of mass transfer coefficient in stirred tank fitted with helical coil. Ind. Eng. Chem. Res., 46:3095–3100.

Nikov I, Delmas H. (1991) The mechanism of liquid–solid mass transfer in three phase fluidized bed reactors. Hung. J. Ind. Chem., 19:311–316.

Nikov I, Delmas H. (1992) Mechanism of liquid–solid mass transfer and shear stress in three phase fluidized beds. Chem. Eng. Sci., 47:673–681.

Ovsenik A. (1982) Optimization of solid particles suspending. In: Proceedings of Fourth European Conference on Mixing, Noordwijkerhout, the Netherlands. p 463–469.

Pangarkar VG, Yawalkar AA, Sharma MM, Beenackers AACM. (2002) Particle–liquid mass transfer coefficient in two-/three-phase stirred tank reactors. Ind. Eng. Chem. Res., 41:4141–4167.

Parthasarathy R, Ahmed N. (1991) Gas holdup in stirred vessels: bubble size and power input effects. In: Proceedings of Seventh European Conference on Mixing, Koninklijke Vlaamse Ingenieursvereniging vzw, Brugge, Belgium. p 295–301.

Ranz WE, Marshall WR. (1952a) Evaporation from drops. Part I. Chem. Eng. Prog., 48(3):141–146.

Ranz WE, Marshall WR. (1952b) Evaporation from drops, Part II. Chem. Eng. Prog., 48(4):173–180.

Sage BH, Galloway TR. (1964) Thermal and material transfer in turbulent gas streams. A method of prediction for spheres. J. Heat Mass Transfer, 10:283–291.

Reynolds O. (1900) Papers on mechanical and physical subjects. 1870–1880. Collected Works. Vol. 1. Cambridge University Press, Cambridge, UK.

Sano Y, Yamagachi N, Adachi T. (1974) Mass transfer coefficient for suspended particles in agitated vessels and bubble columns. J. Chem. Eng. Jpn., 7:255–261.

Schlichting H. (1955) Boundary-layer theory. McGraw-Hill Book Co., New York, USA.

Shaikh A, Al-Dahhan M. (2010) A new methodology for hydrodynamic similarity in bubble columns. Can. J. Chem. Eng., 88:503–517.

Sharma MM, Mashelkar RA. (1968) Absorption with chemical reaction in a bubble columns. In: Pirie JM, editor. Mass transfer with chemical reaction. Proceedings of Symp. Tripartite Chem. Eng. Conf. (Montreal), Inst. Chem. Eng. (London), Symp. Ser., p 10–16.

Schwartzberg HG, Treybal RE. (1968a) Fluid and particle motion in turbulent stirred tanks. Fluid motion. Ind. Eng. Chem. Fundam., 7:1–6.

Schwartzberg HG, Treybal RE. (1968b) Fluid and particle motion in turbulent stirred tanks. Particle motion. Ind. Eng. Chem. Fundam., 7:6–14.

Shinnar R, Church JM. (1960). Statistical theories of turbulence in predicting particle size in agitated dispersions. Ind. Eng. Chem., 52(3):253–256.

Sicardi S, Conti R, Baldi G, Cresta R. (1979) Solid–liquid mass transfer in stirred vessels. In: Proceedings of Third European Conference on Mixing, England. p 217–228.

Snyder WH, Lumley JL. (1971) Some measurements of particle velocity autocorrelation functions in a turbulent flow. J. Fluid Mech., 48:41–71.

Son JE, Hanratty TJ. (1967) Limiting relation for the eddy diffusivity close to a wall. AICHEJ, 13:689–696.

Sreenivasan KR. (1999) Fluid turbulence. Rev. Mod. Phys., 71:S383–S395.

Stewart RW, Townsend AA. (1951) Similarity and self-preservation in isotropic turbulence. Phil. Trans. R. Soc. Lond. Series A, Math. Phys. Sci., 243(867):359–386.

Stolovitzky G, Sreenivasan KR. (1994) Kolmogorov's refined similarity hypothesis for turbulence and general stochastic processes. Rev. Mod. Phys., 66:229–236.

Sykes P, Gomezplata A. (1967) Particle-liquid mass transfer in stirred tank. Can. J. Chem. Eng., 45:189–196.

Taylor GA. (1935) Statistical theory of turbulence. Proc. R. Soc. Lond. A, 151:421–444.

Whitaker S. (1972) Forced convection heat transfer correlations for flow in pipes, past flat plates, single cylinders, single spheres and for flow in packed beds and tube bundles. AICHEJ, 18:361–370.

Wilhelm RH, Conklin LH, Sauer TC. (1941) Rate of solution of crystal. Ind. Eng. Chem., 33:453–457.

Yawalkar AA, Beenackers AACM, Pangarkar VG. (2002a) Gas holdup in stirred tank reactors. Can. J. Chem. Eng., 80:158–166.

Yawalkar AA, Heesink ABM, Versteeg GF, Pangarkar VG. (2002b) Gas–liquid mass transfer coefficient in stirred tank reactor. Can. J. Chem. Eng., 80:840–848.

Zwietering Th N. (1958) Suspending of solid particles in liquids by agitators. Chem. Eng. Sci., 8(3):244–253.

7A

STIRRED TANK REACTORS FOR CHEMICAL REACTIONS

7A.1 INTRODUCTION

Stirred tank reactors are widely used in the chemical and allied industries. They offer excellent flexibility in meeting varying demands of a reaction. The wide range of agitation devices and geometric system parameters available to the chemical engineer allows selection of an agitator–vessel combination precisely suited for a given duty. The impellers available range from low shear–high pumping to high shear–low pumping types. Several excellent texts/handbooks are available on the types of agitators, their characteristics, and selection that can be referred to for more details (Uhl and Gray 1967; Tatterson 1991; Harnby et al. 2000; Hemrajani and Tatterson 2004; Paul et al. 2004). A brief description of the components of the stirred tank reactor is given in Section 7A.1.1.

7A.1.1 The Standard Stirred Tank

Figure 7A.1 shows a standard stirred tank with the major geometric parameters.
The major elements are briefly discussed in the following.

7A.1.1.1 The Tank or Vessel The most common version of the tank or vessel is a vertical cylindrical tank with a dish end at the bottom. In exceptional cases (mostly old designs), the bottom may be conical. In this latter case, there is always the danger of inadequate mixing in the conical part. If a solid phase is an integral part of the

Design of Multiphase Reactors, First Edition. Vishwas Govind Pangarkar.
© 2015 John Wiley & Sons, Inc. Published 2015 by John Wiley & Sons, Inc.

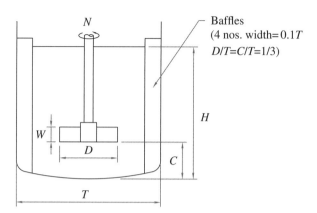

FIGURE 7A.1   Typical stirred tank reactor configuration.

process carried out in the stirred tank reactor, this inadequate mixing in the bottom cone can cause chocking of the underflow outlet. The top of the vessel is closed by a body flange fitted with a dish cover. The cover has a centrally located nozzle for the stuffing box and shaft entry. Nozzles for feed, sight, and light glasses are also provided on the cover. In a majority of applications, height of liquid/dispersion to the tank diameter ratio, (H_D/T), is unity. In this case, the standard configuration is a single impeller mounted at a distance equal to the impeller diameter from the vessel bottom. The diameter of the impeller for this standard configuration is $0.33 \times T$ (McCabe et al. 1986). In some special applications such as fermenters, which require large residence time of the gas phase, the height to diameter ratio is much larger than the conventionally accepted value of unity. In such a case, multiple impellers are employed. This is because a single impeller at the bottom is ineffective for mixing at distances far removed from its location at the bottom. Although there are several impellers located vertically along the axis, the bottom impeller has been found to play a dominant role in dispersion of the gas phase as well as suspension of solids as discussed later.

Baffles: In the conventional stirred tanks, the impeller is mounted on a shaft coincident with the central axis of the tank. Vortex formation is a common phenomenon in a stirred tank, particularly when it is operating at relatively high speeds of agitation. In the case of a vortex, the liquid merely moves in a tangential direction with practically no radial or axial component of velocity. Consequently, this phenomenon results in a state of very poor axial/radial mixing. The vortex, however, can suck substantial amount of gas, which may be one of the ways of surface aeration. To ensure that such a vortex does not form, it is necessary to disrupt the tangential motion of the liquid by providing obstacles in its path. These obstacles are called baffles. The baffles break the tangential component into radial and axial components as shown in Figure 7A.2. The standard *fully baffled* stirred tank has four baffles of width equal to 10% of the diameter of the tank or in some other designs $T/12$. The baffles are secured to the internal surface of the wall of the tank. The baffle wall corners are points of stagnation. In the case when solids are present, these can build up at these corners.

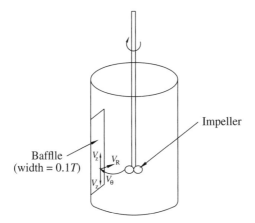

FIGURE 7A.2 Breakup of tangential velocity V_θ into axial and radial velocity components by a baffle.

In such cases, a clearance (approximately 25–50 mm) is maintained between the baffles and the tank wall to allow free movement of the solids through this clearance. Bulk of the tangential component, V_θ, is still disrupted by the rest of baffles.

7A.1.1.2 Impellers Impellers commonly used in stirred tank reactors can be divided into two main categories: (1) radial flow and (2) axial flow impellers. Mixed flow impellers, which develop radial as well as axial flows (Fig. 7A.3b), are also available and are used widely for many applications (the upflow version of such mixed flow impellers will be shown later to perform far better than others). High solidity ratio impellers also discussed later in Chapter 7B are gaining acceptance in special applications. Some of the commonly used conventional impellers are shown in Figure 7A.3.

Propellers and the axial flow hydrofoils of the Lightnin A315 variety belong to the axial flow category. The pitched blade turbines belong to the class of mixed flow impellers. Both the up- and downflow variants in this category develop a greater axial component of velocity as compared to the radial particularly when the agitator diameter to tank diameter ratio is less than 0.4. The most widely used radial flow agitator is the standard Rushton turbine (Fig. 7A.3a). The paddle and curved blade turbines also belong to this class. Middleton (2000) has described the various types of conventional impellers. Collignon et al. (2010) have given details and solid suspension performance of the recently introduced wide blade, low shear–high pumping impellers (Fig. 7B.13).

The variety of impellers available can further be divided into two categories based on whether they create a predominantly shear field or bulk movement. The axial flow propeller, the hydrofoils, and the mixed flow impellers (when $D/T < 0.4$) develop bulk axial patterns. The downflow type in the mixed flow class develops a mean flow directed toward the base of the vessel and is therefore useful for solid suspension in two-phase (solid–liquid) systems. But the same are less efficient in three-phase

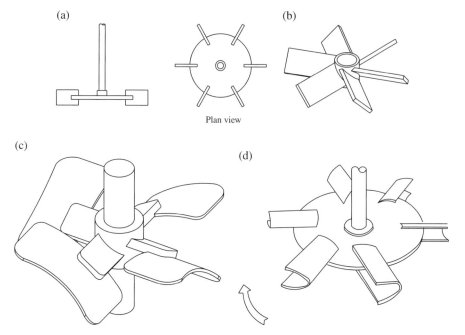

Plan view

FIGURE 7A.3 Some commonly used impellers for two-/three-phase stirred reactors. (a) Standard six-blade Rushton turbine. (b) Six-blade 45° pitched turbine. (c) Lightnin A315®. (d) SCABA 6SRGT. (Reproduced from Middleton 2000 with permission from Elsevier. © 1992.)

(gas–liquid–solid) systems both for gas dispersion and solid suspension. High shear impellers such as the radial flow Rushton turbine generate smaller bubbles with lower rise velocity. However, these impellers consume much higher power. Further, radial and downflow impellers exhibit instability in the presence of the gas phase as discussed later. As a compromise for these two conflicting requirements of low power consumption and gas dispersion, the pitched blade turbine in the upflow mode has been suggested. The latter has been shown to afford stable and efficient operation particularly in three-phase reactors (Nienow 1996; Nienow and Bujalski 2002, 2004). Aubin et al. (2004) have shown that this upflow pitched blade turbine (PTU) yields better performance (36% higher gas holdup) as compared to the downflow variant. Some of the major advantages reported by Nienow and Bujalski (2004) are elaborated in Section 7B.9.1.4. Choice of the impeller for multiphase reactor is complicated by the fact that it has to perform several functions. Majority of investigators have used power input calculated from reported power numbers ($P=N_p\rho_L N^3 D^5$) for quantifying impeller performance. However, as discussed in Section 6.3, this approach has several drawbacks. Machado et al. (2012) investigated the performance of hydrofoil and downflow pitched blade turbine (PTD) impellers with the purpose of suggesting a protocol for impeller selection. Among 14 parameters evaluated, they suggested that the comparison of different impellers should be based on following five parameters:

(1) power number, (2) momentum number, (3) peak rate of dissipation of turbulent kinetic energy, (4) power requirement at the "just"-suspended state, and finally (5) the critical speed for surface aeration. These criteria cover most aspects of gas–liquid–solid contacting, except for gas dispersion. Unfortunately, information on all these criteria for different impellers may not be available under otherwise identical conditions and hence this protocol may be difficult to implement.

7A.2   POWER REQUIREMENTS OF DIFFERENT IMPELLERS

The power required to run a given impeller is dependent on the type of impeller and the geometric configuration of the vessel–impeller combination. Traditionally, the power required by an impeller has been empirically correlated using the dimensionless groups approach. The variables that determine the power required are given by Equation 7A.1:

$P = \varphi \, (\mu_L, \rho_L, D, T, g, N$, impeller geometric parameters such as blade width, blade angle, thickness, and other geometric details relating impeller and vessel dimensions). The usual dimensionless group approach leads to the following functional relationship:

$$\frac{P}{N^3 D^5 \rho_L} = \varphi' \left\{ \frac{ND^2 \rho_L}{\mu} ; \sqrt{\frac{N^2 D}{g}} ; \frac{T}{D} ; \frac{H}{D} ; \frac{W}{D} , \text{etc.} \right\} \qquad (7A.1)$$

In Equation 7A.1, the various dimensionless groups are as follows:

(i) $\dfrac{P}{N^3 D^5 \rho_L}$: power number, N_p; (ii) $\dfrac{ND^2 \rho_L}{\mu}$: impeller Reynolds number; and

(iii) $\sqrt{\dfrac{N^2 D}{g}}$: Froude number. The Froude number is unimportant when sufficient

care is taken to avoid formation of a vortex by providing four baffles of width equal to 10% of the stirred tank reactor diameter. In the latter case, the functional relationship (Equation 7A.1) reduces to

$$\frac{P}{\rho_L N^3 D^5} = \varphi \left\{ \frac{\rho_L ND^2}{\mu} ; \frac{T}{D} ; \frac{H}{D} ; \frac{W}{D} , \text{etc.} \right\} \qquad (7A.2)$$

The relationships in Equation 7A.2 have been used for estimating power requirements in a single-phase (liquid) system. Figure 7A.4 shows typical variation of power number with impeller Reynolds number for this case.

It is evident from Figure 7A.4 that the power number attains a constant value in the turbulent region. For the case of two- (gas–liquid) and three-phase (gas–liquid–solid) systems, additional parameters that represent the gas- and solid-phase holdups have an important bearing on the power required. Hence, these are also included. The effects of introduction of gas and solid phases on the power number are discussed

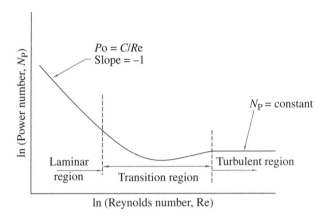

FIGURE 7A.4   Variation of power number with impeller Reynolds number.

separately in Sections 7A.3 and 7A.4. Further, the presence of any internals besides the standard baffles, particularly internal cooling coils, also has an important bearing on the performance of an impeller. The stirred tank reactor is used for fast reactions (involving highly active catalysts) as well as slow reactions. In those cases where the gas-phase reactant is expensive (pure hydrogen/oxygen), the agitation created by the impeller allows the use of relatively low gas velocities to utilize majority of the gas fed. Its use is, however, limited to moderate pressures (<2 MPa) due to the problems associated with sealing of the shaft. Further, it is also limited to moderate heats of reaction ($\Delta H_R < 100$ kJ/mol). For highly endothermic/exothermic reactions, supply/ removal of heat through the jacket may not be sufficient. To cater to the higher heat load, cooling coils can be provided in the stirred tank reactor. However, this causes overall dampening of the turbulence in the stirred tank reactor, and relatively higher power inputs are required to obtain the same results as in the absence of the coils (Nikhade et al. 2005). As an alternative to the internal cooling coils, an external heat exchanger loop may be used. This latter alternative also demands higher overall power input besides the extra capital expenditure on heat exchanger, pump, piping, etc. In view of the above limitations, the venturi loop reactor (Chapter 8) is a better alternative to the stirred tank reactor for highly exothermic reactions that also require higher operating pressures (>2 MPa).

7A.3   HYDRODYNAMIC REGIMES IN TWO-PHASE (GAS–LIQUID) STIRRED TANK REACTORS

Chapman et al. (1983b) have discussed the dispersion of a gas in a stirred vessel. In general, when a gas is introduced into a stirred vessel, it is sucked into the low pressure region at the rear face of the impeller blade. The picture can be viewed from two different perspectives: (1) constant impeller speed and (2) constant gas velocity. Considering the second view, at relatively low gas rates, a large cavity covers the entire rear face of the impeller blade (Fig. 7A.5a). As the speed of the impeller

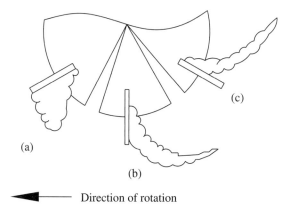

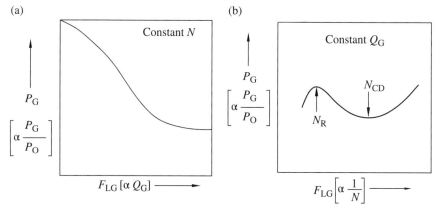

FIGURE 7A.5 Changes in cavity shape with increasing speed of agitation. (Reproduced from Nienow and Wisdom 1974 with permission from Elsevier Ltd. © 1974 Published by Elsevier Ltd.)

FIGURE 7A.6 Plot of P_G versus F_{LG}. (a) Constant impeller speed and (b) constant gas flow rate. (Reproduced from Chapman et al. 1983b with permission from The Institution of Chemical Engineers. © 1983, The Institution of Chemical Engineers. Published by Elsevier B.V. All rights reserved.)

increases, the cavity deforms into a trailing vortex on each blade as shown in Figure 7A.5b. At still higher speeds, the vortex starts detaching from the blade by moving inward to the hub of the impeller (Fig. 7A.5c).

The power drawn by the impeller varies with the progress of the vortices as indicated earlier, and it is generally expressed in terms of a power number, N_{PG}, $(P_G/\rho_L N^3 D^5)$ in the presence of the gas as a function of the flow number, F_{LG} (Q_G/ND^3). Chapman et al. (1983b) have given the general form of variation of N_{PG} with F_{LG} (Fig. 7A.6a and b) for both the cases of constant speed of agitation and constant gas flow rate. The interpretation of the behavior of N_{PG} in Figure 7A.6a for a disc turbine (DT) is as follows (Chapman et al. 1983b).

7A.3.1   Constant Speed of Agitation

With increasing gas flow rate, there is ingress of more gas in the cavity and the pressure on the rear face of the impeller blade rises. This results in lower difference in the pressure between the two faces of the blade. As a consequence, there is a concomitant decrease in the drag and the power drawn. As the gas rate is further increased, the number of cavities increase, which leads to a more streamlined blade + cavity geometry. This phenomenon further reduces the drag. Consequently, the power drawn also decreases. The final stage is reached when all impeller blades are covered by gas cavities and the pumping action of the impeller is severely hampered. Warmoeskerken and Smith (1985) have given a slightly varying description of the decrease in power drawn. According to them, three large cavities form on the rear face of the blade simultaneously, which cause the initial drop in the power drawn. Further drop in the power is a result of the low mean density of the fluid in the vicinity of the impeller.

7A.3.2   Constant Gas Flow Rate

Figure 7A.6b shows typical P_G versus F_{LG} plot at constant gas flow rate for a DT. It can be observed that the power number in the presence of gas exhibits two points of inflection. At low impeller speeds or high F_{LG}, the gas passes vertically upward in the form of plumes as it exits the sparger pipe. The impeller tips are devoid of the gas phase and the impeller action is unaffected by the presence of the gas. Therefore, the power number is practically the same as in the absence of the gas. With increasing speed, the suction of the impeller causes buildup of gas cavities on the rear face of the blades as mentioned earlier. The presence of the cavities causes a drop in N_{PG}. With further increase in the speed, the entire rear blade surface is covered by the gas cavity that causes a minimum in N_{PG}. This is the point (point denoted by N_{CD} in Fig. 7A.6b) at which the gas is considered to be completely dispersed in the vessel (Nienow et al. 1977). At still higher speeds, the impeller starts shedding the gas in the cavities and N_{PG} rises till a maximum in N_{PG} is reached. If the speed is raised higher than this value, recirculation of the gas into the impeller zone starts, thus again lowering fluid density in this region. The consequence is that a small decrease in N_{PG} is witnessed. Different impellers show different drops in the power drawn in the presence of the gas. For example, impellers such as PTD or propeller pump the liquid against the buoyancy of the gas. Dispersion of the gas (point N_{CD} in Fig. 7A.6b) can occur only after the impeller flow momentum overcomes the buoyancy of the gas. It is not easy to define the buoyancy of the gas since the rise velocity of bubbles in the swarm depends on the bubble diameter and the drag it experiences. The latter depends on the level of turbulence to which the bubbles are exposed. These dependences are quite complex. Because of the opposite directions of flow developed by the impeller and buoyancy of the gas, such downward pumping impellers are prone to instability as reflected by sudden variations in the power drawn. Chapman et al. (1983b) found that increasing the number of blades from four to six narrowed the region of instability for the PTD impeller. According to Oyama and Endoh (1955), the extent of decrease in the gassed power is an indication of ability of the impeller to operate in a stable manner in the presence of the gas. The efficiency of the downflow impeller for solid suspension in ungassed (solid–liquid) systems (Rao et al. 1988) had led to the wrong belief that they would be superior for solid suspension in the presence of gas also. However, subsequent investigations clearly brought out the unstable behavior of these downflow impellers. For instance, Fishwick et al. (2005) reported significant reduction in mixing efficiency of six-blade PTD impeller even at relatively very low gas rates. This is in stark contrast to a negligible effect in the case of its upflow counterpart. The original work of Chapman et al. (1983b) had indicated that contrary to the intuitive reaction, the downflow variety is inferior to the upflow impellers. The latter could yield better performance in the presence of gas. This has been attributed to the cooperative action of the upward flow generated by the impeller and the buoyancy-driven gas flow (Nienow and Bujalski 2002). On this count, the different types of upflow impellers caught the attention of various investigators. Nienow's group (Vrabel et al. 2000; Nienow and Bujalski 2004; Nienow 2006), Pangarkar et al. (2002), Fishwick et al. (2005), and other investigators have clearly shown that the upflow impellers are, indeed, far better suited for both dispersion of gas and suspension of solids in three-phase stirred reactors. Further, Oyama and Endoh's (1955) criterion as applied for three-phase stirred vessels shows the superiority of these upflow impellers because the decrease in power drawn by these impellers is very low as compared to the marked drop observed for downflow impellers. Figure 7A.7 from Middleton

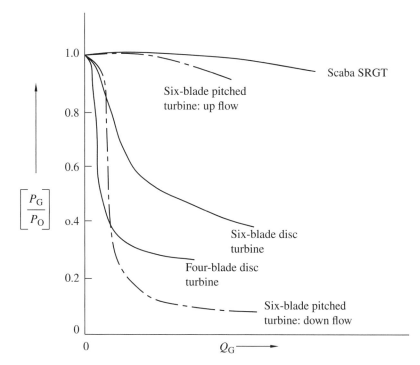

FIGURE 7A.7   Plot of (P_G/P_O) against Q_G for various impellers. (Adapted from Middleton 2000 with permission from Elsevier. © 1997, Elsevier.)

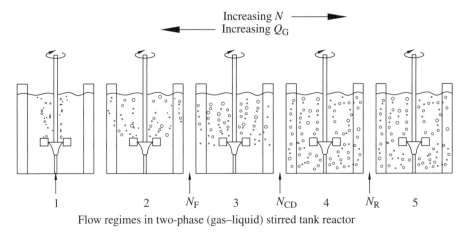

Increasing N ⟶
⟵ Increasing Q_G

1   2 N_F 3 N_{CD} 4 N_R 5

Flow regimes in two-phase (gas–liquid) stirred tank reactor

FIGURE 7A.8 Hydrodynamic regimes in two-phase (gas–liquid) stirred tank reactor. 1, flooding of the impeller; 2, gas dispersion above the impeller; 3, gas circulation above the impeller with marginal dispersion below the impeller; 4, gas circulation both above and below the impeller; 5, recirculation of the gas resulting in the formation of secondary loops besides main discharge streams from the impeller. (Reproduced from Middleton 2000 with permission from Elsevier. © 1997, Elsevier.)

(2000) shows the behavior of various impellers at varying gas flow rates. Cooke and Heggs (2005) have compared the performance of different designs of hollow blade impellers with PTD and aerofoil impellers. Their study supports the superior performance of Scaba 6SRGT, a hollow blade impeller as shown in Figure 7A.7. Further, Cooke and Heggs showed that at high solid loadings, these impellers were most energy efficient with practically no effect of the gas flow rate on the suspension performance. The performance improvement is related to the degree of streamlining incorporated in the impeller blade.

Figure 7A.8 is a visual depiction of the various stages of gas–liquid mixing described earlier with the help of Figures 7A.6a and b as given by Nienow et al. (1977). In Figure 7A.8, the changes in the flow pattern of the gas phase are illustrated for the case of a DT. Initially as expected, at very low speed of agitation, the impeller action is very weak and the gas basically flows in the form of a plume in an area close to the impeller shaft. With increasing speed, the impeller action gains momentum and starts dispersing gas in the same direction as the discharge stream from the impeller. Thus, with increasing impeller speed, the following five broad stages of gas flow pattern can be visualized: (i) practically no gas dispersion or flooding of the impeller (Warmoeskerken and Smith 1985), (ii) dispersion just sufficient such that the zone above the impeller acts as a sparged reactor, (iii) circulation of the gas in the upper part with marginal dispersion below the impeller, (iv) circulation of the gas in both the upper and lower zones, and (v) recirculation of the gas resulting in the formation of secondary loops in addition to the main discharge streams from the impeller. Nienow et al. (1977) suggested that the shift between stages 3 and 4 is the beginning of gas dispersion that is satisfactory from a mass transfer point of view. Alternatively, this transition can be used to derive the minimum impeller speed for satisfactory gas dispersion, N_{CD}. The minimum value of P_G in Figure 7A.6b corresponds to N_{CD}. At this stage (N_{CD}), as explained by Chapman et al. (1983b), the entire rear face of the impeller blade is covered by the gas cavity and the power drawn attains a minimum value.

For applications of a stirred tank as a gas–liquid reactor, it is obvious that the minimum effective speed of the impeller is N_{CD}. For $N > N_{CD}$, practically all the reactor volume is utilized for gas–liquid contacting. If the operating speed of agitation is above N_R, it is safe to assume that the gas phase is back mixed due to significant gas recirculation. Power numbers for standard six-blade Rushton turbine or DT, narrow and wide hydrofoils, and pitched blade turbines for different values of D/T, C/T, and other major geometric parameters (blade width, thickness, etc.) are available (Nienow and Miles 1971; Rewatkar et al. 1990; Aubin et al. 2004; Broz et al. 2004, and http://www.pdhengineer.com/courses/m/BP-10001.pdf). For the upflow, six-blade pitched turbines that will be employed in the solved reactor design problem later in this chapter, the value of N_P for $D/T = 1/3$ is 1.7 (Ibrahim and Nienow 1995). However, for a conservative estimate of the design power rating, a value of 1.9 closer to that given in Gray (1973) will be used.

7A.4 HYDRODYNAMIC REGIMES IN THREE-PHASE (GAS–LIQUID–SOLID) STIRRED TANK REACTORS

Similar to two-phase stirred tank reactors, three-phase stirred tank reactors also exhibit different hydrodynamic regimes. The difference, however, is that the three-phase stirred tank reactor has an additional hydrodynamic regime of just solid suspension that may occur before or after complete dispersion of the gas at N_{CD}. The literature abounds with studies on solid suspension in three-phase stirred tank reactors, and a brief review of the same has been given by Pangarkar et al. (2002). The various regimes can be explained through Figure 7A.9 in a manner analogous to the case of two-phase gas–liquid stirred reactors depicted in Figures 7A.5 and 7A.6. The impeller chosen in this case is the PTD in which solid suspension succeeds gas dispersion ($N_{SG} > N_{CD}$) except for some low density solid particles.

In Figure 7A.10 (Rewatkar et al. 1991), the power number is plotted versus the impeller speed for two- and three-phase stirred reactors separately for PTD. ABCC'D (solid line) in Figure 7A.10 is a qualitative depiction of the variation in N_P with N for low solid loadings (<5 wt.%), which are representative of solid-catalyzed three-phase reactions such as hydrogenations. The broken curve ABCD that is analogous to Figure 7A.6b is a similar qualitative representation of the variation of the power number with the impeller speed for a solid–liquid system. It is evident that similar to Figures 7A.6a and b, Figure 7A.10 also shows a point of inflection in the power number plot. This point corresponds to the critical speed for just suspension of the solid particles, N_{SG}. In a two-phase (solid–liquid) system at low

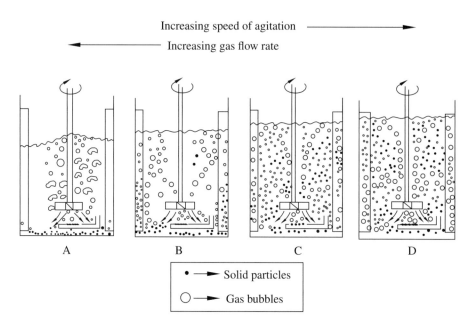

FIGURE 7A.9   Hydrodynamic regimes in three-phase (gas–liquid–solid) stirred tank reactors: downflow pitched turbine. A, no dispersion of gas; solid settled on bottom; B, gas dispersed; beginning of solid suspension; C, gas dispersed; off-bottom suspension of solids; D, recirculation of mixture and possible surface aeration. (Reproduced from Rewatkar et al. 1991 with permission from American Chemical Society. © 1991, American Chemical Society.)

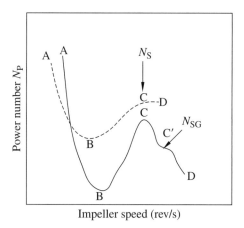

FIGURE 7A.10   Variation of power number with impeller speed for two-phase (gas–liquid) and three-phase (gas–liquid–solid) stirred reactors. Two phase (solid–liquid) system. A, fillet formation; B, disappearance of fillets; C, off-bottom suspension of solids; D, recirculation of mixture. Three phase (gas–liquid–solid). A, no dispersion of gas; solid settled on bottom; B, gas dispersed; beginning of solid suspension; C, gas dispersed; off-bottom suspension of solids; D, recirculation of mixture. (Reproduced from Rewatkar et al. 1991 with permission from American Chemical Society. © 1991, American Chemical Society.)

impeller speeds, N_p decreases with increase in N. This reduction is due to the formation of solid fillets at the center and along the periphery of the bottom. In three-phase systems, there is an additional reason for the reduction in power consumption. This is due to the formation of gas cavities behind the impeller blades. The cavity formation reduces the form drag behind the impeller and the power consumption decreases. The reduction in power consumption due to the cavity formation (in the gas–liquid system) is of greater magnitude than the reduction due to fillet formation (in solid–liquid system). At point B, breakup of cavity starts. The initial stage of particle suspension depends on the physical properties of liquid and solid (particles), impeller diameter, and design and location of the sparger. Along the line BC, more and more gas bubbles are formed and an increasing amount of solids gets suspended. The increasing energy dissipation at the gas–liquid and the solid–liquid interfaces and cavity breakup results in an increase in N_p along the line BC. The rise in power number (in the region BC) depends on the various factors affecting turbulence mentioned earlier in the previous section. At lower solid loading ($X \leq 6.67\%$), recirculation of gas–liquid–solid mixture (point C) becomes prominent before the critical impeller speed for solid suspension (N_{SG}) is reached. The decrease in N_p observed prior to N_{SG} continues beyond N_{SG} as shown in Figure 7A.10.

7A.5 GAS HOLDUP IN STIRRED TANK REACTORS

7A.5.1 Some Basic Considerations

The gas holdup is an important characteristic of any gas–liquid contacting device. It is a direct indication of the effective gas–liquid interfacial area for mass transfer. Several literature studies are reported in the literature on this aspect. Yawalkar et al. (2002a) have reviewed the available literature. Table 7A.1 gives a summary of some important literature investigations on this aspect of stirred reactors. The main factors that have important bearing on the gas holdup are geometric configuration of the impeller–vessel combination, operating parameters, and system physical properties. A brief summary of the same is presented in Section 7A.5.1.1.

7A.5.1.1 *Effect of Geometric Configuration*

1. Tank diameter (T).
2. Type of impeller and its geometric proportions with respect to the vessel. These include the impeller diameter to tank diameter, (D/T), and the impeller bottom clearance to tank diameter, (C/T), ratios (Fig. 7A.1), width of the blade and angle of pitch of the blade in case of pitched blade turbines and propellers, diameter of the hub through which the impeller is connected to the shaft, etc.
3. Size, type, and location of the sparger with respect to impeller.

TABLE 7A.1 Correlations for Gas Holdup in Stirred Tank Reactors Available in the Literature

Reference	Impeller type, system used	Correlation proposed
Calderbank (1958)	6-DT. Air–water	$\varepsilon_G \propto (\varepsilon_M)^{0.4} \, (V_G)^{0.5}$
Machon et al. (1978)	6-DT. Air–aqueous inorganic salt solutions (noncoalescing system)	$\varepsilon_G \propto We \, F_{LG}^{0.36} Y^{-0.56}; Y = f(\psi)$ $\psi = c\left(\dfrac{d\sigma}{dc}\right)^2 \times \phi, \; \phi = \left[1 + \dfrac{d\ln a}{d\ln c}\right]^{-1}$
Loiseau et al. (1977)	1. 6-DT. Air–nonfoaming liquids 2. 6-DT. Air–foaming liquids	1. $\varepsilon_G = 0.011 \times V_G^{0.36} \times \sigma_L^{-0.36} \, \mu_L^{-0.056} + \left[\dfrac{P_O}{V_L} + \dfrac{P_G}{V_L}\right]^{-0.27}$ 2. $\varepsilon_G = 0.051 \times V_G^{0.24} \left[\left(\dfrac{P_O}{V_L}\right) + \left(\dfrac{P_G}{V_L}\right)\right]^{-0.57}$
Hassan and Robinson (1977)	6-DT. Air–deionized water, propionic acid, methyl acetate, ethylene glycol	$\varepsilon_G = 0.113\left(\dfrac{Q_G N^2}{\sigma_L}\right)$
Yung et al. (1979)	6-DT and four-blade paddle. Air–water/acetone/ ethylene glycol/aqueous salt solutions (noncoalescing)	$\varepsilon_G \propto (F_{LG})^{0.5} \, (We)^{0.65} \left(\dfrac{D}{T}\right)^{1.4}$
Warmoeskerken and Smith (1981)	6-DT. Air–water	No large cavity regime: $\varepsilon_G = 0.62 (Q_G)^{0.45} (N)^{1.6}$ Large cavity regime: $\varepsilon_G = 1.167 (Q_G)^{0.75} (N)^{0.7}$
Chapman et al. (1983b)	6-PTD and 6-PTD. Air–water	$\varepsilon_G = 0.086\left(\dfrac{P}{M}\right)^{0.31} (\text{vvm})^{0.67}$
Bujalski et al. (1988)	6-PTD, 6PTU. Air–water 6 PTU alone. Air–water	$\varepsilon_G = 211\left(\dfrac{P}{M}\right)^{0.326} (V_G)^{0.776}$ $\varepsilon_G = 103\left(\dfrac{P}{M}\right)^{0.329} (V_G)^{0.553}$

Greaves and Barigou (1988)	6-DT 1. For Air–water system	$$\varepsilon_G = 4.07(Q_G)^{0.64}(N)^{0.62}\left(\frac{D}{T}\right)^{1.39}$$
	2. For air–aqueous sodium chloride	$$\varepsilon_G = 4.2(Q_G)^{0.52}(N)^{0.79}\left(\frac{D}{T}\right)^{1.92}$$
Smith (1991)	6-DT 1. Air–water	$$\varepsilon_G = 0.85(Re\times Fr\times F_{LG})^{0.35}\left(\frac{D}{T}\right)^{1.25}$$
	2. Air–water with surface tension variation	$$\varepsilon_G = fn\left\{\left(\frac{N^2 Q_G}{\sigma_L}\right)\times\left(\frac{\rho_L^2\,\mu_L}{g^2}\right)^{0.33}\right\}$$
Whitton and Nienow (1993)	6-DT. Air–water	$$\varepsilon_G = 1.28(\varepsilon_M)^{0.26}\times(V_G)^{0.66}$$
Rewatkar et al. (1993)	6-PTD. Air–water	$$\varepsilon_G = 3.54\left(\frac{D}{T}\right)^{2.08}(Fr)^{0.51}(F_{LG})^{0.43}$$
Vrabel et al. (2000)	6-DT, 6-blade Scaba radial turbine; 3- and 4-blade Scaba axial hydrofoil impellers (multiple impellers) Air–water	$$\varepsilon_G = 0.37\left(\frac{P}{V_L}\right)^{0.16}(V_G)^{0.55}$$
Yawalkar et al. (2002a)	6-DT, 6-PTD 1. Air–water, air-aqueous NaCl/Na$_2$SO$_4$/KCl/KOH/KNO$_3$/MgSO$_4$	$$\varepsilon_G = 0.104\times\left[\frac{N}{N_{CD}}\right]^{0.62}(vvm)^{0.64};\quad \varepsilon_G = 0.122\left(\frac{N}{N_{CD}}\right)^{0.64}(vvm)^{0.69}(T)^{0.22}\left(\frac{D}{T}\right)^{0.14}$$
	2. For surface tension factor, STF: 7.9×10^{-8} to 2.07×10^{-7} kg²·m³/kgmole.s⁴	$$\varepsilon_G = 2.97\left(\frac{N}{N_{CD}}\right)^{0.863}(vvm)^{0.7}(STF)^{0.197}$$
	3. For surface tension factor, STF$>2.07\times10^{-7}$ kg²·m³/kgmol·s⁴	$$\varepsilon_G = 15.81\times10^{-2}\left(\frac{N}{N_{CD}}\right)^{0.734}\times(vvm)^{0.85}$$

(Continued)

TABLE 7A.1 (Continued)

Reference	Impeller type, system used	Correlation proposed
Moucha et al. (2003)	6-DT, 6-PTD, and up- and downflow aerofoils 1, 2, and 3 impeller system: $\left(\dfrac{D}{T}\right) = \left(\dfrac{C}{T}\right) = 0.33$, $T = 0.29$. Interimpeller spacing $= T$. Air–aqueous Na_2SO_4	$\varepsilon_G = a \left(\dfrac{P_{Tot}}{V_L}\right)^b (V_G)^c$ Values of a, b, and c are given for different impellers/combinations
Smith et al. (2004)	6-DT. Air–water/electrolyte solutions, N_2/toluene in water	$\varepsilon_G = 69 \times 10^6 (\varepsilon_M)^{0.2} (V_G)^{0.55} (T)^{-3.2}$
Cents et al. (2001, 2005)	6-DT. Air–aqueous K_2CO_3/$KHCO_3$	$\varepsilon_G = 1.29 \times 10^{-2} (N)^{0.87}$
Bao et al. (2006, 2007)	Air–water, glass beads, multi-impeller system	Correlations of the type: $\varepsilon_G = a \left((\varepsilon_M)^\beta (V_G)^\gamma\right)$ (for a variety of impeller combinations)
Zhang et al. (2006)	Multi-impeller system of 6-DT. Air–water/aqueous sugar solutions	$\varepsilon_G = 0.31 (N)^{0.7} (V_G)^{0.52} \left(\dfrac{\mu_L}{\mu_w}\right)^{-0.19}$
Nikhade (2006)	6-DT, 6-PTD, 6PTU: vessel fitted with helical coils. Air–water	$\varepsilon_G = 0.086 \left(\dfrac{N}{N_{CD}}\right)^{0.79} (vvm)^{0.38}$
Bao et al. (2007)	Multi-impeller system: hollow half-elliptical blade dispersing turbine below two up-pumping wide blade hydrofoils, identified as HEDT + 2WHU. Air–hot water (temperature: 355°K), polypropylene particles	$\varepsilon_G = 0.36 (\varepsilon_M)^{0.15} (V_G)^{0.50} (1 + C_V)^{0.6}$ C_V: volume fraction of solids

Bao et al. (2008a)	Multi-impeller system Gas-dispersing disc turbine with concave half-elliptical blades (HEDT) at bottom with two up-pumping (WHU) wide blade hydrofoils above. Air–water (hot and cold) containing 90 μm glass beads: $\rho_S \sim 2500$ kg/m³. Solid loading 3–21 vol.%	Cold: $\varepsilon_G = 0.9(\varepsilon_M)^{0.15}(V_G)^{0.55}(1+C_V)^{-1.77}$ Hot: $\varepsilon_G = 0.48(\varepsilon_M)^{0.15}(V_G)^{0.55}$ No effect of solid phase on ε_G in hot system
Bao et al. (2008b)	Air–water (hot and cold). Multi-impeller system. Gas-dispersing disc turbine with concave half-elliptical blades (HEDT) at bottom with two up-pumping (WHU) wide blade hydrofoils above. Air–water (hot and cold) containing 90 μm glass beads: $\rho_S \sim 2500$ kg/m³. Solid loading 3–21 vol.%	$\varepsilon_G = \alpha'(\varepsilon_M)^{0.16}(V_G)^{0.55}(T)^{m'}$; $\varepsilon_G = \lambda'(\varepsilon_M)^{0.16}(V_G)^{0.55}(1+C_V)^{n'}$ α' and m' are functions of the volumetric solid loading C_V T in K. Values of m', n', α', and λ' are given
Tagawa et al. (2012)	4 flat blade DT. Air–water. Solids used: synthetic adsorbent (SA) ($\rho_S \sim 800$ kg/m³) and polypropylene (PP) ($\rho_S \sim 850$ kg/m³). ε_S: 0–0.5	$\varepsilon_G = 0.25\left(\dfrac{P_G}{V_D}\right)^{0.4}(V_G)^{0.75}(1+\varepsilon_S)^{a_2}$ $a_2 = -1.0$ for SA and -0.7 for PP, respectively

7A.5.1.2 Effect of Operating Conditions Operating conditions in this case include (i) the gas throughput and (ii) speed of agitation. These along with the type of impeller and geometric configuration decide the actual power input to the vessel.

7A.5.1.3 Effect of System Physical Properties Liquid-phase physical properties have a significant effect on the gas holdup in stirred reactors. The main physical properties of interest are (i) surface tension (σ_L), (ii) viscosity (μ_L), and (iii) density (ρ_L) of the liquid phase. The gas holdup in any gas–liquid contactor is intimately related to the average bubble size. The bubble size itself is a complex function of the physical properties of the system and the turbulence prevailing in the contactor. Besides these, the total pressure at which the system is operating is also known to affect the bubble size. In the following discussion, some basic aspects of breakup of bubbles in a turbulent system are presented (Walter and Blanch 1986; Parthasarathy and Ahmed 1991).

A bubble owes its stability to the surface tension forces. It can break when the hydrodynamic stresses are sufficiently large to overcome the forces due to surface tension (Hinze 1955). Therefore, a relative estimate of these two opposing forces is necessary to determine the operating conditions that can cause breakup of bubbles. The discussion is limited to highly turbulent flow fields since industrial operations are invariably in the turbulent region. When the two opposing forces are equal, there is a quasi equilibrium. This situation is quantified as follows:

$$\tau \approx \left(\frac{\sigma}{d_B} \right) \tag{7A.3}$$

Hinze (1975) has suggested that breakup mechanisms can be of two broad types: (1) dominated by viscosity effects and (2) dominated by turbulent phenomena. Hinze derived a critical Weber number from Equation 7A.3 as follows:

$$\text{We}_{\text{Crit}} = \left(\frac{\tau d_M}{\sigma} \right) \tag{7A.4}$$

Critical Weber numbers reported in the literature range from 1 to 5. For turbulent multiphase systems, a much shorter range, 1.2–1.5, has been suggested (Rigby et al. 1997; Lane et al. 2002, 2004; Kerdouss et al. 2006). Besides the excellent discussion on bubble breakup in stirred vessels presented by Walter and Blanch (1986), the studies of Prince and Blanch (1990), Luo and Svendson (1996), and Martin et al. (2010) will be used in the following to rationalize the various mechanisms that lead to bubble breakup. As explained by Walter and Blanch (1986), if the control volume is relatively small, it is safe to assume that local isotropy and homogeneity prevail. Isotropy requires that the microscale of turbulence (l) be very small as compared to the scale of the primary eddies. For stirred vessels, the primary eddies can be assumed to be of the same scale as that of the impeller diameter, D (Moo-Young and Blanch 1981). Walter and Blanch (1986) suggested that a simpler and reasonable procedure would be to assume that the microscale of turbulence is relatively very small as compared to the dimension of a stable bubble for the given configuration of vessel–impeller geometry, physical properties, and the power input. It will be seen later that it is possible to estimate this stable bubble size. Breakage of bubbles is accompanied by an increase in the surface per unit volume. This overall increase in surface area can come through energy expenditure. In a turbulent system, the eddies initiate oscillations/deformation of a bubble. Walter and Blanch (1986) have given photographic evidence that suggests that bubble breakup occurs in stages. Initially, the bubble is stretched into a dumbbell shape. This stretched and deformed shape has two major centers of gravity connected by a thin threadlike link. When this connecting link is severed, the bubbles get detached and form two independent bubbles. This mechanism has been termed "binary splitting." The initial deformation of a free surface (drops/bubbles) is the result of continuous impact of eddies generated by the underlying turbulence. When these eddies are of the desired size range and energy content, the probability of formation of new surface increases. In the absence of a strong disturbance, a bubble can continue to oscillate maintaining a "quasi equilibrium" bubble size implied in Equation 7A.3 (Walter and Blanch 1986). The actual breakage occurs when the shear stress in Equation 7A.3 overcomes the forces stabilizing the bubble. The shear stresses to be considered are (i) inertial and (ii) viscous. In the case of highly turbulent flow, the viscous contribution may be neglected in view of the relatively large size of the bubbles in comparison to the microscale of turbulence (Luo and Svendson 1996). Walter and Blanch also based their analysis on the assumption that the microscale of turbulence is much smaller than the maximum stable bubble size, d_M. The microscale, l_{MK}, was calculated using Kolmogorov's theory of isotropic turbulence:

$$l_{MK} = \left(\frac{\mu^{3/4}}{(P/V)^{1/4} \, \rho^{1/2}} \right) \tag{7A.5}$$

Isotropy was assumed to prevail when $200 \times l_{MK} < d_M$. Prince and Blanch (1990) showed that eddies having size greater than $0.2 \, d_B$ can only cause breakup. The frequency of collisions between bubbles and eddies of similar size is the determining factor in bubble breakup (Wu et al. 1998). Two broad types of eddies are shown in Figure 7A.11: (1) large eddy engulfing the bubble and (2) eddies smaller than the bubble. For relatively large Reynolds numbers, eddies much larger than the bubble size engulf the bubbles and only transport the bubble along their path (Olmos et al. 2001). On the other hand, the smaller eddies that are part of the large eddy packet but smaller than the bubble continue to directly interact with the bubble surface (Fig. 7A.11).

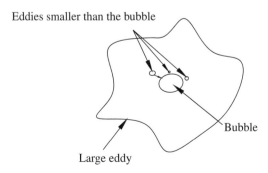

FIGURE 7A.11 Interaction of different eddies with a gas bubble.

In this class of eddies, only the larger size eddies (eddy size $> 0.2\ d_B$) can cause stresses sufficient to deform the bubble (Nienow 2006).

For highly turbulent flow fields such as in an industrial stirred vessel, the forces that are a source of deformation of the free interface are those due to Reynolds stresses. The latter are caused by the fluctuating velocity components. These forces are given by (Walter and Blanch 1986)

$$\tau = \left(\frac{\rho_L u'}{2} \right) \tag{7A.6}$$

$$\text{and } u' = \left(\frac{P}{V} \right)^{1/3} \left(\frac{d_B}{\rho_L} \right)^{1/3} \tag{7A.7}$$

The condition for bubble breakup stipulates that the turbulent stresses should prevail over the surface tension forces that maintain the identity of the bubble (Hinze 1955; Angeli and Hewitt 2000; Kerdouss et al. 2006). Based on this, the following condition for equilibrium between the two forces is realized (Walter and Blanch 1986):

$$\left(\frac{P}{V} \right)^{2/3} \left(\frac{d_B}{\rho} \right)^{2/3} = K \left(\frac{\sigma}{d_B} \right) \tag{7A.8}$$

The maximum stable bubble diameter obtained from the application of the above logic is

$$d_M = K \left(\frac{\sigma^{0.6}}{\left(\dfrac{P}{V} \right)^{0.4} \rho^{0.2}} \right) \tag{7A.9}$$

Predictably, the constant K in Equations 7A.8 and 7A.9 is decided by any additives (surface-active agents) or liquid physical properties (viscosity/surface tension) that tend to stabilize the bubble. The effects of surface-active agents and electrolytes and viscosity are discussed in detail in Section 10.3.2.1 and 10.3.2.1, respectively. Parthasarathy and Ahmed (1991) have given Equation 7A.10 for the Sauter mean bubble diameter based on their photographic measurements in stirred vessels:

$$d_{32} = 2.0 \left(\frac{\sigma^{0.6}}{\left(\dfrac{P}{V} \right)^{0.4} \rho_L^{0.2}} \right) \tag{7A.10}$$

This equation is exactly the same as that given by Walter and Blanch (1986), except that the constant K in Equation 7A.9 is now explicitly given as 2.0 by Parthasarathy and Ahmed (1991). In their study, Parthasarathy and Ahmed covered a wide variety of impellers (six-blade: Rushton, flat blade, and pitched blade turbines and three-blade marine propeller) and spargers. Methyl isobutyl carbinol was added at 50 ppm concentration to obtain a noncoalescing system. In view of this, the constant 2.0 in Equation 7A.10 is valid for noncoalescing media only. The actual local bubble size varies greatly in a stirred tank due to disparate energy dissipation rates in different regions. Bubbles in the vicinity of the impeller are broken up more frequently than those away from the impeller. Martin et al. (2010) indicate that the critical Weber number for a stirred tank is higher than that for a three-phase sparged reactor wherein the energy dissipation is relatively uniform.

The process of coalescence is equally important in deciding the equilibrium bubble size. Two or more bubbles that can coalesce need to come within a collision distance. The latter is decided by the frequency of collisions, which in turn is decided by the number of bubbles in the control volume (or gas holdup). Further, these bubbles in contact can coalesce if the contact time is sufficiently long. During this contact time, the intervening liquid film between these bubbles starts draining and becoming thinner. When a certain critical film thickness is reached, the film is ruptured and the two separate bubbles unite. Wu et al. (1998) have given a general expression for the rate of coalescence.

Effect of Physical Properties on Bubble Breakup and Size High surface tension liquids tend to stabilize bubbles. This is evident from Equation 7A.3, which indicates that a higher surface tension of the liquid requires higher turbulent stress to break the bubble.

Viscosity of the liquid phase also plays an important role in bubble stability. Martin et al. (2010) argued that high viscosity liquids can absorb substantial part of the inertial energy associated with the eddies reaching the bubble surface. Thus, viscosity has a dampening effect on the stress transmitted to the bubble surface, which in turn curtails bubble breakage. For Rushton turbine, Martin et al. (2010) have given Equation 7A.11 for the effect of the liquid viscosity on the critical Weber number:

$$\text{We}_{\text{Crit}} = 1.95 \left(\frac{\mu_{\text{L}}}{\mu_{\text{w}}} \right) + 14 \tag{7A.11}$$

Effect of Gas Sparger on Bubble Size Rewatkar et al. (1993) reported gas holdup data for different types of spargers and six-blade PTD (it is worth repeating here that the downflow type of impellers have serious instability problems). Increase in the dispersion height was used to determine the gas holdup. Using this subjective method, the authors concluded that a ring-type sparger having a diameter equal to $0.8 \times D$ gave 18–25% higher holdup than the other impellers studied. This implied that the sparger does have an effect on the stable bubble size. Parthasarathy and Ahmed's (1991) finding are different than this apparent subjective observation. In their study of effect of bubble size and power input on gas holdup, Parthasarathy and Ahmed

(1991) also used a variety of spargers ranging from coarse ring type to porous sintered glass spargers having different porosities. For the case of the high shear Rushton turbine, Parthasarathy and Ahmed (1991) found that at sufficiently high impeller speeds (or power inputs) approximately close to N_{CD}, there was little difference in the equilibrium bubble size for the ring and the three sintered glass spargers. This was despite the high power input (not included in the impeller power input) and consequent low initial bubble size in the sintered glass spargers. The technique used was more objective than the visual gas holdup measurement technique of Rewatkar et al. (1993). As explained earlier, industrial stirred tanks are designed to operate at speeds of agitation above N_{CD}. The actual operating speed may be 10–20% higher than N_{CD}. The corresponding power input is sufficiently high to lead to the observation of Parthasarathy and Ahmed regarding the ineffectiveness of sintered spargers. It will therefore be fair to conclude that at the relatively high power inputs used in industrial situations, it is futile to use a high energy-consuming sparger that does not give correspondingly higher gas holdup. A simple ring sparger such as that used by Parthasarathy and Ahmed will be far more economical in terms of the gas holdup generated in relation to the overall power input.

7A.5.2 Correlations for Gas Holdup

Some important investigations and the correlations for gas holdup given therein are listed in Table 7A.1.

Most investigators have relied on the power input per unit mass, ε_M, and superficial gas velocity, V_G, as the main variables for correlating the gas holdup in stirred vessels. The other groups of correlations in Table 7A.1 have used dimensionless groups such as the Froude and flow numbers along with important geometrical parameters of the stirred vessel assembly. The gas flow rate is an important variable, and the reported investigations have employed two variations of the same to account for its effect on the fractional gas holdup: (1) superficial gas velocity and (2) gas throughput as volume of gas passed in terms of vessel volume per unit time (vvm). Chapman et al. (1980, 1983b) argued that the use of vvm eliminates the scale-up effect since it is the reciprocal of the residence time, V_D/Q_G, of the gas phase in the reactor. Maintaining a constant superficial gas velocity during scale-up results in increase in gas-phase residence time as the vessel diameter (or volume) increases. In gas–liquid reactions, such as oxidation using air, the reaction consumes oxygen, thereby decreasing the average oxygen partial pressure (and implicitly dissolved concentration). Simultaneously, partial pressure of nitrogen, which is the main inert species in air, increases. In this case also, it has been shown by Schluter and Deckwer (1992) that the use of vvm is advantageous in scaling up of the reactor. Yawalkar et al. (2002a) have analyzed the literature correlations prior to 2002 (Table 7A.1) and have shown that the correlations based on power input or dimensionless numbers have serious limitations when applied to a different geometry/scale of operation or operating conditions. The reasons for these limitations are the same of those discussed in the case of prediction of mass transfer coefficient in Chapter 6, such as disregard of the mean flow strength and direction.

7A.5.3   Relative Gas Dispersion (N/N_{CD}) as a Correlating Parameter for Gas Holdup

The phenomena influencing bubble breakup and gas dispersion in stirred reactors have been discussed in Sections 7A.3 and 7A.4. The shear stresses that cause bubble breakup reduce the average bubble rise velocity through (i) a reduction in bubble size (Walter and Blanch 1986) and (ii) drag on the bubble (Hughmark 1974). Consequently, these smaller bubbles with reduced rising velocity are relatively easily entrained by the mean flow generated by the impeller. Here, it is important to note the difference between the effects of turbulent shear and the mean flow. The former directly affects the bubble size, whereas the latter relates to the travel path of the bubble as well as the momentum of the discharge from the impeller that acts on the buoyant bubbles. Yawalkar et al. (2002a) argued that N_{CD} uniquely characterizes a given impeller–vessel geometric configuration and system physical properties such as surface tension and viscosity of the liquid phase. It was argued that at N_{CD} the mean flow developed by the impeller is one of the factors that overcome the buoyancy of the bubbles and their subsequent dispersion. The other factor is the turbulence associated with the mean flow. N_{CD} also characterizes the turbulence intensity, which is denoted by u'_{CD}. As discussed earlier, u'_{CD} plays an important role in deciding the average bubble size and hence the rise velocity of the bubble. Deshpande (1988) in his investigation of turbulence parameters in stirred reactors had shown that the turbulence intensity, u', is approximately proportional to the speed of agitation, N. Yawalkar et al. (2002a) argued that the speed of agitation, which is readily measured, can replace the rather esoteric turbulence intensity. The implication is that for a given impeller–reactor combination and specific geometric configuration (D/T, C/T, etc.), the turbulence intensity at the completely dispersed condition, u'_{CD}, can be conveniently represented by N_{CD} as a unique value. Yawalkar et al. (2002a) defined a relative gas dispersion parameter in terms of the ratio, N/N_{CD}. A similar approach of relative suspension was used by Jadhav and Pangarkar (1991) and Kushalkar and Pangarkar (1994) prior to the work of Yawalkar et al. (2002a). As discussed in Section 7A.6, this relative suspension approach was found to be very useful in deriving a universal correlation for the solid–liquid mass transfer coefficient, K_{SL}. Yawalkar et al. (2002a) extended this hypothesis of Jadhav and Pangarkar (1991), Dutta and Pangarkar (1994, 1995a, b, 1996), and Kushalkar and Pangarkar (1994) [finally generalized by Nikhade and Pangarkar 2007] to correlate the gas holdup with the relative dispersion parameter, N/N_{CD}. The correlation thus obtained (Table 7A.1) was

$$\varepsilon_G = 0.104 \left(\frac{N}{N_{CD}} \right)^{0.62} (\text{vvm})^{0.64} \qquad (7A.12)$$

Equation 7A.12 was shown to uniquely represent a very large number of literature data covering wide variety of impellers and impeller–vessel combinations. The major feature of Yawalkar et al.'s approach is that (i) it is based on sound logic and (ii) requires variables (N_{CD} and vvm) that are readily estimated and measured, respectively, as compared to turbulence parameters, which N_{CD} represents (Section 7A.3). The R^2 value for the correlation of the type represented by Equation 7A.12 was 0.9, and the standard error was 0.15. To improve on the accuracy of Equation 7A.12, Yawalkar et al. (2002a) proposed Equation 7A.13 ($R^2 = 0.96$ and standard error ~0.01; Table 7A.1) to incorporate the effect of the vessel diameter and D/T:

$$\varepsilon_G = 0.122 \left(\frac{N}{N_{CD}} \right)^{0.62} (\text{vvm})^{0.69} (T)^{0.22} \left(\frac{D}{T} \right)^{0.14} \tag{7A.13}$$

The relatively better fit of Equation 7A.13 is obviously due to inclusion of additional parameters. However, the differences between the predictions of Equations 7A.12 and 7A.13 are not significant. For example with $T = 2$ m, $D/T = 0.33$, vvm $= 0.1$, and $N/N_{CD} = 1.1$, Equation 7A.12 predicts $\varepsilon_G = 0.22$, whereas Equation 7A.13 yields $\varepsilon_G \sim 0.25$. This difference of ~15% between the two predictions of Equations 7A.12 and 7A.13 should be acceptable within the limits of experimental errors. Both Equations 7A.12 and 7A.13 are simple to use and the choice of one of them is a matter of convenience. When a relatively accurate estimate is desired, Equation 7A.13 should be used.

7A.5.4   Correlations for N_{CD}

The aforementioned correlations for the gas holdup require knowledge of N_{CD}. Hydrodynamics in two- and three-phase reactors has been widely investigated as discussed in Sections 7A.3 and 7A.4, respectively. Various investigators have reported reliable correlations for N_{CD} over a wide range of geometric configurations and operating parameters for some widely used impellers. These are:

Six-blade DT (Nienow et al. 1977):

$$N_{CD} = \frac{4Q_G^{0.5}T^{0.25}}{D^2} \tag{7A.14}$$

Six-blade 45° downflow pitched turbine (Rewatkar and Joshi 1993):

$$N_{CD} = 2.143 \times (T)^{1.06}(D)^{-1.88}(V_G)^{0.35} \tag{7A.15}$$

For the most versatile six-blade upflow pitched turbine, Bujalski et al. (1988) gave

$$(F_{LG}) = 12.1 \times 10^4 \times (\text{Fr})^5_{CD} \left(\frac{T}{T_o} \right)^{2.8} \quad (\text{where } T_0 = 1 \text{ m}) \tag{7A.16}$$

7A.6   GAS–LIQUID MASS TRANSFER COEFFICIENT IN STIRRED TANK REACTOR

In Section 2.1, it has been indicated that this book deals with reactions in which the gas-phase reactant is sparingly soluble in the liquid phase and, consequently, the gas film resistance in the overall mass transfer process can be ignored (Figs. 2.1 and 2.2;

Danckwerts 1970). Therefore, the first step in the sequence is gas–liquid mass transfer. Knowledge of volumetric gas–liquid mass transfer coefficient, $k_L \underline{a}$, and the gas–liquid mass transfer rate is an important aspect of design/scale-up of two- and three-phase stirred tank reactors. A detailed discussion on various methods of correlating the various mass transfer coefficients is available in Chapter 6. Therefore, the discussion presented in this chapter will be focused on scale-independent universal correlations that are desirable in any scale-up procedure.

As discussed in Chapter 6, a large number of investigators (Smith et al. 1977; van't Riet 1979; Linek et al. 1987; Hickman 1988; Smith 1991; Whitton and Nienow 1993; Zhu et al. 2001) have used Kolmogorov's theory to correlate various mass transfer coefficients in multiphase reactors. These correlations were of the form

$$k_L \underline{a} = f\left[\left(\frac{P_G}{V_D}\right)^{\alpha} (V_G)^{\beta}\right].$$ The stirred tank reactor has also been treated in this

manner. The advantages of Kolmogorov's theory/power input approach are obvious

for a stirred vessel. The most important of these is that fairly reliable correlations for the power number, N_P/N_{PG}, are available for both two- and three-phase systems. As a result, using the particular correlation for a given system and the speed of agitation, the power input could be readily calculated. Unfortunately, this simple power input approach has the drawback that it ignores the flow patterns (and also the turbulence intensity) developed in different configurations. As a consequence, various similarity regimes such as kinematic, hydrodynamic, and turbulence similarity regimes are likely to be violated. It is therefore not surprising that Kolmogorov's theory approach could not yield a unique correlation for agitator/vessel geometric combinations different than those used for deriving the power input-based correlations. Similarly, correlations based on mechanical similarity using dimensionless groups (Smith and Warmoeskerken 1985; Smith 1991) having the form $k_L \underline{a} = f(\text{Fr}, F_{LG}, D/T, \text{etc.})$ have also failed to represent data over a wide range of geometric configurations and physical properties. A survey of the available literature on $k_L \underline{a}$ in gas-inducing reactors by Markopoulos et al. (2007) has also derived a conclusion that the usual approach based on dimensionless groups does not yield unique correlations. Sardeing et al.'s (2004a) $k_L \underline{a}$ data for DT and A315 (both down- and upflow modes) further suggest that location of the sparger is also an important factor. It was found that sparger located such that the sparged gas is swept by the impeller discharge and cocirculated gives better results. For the impellers they studied, this location is below the impeller level. The diameter of the ring sparger used did not have a significant effect as long as it was at least equal to the impeller diameter. This investigation also supported the argument (Aubin et al. 2001; Nienow and Bujalski 2004; Sardeing et al. 2004b; Section 7A.7) that the upflow variant utilizes the power input better than the downflow type. A recent investigation by Bustamante et al. (2013) has reported and compared $k_L \underline{a}$ for 6-DT and the low shear Elephant Ear impellers (Collignon et al. 2010) in down- and upflow modes. The upflow mode was found to yield higher $k_L \underline{a}$ values as compared to the downflow mode. Evidently, the difference in the performance of up- and downflow modes is due to the different directions of the mean flow from the impeller. This could obviously not be incorporated in either the Kolmogorov's theory-based correlations or those based on mechanical similarity. Although, Bustamante et al.'s study employed relatively small vessel/impeller system ($T=0.17$ m, $D=0.08$ m), the data for the three types of impellers employed were collected under identical conditions. Therefore, this comparison of $k_L a$ values is a reliable reflection of the relative efficacies of the three impellers used. The major conclusion that can be drawn is as follows: the difference in the performance of up- and downflow impellers is due to the different directions of the mean flow from the impeller.

The diversity in correlations reported in the literature can therefore be attributed to differences in the flow patterns, mean flow strengths and direction, turbulence intensity variations, etc. Table 7A.2 lists some important literature studies and the resulting correlations for the volumetric gas–liquid mass transfer coefficient in stirred vessels.

The success of their approach in obtaining a unique scale- and system-independent correlation for the gas holdup (Sections 7A.5.2 and 7A.5.3; Yawalkar et al. 2002a) prompted Yawalkar et al. (2002b) to apply a similar approach for correlating the volumetric gas–liquid mass transfer coefficient in stirred vessels. The argument employed was as follows:

1. N_{CD} represents a unique condition for a given combination of stirred vessel and system properties. At a given superficial gas velocity and the corresponding N_{CD}, for all impeller–vessel geometric combinations and system properties, the fractional gas holdup is the same. The effective gas–liquid interfacial area, a, is related to the gas holdup. Therefore, a correlation based on the relative dispersion inherently includes a through ε_G.
2. As discussed in Section 7A.5.2, N_{CD} represents a unique turbulence state in any given stirred vessel. Further, the detailed discussion in Chapter 6 bears evidence to the fact that the true transfer coefficient in any multiphase system is decided by the turbulence parameters of the system. Therefore, N_{CD} is again a crucial parameter in deciding k_L.

Based on the earlier logic, Yawalkar et al. (2002b) employed the literature data on $k_L a$ in stirred vessels to test the veracity of the relative dispersion approach. A correlation of the type similar to Equation 7A.12, $k_L a = f[(N / N_{CD}),\text{vvm},T,(D/T)]$, was attempted. It was found that $k_L a$ is a weak function of D/T probably because this effect was inherent in N_{CD}. Therefore, the D/T term was eventually not included. The regression results indicated that Equation 7A.17 could unify all the literature data:

$$k_L a = 5.58 \times 10^{-2} \times \left(\frac{N}{N_{CD}} \right)^{1.464} (\text{vvm})(T)^{1.05} \tag{7A.17}$$

Equation 7A.17 is based on a total of 178 data points and has R^2 value of 0.96 with standard deviation of 0.13. The exponent of T in Equation 7A.17 is very close to 1.

TABLE 7A.2 Correlations for $k_L \underline{a}$ in Stirred Tank Reactors Available in the Literature

Reference	Details of equipment and system used	Correlation
Calderbank (1959)	6-DT, $\left(\dfrac{D}{T}\right) = 0.33$; $\left(\dfrac{P}{V_L}\right) = 0.3 - 3.35 \ kW/m^3$. Orifice sparger. Stripping of sparingly soluble organic liquids (benzene, carbon disulfide, ethyl bromide, etc.) from water; absorption/stripping of CO_2, ethylene, butadiene, and oxygen from water for measuring $k_L \underline{a}$. Aqueous glycerol and glycol solutions used to vary liquid viscosity	$k_L \underline{a} = 0.026 \times \left(\dfrac{P}{V_L}\right)^{0.4} (V_G)^{0.5}$
Smith et al. (1977)	6-DT, $\left(\dfrac{D}{T}\right) = 0.33,\ 0.5$; $T = 0.61$. Pipe sparger. Dynamic stripping of oxygen with nitrogen used for measuring $k_L \underline{a}$	$k_L \underline{a} = 0.01 \times \left(\dfrac{P}{V_L}\right)^{0.475} (V_G)^{0.4}$
Van 't Riet (1979)	Literature data of various investigators used to derive correlations	Coalescing system: $k_L \underline{a} = 2.6 \times 10^{-2} \times \left(\dfrac{P}{V_L}\right)^{0.4} (V_G)^{0.5}$ Noncoalescing system: $k_L \underline{a} = 2 \times 10^{-3} \times \left(\dfrac{P}{V_L}\right)^{0.7} (V_G)^{0.2}$
Smith and Warmoeskerken (1985)	6-DT, $\left(\dfrac{D}{T}\right) = 0.4$, $T = 0.44$. Dynamic stripping of oxygen with nitrogen used for measuring $k_L \underline{a}$	BLC[1] regime: $k_L \underline{a} = 1.1 \times 10^{-7} \times (F_{LG})^{0.6} (Re_N)^{1.1} (N)$ ALC[2] regime: $k_L \underline{a} = 1.6 \times 10^{-7} \times (F_{LG})^{0.42} (Re_N)^{1.02} (N)$

(Continued)

TABLE 7A.2 (Continued)

Reference	Details of equipment and system used	Correlation
Linek et al. (1987)	6-DT, $\left(\dfrac{D}{T}\right) = 0.33$, $T = 0.29$. Pipe sparger. Dynamic stripping of oxygen with nitrogen used for measuring $k_L a$	$k_L \underline{a} = 4.95 \times 10^{-3} \times \left(\dfrac{P}{V_L}\right)^{0.593} (V_G)^{0.4}$
Hickman (1988)	6-DT, $\left(\dfrac{D}{T}\right) = 0.33$, $T = 0.6$ and 2 m. Steady-state method based on enzymatic decomposition of H_2O_2. Oxygen produced in the decomposition reaction measured to obtain $k_L a$.	For $T = 0.60$ m: $k_L \underline{a} = 0.043 \times \left(\dfrac{P}{V_L}\right)^{0.4} (V_G)^{0.57}$ For $T = 2$ m: $k_L \underline{a} = 0.027 \times \left(\dfrac{P}{V_L}\right)^{0.54} (V_G)^{0.68}$
Bujalski et al. (1990)	6PTU, 6DT, $\left(\dfrac{D}{T}\right) = 0.5$, $\left(\dfrac{C}{T}\right) = 0.16$. Dynamic stripping of oxygen with nitrogen used for measuring $k_L a$	For 6 PTU: $k_L \underline{a} = 0.34 \times \left(\dfrac{P_{Tot}}{M}\right)^{1.04} \times (V_G)^{0.34}$ For both 6-DT and 6 PTU: $k_L \underline{a} = 2.3 \times \left(\dfrac{P_{Tot}}{M}\right)^{0.7} (V_G)^{0.7} \left(\dfrac{D}{g}\right)^{-0.5}$
Smith (1991)	6-DT, $\left(\dfrac{D}{T}\right) = 0.33 - 0.44$, $T = 0.6 - 2.7$ m. Orifice sparger. Dynamic stripping of oxygen with nitrogen used for measuring $k_L \underline{a}$	$k_L \underline{a} = 1.25 \times 10^{-4} \times \left(\dfrac{D}{T}\right)^{2.8} (Fr)^{0.6} (Re_N)^{0.7} (F_{LG})^{0.45}$
Whitton and Nienow (1993)	6-DT, $\left(\dfrac{D}{T}\right) = 0.3 - 0.33$, $T = 0.61$ and 2.67 m. Ring sparger. Dynamic stripping of oxygen with nitrogen used for measuring $k_L \underline{a}$	$k_L \underline{a} = 0.57 \times \left(\dfrac{P}{M}\right)^{0.4} (V_G)^{0.55}$

Reference	Description	Correlation
Zhu et al. (2001)	6-DT, $\left(\frac{C}{T}\right) = \left(\frac{D}{T}\right) = 0.33$, $T = 0.39$ m $= \left(\frac{D}{T}\right) =$. Ring sparger. Dynamic stripping of oxygen with nitrogen used for measuring $k_L a$	$k_L a = 0.031 \times \left(\frac{P}{V_L}\right)^{0.4} (V_G)^{0.5}$
Yawalkar et al. (2002b)	6-blade DT, A315, 6-PTD. Data from literature covering wide range of geometric configurations and impellers	$k_L a = 3.35 \times \left(\frac{N}{N_{CD}}\right)^{1.464} (V_G)$
Moucha et al. (2003)	Multiple impeller system (1, 2, and 3 impellers): 6-DT, 6-PTD and up- and downflow aerofoils: $\left(\frac{D}{T}\right) = \left(\frac{C}{T}\right) = 0.33$, $T = 0.29$. Interimpeller spacing $= T$. Dynamic step pressure method—oxygen used as the solute	$k_L a = 1.08 \times 10^{-3} \times \left(\frac{P_{Tot}}{V_L}\right)^{1.19} (V_G)^{0.55}$ (relatively high standard deviation of 30%)
Kapic (2005)	6-DT, $\left(\frac{D}{T}\right) = 0.35$, $T = 0.211$ m. Dynamic stripping of oxygen with nitrogen used for measuring $k_L a$	$k_L a = 0.04 \times \left(\frac{P_G}{V}\right)^{0.47} \times (V_G)^{0.6}$. Applicable only for $T = 0.211$ m
Kapic and Heindel (2006)	6-DT, $\left(\frac{D}{T}\right) = 0.35$, $\left(\frac{C}{T}\right) = 0.25$, $T = 0.211$ m. Dynamic stripping of oxygen with nitrogen used for measuring $k_L a$	$k_L a = 1.59 \times \left(\frac{N}{N_{CD}}\right)^{1.34} \left(\frac{T}{D}\right)^{0.41} \times V_G^{0.93}$. Applicable over $0.211 < T < 2.7$ m
Martin et al. (2008)	Multi-impeller system having three 6-DT, $\left(\frac{D}{T}\right) = 0.4$, $T = 0.21$ m, $\left(\frac{H}{T}\right) = 2$. Dynamic stripping of oxygen with nitrogen used for measuring $k_L a$	$k_L a = 1 \times 10^{-3} \times \left(\frac{P_G}{V_L}\right)^{0.37} (V_G)^{0.54}$

(Continued)

TABLE 7A.2 (Continued)

Reference	Details of equipment and system used	Correlation
Yagi and Yoshida (1975); Martin et al. (2010)	6-DT; $\left(\dfrac{D}{T}\right) = \left(\dfrac{C}{T}\right) = 0.4$; $T = 0.2\,\mathrm{m}$ (for Yagi and Yoshida and $T = 0.33$ m for Martin et al.). Aqueous glycerol/millet jelly as Newtonian and aqueous CMC/sodium polyacrylate as non-Newtonian liquids. Dynamic stripping of oxygen with nitrogen used for measuring $k_L a$	$k_L a = 0.06 \times \left(\dfrac{D_M}{D^2}\right)\left(\dfrac{ND^2 \rho_L}{\mu_L}\right)^{1.5}\left(\dfrac{N^2 D}{g}\right)^{0.19}$ $\left(\dfrac{\mu_L}{\rho_L D_M}\right)^{0.5}\left(\dfrac{\mu_L V_G}{\sigma}\right)^{0.6}\left(\dfrac{ND}{V_G}\right)^{0.32}$ $(1 + 2(\lambda N)^{0.5})^{-0.67}$
Tagawa et al. (2012)	Two 4 DT spaced at $\geq 2 \times D$, $T = 0.2\,\mathrm{m}$, $\left(\dfrac{D}{T}\right) = 0.4$; $\left(\dfrac{C}{T}\right) = 0.25$. Solids used: synthetic adsorbent (SA) $(\rho_S \sim 800\,\mathrm{kg/m^3})$ and polypropylene (PP) $(\rho_S \sim 850\,\mathrm{kg/m^3})$. ε_S: 0–0.5; dynamic stripping of oxygen with nitrogen used for measuring $k_L a$	For both Newtonian and non-Newtonian fluids $k_L a = 9 \times 10^{-3} \times \left(\dfrac{P_G}{V_L}\right)^{0.5} \times (V_G)^{0.45}$; $(1 + 3.2\varepsilon_S)^{a_3}$ a_3: SA $= -3.0$ and PP: -0.6
Kielbus-Rapala et al. (2011)	6-DT, A315, $\left(\dfrac{D}{T}\right) = 0.33$, $T = 0.288\,\mathrm{m}$. Distilled water and aqueous NaCl as liquid with air and sea sand as solid phase. Dynamic stripping of oxygen with nitrogen used for measuring $k_L a$	6-DT: distilled water: $k_L a = 0.029 \times \left(\dfrac{P}{V_L}\right)^{0.44} \times (V_G)^{0.53}$ 6-DT: aqueous NaCl (0.4 mol/m³): $k_L a = 0.029 \times \left(\dfrac{P}{V_L}\right)^{0.44} \times (V_G)^{0.53}$

6-DT: aqueous NaCl (0.8 mol/m^3):

$$k_L \underline{a} = 0.454 \times \left(\frac{P}{V_L}\right)^{0.091} \times (V_G)^{0.501}$$

A315: distilled water:

$$k_L \underline{a} = 0.122 \times \left(\frac{P}{V_L}\right)^{0.432} \times (V_G)^{0.79}$$

A315: aqueous NaCl (0.4 mol/m^3):

$$k_L \underline{a} = 0.116 \times \left(\frac{P}{V_L}\right)^{0.375} \times (V_G)^{0.608}$$

A315: aqueous NaCl (0.8 mol/m^3):

$$k_L \underline{a} = 0.128 \times \left(\frac{P}{V_L}\right)^{0.368} \times (V_G)^{0.629}$$

Newtonian liquids:

DT: $k_L \underline{a} = 13.79 \times 10^{-2} (N)^{0.78} (\text{vvm})^{0.69} (\mu)^{-0.62}$

EEID (downflow): $k_L \underline{a} = 1.84 \times 10^{-2} (N)^{0.12} \times (\text{vvm})^{0.585} (\mu)^{-0.521}$

EEIU (upflow): $k_L \underline{a} = 2.47 \times 10^{-3} (N)^{1.7} \times (\text{vvm})^{0.618} (\mu)^{-0.716}$

Non-Newtonian liquids:

DT: $k_L \underline{a} = 43.3 \times 10^{-2} (N)^{1.43} (\text{vvm})^{0.49} (k)^{0.13} (n)^{1.2}$

EEID (downflow): $k_L \underline{a} = 5.02 \times 10^{-3} (N)^{1.52} (\text{vvm})^{0.41} (k)^{-0.53} (n)^{-0.68}$

EEIU (upflow): $k_L \underline{a} = 18.4 \times 10^{-2} (N)^{1.52} (\text{vvm})^{0.46} (k)^{0.12} (n)^{1.33}$

Bustamante et al. (2013)

Two DT spaced at a distance of 1 impeller diameter; $\left(\frac{C}{T}\right) = 0.2$; $\left(\frac{D}{T}\right) \cong 0.45$. Two Elephant Ear impellers (EEI) in the downflow (EEID) and upflow (EEIU) spaced at a distance of 1 impeller diameter; $\left(\frac{D}{T}\right) \cong 0.47$; $\left(\frac{C}{T}\right) = 0.2$. $T = 0.17$ m.

Liquid phase used: aqueous glycerol (Newtonian) and xanthan gum (non-Newtonian). Dynamic step pressure method—oxygen as the solute

^{a} Before large cavity regime.
^{b} After large cavity regime.

Further, $vvm \times T = V_G$. Therefore, Equation 7A.17 can be converted to the form as given in Equation 7A.18:

$$k_L \underline{a} = 3.35 \times \left(\frac{N}{N_{CD}} \right)^{1.464} (V_G) \qquad (7A.18)$$

Equation 7A.18 indicates that for correlating gas holdup, vvm is a better option but the same does not hold for $k_L \underline{a}$. This observation is similar to that of Bujalski et al. (1990). Equation 7A.18 could correlate most of the literature data satisfactorily except for few data points of Smith et al. (1977), which showed ~ ± 20% deviation. It must be stressed here that for many data sets, impeller speeds were not explicitly given, and Yawalkar et al. (2002b) had to resort to estimation of the corresponding values of N for use in the regression to obtain the empirical constants of Equation 7A.17. This was achieved by using Hughmark's (1980) following correlation for the gassed power number, P_G:

$$\left(\frac{P_G}{P} \right) = 0.1 \times \left(\frac{Q_G}{NV_L} \right)^{-0.25} \left(\frac{N^2 D^4}{g W_{blade} V_L^{0.66}} \right)^{-0.2} \qquad (7A.19)$$

There is an inherent inaccuracy in Equation 7A.19, which has a standard deviation of 0.117. Considering these inaccuracies, Equation 7A.18 seems to be the best alternative available to a process designer for getting a relatively reliable estimate of $k_L \underline{a}$ in the least time and efforts.

The most significant aspect of the above N/N_{CD} based correlations is that they do not have any scaling factor such as D, T, D/T, C/T, etc. Further, they also do not refer to a specific impeller type. It has been mentioned earlier that all these factors are incorporated in a single parameter, N_{CD}. Therefore, as long as a reliable estimate of N_{CD} is available, Equations 7A.17 and 7A.18 present the simplest and relatively reliable procedure for predicting $k_L \underline{a}$ in such a complex system as a stirred tank reactor. Correlations for predicting N_{CD} for the commonly used impellers are presented in Section 7A.5.4. For the relatively new types of impellers such as used by Collignon et al. (2010), correlations for N_{CD} are not available. However, in such cases, N_{CD} can be experimentally measured with the simple visual observation technique used by Yawalkar et al. (2002a). These measurements should be made by employing a minimum vessel diameter of 0.5 m. It is known that in smaller diameter vessels, surface aeration could play a significant role particularly at relatively higher speeds of agitation. Therefore, this minimum vessel diameter is necessary to prevent masking of the measured N_{CD} by surface aeration. Kapic and Heindel (2006) also used the concept of relative dispersion proposed by Yawalkar et al. (2002b) and gave a modified correlation for $k_L \underline{a}$:

$$k_L \underline{a} = 1.59 \times \left(\frac{N}{N_{CD}} \right)^{1.34} \left(\frac{T}{D} \right)^{0.41} \times V_G^{0.93} \qquad (7A.20)$$

The exponent on V_G in Kapic and Heindel's above correlation (Equation 7A.20) is not significantly different than that in Yawalkar et al.'s (2002b) correlation. As explained earlier, the latter correlation has an inbuilt T/D effect through N_{CD}, and it can be shown that the inclusion of T/D in Kapic and Heindel's correlation is super-fluous. Indeed, this inclusion causes changes only in the empirical constant and exponent on N/N_{CD} without any significant improvement in the accuracy of predicted $k_L a$. Yawalkar et al.'s (2002b) correlation is simpler than that of Kapic and Heindel and is based on data covering a very wide range of system and operating parameters. Lakkonen et al. (2007) used computational flow dynamics to study the mass transfer process in stirred vessels particularly with the view of assessing the importance of nonhomogeneous mixing in a stirred vessel. Their results point out that there is a significant variation in $k_L a$ at different locations (inhomogeneity in mass transfer). These investigators have attributed the inhomogeneity to (i) spatial variation of bubble size/interfacial area; (ii) spatial variation of gas-phase concentration, which implies that the "completely mixed" model for evaluation of $k_L a$ is not valid; and (iii) spatial variation of energy dissipation. The spatial variations in (i) and (iii) are well known, but the extent of gas-phase mixing may differ. Lakkonen et al. (2007), however, concluded that $k_L a$ correlations based on the relative dispersion parameter as proposed by Yawalkar et al. (2002b) and later extended by Kapic and Heindel (2006) include the effects of nonideal mixing.

7A.7 SOLID–LIQUID MASS TRANSFER COEFFICIENT IN STIRRED TANK REACTOR

As depicted in Figure 2.2, the dissolved gaseous reactant has to overcome the resis-tance offered by the liquid–solid film before it reaches the solid catalyst surface where the reaction between the adsorbed reactive species takes place. The intrinsic capacity of a catalyst is realized when all mass transfer processes are at equilibrium. Therefore, it is required to know the rate of the solid–liquid mass transfer step. Such an estimate should reveal the relative importance of this step and also establish the controlling step in an overall process. In a multiphase system, the mass transfer bet-ween the liquid and particulate phases is considered to be good when there is an intimate mixing between the two phases. In the case of solid–liquid mass transfer, the minimum desirable condition is suspension of the solid in the liquid or in the gas–liquid dispersion as the case may be.

7A.7.1 Solid Suspension in Stirred Tank Reactor

Most catalysts used in industrial processes have a particle density higher than the liquid comprising the reactant. As a result in the absence of agitation, the solid phase forms a settled layer at the bottom of the reactor. The process of solid suspension can be visualized as follows (Nienow 1968): (i) with the commencement of agitation, a mean flow and its associated turbulence manifest; (ii) as the agitation speed is increased, the strength of the mean flow and the level of turbulence increases. This continues till the mean flow and the turbulent eddies reach the settled solids layer. At this point, sudden bursts of eddies that cause some motion in the solid layer can be visually observed; (iii) further increase in the speed results in more activity/movement of the solids on the base of the vessel; (iv) this process continues with increasing speed of agitation, and finally, at a certain speed, N_S, all the solid particles are in motion although some return to the bottom for a very short period. This state is referred to as the "just-suspended" state. (v) Beyond N_S, the spatial variation (particularly, along the reactor height) of solid concentration decreases. At relatively very high speeds, these differences vanish and a uniform catalyst concentration in both radial and axial directions can be obtained. This relatively very high speed is denoted as N_{CS}, the speed at which the solids are "completely" suspended. Different criteria have been used in the literature to experimentally record the "just-suspended" state. These are (i) the 1–2 s criterion of Zwietering (1958) and (ii) sampling and measurement of solid concentration near the base of the vessel (Chapman et al. 1983a). When the solid concentration is plotted against the speed of agitation, there is steady increase in the solid concentration till a point of inflection appears in this plot (Fig. 7A.12). It has been observed that this point of inflection corresponds to N_S (iii) visual method based on measurement of the height and radius of the layer of unsuspended particles at the base of the vessel (Rieger and Ditl 1994) and (iv) more objective instrumental methods: conductivity, turbidimetry/optical density (Sessiecq et al. 1999; Fajner et al. 2008), wall pressure fluctuation method (Sardeshpande et al. 2009), and acoustic emission method (Congjing et al. 2008). The visual method/criterion of Zwietering, however, has been the most widely used due to its simplicity and reproducibility. It is worth noting that the acoustic emission method of Congjing et al. (2008) that is one of the recently developed objective methods yielded results in close agreement with those predicted by the Zwietering correlation. Sardeshpande et al. (2009) compared values of N_S predicted by Zwietering's and other correlations.

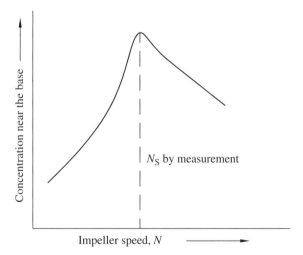

FIGURE 7A.12 Determination of N_S by measurement of solid concentration at one-fifth height above the vessel base. (Reproduced from Chapman et al. 1983a with permission from Elsevier. © Institution of Chemical Engineers 2005 all rights reserved.)

The values of N_S obtained by various instrumental techniques were also compared and found to differ only by 11%. Nienow (2000) has also indicated that a single observer can get results that are reproducible within ±5%. Therefore, despite its subjective nature, Zwietering's criterion and the results/correlations based on the same continue to be widely used for predicting the just-suspended condition.

Based on the physical description of the solid suspension process provided at the beginning of Section 7A.7.1, three basic approaches for modeling solid suspension have been used in the literature (Nienow 2000): (i) models based on drag on the solid and the velocity required to create the necessary upward thrust higher than the particle settling velocity, (ii) approach based on Kolmogorov's theory in which the energy and frequency of the eddies (the "bursts" mentioned previously) decide the suspension, and (iii) analogy based on fluidized beds/sedimentation (reverse of suspension). In the recent past, computational fluid dynamics approach has been advocated (Montante et al. 2008; Shan et al. 2008; Panneerselvam et al. 2009). Most models yield similar functional dependence of N_S on the different variables. However, there are significant differences in the scale-up rules proposed in different studies (Miller 1971; Kraume and Zehner 2002; Montante et al. 2008).

Categories of Suspension   Nienow (2000) and Shaw (1992) have suggested the following three categories depending on the degree of suspension:

1. **Particles in motion but not suspended:** In this case, the particles are in contact with the base of the vessel but are moving. The movement is mainly restricted to the plane of the base of the vessel.
2. **Just off-bottom suspension or "just-suspended" condition:** The particles moving along the base in category 1 are now lifted, and the criterion that is to be applied is that no particle remains on the vessel bottom for more than 1–2 s (Zwietering 1958).
3. **Complete or uniform suspension:** The particle concentration is fairly uniform, except in the immediate vicinity of the top of the liquid surface. This last category of solid suspension requires relatively very high power inputs.

Guha et al. (2007) used noninvasive computer-automated radioactive particle tracking (CARPT) technique to determine the Lagrangian description of the solids flow field for particles heavier than the liquid. The system employed was an agitated vessel equipped with a standard Rushton turbine. The measurements yielded the mean stopover time of the particles at various locations. Of particular interest is the mean stopover time at the base of the vessel. The results showed that the mean stopover time at the base was less than the "one second" criterion of Zwietering (1958). These results suggest that Zwietering's correlation yields a higher value of N_S and therefore is safe for process design. Guha et al. found that the actual N_S value is around 80% of that predicted by Zwietering's criterion/correlation.

Another representation of solid suspension is through the "cloud height" (Kraume 1992; Hicks et al. 1997; Bujalski et al. 1999; Bittorf and Kresta 2003; Takenaka et al.

2005; Sardeshpande et al. 2009; Hosseini et al. 2010). With increase in the speed of agitation although most particles are suspended, a distinct interface that separates a suspension of solid and a clear liquid is observed. The height to which the solids get suspended is termed "cloud height." This technique gives a better idea of the degree of axial homogeneity with respect to the solid concentration.

Shaw (1992) has shown that the average power inputs required in categories 1 and 2 are 10 and 32% of that for uniform suspension (category 3). Therefore, for most common applications, category 2 is the designer's choice since it achieves the minimum desirable result at much lower power inputs. A better estimate of the relative powers in categories 2 and 3 can be obtained from the measurements of Sessiecq et al. (1999). They measured the axial component of the instantaneous fluc-tuating liquid velocity using hot-wire anemometry and axial solid concentration pro-files using *in situ* turbidimetry for a four-blade PTD. Agitation speed for "just" suspension (Zwietering's criterion) and complete suspension were obtained. The authors comment that "complete suspension requires stirring rates slightly larger than the values predicted by Zwietering's equation." However, the corresponding power required is proportional to $(N)^3$. Thus, the power inputs calculated from the respective speeds of agitation suggest that "complete suspension" requires approxi-mately 2.25 times higher power than that for "just suspension." These results also support operation at "just-suspended" condition as a cost-effective alternative. In view of this, in the discussion in this chapter, as well as in Chapter 7B, solid suspension will have the connotation of N_S (critical speed for the "just-suspended" state) rather than N_{CS}. Nienow (2000) and Shaw (1992) have presented excellent discussion on various aspects of solid suspension in both two-phase (solid–liquid) and three-phase (gas–liquid–solid) systems. In addition, there is a large body of literature available on solid suspension for both types of systems. These include Rao et al. (1988), Rewatkar et al. (1989), Dutta and Pangarkar (1995, 1996), Pangarkar et al. (2002), Sharma and Shaikh (2003), Sardeshpande et al. (2009), and Ayranci et al. (2011, 2012). Jafari et al. (2012b) have presented an exhaustive survey of the literature and analysis of the results in the same.

The hydrodynamics of solid suspension in two- and three-phase systems are sig-nificantly different. In view of this, these two systems are discussed separately in Sections 7A.7.1.1 and 7A.7.1.3.

7A.7.1.1   Solid–Liquid (Two-Phase Systems)   The flow fields developed by var-ious types of impellers are different inasmuch as the flow patterns of their discharge are different. Figure 7A.13 shows the flow patterns developed by three widely used impellers for a standard configuration (McCabe et al. 1986) represented by $D/T = C/T = 1/3$.

The last location of solid suspension from the base varies depending on the flow pattern. In the case of radial flow, solid particles are swept from the corner toward the center of the base. Therefore, the "just-suspended" condition represented by the 1–2 s criterion of Zwietering (1958) should now pertain to particles at the center of the base rather than at other locations. In the case of axial flow impellers [for $C/T < 0.35$], the situation is reversed. Final suspension occurs from the corners at the base of the

(a)   (b)   (c)

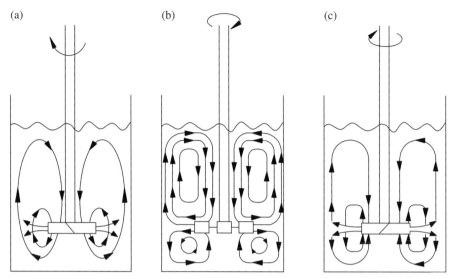

FIGURE 7A.13 Flow patterns developed by various impellers. (a) Pitched blade turbine—downflow (PTD) (Aubin et al. 2001). (b) Disc turbine—radial flow. (c) Pitched blade turbine—upflow (PTU). (Adapted from Aubin et al. 2001 with permission from Elsevier. ©Institution of Chemical Engineers 2001.)

vessel. This difference in the location of final suspension is probably a reason for the divergence of various correlations (Bowen 1989; Sharma and Shaikh 2003) particularly when the "just-suspended" condition is determined by visual observation. The flow pattern of a given type of impeller can be affected by two geometric parameters, D/T and C/T ratios. These effects are noticed for both axial/mixed and radial flow impellers and are discussed with respect to the "just-suspended" condition. Chapman et al. (1983a) have found that although PTU affords higher N_S in solid–liquid system, homogeneity in axial solid-phase concentration is better with PTU than PTD. This must be attributed to the upward lift caused by the flow pattern created by PTU. Once the particles are suspended, the flow pattern propels them such that they come under the influence of the upward impeller discharge. As a result, the overall "cloud height" for PTU is higher than that for PTD.

Effect of C/T

AXIAL/MIXED FLOW IMPELLERS   For the more common application using solid particles denser than the liquid, under conditions such that the actual speed of agitation $N < N_S$, the particles settle on the bottom of the vessel. Hence, an impeller that discharges a stream directed toward the bottom of the stirred tank reactor (axial flow) is more effective. Sharma and Shaikh (2003) in their study of suspension of glass beads by PTDs have found significant and peculiar effect of C/T on N_S. With respect to C/T, these investigators observed three different regions of behavior: (1) in region 1, $C/T < 0.1$; N_S was independent of C/T. (2) In region 2, $0.1 < C/T < 0.35$; N_S starts increasing with C/T. And (3) in region 3, $C/T > 0.35$; N_S becomes a relatively strong function of C/T. According to Sharma and Shaikh, in region 1, the small clearance from the base delivers an axial discharge to the vessel bottom without much loss of momentum. This momentum is likely to be continued as the axial flow spreads along the base to reach the corners from where the last portion of the solid particles is suspended. Sharma and Shaikh analyzed Rao et al.'s (1988) data and showed that region 1 exists in the latter's results also. In region 2, with increasing clearance, the loss of momentum is appreciable. The rapid increase in N_S observed in region 3 was attributed by the authors to a change in the flow pattern. Several studies, starting with Gray (1987), Jaworski et al. (1991, 2001), Bakker et al. (1994), and Myers et al. (2013), observed a flow reversal in the case of a PTD around $C/T \sim 0.35$. Below this clearance, the flow is mainly axial and directed to the base of the tank. Above $C/T \sim 0.35$, the impeller discharge is directed toward the wall, which then returns to the impeller at or above the impeller height. The predominantly downward axial flow is now transformed into radial flow typical of an impeller such as a DT. In addition to this primary loop, a weak secondary loop exists in the region near the base. This loop is directed inward from the wall and returns to the impeller near the axis of the tank with an upward momentum. The particles are forced to the center of the base of the vessel by the secondary loop. These form a heap that is trapped in a whirling eddy below the impeller. The movement becomes erratic, and the Zwietering 1–2 s criterion cannot rightly describe the "just-suspended" condition. The weak secondary loop is responsible for particle suspension, and it must attain sufficient strength to lift the particles. This requires additional power, and therefore, N_S rises sharply. Sharma and Shaikh used this physical picture to rationalize the different dependences of N_S on D/T and C/T observed in the literature. Armenante and Nagamine (1998) have studied the dependence of N_S on C/T for various impellers. An exponential relationship of the type $N_S \propto e^{(\alpha(C/T))}$ was found to best represent the dependence. Empirical constant α was found to be a weak function of D/T. The values of α were less than 1 for PTD and HE-3 impellers. This indicates that the dependence of N_S on C/T for the mixed flow (PTD) and axial flow (HE-3) is weak. The data collected were in the range of $0.06 < C/T < 0.26$. This overlapped region 1 [$C/T < 0.1$; N_S independent of C/T] and region 2 [$0.1 < C/T < 0.35$; N_S weak function of C/T] were described by Sharma and Shaikh (2003). Therefore, major part of the weak dependence of N_S on C/T observed by Armenante and Nagamine must be ascribed to the fact that data for the two ranges are combined to derive a single correlation.

As explained earlier, a downflow impeller is not efficient for solid suspension for $C/T > 0.35$. Indeed, for this sole reason, Sharma and Shaikh (2003) did not consider region 3 worthy of further studies.

RADIAL FLOW IMPELLERS   Rushton turbine (DT) is the most widely used in this category. Therefore, the following discussion is limited to its performance. As shown in Figure 7A.13, for the standard [$C/T = D/T = 1/3$] configuration, DT discharges a liquid stream directed radially to the vessel wall. This stream divides into two portions at the vessel wall—one portion directed down toward the base, while the rest travels upward. The lower flow loop has to travel considerable distance comprising three sections: (1) outward along the vessel radius from the impeller tip, (2) downward at the wall, and

(3) radially inward from the wall to the center of the base of the vessel. This involves two changes in direction. The turbulence required to lift the particles from the bottom gradually dissipates as the stream flows along the flow path described earlier. Therefore, the strength of the flow and associated turbulence is significantly low when it reaches the center of the base from where the last particles are suspended. Thus, under comparable power inputs, the radial flow impeller exhibits relatively poor solid suspension. Alternatively, for solid–liquid systems, radial flow impellers require greater power for solid suspension than the downward-directed axial (propeller) or mixed flow impellers (PTD). Several investigators have studied the effect of C/T on the flow pattern generated by a DT (Yianneskis et al. 1987; Rutherford et al. 1996; Montante et al. 1999, 2001; Jafari et al. 2012a, b, c). Montante et al.'s (2001) LDA measurements showed that with decreasing C/T the two-loop "8"-shaped flow changes to a single "8" loop, which is a salient feature of axial flow impellers. Therefore, in this region, radial flow impellers should also afford facile solid suspension. The transition in the flow pattern takes place at C/T between 0.15 and 0.2. Armenante and Nagamine (1998) studied the effect of C/T on N_S in the single "8" region. Similar to axial and mixed flow impellers, the correlation employed was $N_S \propto e^{(\alpha(C/T))}$. The values of α were ~2 for both DT and FBT. These relatively high values of α indicate that radial flow impellers are more affected by variations in C/T as compared to the axial/mixed flow impellers. For $C/T>0.2$, the double "8" pattern exists. Chapman et al. (1983) measured N_S for $C/T>0.2$ and found that increasing the clearance results in higher values of N_S but no explicit relationship was reported. These authors suggest that large DT impellers located at relatively low clearance yield solid suspension at relatively low power input. The empirical constant α was found to be a weak function of D/T.

Effect of D/T   The effect of D/T on N_S cannot be described by a single parameter in the same manner as the effect of C/T. This is because of the interaction between the two terms involving the impeller diameter in the correlations for N_S (Table 7A.3). These are D alone and as $(D/T)^n$. The general trend reported is (i) at constant T, increase in D causes decrease in N_S and (ii) at constant D, increase in T results in higher N_S. In the absence of an explicit relationship between N_S and D/T, information on the flow pattern at different D/T available in the literature can give some valuable inputs. This information is important inasmuch as the flow pattern has a strong bearing on N_S as discussed earlier. In general, $D/T>0.33$ is not used since at the same power input, a higher D/T implies lower agitation speeds and higher torque. The gearbox (speed reducer) required is correspondingly expensive (Shaw 1992).

AXIAL/MIXED FLOW IMPELLERS   Flow patterns of propeller (axial flow) and pitched blade turbines (mixed flow) have been widely studied in the literature. For these impellers, at low D/T values (~<0.25), the radial component is much less as compared to the axial component. For larger $[D/T~0.4]$ values, there is a substantial radial component for both axial flow (propeller; Zwietering 1958) and mixed flow (Nienow and Bujalski 2004) impellers. Ayranci et al. (2012) also found a distinct difference in the behavior of Lightnin A310 impellers of $D/T=0.33$ and 0.5. For the $D/T=0.33$ impeller, it was argued that turbulence plays an important role in solid

TABLE 7A.3 Summary of Important Literature Studies on Solid Suspension in Two-phase (solid–liquid) Systems

Reference	Impeller, system	Correlation
Zwietering (1958)	Paddles, $D/W=2$; paddles, $D/W=4$; flat blade turbines; vaned discs and propellers. Sand, NaCl–water/ acetone/CCl_4/K_2CO_3/oil	$N_S = S \times \dfrac{v^{0.1} d_P^{0.2} \left(\dfrac{g(\rho_S - \rho_L)}{\rho_L} \right)^{0.45} X^{0.13}}{D^{0.85}}$
Rao et al. (1988)	6-PTD. Quartz spheres–water ($\rho_S = 2520$ kg/m³)	$N_S = 3.3 \times \left(\dfrac{v^{0.1} d_P^{0.11} \left(\dfrac{g(\rho_S - \rho_L)}{\rho_L} \right)^{0.45} X^{0.1} T^{0.31}}{D^{1.16}} \right)$
Rieger and Ditl (1994)	6-PTD. Glass ballotini–water/aqueous glycerol	$\left(\dfrac{N^2 D \rho_L}{(\rho_S - \rho_L) \times g} \right) = 0.855^a; \ \dfrac{N^2 D \rho_L}{(\rho_S - \rho_L) \times g} = 16.4 \left(\dfrac{d_P}{D} \right)^{0.6}$
Ibrahim and Nienow (1996)	6-DT/PTU/PTD; Lightin A-310/Chemineer HE-3 hydrofoils; Ekato Intermig (one and two). Blue glass ballotini ($\rho_S = 2950$ kg/m³), bronze shots ($\rho_S = 8450$ kg/m³)–water	Values of "S" in Zwietering's (1958) correlation are given
Aravinth et al. (1996)	6 FBT—various D/T and C/T. Sand–water	6 FBT: $S = 0.135 \left(\dfrac{T}{D} \right)^{2.6} \left(\dfrac{C}{T} \right)^{0.15} \times \left[\dfrac{v^{0.1} d_P^{0.2} \left(\dfrac{g\Delta\rho}{\rho_L} \right)^{0.45} X^{0.13}}{D^{0.85}} \right]$

Armenante and Nagamine (1998)

6-DT ($0.217 < D/T < 0.348$); 6 FBT ($0.217 < D/T < 0.348$); 6-PTD ($0.217 < D/T < 0.348$); Chemineer HE-3 ($0.304 < D/T < 0.391$); $0.041 < C/T < 0.25$. Glass beads: 60–300 μm ($\rho_s = 2500$ kg/m³), ion exchange resins ($\rho_s = 1375$ kg/m³)—water and aqueous sugar solutions. Solid loading: 0.5–1.5 wt.%

6-DT: $N_s = \left[0.99 \times \left(\dfrac{T}{D} \right)^{1.4} \times \exp\left(2.18 \dfrac{C}{T} \right) \right] \times \left[\dfrac{v^{0.1} d_P{}^{0.2} \left(\dfrac{g\Delta\rho}{\rho_L} \right)^{0.45} X^{0.13}}{D^{0.85}} \right]$

6 FBT: $N_s = \left(1.43 \times \left(\dfrac{T}{D} \right)^{1.2} \times \exp\left(1.95 \dfrac{C}{T} \right) \right) \times \left[\dfrac{v^{0.1} d_P{}^{0.2} \left(\dfrac{g\Delta\rho}{\rho_L} \right)^{0.45} X^{0.13}}{D^{0.85}} \right]$

6-PTD: $N_s = \left(2.28 \times \left(\dfrac{T}{D} \right)^{0.83} \times \exp\left(0.65 \dfrac{C}{T} \right) \right) \times \left[\dfrac{v^{0.1} d_P{}^{0.2} \left(\dfrac{g\Delta\rho}{\rho_L} \right)^{0.45} X^{0.13}}{D^{0.85}} \right]$

HE-3: $N_s = \left(3.49 \times \left(\dfrac{T}{D} \right)^{0.79} \times \exp\left(0.66 \dfrac{C}{T} \right) \right) \times \left[\dfrac{v^{0.1} d_P{}^{0.2} \left(\dfrac{g\Delta\rho}{\rho_L} \right)^{0.45} X^{0.13}}{D^{0.85}} \right]$

(Continued)

TABLE 7A.3 (Continued)

Reference	Impeller, system	Correlation
Wu et al. (2001a, b)	2 to 6-DT, pitched blade turbines with varying angle of pitch (20°–45°), curved blade DT; Lightnin A310; D/T: 0.41. Glass ballotini–tap water	$\left(\dfrac{N_s \times Q}{ND^3}\right) = $ Constant $\left(\dfrac{Q}{ND^3} \times S\right) = K$; K is independent of density difference. Impeller type, $\left(\dfrac{C}{T}\right), \left(\dfrac{D}{T}\right)$ incorporated in S from Zwietering's (1958) results
Sharma and Shaikh (2003)	6-PTD; $0.025 < C/T < 0.35$; $D/T < 0.35$. Glass beads: 130–850 µm ($\rho_S = 2400$–2600 kg/m³)	$N_S = 1.16 \left[e^{(0.65C/T)} \times \left[\dfrac{v^{0.1} d_P^{0.2} \left(\dfrac{g\Delta\rho}{\rho_L}\right)^{0.45} X^{0.13} \times \left(\dfrac{D}{T}\right)^{-1.15}}{D^{-0.85}} \right] \right]$
Ibrahim and Nienow (2004)	6-PTD; single Intermig, HE-3.Cytodex beads ($\rho_S = 1040$ kg/m³)–phosphate buffer solution	Values of "S" in Zwietering's (1958) correlation are given and compared with high density particles
Jafari et al. (2012a)	4 PTD, 6-DT, 6 CBT; D/T = 0.33 Water–sand ($\rho_S = 2650$ kg/m³)	$N_S = \left(a + b\left(\dfrac{C}{T}\right)\right)^{\alpha} \left(\dfrac{\mu_L}{\rho_L}\right) \left(\dfrac{g(\rho_S - \rho_L)}{\rho_L}\right)^{\beta} (d_P)^{\gamma} (D)^{\delta} (X)^{\varphi}$: $\alpha = 0.1$, $\beta = 0.45,\ \gamma = 0.2,\ \delta = -0.85$ 6-DT: $0.1 < (C/T)$: $a = 4.7$; $b = 1.1$; $\phi = 0.12$. 4-PTD: $0.1 < \left(\dfrac{C}{T}\right) < 0.35$, $a = 3.47$; $b = 1.35$; $\phi = 0.22$; 6CBT: $0.1 < \left(\dfrac{C}{T}\right)$: $a = 5.4$; $b = 0.98$; $\phi = 0.12$

suspension, whereas as for $D/T=0.5$, both mean flow and turbulence contribute to solid suspension. For the larger D/T impeller, the presence of a significant radial component was presumed to result in a higher power requirement for solid suspension. Sharma and Shaikh (2003) have analyzed the literature data on the effect of D/T on N_S for $C/T<0.35$. They found that an ln (N_S) versus ln (D/T) plot is linear up to $D/T=0.35$. At higher values, a constant value of N_S is indicated. Assuming that T is held constant, this implies that although power input increases significantly with increase in D $(P_O \propto D^5)$, there is no concomitant decrease in N_S. This inefficiency in utilization of power input can again be attributed to the presence of a significant radial component. It can therefore be concluded that to obtain energy-efficient suspension, the value of D/T should not exceed 0.35.

RADIAL FLOW IMPELLERS   As discussed earlier, the standard Rushton turbine (DT) exhibits different flow patterns for $C/T<0.2$ and $C/T>0.3$. The transition in the flow pattern from double "8" to single "8" or *vice versa* is affected by D/T ratio. With increasing D/T, the critical value of C/T at which the flow pattern changes to single "8" is reduced. The effect of D/T needs to be considered separately for the two flow patterns:

1. $C/T<0.2$: The flow pattern is axial rather than radial. The final point of solid suspension moves toward the corner of the base. Further, as D/T increases, the flow strength $(\propto ND^3)$ increases. The overall result is a better utilization of the energy input for solid suspension, and N_S for DT decreases with increasing D/T.

2. $C/T>0.3$: In this case, the flow pattern is the classical two "8"s. The lower "8" is responsible for solid suspension. The location of final solid suspension shifts to the center of the base. Increasing D/T at constant T reduces distance from the tip of the impeller to the center of the base from where the final solid suspension occurs. Additionally, the flow strength increases. Thus, similar to $C/T<0.2$, in this case also, there is a decrease in N_S with increasing D/T. In the absence of specific data for this region, a qualitative inference based on the ineffectiveness of radial flow pattern indicates that increase in D/T (or power input) is unlikely to yield proportionate decrease in N_S.

Considering the earlier discussion on D/T and C/T effects, it is not surprising that a standard text in chemical engineering (McCabe et al. 1986) suggests the configuration shown in Figure 7A.1. Fortunately, most stirred vessels are designed such that neither C/T nor D/T is >0.33, and therefore, most correlations for N_S obtained disregarding the changes in regions (1, 2, and 3 as defined by Sharma and Shaikh 2003) do not have a serious impact on the prediction.

Rewatkar et al. (1989) have reviewed the literature on solid suspension in two-phase systems. Subsequently, a large body of information has been generated, which rationalizes the earlier data as discussed earlier. It should be noted here that most investigators (e.g., Rewatkar et al. 1989; Nikhade and Pangarkar 2005) used vessels with flat bottoms. In such vessels, fillets of solids tend to form at the vessel wall base corners. For larger vessels with dish end bottoms, these fillets are found predominantly below the impeller or halfway between the vessel axis and periphery.

The N_S values for the flat bottom vessels are typically 10–20% higher than those for vessels with dish ends. Thus, N_S predicted from correlations of data in flat bottom vessels are conservative (Jafari 2010).

Effect of Solid Loading Most investigators have reported agreement with Zwietering's correlation in terms of the effect of solid loading on N_S. Myers et al. (2013) recently investigated this effect over a relatively wide range as compared to previous studies. The range covered varied from practically no solids ($X \to 0$) to as high as 67 wt.%. It should be noted that the power law-type correlation of Zwietering implies that for $X=0$, $N_S=0$. Myers et al. showed that N_S approaches a finite, nonzero value for relatively very low solid loading. They found that for very low loading such as in terms of number of particles (rather than wt.% or $X \to 0$) $= 25$, N_S remains finite. These investigators used downflow impellers (HE-3 and PTD). They found that the following ratio $\dfrac{N_S(X \to 0)}{N_S(X \to 1)}$ is $=0.72$ and 0.73 for HE-3 and PTD, respectively, with a very low statistical average error. A correlation covering the entire range from $X=0$ to 67 wt.% yielded $N_S \propto (X)^{0.17}$. However, this single correlation gave significant negative deviations for $X < 20$ wt.% and positive deviations for the higher range ($X > 35$ wt.%). In view of the inability to derive a single correlation, Myers et al. used three separate loading ranges: (1) $0.25 \le X \le 5$, (2) $5 \le X \le 25$, and (3) $25 \le X \le 67$. Such a division gave much better fit of the data. Nevertheless, for the lighter resin particles [ρ_S: 1080–1180 kg/m^3], the exponent on X was much lower than the heavier sand and glass particles [ρ_S: 2440–2780 kg/m^3] for the low (i) and intermediate (ii) solid loading ranges. On the other hand, the same low density particles exhibited substantially higher exponents for the highest range of solid loadings. These authors attributed the differences as possibly due to different particle size, shape, and size distribution. Myers et al. used a modified method for determining the just-suspended condition. This method involved simultaneous visual comparison of a fully suspended reference tank and the experimental tank by two independent observers. Although this method was claimed to be an improvement over the Zwietering criterion used by most investigators, it still had elements of subjectivity that also could be responsible for the differences observed. For instance, the two observers may not agree on the precise degree of incomplete suspension. Additionally, the difference between N_S measured by Myers et al.'s method and Zwietering's data for $0 \le X \le 18$ is practically negligible. The exponents on X for Myers et al.'s and Zwietering's correlations (0.12 and 0.13, respectively) also confirm the reliability of Zwietering's time-tested "1–2 s" criterion. It should be noted that except for some special cases (such as oxidation of *p*-xylene to PTA or Fischer–Tropsch synthesis discussed in Section 3.4.2), the solid loading in most three-phase catalytic reactions is rarely higher than the maximum value in Zwietering's study. Nevertheless, for higher solid loadings, relying on observations of two persons along with a reference "just-suspended" state is an improvement.

Complete Suspension versus Just Suspension   It has been mentioned that for most situations, operation at the just-suspended condition is satisfactory. However, it must be pointed out that at N_S, substantial variations in solid-phase concentration along the

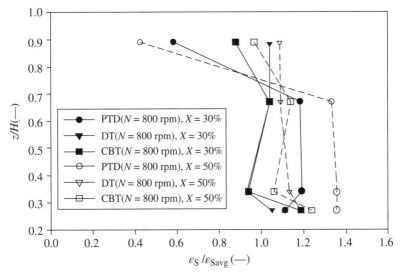

FIGURE 7A.14 Axial variation of solid concentration for various impellers with C/T as a parameter. (Figure 6 from Jafari et al. 2012c: Reproduced with permission from Elsevier. © 2011 The Institution of Chemical Engineers. Published by Elsevier B.V. All rights reserved.)

vessel axis have been observed (Buurman et al. 1986; Hicks et al. 1997; Jafari 2010). Figure 7A.14 from Jafari et al. (2012c) gives the axial solid concentration profiles for DT and PTD impellers at different clearances.

It is evident that whereas DT shows relative homogeneity, there is substantial variation in the normalized concentration along the vessel height particularly for PTD. In the case of PTD, as expected, increase in C/T results in a higher solid concentration near the base of the vessel. However, there is a marginal improvement in the homogeneity near the liquid surface. Figure 7A.14 indicates that DT at $C/T = 0.33$ gives the best results. However, assuming that all other conditions are exactly identical, the

relative power requirement is $\left(\dfrac{(N_P)_{DT}}{(N_P)_{PTD}} \right)$ or about 3.

A summary of the important studies pertinent to prediction of the "just-suspended" condition is given in Table 7A.3.

7A.7.1.2 Scale-Up Criterion for Solid Suspension This subject has received far more attention in the literature than the corresponding condition of gas dispersion. Different criteria have been employed by various investigators for predicting the scale-up rule for just solid suspension. The two main criteria employed are as follows:

1. Constant power input per unit volume or per unit mass, $\left(\dfrac{N_P N^3 D^5 \rho_L}{V_L} \right)$
2. Constant tip speed, (ND^n)

These criteria are interrelated and are qualitatively similar. They differ only in nomenclature/magnitude since for a given system (ρ_L in particular) and impeller (given N_p), the power input depends mainly on N and D.

For geometrically similar systems, Nienow (2000) has suggested Equation 7A.21:

$$\left(\varepsilon\right)_{NS} \propto \text{scale factor}^{b} \tag{7A.21}$$

Using the approach proposed by Shamlou and Koutsakos (1989) and $b=0.85$, Equation 7A.21 was modified as (Nienow 2000)

$$\left(\varepsilon\right)_{NS} \propto \text{scale factor}^{2-3b} \tag{7A.22}$$

According to Equation 7A.22, ε_{NS} decreases for values of $b>2/3$, remains constant for $b=2/3$, and increases when $b<2/3$. Nienow (2000) pointed out that the large variations in ε_{NS} on scale-up arise out of its extreme sensitivity to the experimentally determined value of b. Figure 16.6 of Nienow (2000) shows that the variation in scale-up rule can indeed be very dramatic.

Kraume and Zehner (2002) have given details of the work subsequent to Nienow's (2000) review of solid suspension in two-phase stirred tank reactors. This later survey has also shown a wide variance in the criterion proposed by different investigators. It also confirmed that the value of "n" in the constant power input criterion obtained by different investigators varies as widely as from -1 to 0.5. The differences in the various criteria were attributed to varying vessel–impeller geometries/combinations and most importantly to the different definitions of the suspension criterion itself. Kraume and Zehner (2002) developed a model that could explain the variation in the solid suspension criteria employed in the literature. Their model was based on a cloud height criterion that assumes that the solids are suspended when the slurry level reaches a height equal to 90% of the liquid level in the stirred vessel. This model always predicts a conservative estimate of the power required for suspension as against 30% lower predictions based on $n=-0.5$ from Zwietering's (1958) and Nienow's (1968) correlations. However, the procedure suggested by Kraume and Zehner (2002) requires experimental determination of the parameters for the geometry of interest, which includes the type of impeller, C/T, shape of the vessel bottom, etc. In the case of suspension of multisized particles, conservative design would call for use N_S for the largest particle diameter. Ayranci et al. (2011) have indicated that this rule is a safe practice when the smaller particle has $Ar<40$ and the mixture does not contain appreciable proportion of the second type of particles. For large solid loadings, these investigators have shown that N_S actually decreases because of particle–particle interactions. Fortunately, in the case of three-phase reactions, the solid catalyst loading is generally <10 wt.%. Further, the catalyst particles employed have the same density and also do not differ appreciably in size.

For a particular case with higher solid loadings, Rieger et al. (1988) applied the constant tip speed $[ND]^n$ criterion for scale-up. They proposed two criteria: $n=0.67$ for the just-suspended situation and $n=1$ for the "complete" or "homogeneous" suspension condition. Montante et al. (2008) studied solid suspension in multi-impeller stirred vessels. These investigators also used the constant tip speed approach. They argued that intermittent turbulent fluctuations stronger than the average fluctuations were responsible for maintaining the solid in suspension. Their analysis yielded a value of $n=0.93$.

Risking repetition, it should be pointed out that the "just-suspended" condition is the most cost-effective operating condition (Shaw 1992; Pangarkar et al. 2002). The design speed of agitation should be 10–20% higher than the estimated N_S value to account for minor deficiencies/deviations in the correlation used (Table 7A.3).

7A.7.1.3   Gas–Liquid–Solid (Three-Phase Systems)   When a gas is introduced in a stirred tank reactor, the buoyancy of the gas dampens the liquid flow directed toward the bottom. It has also been noted that the power drawn by the impeller decreases depending on the type of the impeller. For downflow impellers and Rushton turbine, the mean flow at the base is reduced due to the introduction of the gas. Visual observations indicate that for these impellers solid suspension is preceded by the impeller flow overcoming the buoyancy of the gas. This condition as noted earlier coincides with minimum speed for gas dispersion, N_{CD} (Fig. 7A.10). Subsequent to N_{CD}, when the bursts of the two-phase flow reach the base of the vessel, movement in the settled solid particles is noticed (Nikhade et al. 2005). As the speed of agitation increases, the energy and frequency of these bursts increase, and at N_{SG}, the just-suspended condition is achieved. For the majority of three-phase solid-catalyzed reactions, increasing the speed beyond N_{SG} has little justification since K_{SL} is almost independent of N as discussed in Section 6.4.1 and shown in Figure 6.3 (Nienow 1969). According to the physical picture described earlier for downflow and radial flow impellers, additional power is required to propel the gas in the zone below the impeller. For the upflow impellers, the picture is quite different. As mentioned earlier, the upflow impeller generates two loops below and above the level of the blades. The direction of impeller discharge and gas flow is same. There is no gas buoyancy that needs to be overcome. The two flows, indeed, cooperate in creating a higher net upward flow, which in turn generates a stronger recirculation since flow continuity cannot be violated. This stronger recirculating flow returns to the base to close the circulation loop (Figure 3 of Nienow and Bujalski 2004). Therefore, the mean flow near the base is relatively high as compared to PTD. This higher strength of mean flow facilitates suspension.

In all these cases, there is an increase in the critical speed for suspension of solids when a gas is introduced. Most investigators have accounted for this increase through an incremental speed of agitation, ΔN_{SG}. Thus, the minimum speed for just suspension, N_{SG}, is given by

$$N_{SG} = N_S + \Delta N_{SG} \qquad (7A.20)$$

Table 7A.4 gives a summary of the literature on solid suspension in three-phase stirred reactors.

Chapman et al. (1983c) found that in the case of the Rushton turbine, the value of critical impeller speed for solid suspension (N_{SG}) is higher than the impeller speed for complete dispersion of the gas phase (N_{CD}), except for some low density particles such

TABLE 7A.4 Correlations for N_{SG} Available in the Literature

Reference	Impeller/system used	Correlation
Chapman et al. (1983c)	6-DT. Deionized water–various solids ($1050 < \rho_P < 2900$ kg/m³)	$\Delta N_{SG} = 0.94 \times Q_G$
Wong et al. (1987)	Propeller, 4 PTD, 6-DT. Water–various solids ($2514 < \rho_P < 8642$ kg/m³)	$N_{SG} = \text{Constant} \times Q_G$
Rewatkar et al. (1991)	6-DT, 6-PTD, 6PTU. Water/aqueous i-propanol–spherical glass particles ($\rho_P = 2520$ kg/m³)	$\left(\dfrac{V_{Sxo}^{0.5} - a V_C}{N_{SG} \times D} \right) \left(\dfrac{T}{D} \right) X^{0.1} V_G^{0.14} = 0.18$ $a = 0.9 (V_{Sxo})^{1.27}$
Kawase et al. (1997)	6-DT. Water/aqueous CMC/xanthan gum–various solids ($1030 < \rho_P < 1430$ kg/m³)	$\Delta N_{SG} = 0.94 \times Q_G$ Correlation of Chapman et al. (1983c) extended to non-Newtonian fluids. No effect of non-Newtonian nature of the liquid on N_{SG}
Nienow and Bujalski (2002)	6-DT, 6-PTD, 6PTU, Scaba 6SRGT, Chemineer HE-3(D[a]), Lightnin A315(U[a]). Water–glass ballotini ($\rho_P = 2520$ kg/m³)	6-DT: $N_{SG} = N_S + 0.85 \times \text{vvm}$ 6-PTD: $N_{SG} = N_S (0.83 + 0.31 \times \text{vvm})$ 6PTU: $N_{SG} = N_S (1 + \text{vvm})^{0.11}$ Values of "S" in Zwietering's correlation for N_S
Zhu and Wu (2002)	6-DT. Tap water–glass ballotini ($\rho_P = 2520$ kg/m³)	$\left(\dfrac{N_{SG}}{N_S} \right) = 1 + 2.6 \left(\dfrac{Q_G}{N_S D^3} \right)_S^{0.7}$
Nikhade et al. (2005)	6-DT, 6-PTD, 6 PTU with helical coils; water–spherical glass particles ($\rho_P = 2520$ kg/m³)	6-DT: $\Delta N_{SG} = 601 (V_G)^{0.01} (d_P)^{0.59} (X)^{0.25} (S_{Coil})^{0.61}$ 6-PTD: $\Delta N_{SG} = 1119 (V_G)^{0.27} (d_P)^{0.55} (X)^{0.03} (S_{Coil})^{0.3}$ 6PTU: $\Delta N_{SG} = 13.86 (V_G)^{0.09} (d_P)^{0.11} (X)^{0.11} (S_{Coil})^{0.70}$

[a]D, Downflow; U, upflow.

as polystyrene. For the downflow pitched turbine, the relative values of N_{CD} and N_{SG} depended on D/T ratio as well as the gas flow rate (vvm). This impeller shows significant instability with the introduction of gas, and the instability increases with increase in gas flow rate (Fig. 7A.7). For the system with $D/T=0.5$, solid suspension was found to occur prior to gas dispersion ($N_{SG}<N_{CD}$). The effect of gas flow rate was somewhat different. For vvm≥ 0.25, due to the inability of this impeller to handle higher gas loads, gas dispersion becomes the major concern, and it was found that $N_{SG}\sim N_{CD}$. Thus, the downflow pitched turbine in three-phase systems needs to be used with prudence. However, as observed in Figure 7A.7, the upflow pitched turbine is relatively insensitive to the introduction of the gas and its flow rate. Therefore, the difference between N_{SG} and N_{S} (ΔN_{SG}) is relatively low. Bujalski et al. (1990) showed that even at a relatively high vvm value of 3.5, N_{SG} is only 15% higher than N_{S}. The correlations of Nienow and Bujalski (2002) presented in Table 7A.4 can be used to further substantiate the inferior performance of PTD in comparison to DT and PTU in terms of the extent of increase in N_S due to the introduction of the gas phase. Considering 0.1 vvm as a representative value of the gas flow rate, the increments (ΔN_{SG}) are as follows: PTU, 0.01; DT, 0.085; and PTD, 0.86 rev/s. Evidently, PTD exhibits the highest ΔN_{SG} under otherwise the same conditions. This is also clear because PTD also suffers from the highest reduction in power in the presence of gas (Fig. 7A.7).

The following conclusions can be drawn from the foregoing discussion: (i) DT requires the highest power for solid suspension in three-phase systems not so much as due to high N_{SG} but because of its significantly higher power number; (ii) both DT and PTD suffer from serious instabilities in the presence of gas. These instabilities increase with increase in the gas flow rate. (iii) The six-blade upflow pitched turbine yields most stable operation and requires relatively low power inputs as compared to other impellers. Therefore, for duties that require solid suspension in the presence of gas, the six-blade upflow pitched turbine should be preferred (Nienow and Bujalski 2004). This advantage is also discussed and applied in Chapter 7B.

7A.7.2 Correlations for Solid–Liquid Mass Transfer Coefficient

As mentioned earlier, Pangarkar et al. (2002) have critically examined the available information on correlations for predicting K_{SL} in stirred tank reactors. It was concluded in Section 6.4.2 that an approach based on the analogy between momentum and mass transfer can yield simple yet theoretically sound and reliable correlations for the mass transfer coefficient in several cases. Before this approach is applied to prediction of K_{SL} in stirred tank reactors, it is desirable to reconcile the available literature on the effects of various relevant parameters on K_{SL}. The conclusions derived by Pangarkar et al. (2002) on examination of data from their group as well as other literature data are as follows:

1. For $d_p<1100$ μm, $K_{SL}\propto (d_p)^{-0.5}$ at constant N. However, the literature data show a variation in the exponent on d_p ranging from -0.7 to 0.33.
2. K_{SL} varies approximately linearly with N at constant d_p. The literature data however, show a variation ranging from $K_{SL}\propto (N)^{0.27}$ to $(N)^{1.28}$.

3. K_{SL} increases with decreasing clearance of the impeller from the bottom at constant N in the range of C/T from 0.17 to 0.37.

4. K_{SL} increases with increasing diameter of the impeller. Literature data show a varying dependence of $K_{SL} \propto (D)^{0.17}$ to $(D)^{1.67}$.

5. $K_{SL} \propto (D_M)^{0.47}$ to $(D_M)^{0.55}$. Literature reports show wider variation: $K_{SL} \propto (D_M)^{0}$ to $(D_M)^{0.8}$.

6. Dependence of K_{SL} on vessel diameter varied from $K_{SL} \propto (T)^{-0.22}$ to $(T)^{0.74}$.

7. *Apparent* dependence of K_{SL} on viscosity of the liquid in the literature ranges from $K_{SL} \propto (\mu_L)^{-0.67}$ to $(\mu_L)^{0.12}$.

It is evident that the experimental techniques employed were sound and could not have been the cause of such widely varying dependences on key parameters. For example, the effects of D/T and C/T were significant only in the region A–B–C in Figure 6.3. Once point C, which represents N_S or N_{SG}, is passed, these effects vanished (Harriott 1962; Nienow 1969, 1975; Pangarkar et al. 2002). Evidently, the D/T and C/T effects were confined to the just-suspended condition. It can be similarly argued that the effects of d_p, X, and $\Delta\rho$ also relate to N_S and hence vanish once the just-suspended condition is achieved. The varying effects of physical properties $(D_M, \rho,$ and $\mu_L)$ listed earlier are also likely to be due to different hydrodynamic situations ranging from stagnant particles to particles suspended in a turbulent liquid (Keey and Glen 1964, 1966). The effect molecular diffusivity is simpler to analyze since this property is involved only in the Schmidt number. Under suspended conditions, the particles experience a highly turbulent flow field and a dependence closer to that predicted by the penetration model is justified. Viscosity and density of the liquid have complex effects on N_S through the kinematic viscosity and $\Delta\rho$ (which affect the settling velocity) besides the value of the Schmidt number. Therefore, the dependence reported in the literature data should be termed *apparent* dependence as listed earlier (item 7).

In their exhaustive review, Pangarkar et al. (2002) included the pertinent literature on K_{SL} in stirred tank reactors equipped with different types/sizes of impellers. This review, as well as other papers from the same group (Jadhav and Pangarkar 1991; Kushalkar and Pangarkar 1994, 1995; Dutta and Pangarkar 1994, 1995a, b, 1996), has clearly brought out the importance of solid suspension in stirred tank reactors employing a solid phase as a catalyst or reactant. These aforementioned studies focused on the "relative suspension parameter" and obtained scale-/geometry-independent correlations for K_{SL}. This approach was formally hypothesized by Nikhade and Pangarkar (2007) in the form of a theorem of corresponding hydrodynamic states, which states that "The particle-liquid mass transfer coefficient in a stirred reactor is the same for different geometrical and operating conditions as long as these conditions result in the same corresponding hydrodynamic state." The corresponding hydrodynamic state was defined in the form of the relative suspension parameter: N/N_S or N/N_{SG} as the case may be.

Equations 7A.23, 7A.24, and 7A.25 were found to fit a very large number of data from the literature.

7A.7.2.1 Newtonian Liquids

1. Two-phase (solid–liquid) system: with single impeller (Jadhav and Pangarkar 1991):

$$K_{SL} = 1.72 \times 10^{-3} \left(\frac{N}{N_S} \right)^{1.16} \left(\frac{\mu_L}{\rho_L D_M} \right)^{-0.53} \tag{7A.23}$$

2. Three-phase gas–liquid–solid system with single impeller (Pangarkar et al. 2002):

$$K_{SL} = 1.8 \times 10^{-3} \left(\frac{N}{N_{SG}} \right) \left(\frac{\mu_L}{\rho_L D_M} \right)^{-0.53} \tag{7A.24}$$

3. Three-phase multi-impeller systems with six-blade DT and 45° PTD spaced at a distance equal to the vessel diameter (Dutta and Pangarkar 1996):

$$K_{SL} = 1.02 \times 10^{-3} \left(\frac{N}{N_S} \right)^{0.97} \left(\frac{\mu_L}{\rho_L D_M} \right)^{-0.45} \tag{7A.25}$$

In Equations 7A.23, 7A.24, and 7A.25, the dependence of K_{SL} on the molecular diffusivity varies in the relatively narrow range of $(D_M)^{0.47}$ to $(D_M)^{0.55}$, which can be averaged as $(D_M)^{0.5}$ as compared to the very wide range of dependence on N, $K_{SL} \propto (N)^{0.27}$ to $(N)^{1.28}$ revealed by the literature review of Pangarkar et al. (2002). The square root dependence of K_{SL} on D_M implies a surface renewal mechanism in a highly turbulent medium as encountered in industrial stirred reactors. The minor difference in dependence (0.47–0.55) can be attributed to variations caused by disparities, for example, in the values of viscosity and diffusivity employed or the estimation of the diffusivity. The fact that the difference in the dependence on D_M has narrowed down to a marginal level also supports the earlier observation that the wide divergence reported in the literature was mainly due to varying hydrodynamic conditions or degrees of suspension. There are no explicit dependences on D/T, C/T, and other geometric parameters as well as the type of impeller in Equations 7A.23, 7A.24, and 7A.25. All these are all included in N_S or N_{SG} depending on whether the correlation is for two- or three-phase system.

7A.7.2.2 Non-Newtonian (Power Law Liquids)

1. Single-impeller three-phase system (Kushalkar and Pangarkar 1995):

$$K_{SL} = 1.19 \times 10^{-3} \left(\frac{N}{N_{SG}} \right)^{1.15} \left(\frac{\mu_{Eff}}{\rho_L D_M} \right)^{-0.47} \tag{7A.26}$$

Equations 7A.23, 7A.24, 7A.25, and 7A.26 are remarkably similar in terms of the empirical constants involved although they are derived for very widely differing conditions both in terms of type of systems (two- or three-phase, single or multiple impellers, Newtonian or non-Newtonian liquids, etc.). This observation buttresses the theoretical approach outlined in Section 6.4.2. As regards the energy efficiency and operational stability, the PTU yields significantly better performance since it does not suffer from instabilities discussed in Section 7A.3 (Pangarkar et al. 2002). This will be evident from the design examples illustrated in Section 7A.10.

7A.8   DESIGN OF STIRRED TANK REACTORS WITH INTERNAL COOLING COILS

In many industrial situations, a stirred tank reactor designed for specific reaction and specific heat and mass transfer duties may have to be adopted for another set of duties. If the heat transfer duty is much higher than that can be matched by the available jacket area, internal cooling coils may be used to provide the additional heat transfer duty for the new application (Ramachandran and Chaudhari 1983). This seemingly innocuous modification can have grave consequences for hydrodynamics in a stirred reactor. Nikhade et al. (2005) and Nikhade and Pangarkar (2007) investigated the phenomena resulting from the addition of internal cooling coils. They have shown that the critical speeds of agitation for solid suspension, N_{SG}; complete gas dispersion, N_{CD}; the gas holdup, ε_G; and K_{SL} are greatly affected by the presence of the cooling coils. In the case of N_{SG}, there is an increase due to the presence of the coil that dissipates some portion of the momentum of the stream generated by the impeller. Since there is a close relationship between N_{SG} and K_{SL}, the latter needs to be evaluated using N_{SG} in the presence of the coil. On the other hand, the presence of coils has a positive effect on ε_G. The coils break up the bubbles, thereby generating smaller bubbles and increasing ε_G. Based on the work in a 0.57 m diameter stirred tank reactor fitted with three different coils and various impellers, correlations were derived for these important parameters as discussed in Section 7A.8.1.

7A.8.1   Gas Holdup

Nikhade (2006) has determined the gas holdup in 0.57 m diameter stirred tank reactors equipped with different impellers and internal coils. The investigation covered three impellers (PTD, DT, and PTU) and three internal coils with different surface areas. He could successfully obtain a unique correlation for the gas holdup data for three different impellers and coils using the approach proposed by Yawalkar et al. (2002a). The correlation obtained was

$$\varepsilon_G = 0.0857 \left(\frac{N}{N_{CD}} \right)^{0.79} (\text{vvm})^{0.38} \qquad (7A.27)$$

If Equation 7A.27 is compared with the correlation for ε_G without coils (Table 7A.1), it is evident that the gas holdup is a relatively strong function of N/N_{CD} but a relatively weak function of V_G when internal coils are present. The coils tend to break the bubbles. The resulting smaller bubbles are easily dragged down below the impeller, and hence, ε_G is higher in the presence of internal coils.

7A.8.2   Critical Speed for Complete Dispersion of Gas

There is absolutely no information in the literature on the critical speed for complete dispersion of the gas phase, N_{CD}, in stirred tank reactors fitted with helical coils. The only work reported so far is that of Nikhade (2006) in the 0.57 m diameter stirred tank reactor referred to earlier. The correlations obtained for the increase in N_{CD} and ΔN_{CD} were (Nikhade and Pangarkar 2006)

$$[\Delta N_{CD}]_{Coil} = 536(V_G)^{0.19}(S_{Coil})^{0.17}: 6\text{-PTD} \tag{7A.28}$$

$$[\Delta N_{CD}]_{Coil} = 4.8(V_G)^{0.05}(S_{Coil})^{0.57}: 6\text{-DT}\ (\text{Rushton turbine}) \tag{7A.29}$$

$$[\Delta N_{CD}]_{Coil} = 14.2(V_G)^{0.08}(S_{Coil})^{1.15}: 6\text{-PTU} \tag{7A.30}$$

$$(N_{CD})_{Coils} = \Delta N_{CDCoil} + N_{CD} \tag{7A.31}$$

N_{CD} (without coils) can be estimated by using the corresponding correlation in Section 7A.5.4. The above correlations indicate that N_{CD} increases with increase in the coil area. In situations which require substantial heat removal through the coils and thereby substantial coil area $(\Delta N_{CD})_{Coil}$ will be relatively high. The extra power required for gas dispersion will be correspondingly very high in view of $[\Delta Power]_{Coil} \propto [(\Delta N_{CD})_{Coil}]^3$ Clearly, the presence of internal cooling coils calls for significantly higher power input and hence will be counter productive.

7A.8.3 Critical Speed for Solid Suspension

Similar to N_{CD}, there is absolutely no information in the literature on the critical speed for suspension of solids, N_{SG}, in stirred tank reactors fitted with cooling coils. Nikhade et al. (2005) have provided Equations 7A.32, 7A.33, and 7A.34 for ΔN_{SGCoil} for the three impellers referred to earlier. ΔN_{SGCoil} is defined as the difference between the value of N_{SG} with and without a coil of surface area, S:

$$[\Delta N_{SG}]_{Coil} = 1119(V_G)^{0.27}(d_p)^{0.55}(X)^{0.03}(S_{Coil})^{0.3}: 6\text{-PTD} \tag{7A.32}$$

$$[\Delta N_{SG}]_{Coil} = 601(V_G)^{0.01}(d_p)^{0.59}(X)^{0.25}(S_{Coil})^{0.61}: 6\text{-DT}\ (\text{Rushton turbine}) \tag{7A.33}$$

$$[\Delta N_{SG}]_{Coil} = 13.86(V_G)^{0.09}(d_p)^{0.11}(X)^{0.11}(S_{Coil})^{0.7}: 6\text{-PTU} \tag{7A.34}$$

The dependence of ΔN_{SG} on the various operating parameters is different for the three types of impellers studied, which has been rationally explained by Nikhade et al. (2005). It has been shown in section 7A.8.2 that the presence of an internal coil leads to relatively very high power requirement for gas dispersion. This argument is equally valid for the case of solid suspension. Overall, an internal coil is counter productive particularly when the coil area required is high due to a substantial heat load.

7A.8.4 Gas–Liquid Mass Transfer Coefficient

There is absolutely no reported study on $k_L a$ in stirred tank reactors containing internal coils. The work of Nikhade (2006) indicates that the gas holdup in such stirred tank reactors increases due to the presence of the coils. This observation implies that the effective gas–liquid interfacial area is also correspondingly higher. However, because of the dampening of the turbulence caused by the coil, the true k_L

should decrease. Thus, in the absence of any systematic study, it is reasonable to believe that the volumetric mass transfer coefficient, $k_L a$, should be the same as that for a stirred tank reactor without coils. In view of this, Equation 7A.6 is recommended for design purpose. However, N_{CD} to be used in Equation 7A.6 should be obtained from Equation 7A.31 using the corresponding correlations for ΔN_{CD} for a stirred tank reactor with internal coil (Eqs. 7A.28, 7A.29, and 7A.30).

7A.8.5 Solid–Liquid Mass Transfer Coefficient

Nikhade and Pangarkar (2006) employed the benzoic acid dissolution method to determine K_{SL} in the stirred tank reactor described in Sections 2.7.1, 7A.7.2, 7A.7.3, and 7A.7.4. Using an approach exactly analogous to that of Jadhav and Pangarkar (1991) and Pangarkar et al. (2002), these investigators obtained the following unique correlation satisfying the various system configurations and operating variables covered:

$$K_{SL} = 1.92 \times 10^{-3} \left(\frac{N}{N_{SGCoil}} \right)^{0.9} (Sc)^{-0.53} \tag{7A.35}$$

The critical speed for solid suspension in the presence of a coil was obtained by an equation similar to Equation 7A.31: $N_{SGCoil} = N_{SG\ withoutcoil} + \Delta N_{SGcoil}$. ΔN_{SGcoil} was obtained from Equations 7A.32, 7A.33, and 7A.34. Equation 7A.35 is also strikingly similar to Equations 7A.7, 7A.8, 7A.9, and 7A.10 with only minor differences in the empirical constants. The hydrodynamics situations covered in configuration resulting into these correlations are vastly different. Thus, it can be again argued that this similarity is not fortuous but arises out of the sound basis outlined in Section 6.4.2.

7A.9   STIRRED TANK REACTOR WITH INTERNAL DRAFT TUBE

In some multiphase reactions, the solid phase may be lighter than the liquid phase. In such a case, instead of settling at the bottom, floating of the solid phase has to be prevented. In this event, it is advantageous to use a draft tube with an axial flow impeller (Fig. 7A.15).

The solid phase (catalyst or reactant) can be added at the top of the draft tube in the vicinity of the eye of the impeller. The axial flow impeller sucks the fluid from the annulus into the draft tube and then back into the annulus. With proper design and appropriate speed of agitation, the well-directed axial flow generated can overcome the buoyancy of the solid. The low density particles are propelled downward into the draft tube and thereafter into the annulus. The circulation pattern is shown in Figure 7A.15. The draft tube crystallizer and draft tube fermenter (also called propeller loop reactor) are popular applications of a standard stirred tank reactor with draft tube. Another major class of application is ion exchange resin-catalyzed esterifications and transesterification. The latter is widely used for converting triglycerides (vegetable oils) to biodiesel (fatty acid methyl ester, FAME) by transesterification with methanol. In the initial phase

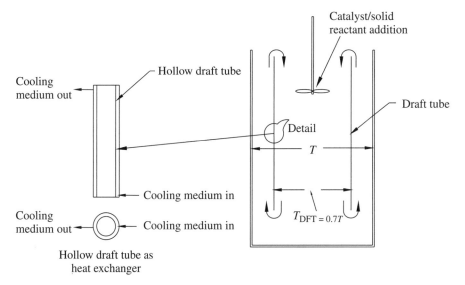

FIGURE 7A.15　Stirred reactor with draft tube cum heat exchanger.

of the reaction, the pores of the resin are filled with air, and it is lighter than the oil phase, which forms the continuous phase. In this stage, the stirred tank reactor with a draft tube containing a low shear axial flow impeller is a better option than the standard stirred tank reactor. In the propeller loop reactor used in many industrial fermentations, the objective is to recirculate the gas in a manner analogous to the circulation of the solid particles described earlier. Chisti and Jauregui-Haza (2002) have used a loop reactor equipped with axial flow (Prochem®) impeller with gas sparging in the annulus. The impeller discharge from the draft tube into the annulus and the direction of gas flow in the annulus is same. Therefore, the possibility of flooding of the impeller is eliminated.

The propeller loop reactor with internal draft tube cum heat exchanger configuration has also been used by the author for an extremely exothermic (almost explosive) solid–liquid reaction:

$$A \text{ (light solid)} + B \text{ (heavy liquid)} \rightarrow \text{product } C + \text{by-product } D \text{ (gaseous)} + (\text{very high} - \Delta H_R)$$

The solid-phase reactant is relatively light as compared to the liquid. In the absence of a well-directed downflow, it tends to float on the surface, resulting into highly localized exotherm and "popcorn"-type bursting followed by gas emission. The solid-phase reactant was fed into the suction of the propeller (Fig. 7A.15). For the case of three-phase sparged reactors with internal draft tube, it will be seen later that T/T_{DFT} ratio of 2 gives the maximum circulation velocity (Section 10.10.2). This ratio implies that the draft tube and annulus areas are 25 and 75%, respectively, of the total vessel cross-sectional area. In the present case, because of the relatively high density difference between the solid and liquid, the buoyancy of the solid is significantly high. Therefore, the downward discharge of the propeller must be such that it overcomes the buoyancy or Q_L ($\propto ND^3$) must be high. It was found that a draft tube designed such that the draft tube and annulus area are equal ($T_{DFT} = 0.7\ T$) provided the optimum configuration. Power measurements indicated that the chosen propeller (confined in the draft tube designed using the equal area logic) afforded approximately 30% higher power number. The other parameters such as clearance of the draft tube were chosen from the optimum value reported later in Section 10.10.3b. Implementation of the scheme depicted in Figure 7A.15 coupled with a carefully chosen speed of agitation resulted in uniform distribution of the lighter solid by the focused and strong circulation. Another advantage of this arrangement was that the gaseous by-product produced in the reaction was also propelled into the annulus wherein its evolution and eventual exit was helped by the upward flow in the annulus. The draft tube that is directly in contact with the zone of the reaction and heat evolution on both the inside and outside surfaces was also used as a heat exchanger. This arrangement was found to yield stable and safe operation with normal cooling tower water. In the previous version using a conventional stirred reactor without draft tube, the relatively very high heat of reaction was removed only through the jacket of the reactor. The latter having a limited area (besides being far away from the main reaction zone) required use of chilled water. The operating cost of the chilled water unit contributed significantly to the overall processing cost. With the modifications shown in Figure 7A.15, the reaction could be carried out smoothly using cooling tower water for both the draft tube and vessel jacket. Temperature measurements and a heat balance showed that the draft tube was extracting a substantial portion of the extremely high heat of reaction. Thus, the refrigeration cost component was completely eliminated, resulting in not only 50% lower operating cost and proportionately higher profit but also a safe, hazard-free operation.

7A.10   WORKED EXAMPLE: DESIGN OF STIRRED REACTOR FOR HYDROGENATION OF ANILINE TO CYCLOHEXYLAMINE (CAPACITY: 25,000 METRIC TONNES PER YEAR)

Calculations for design of stirred reactor for manufacturing of 25,000 metric tonnes per year of cyclohexylamine by hydrogenation of aniline are presented in a separate Excel file. This file works in all Excel versions of Microsoft Office Excel 97-2003 and higher.
    The procedure is as follows:

1. Choose an annual capacity of cyclohexylamine and feed it in cell #C2.
2. To calculate the reactor volume required, feed an initial guess (e.g., 10 m³) in cell #C29.
3. Press "Calculate" button in cell G27. This initiates iteration-enabled macro. The calculations return the value (in cell C29) that balances the kinetic and hydrodynamic conditions for the reaction.

Output of Excel spreadsheet for design of stirred reactor for 25,000 metric tonnes per year of cyclohexylamine is given in Figure 7A.16.

Basis	Capacity tonnes of cyclohexyl amine (CHA) per annum	25,000		TPA

	Aniline	→	CHA
Reaction	$C_6H_5NH_2$	$3H_2$	$C_6H_{13}N$
Molecular weight	93.13	3 X 2	99

Working days/year	300	days/year
Capacity of CHA production required per hour	3.47	Tons/hr
Capacity of CHA production required	0.965	kg/s
Molar flow rate of CHA	0.00974	kmol/s
Desired conversion of Aniline	90%	
Aniline molar flow rate	0.01082	kmol/s
Aniline molar flow rate	38.97	kmol/hr
Aniline mass flow rate	3629.27	kg/hr
Density (Aniline)	900	kg/m³
Aniline vol flow rate	4.03	m³/hr

Stoichiometric factor, Z	3

Initial Aniline concentration	9.66	kmol/m³
Final Aniline concentration	0.97	kmol/m³
Conversion	90%	

Total Volumetric flow rate	4.03	m³/hr
Average Density	900.00	kg/m³

				Calculate

Dispersion (working) volume (assumed)	1627	m³
Total volume (40% extra for safety)	2277	m³

Calculated Volume	Error
1626.7	0.000

H/T Ratio	1	
Tank Diameter (T)	12.75	m
Height (H)	12.75	m
C.S area	127.61	m²
Residence time	403.39	hr
Molar flow rate of H2 required	105.22	kmol/hr
Molar flow rate of H2 required	0.0292	kmol/s
Temp	403	Deg K
P	1,000,000	N/m²
Gas constant, R	8314	J/kmol-K
Hydrogen molar Density	0.298	kmol/m³
Flow rate of Hydrogen required	352.54	m³/hr
Excess Hydrogen	20%	
Using excess hydrogen	423.05	m³/hr
QGin	423.05	m³/hr
QGout	70.51	m³/hr

QGin	0.118	m³/s

Superficial gas velocity	0.00092	m/s
Operating superficial gas velocity	0.00092	m/s

Hydrogen flowrate vessel volume per minute	0.0043	vvm

FIGURE 7A.16 Output of Excel spreadsheet for design of stirred reactor for 25,000 metric tonnes per year of cyclohexylamine and explanation.

Physical Properties:

Catalyst particle density, r_P	1750	kg/m^3
Catalyst particle diameter, d_p	1E-05	m
Solution viscosity	0.5	cp
Solution viscosity	5E-04	Pa s
Diffusivity of Hydrogen in reaction mixture	1.16E-09	m^2/s
Catalyst loading, X	20.0	kg catalyst/m^3 liquid
w	32533.29	kg
Impeller dia (D)	4.249	m

Calculation of stirring speeds

	6 Blade down flow pitched turbine (6PTD)	6 Blade up flow pitched turbine (6PTU)	
Critical speed for dispersion of gas, N_{cD}	0.182	0.147	rev/s
Critical speed for just suspension solid (2 phase), Ns	0.338	0.193	rev/s
Change in Ns due to presence of gas phase, DNs	0.001		rev/s
Critical speed for just suspension solid (3phase), N_{SG}	0.340	0.193	rev/s
Critical speed, $(max(N_{cD}, N_{SG}))$	0.340	0.193	rev/s
Ratio - operating speed/critical speed	1.2	1.2	
Operating speed, N	0.408	0.231	rev/s
Total Catalyst surface area, a_p	1.115E+07	1.115E+07	m^2
a_p/V (m^2/liq Vol.)	6857.14	6857.14	m^2 / m^3
Schmidt number	478.93	478.93	
k_{sL}	8.202E-05	8.202E-05	
$k_{sL}*a_p/V$	0.562	0.562	1/s
kla	0.0040	0.0040	1/s
Rate Constant	0.053	0.053	m^3/kg Catalyst /s
first order rate constant	1.0698	1.0698	1/s
k_2*C_b	1.0698	1.0698	1/s
Solubility of Hydrogen	4.46E-03	4.46E-03	$kmol/m^3$
Resistance comparison	**Resistances %**	**Resistances %**	
$k_{sL}*a_p/V$	0.71%	0.71%	
k_{la}	98.92%	98.92%	
$k_2 C$	0.37%	0.37%	
Total	100.00%	100.00%	
	As ksL*ap/V >> kla and k2*Cb >> kla, contribution of kla will be considered only	As ksL*ap/V >> kla and k2*Cb >> kla, contribution of kla will be considered only	
Reactor Volume	1626.664	1626.664	m^3

Power Required for the Impeller

$P_G/P_0 = 0.434 (Q_G/ND^3)^{-\alpha} (ND^2/g)^{-\beta} (W/D)^{0.174} (D/T)^{-0.064}$

W/D	0.300	
$\alpha = 0.064 T^{-0.94}$	0.006	
$\beta = 0.148 T^{-1.42}$	0.004	
P_G/P_0	0.391	0.90
Power Number, N_p	1.930	1.93
$P_0 = \rho_L \times N_P \times N^3 \times D^5$	163208	29747	W
P_G	63750	26772	W
P_G	63.75	26.77	kW

It is evident that 6-PTU requires much lower power for the same duty. Therefore, 6-PTU is chosen.

Power requied for agitation	26.77	kW
Volume of reactor	1627	m3

FIGURE 7A.16   (Continued)

Power calculations for Hydrogen compressor

The unreacted hydrogen need to be recirculated by compressor, which will overcome the pressure drop across the pool of liquid

Flowrate of unused hydrogen	0.0196	m^3/s
Height of reaction liquid	12.75	m
Density of liquid	900.00	kg/m^3
Liquid head	112,542	N/m^2
Assumng 20% more pressure drop considering the losses	135,050	N/m^2
Density of hydrogen	0.597	kg/m^3
Power required for compressor	1578.870	W
Power required for compressor	1.579	kW

Estimation of heat of reaction

Heat of formation of CHA	-115.8	kJ/mol	Reference: Yaws et al (1999)
Heat of formation of aniline	-75.39	kJ/mol	
Heat of reaction per mol of CHA	-40.41	kJ/mol	
Heat of reaction per kmol of CHA	-40410	kJ/kmol	
Rate of CHA foramtion	35.07	kmols/hr	
Heat evolved during reaction	1,417,298	kJ/hr	

Energy balance for reactor

Assume U, overall heat transfer coefficient (Benz, 2011)	250.00	$W/m^2 {}^{0}C$
Reactor dia	12.75	m
Height	12.75	m
Reactor surface area	510.452	m^2
LMTD	95	deg K
Heat that can be extracted by jacket	12,119,213	W
Heat that can be extracted by jacket	43,629,167	kJ/hr
Heat of reaction	1,417,298	kJ/hr

It is evident that, the jacket is more than sufficint to take heat load.

Total power required (compressor + agitation)	28.35	kW
The total Power per unit volume	0.0174	kW/m^3
Energy input per unit mass of product	29.39	kJ/kg

Energy requirement is higher than ventury loop reactor

Water out at 311°K

Reactor, T=403°K

Water in at 305°K

FIGURE 7A.16 (Continued)

7A.10.1 Elucidation of the Output

Comparison of the resistances reveals that the overall process is governed by gas–liquid mass transfer. Thus, the intrinsic productivity of the reactor as defined by the kinetics cannot be achieved. Further, in this case, the hydrogen concentration on the catalyst surface is negligible. It is important to note that if the feed rate of hydrogen is taken at 20% excess over the stoichiometric rate, the reactor volume required is astronomically high. This is because $k_L a \propto V_G$ and since V_G is very low, $k_L a$ is also very low. Increase in $k_L a$ can lower the reactor volume. Two options are available for this objective: (1) increase in V_G through increase in hydrogen feed rate and (2) increase in operating speed or higher N/N_{SG}. The design calculations were repeated

TABLE 7A.5 Effect of % Excess Hydrogen Feed Rate on Reactor Dimensions and Operating Cost: $N/N_{SG} = 1.2$

% excess H_2	V_G (m/s)	$k_L a$ ($1/s$)	Dispersion volume (m³)	Impeller power (kW)	H_2 unreacted (kg/s)	Power for H_2 recycle (kW)	Energy required per kg CHA (kJ/kg)
20	9×10^{-4}	4×10^{-3}	1627	27	0.011	1.6	29.4
100	4×10^{-3}	1.9×10^{-2}	351	7.73	0.06	4.7	13
200	1.3×10^{-2}	5.6×10^{-2}	123	3.4	0.118	6.68	10.4
300	2.1×10^{-2}	9.4×10^{-2}	87	2.65	0.176	8.92	12
400	3.3×10^{-2}	1.5×10^{-1}	62	2.11	0.235	10.6	13.2

TABLE 7A.6 Effect of N/N_{SG} on Reactor Dimensions and Operating Costs: Excess Hydrogen 20%

N/N_{SG}	V_G (m/s)	$k_L a$ ($1/s$)	Dispersion volume (m³)	Impeller power (kW)	H_2 unreacted (kg/s)	Power for H_2 recycle (kW)	Energy required per kg CHA (kJ/kg)
1.2	9.2×10^{-4}	4×10^{-3}	1627	26.8	0.011	1.6	29.4
1.5	1.8×10^{-3}	1×10^{-2}	610	23.54	0.011	1.14	25.6
2	4.1×10^{-3}	3.8×10^{-2}	173	20.07	0.011	0.75	21.6
2.5	7.9×10^{-3}	1×10^{-1}	65	18	0.011	0.54	19.2
3	8.7×10^{-3}	1.4×10^{-1}	56	27	0.011	0.51	29.1

for 100, 200, 300, and 400% excess hydrogen feed rate at constant $N/N_{SG} = 1.2$ and $N/N_{SG} = 1.5$, 1.8, 2.0, 2.5, and 3 at 20% excess hydrogen. The results are given in Tables 7A.5 and 7A.6.

It can be seen that increasing the hydrogen flow rate gives better results in terms of the energy required per unit mass of the product. However, even with 400% excess hydrogen supply, the reactor volume is still very high as compared to the venturi loop reactor in Section 8.13. A large excess has impact on the capital cost of the hydrogen recycle compressor. It should also be noted that it will be desirable to remove the condensable species present with unreacted hydrogen before recompression to the desired inlet pressure. The resulting relatively very high overall capital cost of the high pressure reactor and recycle compressor does not justify the use of a stirred reactor for this comparatively fast reaction.

These results will be compared with those for the venturi loop reactor in Section 8.13.

NOMENCLATURE

$[A^*]$:	Interfacial concentration of solute A (mol/m³)
$[A]$:	Concentration of solute A (mol/m³)
a:	Activity of solute (—)
$[A_S]$:	Concentration of species A at the catalyst surface (mol/m³)
A_C:	Total heat transfer area of the cooling coil (m²)
A_J:	Available jacket surface area (m²)
a_P or $\left(\dfrac{A_P}{V_D}\right)$:	Specific particle surface area (m²/m³)
A_P:	Total particle/catalyst surface area (m²)
Ar:	Archimedes number $\left(\dfrac{d_P^{\,3} g(\rho_S - \rho_L)}{\mu^2 \rho_L}\right)$ (—)
C:	Impeller clearance from the bottom of the tank (m)
c:	Concentration of electrolyte (mol/m³)
C_V:	Volume % of solid in the three-phase system (%)
D:	Diameter of the impeller (m)
d_{3-2}:	Sauter mean bubble diameter (m)
d_B:	Bubble diameter (m)
d_M:	Maximum stable bubble diameter used by Walter and Blanch (1986) to define critical Weber number (m)
D_M, D_A:	Molecular diffusivity of solute (m²/s)
d_O:	Outside diameter of the pipe (m)
d_P:	Solid particle diameter (m)
F_{LG}:	Impeller flow number $\left(\dfrac{Q_G}{ND^3}\right)$ (—)

Fr:	Froude number $\sqrt{\dfrac{N^2 D}{g}}$ (—)
Fr_{CD}:	Froude number at $N=N_{CD}$, $\sqrt{\dfrac{N_{CD}^2 D}{g}}$ (—)
g:	Gravitational constant, 9.81 (m/s^2)
H:	Height of liquid/dispersion in the tank (m)
ΔH_R:	Heat of reaction (kJ/mol or kcal/mol)
k_L:	True gas–liquid mass transfer coefficient (m/s)
$k_L a$:	Volumetric gas–liquid mass transfer coefficient (1/s)
K_{SL}:	True solid–liquid mass transfer coefficient (m/s)
$K_{SL}\,(A_P/V_D)$:	Volumetric solid–liquid mass transfer coefficient (1/s)
l_M:	Microscale of turbulence (m)
l_{MK}:	Microscale of turbulence based on Kolmogorov's theory of iso-
	tropic turbulence (Walter and Blanch 1986), $\left(\dfrac{\mu^{3/4}}{(P/V)^{1/4}\,\rho^{1/2}}\right)$ (m)
M:	Mass of liquid in the reactor (kg)
N:	Speed of agitation of the impeller (rev/s)
N_{CD}:	Critical speed for complete dispersion of the gas phase (rev/s)
$(N_{CD})_{Coils}$:	Critical speed of agitation for gas dispersion in the presence of helical coil (rev/s)
N_{CS}:	Critical speed for "complete" suspension of solid in two-phase (solid–liquid) system (rev/s)
ΔN_{CDCoil}:	Difference between the speed of agitation for dispersion of gas in the presence of helical coil and in its absence (rev/s)
N_P, N_{PO}:	Power number in the absence of gas phase $\left(\dfrac{P}{\rho_L N^3 D^5}\right)$ (—)
N_{PG}:	Power number in the presence of gas $\left(\dfrac{P_G}{\rho_L N^3 D^5}\right)$ (—)
N_R:	Critical speed for recirculation of the gas phase (rev/s)
N_S:	Critical speed for "just" suspension of solid in two-phase (solid–liquid) system (rev/s)
N_{SG}:	Critical speed for "just" suspension of solid in three-phase (gas–liquid–solid) system (rev/s)
ΔN_{SG}:	Difference in the critical speed for "just" suspension of solid in three-phase (gas–liquid–solid) and two-phase (solid–liquid) system (rev/s)
ΔN_{SGCoil}:	Difference in the minimum speed for just suspension of the solid in three-phase system in the presence and absence of helical coil (rev/s)
P:	Power input (W, kW)
P_G:	Power input in the presence of gas (W, kW)
P_O:	Power input for single (liquid phase) (W, kW)

P_{Tm}:   Mean total specific energy dissipation rate (W/kg)

ΔP: Pressure drop (N/m^2)

Q: Total heat to be removed (kJ/s or kcal/s)

Q_G: Volumetric flow rate of gas (m^3/s)

Q_L: Volumetric flow rate of liquid (m^3/s)

R: Ideal gas law constant, 8.314 (J/mol °K)

Re_I: Impeller Reynolds number $\left(\dfrac{ND^2 \rho_L}{\mu_L} \right)$ (—)

S: Constant in Zwietering's correlation $\left(\dfrac{ND^{0.85}}{v^{0.1} d_P^{\,0.2} \left(\dfrac{g\Delta\rho}{\rho_L} \right)^{0.45} X^{0.13}} \right)$ (—)

Sc: Schmidt number $[\mu/\rho D_A]$ (—)

S_{Coil}:   Surface area of helical coil (m^2)

STF: Surface tension factor, $(c/z)\,(d\sigma/dc)^2$ (kg^{2}m^3/kmol, s^4)

T: Diameter of the stirred reactor (m)

T_{DFT}:   Diameter of draft tube (m)

U: Overall heat transfer coefficient (W/m^2 °K)

u':   Turbulence intensity at a given speed of agitation N (m/s)

u'_{CD}:   Turbulence intensity at $N=N_{CD}$ (m/s)

V: Control volume (m^3)

V_C: Liquid circulation velocity (m/s)

V_D: Volume of the dispersion in the stirred tank (m^3)

V_G: Superficial gas velocity (m/s)

V_L: Volume of liquid in the tank (m^3)

V_R: Volume of reactor (m^3)

v_R: Radial component of fluid velocity (m/s)

$v\theta$:   Tangential component of fluid velocity (m/s)

v_Z: Axial component of fluid velocity (m/s)

$V_{S\infty}$, V_T: Terminal settling velocity of the solid (m/s)

vvm: Gas flow rate in terms of vessel volume per minute (m^3/(m^3/min)) $\left(\dfrac{Q_G \times 60}{V_D} \right)$

W, W_{Blade}:   Width of impeller blade (m)

We: Weber number (Walter and Blanch 1986) $\left(\dfrac{\tau d_B}{\sigma} \right)$ (—)

We_{crit}:   Critical Weber number (Walter and Blanch 1986) $\left(\dfrac{\tau d_M}{\sigma} \right)$ (—)

X: Solid loading (wt.%)

z: Number of ions formed by dissociation of an electrolyte molecule

Greek Letters

α, β: Empirical constants

ε_G: Fractional holdup of the gas phase (—)

ε_{GCD}:   Fractional holdup of the gas phase at the critical speed for gas dispersion (—)

ε: Power input per unit volume (kW/m^3)

ε_M: Power dissipated per unit mass (W/kg)

ε_{NS}:   Power input for just suspending the solids (W/kg)

ε_S: Fractional holdup of the solid phase (—)

μ_L: Liquid-phase viscosity (Pa s)

μ_W: Viscosity of water (Pa s)

ν: Kinematic viscosity (m^2/s)

ρ_L: Density of the liquid phase (kg/m^3)

ρ_S, ρ_P: Density of solid/particulate phase (kg/m^3)

σ_L: Surface tension liquid phase (N/m)

τ: Shear stress $\left(\dfrac{\sigma}{d_B}\right)$; $\left(\dfrac{\rho u'}{2}\right)$(Pa)

$\Delta\rho$:   Density difference between solid and liquid phases (kg/m^3)

Subscripts

G: Gas phase

L: Liquid phase

P, S: Particle/solid phase

Abbreviations

DT: Standard Rushton disc turbine

FBT: Flat blade turbine

LMTD: Log mean temperature difference (°K)

PTD: Pitched blade turbine—downflow

PTU: Pitched blade turbine—upflow

TPA: Tonnes per annum

REFERENCES

Angeli P, Hewitt GF. (2000) Drop size distributions in horizontal oil–water dispersed flows. Chem. Eng. Sci., 55:3133–3143.

Aravinth S, Gangadhar Rao P, Murugesan T. (1996) Critical impeller speed for solid suspension in turbine agitated contactors. Bioprocess Eng., 14:97–99.

Armenante PM, Nagamine EU. (1998) Effect of low off-bottom impeller clearance on the minimum agitation speed for complete suspension of solids in stirred tanks. Chem. Eng. Sci., 53:1757–1775.

Aubin J, Mavros P, Fletcher DF, Bertrand J, Xuereb C. (2001) Effect of axial agitator configuration (up-pumping, down-pumping, reverse rotation) on flow patterns generated in stirred vessels. Chem. Eng. Res. Des., Trans. IChemE, 79 (Part A):845–856.

Aubin J, Fletcher DF, Xuereb C. (2004) Modelling turbulent flow in stirred tanks with CFD: the influence of the modelling approach, turbulence model and numerical scheme. Thermal Fluid Sci., 28:431–445.

Ayranci I, Kresta SM. (2011) Design rules for suspending concentrated mixtures of solids in stirred tanks. Chem. Eng. Res. Des., 89:1961–1971.

Ayranci I, Machado MB, Madej AM, Derksen JJ, Nobes DS, Kresta SM. (2012) Effect of geometry on the mechanisms for off-bottom solids suspension in a stirred tank. Chem. Eng. Sci., 79:163–176.

Bakker A, Fasano JB, Myers KJ. (1994, September 21–23) Effects of flow pattern on the solids distribution in a stirred tank. In: Presented at the Eighth European Conference on Mixing (IChemE Symposium Series No. 136, L8), Cambridge, UK.

Bao Y, Hao Z, Gao Z, Huang X, Shi L, Smith JM, Thorpe RB. (2006) Gas dispersion and solid suspension in a three-phase stirred tank with multiple impellers. Chem. Eng. Commun., 193:801–825.

Bao Y, Gao Z, Huang X, Shi L, Smith JM, Thorpe RB. (2007) Gas–liquid dispersion with buoyant particles in hot sparged stirred tanks. Ind. Eng. Chem. Res., 46:6605–6611.

Bao Y, Zhang X, Gao Z, Chen L, Chen J, Smith JM, Kirkby NF. (2008a) Gas dispersion and solid suspension in a hot sparged multi-impeller stirred tank. Ind. Eng. Chem. Res., 47:2049–2055.

Bao Y, Chen L, Gao Z, Zhang X, Smith JM, Kirkby NF. (2008b) Temperature effects on gas dispersion and solid suspension in a three-phase stirred reactor. Ind. Eng. Chem. Res., 47:4270–4277.

Benz GT. (2011) Bioreactor design for chemical engineers. Chem. Eng. Prog., 107:21–26.

Bittorf KJ, Kresta SM. (2003) Prediction of cloud height for solid suspensions in stirred tanks. Trans IChemE., 81(Part A):568–577.

Bowen RL Jr. (1989) Letter to the editor on the paper by Rao et al., 1988. AICHEJ, 35:1575–1576.

Brož J, Fořt I, Sperling R, Jambere S, Heiser M, Rieger F. (2004) Pumping capacity of pitched blade impellers in a tall vessel with a draught tube. Acta Polytechnica, 44(4):48–53.

Bujalski W. (1987) Three phase mixing: studies of geometry, viscosity and scale. Ph.D. Thesis, University of Birmingham cf: Bujalski et al. (1990).

Bujalski W, Konno M, Nienow AW. (1988) Scale up of 45° pitched blade agitators for gas dispersion and solid suspension. In: Proceedings of Sixth European Mixing Conference, Italy, AIDIC-BHRA Fluid Engineering Center, Cranfield, UK. p 389–398.

Bujalski W, Nienow AW, Huoxing L. (1990) The use of upward pumping 45° pitched blade turbine impellers in three phase reactor. Chem. Eng. Sci., 45:415–421.

Bujalski W, Takenaka K, Paolini S, Jahoda M, Paglianti A, Takahashi K, Nienow AW, Etchells AW. (1999) Suspension and liquid homogenization in high solids concentration stirred chemical reactors. Trans. Inst. Chem. Eng., 77:241–247.

Bustamante MCC, Cerri MO, Badino AC. (2013) Comparison between average shear rates in conventional bioreactor with Rushton and elephant ear impellers. Chem. Eng. Sci., 90:92–100.

Buurman C, Resoort G, Plaschkes A. (1986) Scaling-up rules for solids suspension in stirred vessels. Chem. Eng. Sci., 41:2865–2871.

Calderbank PH. (1958) Physical rate processes in industrial fermentation. Part I. The interfacial area in gas–liquid contacting with mechanical agitation. Trans. Inst. Chem. Eng., 36:443–463.

Calderbank PH. (1959) Physical rate process in industrial fermentation. Part II. Mass transfer coefficients in gas–liquid contacting with and without mechanical agitation. Trans. Inst. Chem. Eng. (UK), 37:173–185.

Calderbank PH, Jones SRJ. (1961) Physical rate processes in industrial fermentation. Part III. Mass transfer from fluids to solid particles suspended in mixing vessels. Trans. Inst. Chem. Eng. (UK), 39:363–368.

Cents AHG, Brillman DWF, Versteeg GF. (2001) Gas absorption in an agitated gas–liquid–liquid system. Chem. Eng. Sci., 56:1075–1083.

Cents AHG, Brillman DWF, Versteeg GF. (2005) Ultrasonic Investigation of hydrodynamics and mass transfer in a gas–liquid (–liquid) stirred vessel. Int. J. Chem. Reactor Eng., 3(A19):1–32.

Chapman CM, Nienow AW, Middleton JC. (1980) Surface aeration in a small agitated and sparged vessel. Biotechnol. Bioeng., 22:981–994.

Chapman CM, Nienow AW, Cooke M, Middleton JC. (1983a) Particle–gas–liquid mixing in stirred vessels. Part I. Particle–liquid motion. Chem. Eng. Res. Des., 61(2):71–81.

Chapman CM, Nienow AW, Cooke M, Middleton JC. (1983b) Particle–gas–liquid mixing in stirred vessels. Part II. Gas–liquid mixing. Chem. Eng. Res. Des., 61(2):82–95.

Chapman CM, Nienow AW, Cooke M, Middleton JC. (1983c) Particle–gas–liquid mixing in stirred vessels. Part III. Three phase mixing. Chem. Eng. Res. Des., 61(3):167–181.

Chapman CM, Nienow AW, Cooke M, Middleton JC. (1983d) Particle–gas–liquid mixing in stirred vessels. Part IV. Mass transfer and final conclusions. Chem. Eng. Res. Des., 61(3):182–185.

Chisti Y, Jauregui-Haza UJ. (2002) Oxygen transfer and mixing in mechanically agitated airlift bioreactors. Biochem. Eng. J., 10:143–153.

Collignon M-L, Delafosse A, Crine M, Toye D. (2010) Axial impeller selection for anchorage dependent animal cell culture in stirred bioreactors: methodology based on the impeller comparison at just-suspended speed of rotation. Chem. Eng. Sci., 65:5929–5941.

Congjing R, Xiaojing J, Jingdai W, Yongrong Y, Xiaohuan Z. (2008) Determination of critical speed for complete solid suspension using acoustic emission method based on multiscale analysis in stirred tank. Ind. Eng. Chem. Res., 47:5323–5327.

Cooke M, Heggs PJ. (2005) Advantages of the hollow (concave) turbine for multi-phase agitation under intense operating conditions. Chem. Eng. Sci., 60:5529–5543.

Danckwerts PV. (1970) Gas–liquid reactions. McGraw Hill Book Company, New York, USA.

Deshpande GB. (1988) Fluid mechanics in mechanically agitated contactors, M. Chem. Eng. Thesis, University of Bombay, Bombay, India.

Dutta NN, Pangarkar VG. (1994) Particle-liquid mass transfer in multi-impeller agitated contactors. Chem. Eng. Commun., 129:109–121.

Dutta NN, Pangarkar VG. (1995a) Critical impeller speed for solid suspension in multi-impeller agitated contactors. Solid–liquid system. Chem. Eng. Commun., 137:135–146.

Dutta NN, Pangarkar VG. (1995b) Critical impeller speed for solid suspension in multi-impeller three phase agitated contactors. Can. J. Chem. Eng., 73(3):273–283.

Dutta NN, Pangarkar VG. (1996) Particle-liquid mass transfer in multi-impeller agitated three phase reactors. Chem. Eng. Commun., 146:65–84.

Fajner D, Pinelli D, Ghadge RS, Montante G, Paglianti A, Magelli F. (2008) Solids distribution and rising velocity of buoyant solid particles in a vessel stirred with multiple impellers. Chem. Eng. Sci., 63:5876–5882.

Fishwick RP, Winterbottom JM, Parker DJ, Fan XF, Stitt EH. (2005) Hydrodynamic measurements of up- and down-pumping pitched-blade turbines in gassed, agitated vessels, using positron emission particle tracking. Ind. Eng. Chem. Res., 44:6371–6380.

Gray JB. (1973) Agitation of low viscosity particle suspensions. In: Perry RH, Chilton CH, editors. Chemical engineers' handbook. 5th ed. International Student Edition, McGraw-Hill, Kogakusha Ltd., Tokyo. p 19–8.

Gray DJ. (1987) Impeller clearance effect on off bottom particle suspension in agitated vessels. Chem. Eng. Commun., 61:151–158.

Greaves M, Barigou M. (1988) Estimation of gas hold up and impeller power in a stirred vessel reactor. Fluid Mixing III. Chem. Eng. Symp. Series No. 108, p 235–256.

Guha D, Ramachandran PA, Dudukovic MP. (2007) Flow field of suspended solids in a stirred tank reactor by Lagrangian tracking. Chem. Eng. Sci., 62:6143–6154.

Harriott P. (1962) Mass transfer to particles. Part I. Suspended in agitated vessels. AICHEJ, 8:93–101.

Harnby N, Edwards MF and Nienow AW. (2000) Mixing in the process industries. 2nd ed. Butterworth-Heinemann, Oxford, UK.

Hassan ITM, Robinson CW. (1977) Stirred-tank mechanical power requirement and gas holdup in aerated aqueous phases. AICHEJ, 23:48–56.

Hemrajani RR, Tatterson GB. (2004) Mechanically stirred vessels: chapter 6. In: Paul EL, Atiemo-Obeng VA, Kresta SM, editors. Handbook of industrial mixing: science and practice. John Wiley & Sons, Inc., Hoboken, NJ, USA.

Hicks MT, Myers KJ, Bakker A. (1997) Cloud height in solids suspension agitation. Chem. Eng. Commun., 160:137–155.

Hickman AD. (1988) Gas–liquid oxygen transfer and scale-up. A novel experimental technique with results for mass transfer in aerated agitated vessels. In: Proceedings of Sixth European Conference on Mixing, May 24–26, 1988, Pavia, Italy. p 369–374.

Hinze JO. (1975) Turbulence. 2nd ed. McGraw-Hill Inc., New York, USA. p 460–471.

Hosseini S, Patel D, Ein-Mozaffari F, Mehrvar M. (2010) Study of solid–liquid mixing in agitated tanks through computational fluid dynamics modelling. Ind. Eng. Chem. Res., 49:4426–4435.

Hughmark GA. (1974) Hydrodynamics and mass transfer for suspended solid particles in a turbulent liquid. AICHEJ, 20:202–204.

Hughmark GA. (1980) Power requirements and interfacial area in gas–liquid turbine agitated systems. Ind. Eng. Chem. Process Des. Dev., 19:638–641.

Ibrahim S, Nienow AW. (1996) Particle suspension in the turbulent regime: the effect of impeller type and impeller/vessel configuration. Trans. IChemE, 74(Part A):679–688.

Ibrahim S, Nienow AW. (2004) Suspension of microcarriers for cell culture with axial flow impellers. Trans. IChemE, Part A, Chem. Eng. Res. Des., 82(A9):1082–1088.

Jadhav SV, Pangarkar VG. (1991) Particle–liquid mass transfer in mechanically agitated contactors. Ind. Eng. Chem. Res., 30:2496–2503.

Jafari R. (2010) Solid suspension and gas dispersion in mechanically agitated vessels. Thèse présentée en vue de l'obtention Du diplôme de Philosophiae Doctor (Ph.D.) (Génie chimique) Département de Génie Chimique École Polytechnique de Montréal, Canada.

Jafari R, Tanguy PA, Chaouki J. (2012a) Characterization of minimum impeller speed for suspension of solids in liquid at high solid concentration, using gamma-ray densitometry. Int. J. Chem. Eng., 2012, Article ID 945314:15. doi:10.1155/2012/945314.

Jafari R, Chaouki J, Tanguy PA. (2012b) A comprehensive review of just suspended speed in liquid–solid and gas–liquid–solid stirred tank reactors. Int. J. Chem. Reactor Eng., 10(Review R1):1–31.

Jafari R, Tanguy PA, Chaouki J. (2012c) Experimental investigation on solid dispersion, power consumption and scale-up in moderate to dense solid–liquid suspensions. Chem. Eng. Res. Des., 90:201–212.

Jaworski A, Nienow AW, Koutsakos E, Dyster K, Bujalski W. (1991) An LDA study of turbulent flow in a baffled vessel agitated by a pitched blade turbine. Trans. Inst. Chem. Eng. (UK), 69A:313–320.

Jaworski Z, Dyster KN, Nienow AW. (2001) The effect of size, location and pumping direction of pitched blade turbine impellers on flow patterns: LDA measurements and CFD predictions. Trans. IChemE, Part A, Chem. Eng. Res. Des., 79:887–894.

Kapic A. (2005) Mass transfer measurements for syngas fermentation, M.S. Thesis, Iowa State University, Ames, IA, USA (cf. Kapic and Heindel, 2006).

Kapic A, Heindel TJ. (2006) Correlating gas–liquid mass transfer in a stirred-tank reactor. Trans. IChemE, Part A, Chem. Eng. Res. Des., 84(A3):239–245.

Keey RB, Glen JB. (1964) Mass transfer from solid spheres. Can. J. Chem. Eng., 42(5): 227–235.

Keey RB, Glen JB. (1966) Mass transfer from fixed and freely suspended particles in an agitated vessel. AICHEJ, 12:401–403.

Kerdouss F, Bannari A, Proulx P. (2006) CFD modeling of gas dispersion and bubble size in a double turbine stirred tank. Chem. Eng. Sci., 61:3313–3322.

Kielbus-Rapala A, Karcz J, Cudak M. (2011) The effect of the physical properties of the liquid phase on the gas–liquid mass transfer coefficient in two- and three-phase agitated systems. Chem Papers, 65(2):185–192.

Kraume, M, Zehner P. (2002) Concept for scale-up of solids suspension in stirred tanks. Can. J. Chem. Eng., 80(4):1–8.

Kraume M. (1992) Mixing times in stirred suspension. Chem. Eng. Technol., 15:313–318.

Kushalkar KB, Pangarkar VG. (1994) Particle-liquid mass transfer in mechanically agitated three phase reactors. Ind. Eng. Chem. Res., 33:1817–1820.

Kushalkar KB, Pangarkar VG. (1995) Particle-liquid mass transfer in three phase mechanically agitated contactors: power law fluids. Ind. Eng. Chem. Res., 34:2485–2492.

Lakkonen M, Moilanen P, Alopaeus V, Aittamaa J. (2007) Modeling local gas–liquid mass transfer in agitated vessels. Trans. IChemE, Part A, Chem. Eng. Res. Des., 85(A5): 665–675.

Lane GL, Schwarz MP, Evans GM. (2002) Predicting gas–liquid flow in a mechanically stirred tank. Appl. Math. Model., 26:223–235.

Lane GL, Schwarz MP, Evans GM. (2004) Numerical modeling of gas–liquid flow in stirred tanks. Chem. Eng. Sci., 60:2203–2214.

Linek V, Vacek V, Benes P. (1987) A critical review and experimental verification of the correct use of the dynamic method for the determination of oxygen transfer in aerated agitated vessels to water, electrolyte solutions and various viscous liquids. Chem. Eng. J., 34:11–34.

Loiseau B, Midoux N, Charpentier JC. (1977) Some hydrodynamics and power input data in mechanically agitated gas–liquid contactors. AICHEJ, 23:931–935.

Luo H, Svendson F. (1996) Theoretical model for drop and bubble breakup in turbulent dispersions. AICHEJ, 42:1225–1233.

Machado MB, Nunhez JR, Nobes D, Kresta SM. (2012) Impeller characterization and selection: balancing efficient hydrodynamics with process mixing requirements. AIChEJ, 58: 2573–2588.

Machon V, Vleck J, Kudrna V. (1978) Gas hold-up in agitated aqueous solutions of strong inorganic salts. Collect. Czech. Chem. Commun., 43:593–603.

Markopoulos J, Christofi C, Katsinaris I. (2007) Mass transfer coefficients in mechanically agitated gas–liquid contactors. Chem. Eng. Technol., 30:829–834.

Martin M, Montes FJ, Galan MA. (2008) On the contribution of the scales of mixing to the oxygen transfer in stirred tanks. Chem. Eng. J., 145:232–241.

Martin M, Montes FJ, Galan MA. (2010) Mass transfer rates from bubbles in stirred tanks operating with viscous fluids. Chem. Eng. Sci., 65:3814–3824.

Maxey MR, Riley JJ. (1983) Equation of motion for a small rigid sphere in a nonuniform flow. Phys. Fluids, 26:883–889.

McCabe WL, Smith JC, Harriott P. (1986) Unit operations of chemical engineering. IV International Student Edition, McGraw-Hill Chemical Engineering Series, McGraw-Hill Book Co., New York, USA.

Middleton JC. (2000) Gas–liquid dispersion and mixing .Chapter 15. In: Harnby N, Edwards MF, Nienow AW, editors. Mixing in the process industries. 2nd ed. Butterworth-Heinemann, Oxford, UK. p 322–363.

Miller DN. (1971) Scale-up of agitated vessels: gas–liquid mass transfer. AIChEJ, 20:445–453.

Montante G, Lee KC, Brucato A, Yianneskis M. (1999) An experimental study of double-to-single-loop transition in stirred vessels. Can. J. Chem. Eng., 77(4):649–659.

Montante G, Lee KC, Brucato A, Yianneskis M. (2001) Numerical simulations of the dependency of flow pattern on impeller clearance in stirred vessels. Chem. Eng. Sci., 56:3751–3770.

Montante G, Bourne JR, Magelli F. (2008) Scale-up of solids distribution in slurry, stirred vessels based on turbulence intermittency. Ind. Eng. Chem. Res., 47:3438–3443.

Moo-Young M, Blanch HW. (1981) Design of biochemical reactors: mass transfer criteria for simple and complex systems. Adv. Biochem. Eng., 19:1–69.

Moucha T, Linek V, Prokopova E. (2003) Gas hold-up, mixing time and gas–liquid volumetric mass transfer coefficient of various multiple-impeller configurations: Rushton turbine, pitched blade and techmix impeller and their combinations. Chem. Eng. Sci., 58:1839–1846.

Myers KJ, Janz EE, Fasano JB. (2013) Effect of solids loading on agitator just-suspended speed. Can. J. Chem. Eng., 91(9):1508–1512.

Nienow AW. (1968) Suspension of solid particles in turbine agitated baffled vessels. Chem. Eng. Sci., 23:1453–1459.

Nienow AW. (1969) Dissolution mass transfer in a turbine agitated baffle vessel. Can. J. Chem. Eng., 47(3):248–258.

Nienow AW, Miles D. (1971) Impeller power numbers in closed vessels. Ind. Eng. Chem. Process Des. Develop., 10:41–43.

Nienow AW, Wisdom DJ. (1974) Flow over disc turbine blades. Chem. Eng. Sci., 29:1994–1997.

Nienow AW. (1975) Agitated vessel particle–liquid mass transfer: a comparison between theories and data. Chem. Eng. J., 9:153–160.

Nienow AW, Wisdom DJ, Middleton JC. (1977) The effect of scale and geometry on flooding, recirculation and power in gassed stirred vessels. In: Proceedings of Second European Conference on Mixing, Cambridge, UK, BHR Group, Cranfield, UK. p F1-1–F1-16.

Nienow AW. (1996) Gas–liquid mixing studies: a comparison of Rushton turbines with some modern impellers. Trans. IChemE, Part A, Chem. Eng. Res. Des., 74:417–423.

Nienow AW. (2000) The suspension of solid particles. Chapter 16. In: Harnby N, Edwards MF, Nienow AW, editors. Mixing in the process industries. 2nd ed. Butterworth-Heinemann, Oxford, UK. p 364–393.

Nienow AW, Bujalski W. (2002) Recent studies on agitated three-phase (gas–solid–liquid) systems in the turbulent regime. Trans. IChemE, Part A, Chem. Eng. Res. Des., 80:832–838.

Nienow AW, Bujalski W. (2004) The versatility of up-pumping hydrofoil agitators. Trans. IChemE, Part A, Chem. Eng. Res. Des., 82(A9):1073–1081.

Nienow AW. (2006) Reactor engineering in large scale animal cell culture. Cytotechnology, 50:9–33.

Nikhade BP, Moulijn JA, Pangarkar VG. (2005) Critical impeller speed (N_{SG}) for solid suspension in sparged stirred vessels fitted with helical coils. Ind. Eng. Chem. Res., 44:4400–4405.

Nikhade BP. (2006) Hydrodynamics, modeling and process intensification in multiphase systems. Ph.D. (Tech) Thesis, University of Mumbai, Mumbai, India.

Nikhade BP, Pangarkar VG. (2006) Gas dispersion and hold up in stirred tanks fitted with helical coils (unpublished work).

Nikhade BP, Pangarkar VG. (2007) A theorem of corresponding hydrodynamic states for estimation of transport properties: case study of mass transfer coefficient in stirred tank fitted with helical coil. Ind. Eng. Chem. Res., 46:3095–3100.

Olmos E, Gentric C, Vial C, Wild G, Midoux N. (2001) Numerical simulation of multiphase flow in bubble column reactors. Influence of bubble coalescence and break-up. Chem. Eng. Sci., 56:6359–6365.

Oyama Y, Endoh K. (1955) Power characteristics of gas–liquid contacting mixers. Chem. Eng. (Japan), 9:2–11.

Pangarkar VG, Yawalkar AA, Sharma MM, Beenackers AACM. (2002) Particle–liquid mass transfer coefficient in two/three-phase stirred tank reactors. Ind. Eng. Chem. Res., 41:4141–4167.

Panneerselvam R, Savithri S, Surender GD. (2009) Computational fluid dynamics simulation of solid suspension in a gas–liquid–solid mechanically agitated contactor. Ind. Eng. Chem. Res., 48:1608–1620.

Parthasarathy R, Ahmed N. (1991) Gas holdup in stirred vessels: bubble size and power input effects. In: Proceedings of Seventh European Conference on Mixing, Koninklijke Vlaamse Ingenieursvereniging vzw, Brugge, Belgium. p 295–301.

Paul EL, Atiemo-Obengo VA and Kresta SM, editors. (2004) Handbook of industrial mixing: Science and practice. Wiley-Interscience, Hoboken, New Jersey, USA.

Prince MJ, Blanch HW. (1990) Bubble coalescence and break-up in air-sparged bubble columns. AICHEJ, 36:1485–1499.

Ramachandran PA, Chaudhari RV. (1983) Three phase catalytic reactors. Gordon and Breach Science Publishers Inc., New York, USA.

Rao KSMS, Rewatkar VB, Joshi JB. (1988) Critical impeller speed for solid suspension in mechanically agitated contactors. AICHEJ, 34:1332–1340.

Rewatkar VB, Rao KSMS, Joshi JB. (1989) Some aspects of solid suspension in mechanically agitated reactors. AICHEJ, 35:1577–1580.

Rewatkar VB, Rao KSMS, Joshi JB. (1990) Power consumption in mechanically agitated contactors using pitched bladed turbine impellers. Chem. Eng. Commun., 88(1): 69–90.

Rewatkar VB, Rao KSMS, Joshi JB. (1991) Critical impeller speed for solid suspension in mechanically agitated three phase reactors. Part I. Experimental part. Ind. Eng. Chem. Res., 30:1770–1784.

Rewatkar VB, Joshi JB. (1993) Role of sparger design in gas dispersion in mechanically agitated gas/liquid reactors. Can. J. Chem. Eng., 71(2):278–291.

Rewatkar VB, Deshpande AJ, Pandit AB, Joshi JB. (1993) Gas hold-up behavior of mechanically agitated gas–liquid reactors using pitched blade down flow turbines. Can. J. Chem. Eng. 71(2):226–237.

Rieger F, Ditl P, Havelkova O. (1988) Suspension of solid particles—concentration profiles and particle layer on the vessel bottom. In: Proceedings of Sixth European Conference on Mixing, Pavia, Italy, 24–26 May, BHRA, Cranfield, UK. p 251–258.

Rieger F, Ditl P. (1994) Suspension of solid particles. Chem. Eng. Sci., 49:2219–2227.

Rigby GD, Evans GM, Jameson GJ. (1997) Bubble breakup from ventilated cavities in multiphase reactors. Chem. Eng. Sci., 52(21–22):3677–3684.

Rutherford K, Lee KC, Mahmoudi SMS, Yianneskis M. (1996). Hydrodynamic characteristics of dual Rushton impeller stirred vessels. AICHEJ, 42:332–346.

Sessiecq P, Mier P, Gruy F, Cournil M. (1999) Solid particles concentration profiles in an agitated vessel. Trans. IChemE, 77(Part A):741–746.

Sardeing R, Aubin J, Poux M, Xuereb C. (2004a) Gas–liquid mass transfer: influence of sparger location. Trans. IChemE, Part A, Chem. Eng. Res. Des., 82(A9):1161–1168.

Sardeing R, Aubin J, Xuereb C. (2004b) Gas–liquid mass transfer: a comparison of down- and up-pumping axial flow impellers with radial turbines. Trans. IChemE, 82(A12): 1589–1596.

Sardeshpande MV, Sagi AR, Juvekar VA, Ranade VV. (2009) Solid suspension and liquid phase mixing in solid–liquid stirred tanks. Ind. Eng. Chem. Res., 48:9713–9722.

Schluter V, Deckwer W-D. (1992) Gas/liquid mass transfer in stirred vessels. Chem. Eng. Sci., 47:2357–2362.

Shamlou PA, Koutsakos E. (1989) Solids suspension and distribution in liquids under turbulent agitation. Chem. Eng. Sci., 44:529–542.

Shan X, Yu G, Yang C, Mao Z-S, Zhang W. (2008) Numerical simulation of liquid–solid flow in an unbaffled stirred tank with a pitched-blade turbine down flow. Ind. Eng. Chem. Res., 47:2926–2940.

Sharma RN, Shaikh AA. (2003) Solid suspension in stirred tanks with pitched blade turbines. Chem. Eng. Sci., 58:2123–2140.

Shaw JA. (1992) Succeed at solid suspension. Chem. Eng. Prog., 88(5):34–41.

Smith JM, Van't Riet K, Middleton JC. (1977) Scale-up of agitated gas–liquid reactors for mass transfer. In: Proceedings of Second European Conference on Mixing, Cambridge, UK. p F4-51–F4-66.

Smith JM, Warmoeskerken MMCG. (1985) The dispersion of gases in liquids with turbines. In: Proceedings of Fifth European Conference on Mixing, June 10–12, Wurzburg, West Germany, BHRA Fluid Eng., Cranfield, UK. p 115–126.

Smith JM. (1991) Simple performance correlations for agitated vessels. In: Proceedings of Seventh European Conference on Mixing, Brugge, Belgium, Koninklijke Vlaamse Ingenieursvereniging vzw. p 233–241.

Smith JM, Gao Z, Steinhagen HM. (2004) The effect of temperature on the void fraction in gas–liquid reactors. Exp. Therm. Fluid Sci., 28:473–478.

Tagawa A, Dohi N, Kawase Y. (2012) Volumetric gas–liquid mass transfer coefficient in aerated stirred tank reactors with dense floating solid particles. Ind. Eng. Chem. Res., 51:1938–1948.

Takenaka K, Takashi K, Bujalski W, Nienow AW, Paolini S, Paglianti A, Etchells A. (2005) Mixing time for different diameters of impeller at a high solid concentration in an agitated vessel. J. Chem. Eng. Jpn., 38(5):309–316.

Tatterson GB. (1991) Fluid mixing and gas dispersion in agitated tanks. McGraw-Hill Inc., New York, USA.

Uhl VW, Gray JB. (1966) Mixing: theory and practice. Vol. 1. Academic Press, New York, USA.

Uhl VW, Gray JB. (1967) Mixing: theory and practice. Vol. 2. Academic Press, New York, USA.

Uhl VW, Gray JB. (1986) Mixing: theory and practice. Vol. 3. Academic Press, New York, USA.

Van't Riet K. (1979) Review of measuring methods and results in non-viscous gas–liquid vessel reactor. Ind. Eng. Chem. Proc. Des. Dev., 18:357–364.

Vrabel P, van der Lans RGJM, Luyben KCAM, Boon L, Nienow AW. (2000) Mixing in large-scale vessels stirred with multiple radial or radial and axial up-pumping impellers: modelling and measurements. Chem. Eng. Sci., 55:5881–5896.

Walter JF, Blanch HW. (1986) Bubble break-up in gas–liquid bioreactors: break-up in turbulent flows. Chem. Eng. J., 32(1):B7–B17.

Warmoeskerken MMCG, Smith JM. (1981) Hydrodynamics and mass transfer in agitated vessels. In: Proceedings of Seventh International Congress of Chemical Engineers, Prague, 1981. Paper B 3.4.

Warmoeskerken MMCG, Smith JM. (1985) Flooding of disk turbines in gas–liquid dispersions: a new description of the phenomenon. Chem. Eng. Sci., 40:2063–2071.

Whitton MJ, Nienow AW. (1993) Scale up correlations for gas hold up and mass transfer coefficients in stirred tank reactors. In: Nienow AW, editor. Proceedings of Third International Conference on Bioreactor and Bioprocess Fluid Dynamics, Cambridge, BHR Group/MEP, London, UK. p 135–149.

Wong CW, Wang JP, Huang ST. (1987) Investigations of fluid dynamics in mechanically stirred aerated slurry reactors. Can. J. Chem. Eng., 65(3):412–419.

Wu Q, Kim S, Ishii M, Beus SG. (1998) One-group interfacial area transport in vertical bubbly flow. Int. J. Heat Mass Transfer, 41:1103–1112.

Wu J, Zhu Y, Pullum L. (2001a) Impeller geometry effect on velocity and solids suspension. Trans. IChemE, 79(Part A):989–997.

Wu J, Zhu Y, Pullum L. (2001b) The effect of impeller pumping and fluid rheology on solids suspension in a stirred vessel. Can. J. Chem. Eng., 79(2):177–186.

Yagi H, Yoshida F. (1975) Gas absorption by Newtonian and non-Newtonian fluids in sparged agitated vessels. Ind. Eng. Chem. Process. Des. Dev., 14:488–493.

Yaw CL. (1999) CRC Handbook of Chemicals and Physics, various editions; and Chemical Properties Handbook. McGraw-Hill Publishing Company, New York.

Yawalkar AA, Beenackers AACM, Pangarkar VG. (2002a) Gas holdup in stirred tank reactors. Can. J. Chem. Eng., 80(1):158–166.

Yawalkar AA, Heesink ABM, Versteeg GF, Pangarkar VG. (2002b) Gas–liquid mass transfer coefficient in stirred tank reactor. Can. J. Chem. Eng., 80(5):840–848.

Yianneskis M, Popiolek Z, Whitelaw JH. (1987) An experimental study of the steady and unsteady flow characteristics of stirred reactors. J. Fluid Mech., 175:537–555.

Yung CN, Wong CW, Chang CL. (1979) Gas hold-up and aerated power consumption in mechanically stirred tanks. Can. J. Chem. Eng., 57(6):672–676.

Zhang X, Pan Q, Rempel GL. (2006) Liquid phase mixing and gas hold-up in a multistage-agitated contactor with co-current up flow of air/viscous fluids. Chem. Eng. Sci., 61:6189–6198.

Zhu Y, Bandopadhayay PC, Wu J. (2001) Measurement of gas–liquid mass transfer in an agitated vessel—a comparison between different impellers. J. Chem. Eng. (Japan), 34:579–584.

Zhu Y, Wu J. (2002) Critical impeller speed for suspending solids in aerated agitation tanks. Can. J. Chem. Eng., 80(4):682–687.

Zwietering Th N. (1958) Suspending of solid particles in liquids by agitators. Chem. Eng. Sci., 8(3):244–253.

7B

STIRRED TANK REACTORS FOR CELL CULTURE TECHNOLOGY

7B.1 INTRODUCTION

Biopharmaceuticals are therapeutic drugs produced using biotechnology for the treatment of a variety of diseases. They include proteins (including antibodies), nucleic acids (DNA, RNA), or antisense oligonucleotides used for therapeutic or *in vivo* diagnostic purposes. They are produced by means other than direct extraction from a native (nonengineered) biological source (Walsh 2003).

Biopharmaceuticals are produced by genetic engineering or hybridoma technology or through biopharmaceutical techniques such as recombinant DNA technology, gene transfer, and antibody production processes. Genetic engineering involves use of recombinant DNA technology to selectively produce proteins of interest by a host of organisms. This process is now commonly referred to as cell culture technology. The host organism could be a microbial cell (e.g., *Escherichia coli* or yeast), a mammalian cell, or a plant cell.

Cell culture technology had a modest beginning (Table 7B.1). Its roots can be traced to the production of viral vaccines and other therapeutically important products secreted by human and primate cells such as those listed in Table 7B.1 (Karkare 2004; Griffiths 2007):

Starting with this modest beginning of vaccine for the foot-and-mouth disease in 1962, cell culture technology has now progressed to a level where revolutionary treatment for malignant disorders, diabetes, etc. is available. The most common example is that of bone marrow transplant for the treatment of leukemia, other

Design of Multiphase Reactors, First Edition. Vishwas Govind Pangarkar.
© 2015 John Wiley & Sons, Inc. Published 2015 by John Wiley & Sons, Inc.

TABLE 7B.1 Historical Products of Cell Culture Technology[a]

Product	Cell line	Processes	Year introduced and reference
Foot-and-mouth disease vaccine	Baby hamster kidney	SC	Capstick et al. (1962)
Rabies vaccine	Dog kidney	SM	van Wezel and Steenis (1978)
Interferon	Human Namalwa	SC	Johnstone et al. (1979)
Polio vaccine	Monkey kidney	SM	van Wezel et al. (1980)

FMD, foot-and-mouth disease; SC, suspension culture; SM, microcarrier supported.
[a]Reproduced from Karkare (2004) with permission from John Wiley & Sons, Inc., New York, USA. © 2001.

varieties of cancer, as well as blood disorders (Singec 2007; Sekhon 2010). Some of the milestones in cell culture technology are as follows: (i) Bernard's proposition that the physiological functions of an organism can be sustained even after the death of the organism; (ii) Harrison's (1907, 1910) *in vitro* development of frog nerve fibers starting with neurons in the explanted tissue. He found that the cells could live without any functions but when provided with a substrate (a protein matrix derived from clotted lymph in his case) started physiologically functioning as observed from motility and growth. (iii) Carrel's (1923) introduction of the "D flasks" that facilitated feeding and allowed maintaining the cultures for weeks as compared to "days" earlier. At this stage, the term tissue culture was introduced and applied to cultures that could be maintained for more than 24 hr. (iv) First instance of development of a vaccine (vaccinia) by Carrel and Rivers (1927) and (v) Lindbergh's (1939) perfusion flask for cell culture, which demonstrated practicality of perfusion cultures. Clearly, this was the precursor of the contemporary perfusion devices discussed at length in Sections 7B.4.3 and 7B.5. (vi) Development of the spinner flasks (Owens et al., 1953), which were the forerunners of the giant modern cell culture reactors (Nienow 2006).

Major applications of cell culture technologies have been in the production of monoclonal antibodies (MAbs), recombinant therapeutics, and vaccines. "Highlights from a Jan.2011 report of bcclResearch Market Forecasting estimate the global biologics market at 149 billion US $ in 2010. The forecast indicates that by 2015 the global market will be valued at 239 billion US $. This indicates a compound annual growth rate (CAGR) of 9.9% for the period 2010-2015" (http://www.bccresearch.com/report/biologic-therapeutic-drugs-bio079a.html; BIOTECHNOLOGY Report Code: BIO079A. © 2011. Published: January 2011). This report has detailed the main drivers for this rapid growth. Some of these are the necessity for (i) more drug candidates in development stage, (ii) more biosimilars, (iii) attractive targets against challenging diseases, (iv) personalized medicine using pharmacogenomics (Henry 2000), (v) regenerative therapy (Henry 2000), and (vi), finally, development of better manufacturing technologies that include processes/hardware/engineering designs. The aim is not only to cut the overall manufacturing costs and increase profits but more importantly to make the drugs affordable. The last point is important in view of the debate on biologics and healthcare. The report further gives sector-wise market share data for 2009: (i) monoclonal antibodies had 31.9% at 48 billion U.S.$ with an anticipated growth to 86 billion U.S.$ by 2015, and (ii) therapeutic proteins had the largest share (48.8%) at 66 billion U.S.$ in 2009. Market projections for this sector are: U.S.$ 72 billion in 2010 rising eventually to U.S.$ 107 billion in 2015. These data indicate that monoclonal antibodies will be the fastest growing sector at a compound annual growth of 12.4%, followed by therapeutic proteins at 8.2%. Among current bestseller drugs, four have been obtained from the biotechnology segment, while only one belongs to the classical small molecule type (Kayser and Warzecha 2012). Hou et al. (2011) have reviewed the developments in this field. Gonzalez-Valdez et al. (2013) have given an exhaustive discussion on the state of development in the field of vaccine development. They have pointed out that the recent occurrence of A/H1N1/2009 influenza virus pandemic calls for large-scale, worldwide availability of relevant vaccines. These authors have also illustrated the advantages of DNA-based vaccines over first-generation (comprising attenuated (AV) or inactivated viruses (IV)) and second-generation (recombinant) vaccines.

Several important developments, such as the "abbreviated approval pathway for follow-on biologics" and patent laws, have made the biologics market exceptionally attractive for pharmaceutical companies: (http://thomsonreuters.com/content/press_room/science/biologics_US).

The majority of the currently used conventional drugs are small molecules that are synthesized with conventional synthetic chemistry. These lack specificity. As a result, they tend to bind to different unintended targets in the body. A greater part of the side reactions associated with such small molecule drugs is because of this "spill-over" action. Contrary to these conventional drugs, therapeutic biologics are derived from proteins. These are larger molecules produced in living systems (Projan et al. 2004). Biologics are designed to mimic human body's natural processes and to interact specifically with the target molecule in the body. **Majority** of biologics exhibit high specificity and therefore bind to the precise target. An **ideal** biologic would therefore have no side effects in contrast to the conventional (small molecule) chemically synthesized variety. However, these biologics are more complicated molecules due to their size and elaborate folding. These properties give rise to several forms of the active biologic molecule, not all necessarily desirable. Some of these other forms are likely to be present along with the desired form. Further, these undesired forms may have different pharmacological properties as compared to the active form of interest. Thus, it will be wrong to presume that **all** biologics have high specificity. The tragic results of a "first in human" clinical study of TGN1412, an anti-CD28 monoclonal antibody, have brought into focus the need for additional safeguards required in the appraisal of biologics before use in humans (Marshall 2006 a, b; Suntharalingam et al. 2006; Hansen and Leslie 2006; Stebbings et al. 2007). Table 7B.2 gives the salient differences between conventional chemically synthesized drugs and biologics.

Biopharmaceuticals require extensive research and development work that is far more complex in nature as compared to chemically derived drugs. The entirely new nature of these drugs is also likely to take relatively longer time for statutory approvals. Biopharmaceuticals are also expected to be relatively expensive as

TABLE 7B.2 **Salient Differences Between Conventional Drugs and Biologics**[a]

Property	Conventional	Therapeutic biologic
Order of magnitude molecular weight	10^2, single molecule	10^4 and above, mixture of related molecules
Manufacturing process	Straightforward chemical reactions or well-defined biotransformation	Highly complex
Side effects (arising out of nonspecific binding)	Common (may not be critical)	Well-developed biologics have very high specificity
Clinical tests	Potencies are often similar in animals and humans	Binding in homologous animal counterpart could be different. Comparative binding affinities in primate and human antigen essential to avoid unforeseen situations (Marshall 2006a, b; Suntharalingam et al. 2006; Hansen and Leslie 2006; Stebbings et al. 2007)
Cost of treatment	Low: covered by conventional health insurance	High to very high. Health insurance policies need clarity
Risk involved	Low and identified in established drugs	Not established
Post patent scenario	Generics: much lower cost than original (Frank 2007)	Biosimilars/follow-on biologics: costs may not reduce as much as in the case of generics (Frank 2007; Hou et al. 2011)

[a]Sources of information for the above comparison: (i) Biologics: a different class of medications that makes a difference for our patients. White paper: National Physicians Biologics Working Group. http://safebiologics.org/pdf/NPBWGWhitePaper-1.pdf (accessed March 2014). (ii) How do Drugs and Biologics Differ? (2010) Biotechnology Industry Organization. http://www.bio.org/articles/how-do-drugs-and-biologics-differ (accessed March 2014). (iii) Small molecule versus biological drugs: Generics and Biosimilars Initiative (2012): http://www.gabionline.net/Biosimilars/Research/Small-molecule-versus-biological-drugs (accessed March 2014); (iv) Samanen (2013).

compared to the conventional variety because of the profound associated development costs and longer development life cycles for the former. The principal organizations involved in the research and development programs are the leading drug companies although there are some smaller companies that are specifically devoted to biopharmaceuticals. Some of these biopharmaceutical companies have sales exceeding US$ 1 billion per annum. This large and rapidly increasing market is attracting significant investment in both research and industrial exploitation. Development of generic copies of biopharmaceutical products, referred to as "biosimilars," is one such attractive area.

As the demand and approvals for biopharmaceuticals grow, the scale of production has to keep pace. Figure 7B.1 shows the steady progress made by different products derived from cell culture technology.

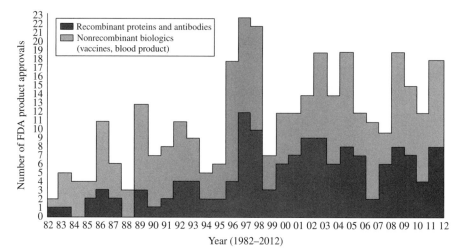

FIGURE 7B.1 FDA approvals for new biopharmaceutical products, 1982–2012. (Reproduced with permission from Biotechnology Information Institute. © 2013, Biotechnology Information Institute.)

List of U.S. Food and Drug Administration (FDA) biopharmaceutical approvals (2010) is given in Table 7B.3.

The links to other biopharmaceutical approvals/applications reproduced in the following indicate the vast range of applications of this class of drugs.

Links on this page:

1. http://www.fda.gov/BiologicsBloodVaccines/BloodBloodProducts/Approved Products/LicensedProductsBLAs/FractionatedPlasmaProducts/ucm217877.htm

2. http://www.fda.gov/BiologicsBloodVaccines/BloodBloodProducts/Approved Products/LicensedProductsBLAs/BloodDonorScreening/InfectiousDisease/ucm210158.htm

3. http://www.fda.gov/BiologicsBloodVaccines/CellularGeneTherapyProducts/ApprovedProducts/ucm210012.htm

4. http://www.fda.gov/BiologicsBloodVaccines/BloodBloodProducts/Approved Products/LicensedProductsBLAs/FractionatedPlasmaProducts/ucm207482.htm

5. http://www.fda.gov/BiologicsBloodVaccines/BloodBloodProducts/Approved Products/LicensedProductsBLAs/FractionatedPlasmaProducts/ucm202630.htm

6. http://www.fda.gov/BiologicsBloodVaccines/Vaccines/ApprovedProducts/ucm201667.htm

7. http://www.fda.gov/BiologicsBloodVaccines/Vaccines/ApprovedProducts/ucm201342.htm. Last Updated: 08/05/2010

(Source:http://www.fda.gov/BiologicsBloodVaccines/DevelopmentApproval Process/BiologicalApprovalsbyYear/ucm201369.htm)

Excellent reviews on various aspects of biotherapeutics are available in the recent literature (Chen and Oh 2010; Allison et al. 2007).

TABLE 7B.3 List of U.S. FDA Biopharmaceutical Approvals (2010)[a]

Trade name/proper name	Indication for use	STN	Manufacturer/license number	Approval date
GLASSIA Alpha1-Proteinase Inhibitor (Human)	Treatment of chronic augmentation and maintenance therapy in individuals with emphysema due to congenital deficiency of alpha-1-proteinase inhibitor (alpha1-PI), also known as alpha1-antitrypsin	125325/0	Kamada Ltd. Beit Kama, 7 Sapir St Kiryat Weitzmann, Science Park P.O. Box 4081 Ness Ziona 74140 Israel License 1826	7/1/2010
ABBOTT PRISM Chagas *Trypanosoma cruzi* (*E. coli*, Recombinant) Antigen	To screen individual human donors, including volunteer donors of whole blood and blood components and other living donors, for the presence of antibodies to *T. cruzi*. The assay is also intended for use in testing serum and plasma specimens to screen organ donors when specimens are obtained while the donor's heart is still beating, in testing blood specimens to screen cadaveric (non-heart-beating) donors. It is not intended for use in testing cord blood specimens	125361/0	Abbott Labs 100 Abbott Park Rd D-49C, AP6C/2 Abbott Park, IL 60064, USA License 0043	4/30/2010
Provenge Sipuleucel-T	Treatment of men with asymptomatic or minimally symptomatic metastatic castrate-resistant (hormone refractory) prostate cancer	125197/0	Dendreon Corp 3005 First Ave Seattle, WA 98121, USA License 1749	4/29/2010

(Continued)

TABLE 7B.3 (Continued)

Trade name/proper name	Indication for use	STN	Manufacturer/license number	Approval date
TachoSil Fibrin Sealant Patch	Fibrin sealant patch indicated for use as an adjunct to hemostasis in cardiovascular surgery when control of bleeding by standard surgical techniques, such as suture, ligature, or cautery, is ineffective or impractical	125351/0	Nycomed Danmark ApS Langebjerg1 DK-4000 Roskilde Denmark License 1825	4/5/2010
Fibrinogen (for further manufacturing use)	Fibrinogen (human) active substance supplied by CSL Behring for further manufacturing of fibrin sealant patch by Nycomed	125356/0	CSL Behring GmbH 1020 First Ave P.O. Box 61501 King of Prussia, PA 19406, USA License 1765	4/2/2010
Thrombin (Human) (for further manufacturing use)	Thrombin (human) active substance supplied by CSL Behring for further manufacturing of fibrin sealant patch by Nycomed	125357/0	CSL Behring GmbH 1020 First Ave P.O. Box 61501 King of Prussia, PA 19406, USA License 1765	4/2/2010
Hizentra Immune Globulin Subcutaneous (Human), 20% Liquid	Treatment of primary immunodeficiency (PI)	125350/0	CSL Behring AG1020 First Ave. P.O. Box 61501 King of Prussia, PA 19406-0901, USA License 1766	3/4/2010

Product	Description	License No.	Manufacturer	Date
Prevnar 13 Pneumococcal 13-valent Conjugate Vaccine (Diphtheria CRM197 Protein)	For active immunization for the prevention of invasive disease caused by *Streptococcus pneumoniae* serotypes, 1, 3, 4, 5, 6A, 6B, 7F, 9V, 14, 18C, 19A, 19F, and 23F, for use in children 6 weeks through 5 years of age; for the prevention of otitis media caused by *S. pneumoniae* serotypes 4, 6B, 9V, 14, 18C, 19F, and 23F	125324/0	Wyeth Pharmaceuticals, Inc 401 North Middleton Road Pearl River, NY 10965, USA License 0003	2/24/2010
Menveo Meningococcal [Groups A, C, Y, and W-135] Oligosaccharide Diphtheria CRM197 Conjugate Vaccine	For active immunization to prevent invasive meningococcal disease caused by *Neisseria meningitidis* serogroups A, C, Y, and W-135 when administered to individuals 11 through 55 years of age	125300/0	Novartis Vaccines and Diagnostics, Inc. 4560 Horton St Emeryville, CA 94608-2916, USA License 1751	2/19/2010
Enbrel	For reducing joint inflammation by blocking chemical tumor necrosis factor (TNF)		Amgen	
Humira	TNF blocker for moderate-to-severe rheumatoid arthritis, juvenile idiopathic arthritis, psoriatic arthritis, ankylosing spondylitis, moderate-to-severe Crohn's disease in adults who have not responded well to other treatments, and moderate-to-severe chronic (lasting a long time) plaque psoriasis		Abbot	

Source: U.S. Federal Drug Administration.

[a]Reproduced from http://www.fda.gov/BiologicsBloodVaccines/DevelopmentApprovalProcess/BiologicalApprovalsbyYear/ucm201368.htm (accessed November 2010).

7B.2 THE BIOPHARMACEUTICAL PROCESS AND CELL CULTURE ENGINEERING

A biopharmaceutical process begins with production of cell cultures in the laboratory using T-flasks to multilayer flasks. As an example, a 10-layer flask can provide surface area of the order of 6000 cm^2 and yield a cell harvest ranging from 150 to 800 million cells per flask (Rowley 2010). The cell number is increased in steps till a sufficiently large quantity is available as a feed to the production-scale reactor. Each such step is part of the overall scale-up for commercial-scale production of biopharmaceuticals. Evidently, the scale-up of a bioprocess needs to be carried at three levels: (i) laboratory scale in small flasks for preliminary studies to obtain basic conditions for the metabolism desired, (ii) pilot plant in which the optimum parameters for the desired metabolism are established, and (iii) commercial production. The culture required for the last stage also has to go through the three stages but more so in terms of its quantity (Garcia-Ochoa and Gomez 2009). Evidently, the scale-up factor in this case (Table 2.1) is relatively very low as compared to chemical reactors. The overall process is defined by conversion of laboratory protocols to a process for the final product. This includes cell development and characterization, optimization of the medium for cultivation, as well as the succeeding processes for cell separation and product recovery (Sharma et al. 2011; Fernandes et al. 2013). The segment of the plant wherein cells are multiplied and allowed to produce the desired product is called the upstream section. The basic unit in this section is a reactor in which conditions that are amenable to the growth (and modification) and expression of the basic cells are maintained. The segment in which products from the upstream section are subsequently processed for purification to the desired level is termed downstream processing section. Chemical engineering is relevant to the reactor part in the upstream and separation part in the downstream sections. This downstream section is the last part of the overall process sequence and includes various steps that start with product harvesting, followed by separation by both conventional and special techniques such as affinity separations and chromatography. Downstream processing is not included in the scope of this book. The recent review of Fernandes et al. (2013) covers several aspects in both upstream and downstream processings, product validation for clinical trials, regulatory guidelines, quality control, as well as clinical trial pipeline.

7B.2.1 Animal Cell Culture *vis-à-vis* Microbial Culture

After the initial fervor about microbial expression of recombinant DNA (rDNA) proteins, it was realized that this method was not suitable for large proteins that needed to be correctly folded to ensure desired activity. Animal cells, on the other hand, have this ability to fold and process such extremely intricate proteins (Lubiniecki 1998). For instance, proteins that require substantial glycosylation as well as appropriate folding or need a proper folding environment owing to the abundance of disulfide bonds are best derived from animal cells. Bacterial cellular machinery does not have the ability to make complex biomolecules such as proteins, which require glycosylation or posttranslational modifications such as phosphorylation and acetylation. In contrast, animal cells secrete the desired product and perform both glycosylation and phosphorylation competently (Yang and Basu 2006). Mammalian cells have by far the most advanced and complex cellular machinery. They can perform most functions and can mimic human cells more closely. Under favorable conditions, stem cells can divide for relatively longer periods of time both *in vivo* and *in vitro*. *In vitro* stem cell culture is relatively complicated. However, comparatively much better protocols that afford reliability have been devised over the last decade.

Against these advantages, there were several major reservations concerning animal cell-derived products that prevailed in the 1980s. These related to possible tumorigenicity of persons receiving products of animal cell origin. The example of human immunodeficiency virus (HIV) transmitted through blood/plasma products (even contaminated syringes) was a major deterrent. As pointed out by Lubiniecki, a clear regulatory roadmap leading to maturity of biological product for human use was lacking. Subsequent research has resulted in allaying some of these major concerns. As a result, a large number of animal cell-derived products have been approved or are in clinical trials (Table 7B.3). Walsh's (2010) review of product approvals for 2006–2010 shows that 32 out of the 58 products approved were based on mammalian cell lines. This is an indication of the promise of animal cell culture products.

7B.2.2 Major Improvements Related to Processing of Animal Cell Culture

The development of several processing tools has been critical to the progress of cell culture technology. Lubiniecki (1998) has given a historical perspective of cell culture engineering including major developments up to 1998. One of these major developments cited by Lubiniecki was (i) the availability of reliable 0.1 μm filters to effectively eliminate mycoplasma contamination from animal serum. Prior to this, contamination was a major problem and required complex methods such as irradiation (found to be ineffective) or alkylation agents (effective but risky). These filters could also effectively prevent contamination due to bacteria such as pseudomonads. The next major advances were (ii) the development of serum-free media to allow large-scale cultivation of hybridomas and Chinese hamster ovary (CHO) cells; (iii) development of sanitary designs such as diaphragm valves along with steam block devices and steam in place (SIP) and clean in place (CIP) equipment, which lent confidence for effective batch processing without contamination from a previous batch; (iv) extensive use of affinity chromatography and ion exchangers that could be base sanitized; and finally (v) availability of sophisticated analytical tools for proteins. All these have allowed consistency in processes and their products (Lubiniecki 1998). The approach to large-scale bioprocessing of animal cell culture has consequently changed from more or less empirical to relatively rational. Sophisticated techniques that distinguish large-scale cell state coupled with innovative procedures for host cell modeling are likely to yield better optimization of next-generation bioprocesses (Carinhas et al. 2012). Downstream processing has also evolved over the years to yield more mature technologies. Hubbard (2009 *cf* Monk 2009) has discussed related developments in separation technologies in this respect.

7B.2.3 Reactors for Large-Scale Animal Cell Culture

The requirements of reactors for cell culture processes are significantly different from conventional reactors for chemical reactions. In view of this, special attention needs to be given while designing these (Monk 2010). Close interaction of chemical engineers with experts in molecular biology is necessary for development of production processes for biopharmaceuticals. In terms of scale-up of a reactor for biopharmaceuticals, besides other considerations, the viability of the cell, its physiological functions, etc. must be maintained on the larger scale. In Section 5.2.5, the additional concept of physiological similarity was introduced for bioreactors. This similarity implies that the microenvironment of the living cell must be identical to reproduce the same physiological functions (e.g., growth, substrate consumption or product formation rates, etc.) in the large-scale bioreactor and the laboratory model. Since cell cultivation, multiplication, and product expression form a highly complicated sequence, the scale-up needs to be stagewise, as mentioned earlier and reflected in Table 2.1. A major difference between chemical reactors and bioreactors is the relative sensitivity of the respective reactants/products. Chemical reactants and products can be affected by severity of conditions of which temperature is a major factor. Techniques for control of reactor temperature that prevent degradation of the chemical molecule are well established. On the other hand, living cells, and in particular mammalian cells, are highly sensitive to the micro- and macroenvironment defined by supply of nutrients, removal of toxins, pH, etc. Indeed, viability of mammalian cells is affected by factors that may be innocuous to chemical molecules. As an example, simple gas sparging/bubble bursting has been found to be the main reason for damage to sensitive mammalian cells, as discussed in Section 7B.8. This chapter is meant to give a chemical engineer's perspective and approach to bioreactor design.

7B.2.3.1 Stem Cell Engineering   In the overall field of biotherapeutics, stem cell engineering is a major area of current research. In view of its growing importance, a brief overview of this subject is presented in the following.

Stem cell therapy involves treatment of a disease through introduction of specific cells into the tissues. Exhaustive contemporary treatment covering basic biology, laboratory methods, stem cell therapy, stem cells, and diseases is available in a recent two-volume compilation edited by Meyer (2013). Stem cells are so named because the other cells in the human body "stem" from them. Commonly used stem cell therapies are based on (i) bone marrow stem cells, (ii) peripheral blood stem cells, and (iii) umbilical cord blood stem cells. The last type is unique to a person and is expected to have far-reaching implications for developments in stem cell therapies. Umbilical cord storage in banks is being already practiced in societies that can afford the cost of such storage. Stem cell treatment uses regenerative cell-based therapies, disease models, and drug screening platforms (Cortes-Caminero 2010). A number of these therapies will be aimed at a variety of afflictions that are challenging the human race. The list includes regeneration of failing organs, cardiovascular conditions, diabetes, multiple sclerosis, Parkinson's and celiac disease, muscle cartilage and bone damage (Porter et al. 2009), etc., besides the omnipresent cancer threat (Walsh

2004; Malda et al. 2004; Murphy et al. 2010; Singec et al. 2007). Bone marrow transplant to treat different types of blood related diseases, including leukemia, is a very well known and successful therapy. *Osteogenesis imperfecta*, a disease caused by formation of abnormal type I collagen, can lead to slow bone development, frequent fractures, and bone distortion. Horwitz et al. (1999) showed that transplantation of mesenchymal stem cells (MSCs) resulted in normal osteoblasts. This in turn resulted in rapid bone growth and reduced frequency of fracture. The use of purified allogenic bone marrow MSCs gave further improved therapeutic results (Horwitz et al., 2002). Monk (2010) has reviewed the application of stem cell engineering for treatment of variety of diseases and malignant conditions. It is evident that rapid strides are being made in the development of stem cell therapies. Eventually, stem cell-based therapies are expected to change the currently practiced treatments for several diseases/disabilities in the near future (Fernandes et al., 2013). As mentioned earlier, high cost of such therapies is a major hurdle. Kelley's (2009) analysis reveals that introduction of large production capacities that use the innovative methods described earlier have led to increase in titers and significant reduction in costs.

The major difference between stem cell bioprocess and conventional processes (such as those for protein production) is in terms of the end product. In the former, the cell is the end product, whereas in the latter the substances expressed by the cell are the focus. Another dissimilarity is that stem cells are generally grown on suitable compatible microcarrier surface for proliferation (Section 7B.8.1.2) (Steiner et al. (2010) and Olmer et al. (2012), however, suggest that stem cell expansion is also feasible in large-scale suspension cultures). On the other hand, in the case of recombinant protein production, suspension-type cultures are preferred although the host cells in this case also exhibit good adherence. Chen and Oh (2010) and Chen et al. (2013) have discussed the advantages and drawbacks of microcarrier-based expansion. For most autologous applications that are patient specific, laboratory-scale cultures will be sufficient. In some applications (beta cell replacement for type I diabetes, liver cell transplantation, treatment of spinal cord injury, retinal pigment epithelial transplantation, etc.), however, relatively large quantities of cells (of the order of 10^{10}–10^{12}) in each defined batch may be needed for clinical applications (Sharma et al. 2011; Chen et al. 2013; Fernandes et al. 2013). This will require production-scale plants rather than laboratory-scale cultivation. Sharma et al. (2011) have discussed various aspects of process scale-up in stem cell engineering. Basic aspects of stem cell engineering as outlined by Monk (2010) are covered in the following discussion. The objective here is to familiarize chemical engineers with the subject and its scope as related to their discipline.

Types and Sources of Stem Cells

EMBRYONIC STEM CELLS   A fertilized egg divides to form two cells. This binary division process continues till a mass of ~150 cells is formed. This mass called blastocyst consists of two regions: the outer shell and an inner cell mass encapsulated by the outer shell. It is possible to extract this inner cell mass. The extracted mass can be cultured to create stem cells. These stem cells are "pluripotent," a term indicative of their multiple potencies for forming "any cell type" under the proper impulse.

Another method for obtaining such pluripotent stem cells is through what is known as somatic-cell nuclear transfer (SCNT). This technique involves the separation of the nucleus from an unfertilized egg. The vacancy thus created is filled with the nucleus of a mature donor cell. The donor cell contains the basic genetic information. This combination of an unfertilized egg coupled with an external donor cell divides under suitable conditions to yield a blastocyst. These blastocysts are then processed to extract the pluripotent stem cells. Because of their origin, such stem cells are called embryonic stem cells. The major advantages of SCNT method are (i) superior therapeutic potential arising out of genetic similarity with the mature donor cell and (ii) transplantation of the stem cells/their tissue derivatives into the original donor without problems of rejection/immunosuppression that are common to transplants from dissimilar donors.

ADULT STEM CELLS   These are present all over the body deep inside many tissues and organs that require a continuous replenishment of new cells or replacement of old dying cells. Few examples are blood and skin. Adult stem cells, in simple language, are the source of continuing life. These stem cells are not pluripotent in the sense that they belong to a specialized class, which can perform limited functions such as differentiating into a limited type of cells. They clearly lack the versatility of embryonic stem cells. Their propagation is also relatively slow. Therefore, it is difficult to maintain them and promote their growth in cultures. Adult stem cells are closer to clinical trials and commercial applications than embryonic or induced pluripotent cells (Lapinskas 2010).

INDUCED PLURIPOTENT (iPS) CELLS   These cells are mature, somatic, non-germline-type cells. They can be derived from, for example, skin on the hand. In culture systems, these cells can be converted to the stem cell variety. Once converted to the latter state, they can differentiate into the desired mature cell type of the body. Reprogramming with three or four properly chosen genes can induce pluripotency in the mature stem cells. Successful research in this area has been acknowledged through the Nobel Prize for 2012 in physiology or medicine jointly awarded to John B. Gurdon and Shinya Yamanaka (http://www.nobelprize.org/nobel_prizes/medicine/laureates/2012/press.html) for the discovery that mature cells can be reprogrammed to become pluripotent. A major advantage of these nonembryonic pluripotent cells route is that they can overcome the strong ethical reservations concerning use of stem cells derived from embryos. Further, since these iPS cells are derived from the patients themselves, the chance of rejection in organ transplants is greatly reduced. iPS cells afford an added advantage that they can be used to develop patient-specific disease models using susceptible cells and understand the genetic causes of the disease. Such models can lead to better therapies. The major drawback of iPS cells is that some of them are not truly pluripotent because they cannot differentiate into specific classes of tissues. In addition, it has been found that several genes used in converting mature cells into the "pseudo" pluripotent cells have links to cancer. Obviously, considerable research efforts are needed to overcome the drawbacks and reap the advantages of the iPS variety.

The expansion and differentiation of stem cells is a very complex process. The differentiation of stem cells into specific lineages is brought about by a combination of various stimuli, which include molecular and cellular signals, physical stresses, and electrical impulses (Monk 2010). Careful control of the conditions that allow differentiation of the cells into a specific type of desired cell is referred to as **engineering of cell fate** (Monk 2010; Fernandes et al. 2013).

7B.3 TYPES OF BIOREACTORS

Basically, two types of bioreactors are used in biopharmaceutical applications. These are (i) sparged reactors and (ii) stirred reactors. A detailed discussion on the afore-mentioned reactors is provided in Chapters 7A and 10, respectively. However, these chapters deal mainly with the use of these reactors for chemical reactions in which the compounds involved are not biologically active, fragile, or sensitive to shear. Therefore, a brief discussion on the relative merits of sparged and stirred reactors for the specific applications pertaining to this chapter is in order. Sparged reactors have the obvious advantage of simplicity of construction and operation. This advantage is particularly important for large-scale operation where mechanical problems associated with relatively large agitation systems need to be addressed. Sparged reactors do not have a rotating part that requires sealing. The importance of sealing in a cell culture stirred reactor is with reference to a reason much different from, for example, that in a stirred reactor for hydrogenation (Chapter 7A). This aspect of mechanical seals is discussed in Section 7B.16.4. In the early stages of mammalian cell culture technology, it was believed that the fragile cells are damaged due to the mechanical shear of the impeller. Subsequent investigations have found this to be erroneous (Section 7B.8). Among sparged reactors, the airlift variety offers better directed liquid circulation than a plain bubble column. Two subtypes in this class are (i) external circulation loop through a downcomer and (ii) internal circulation loop created by a draft tube. The performance of sparged reactors may depend on the type of gas sparger used. Sintered/porous spargers afford relatively high values of the effective interfacial area and hence gas–liquid mass transfer coefficient $k_L a$. Stirred reactors, on the other hand, generally operate at lower gas flow rates, use low pressure drop pipe orifice spargers, and yet yield equivalent performance, particularly at low sparging rates. It is now customary to prefer a stirred reactor in almost all new designs for mammalian cell culture systems. Further, in view of the evidence on stability of the cells to mechanical shear, the upper limit of power input in stirred reactors can be relaxed. This aspect is important in operation with high cell densities and/or viscosity of the media. Thus, stirred bioreactors are the current favorites for this application (Boraston et al. 1984; Varley and Birch 1999; Nienow 2006; Fenge and Lullau 2006; Eibl and Eibl 2009). This matter is discussed in detail in Section 7B.8. Some of the functions in this application are similar to stirred reactors for chemical reactions. These are (i) gas–liquid mass transfer, (2) mass transfer of dissolved gas to the liquid and then to the cells (suspended or anchored on microcarriers), and (iii) homogeniza-tion of the liquid phase for mixing of the nutrients/additives to control pH. Over the last decade, several studies have brought out the merits of upflow impellers (Nienow and Bujalski 2004; Nienow 2006) along with high solidity ratio hydrofoil impellers (McFarlane and Nienow 1996; Collignon et al. 2010; Bustamante et al. 2013). These impellers yield stable operation unlike the downflow impellers that were considered to be superior earlier.

7B.3.1 Major Components of Stirred Bioreactor

The major components of a stirred bioreactor are:

1. A jacketed stainless steel vessel without any sharp corners. Sharp corners facilitate growth/fouling, which is difficult to remove. Standard flanged dish ends are used as a cover with provision for agitation assembly, feed/outlet nozzles, etc. as per the requirement of the process planned. The relevant details are provided in Section 7B.16.
2. Suitable impeller system with variable speed drive, which gives the desired gas dispersion/solid suspension/liquid-phase mixing.
3. Properly designed gas sparger with additional controls for adjusting the head-space gas composition. Gas sparging system needs to be flexible with facility for using air/pure oxygen and also nitrogen/carbon dioxide when required.
4. Cleaning and sterilizing system.
5. Dissolved oxygen measurement and control system.
6. pH measurement and control system.
7. Temperature measurement and control system using the jacket as the heat transfer area. Internal cooling coils should be avoided as far as possible. Surface deposition on internal cooling coils can be a serious problem. These cooling coils may be difficult to clean in place resulting in problems with the succeeding batch. In addition, as discussed in Chapter 7A (Sections 7A2, 7A.8, and 7A.10.8), internal cooling coils suppress turbulence and consume relatively greater power for the same overall mass transfer and mixing effect.
8. Facility for providing injection grade water for use in preparing start-up solutions as well as nutrients, pH control solutions, etc.

Section 7B.16 gives a more detailed discussion pertaining to special precautions that are needed in specifying bioreactor components.

7B.4   MODES OF OPERATION OF BIOREACTORS

Three possible modes available for operation of the bioreactor are shown in Figure 7B.2.

The decision on the mode to be used is based on the type of product to be derived from the process and process requirements in terms of growth and production of the desired product (Bartow and Spark 2004). As noted earlier, the market for

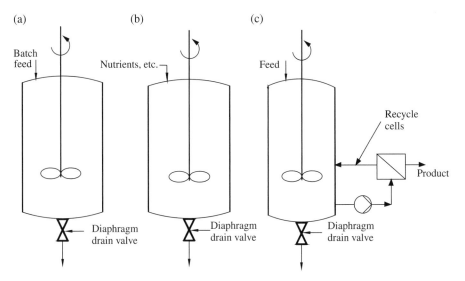

FIGURE 7B.2   Different modes of operation of bioreactors.

biotherapeutics is growing rapidly. Therefore, the scale of operation and the corresponding production facilities required must match this growth. Another reason for the growing demand is greater requirements of products brought about by the order of magnitude higher dosages (grams against milligrams previously). In view of the earlier discussion, the size of bioreactors is steadily growing (Werner et al. 1992; Griffiths 1992; Lubiniecki and Lupker 1994). Contemporary commercial reactors having working volumes of 20 m^3 or larger are common (Kadouri and Spier 1997; Hoeks et al. 2004; Nienow 2006).

7B.4.1   Batch Mode

This mode has been traditionally favored largely because of the flexibility associated with it. All the ingredients are fed together at the start of the operation. The contents are withdrawn and processed at the end of the operation. It is relatively easy to maintain aseptic conditions in this mode because the number of additions is limited. The cultivation of secondary products, which are nongrowth associated, is also relatively easy in batch reactors. Regulatory bodies such as the FDA require each unit of dosage form to be identified by a batch number for tracking and quality assurance purposes (Karkare 2004). This is certainly one of the reasons why most Active Pharmaceuticals Ingredients are manufactured by the batch process. The other obvious reason is their "low volume/high value product" character that may call for campaign-based manufacturing. In view of the preceding text, the majority of biopharmaceuticals are manufactured using either batch or fed-batch mode (discussed later). The general experience that simple technologies are relatively robust and dependable on large scale as compared to intricate technologies has also favored batch processes. Thus, notwithstanding significant advantages and progress made in continuous processing, the simpler batch suspension culture mode may still be an attractive alternative (Lubiniecki 1998).

7B.4.2 Fed-Batch or Semibatch Mode

In this mode, the operation starts in a manner similar to the batch mode, but nutrients are fed according to a predecided program. The withdrawal can be intermittent or at the end of the batch time. Periodic addition can create problems in maintaining an aseptic operation. However, fed-batch mode has the advantage that it allows better control on the availability of the nutrients. An example is of cells that produce inhibitory metabolites in the presence of high glucose concentration levels. Curtailing the temporal availability of glucose can prevent any increase in the concentration of inhibitory species and consequently allow a smoother biogrowth.

Whitford (2006) has listed the following advantages of the fed-batch mode:

1. Variant of batch mode, operated with cells that are already well characterized for the specific application.
2. Rapid, inexpensive and easy implementation relative to other systems.
3. Good reproducibility and predictability of large-scale system modes (after batch).
4. Flexibility in materials employed, process design, adaptation, and optimization of the process to suit the requirements.
5. Affords production capacities higher than the batch mode and also comparable to the best available.
6. Controlled and regulated addition of important ingredients like glucose at appropriate stages of the batch. This provides control over absolute as well as relative nutrient concentrations and consequently affords **reliable production of consistent lots.**
7. **Advantages mentioned in 6 above provide relative ease of characterization and validation that is vital from the commercial and regulatory angle.**
8. Use of specific optimization / implementation strategies available in the public domain is possible.
9. Adapts quite well to production of secondary metabolites that may include non–growth-associated products like recombinant proteins.
10. **Affords a smaller footprint, uses media and other feedstock efficiently. In addition this mode produces lower quantities of waste materials/stream as compared to perfusion.**
11. **Easy and efficient harvest of raw products as well as downstream processing for obtaining the final product and its approval.**

(Whitford 2006)

Aspects relevant to sustainable processing and regulatory requirements, which are of paramount importance in the current scenario, are highlighted in the above list.

7B.4.3   Continuous Mode (Perfusion)

The advantages of semibatch processing do not imply that continuous processing has been neglected as a means of achieving higher production rates at lower costs. On the other hand, considerable research is being devoted to make continuous operation feasible in the near future.

By definition, this mode implies continuous addition of nutrients and removal of products. In the biotechnology industry, this type of operation is popularly referred to as perfusion. A higher yield of primary products is the major advantage of continuous operation. It is evident that continuous addition/removal can impose constraints on aseptic operation. Advantages of continuous operation are (i) sound theory and (ii) high productivity, especially in dedicated manufacture of limited number of products or for producing labile products (Whitford 2006; Karkare 2004). The low residence time, particularly in high-density culture systems, restricts the contact of the product with enzymes capable of degrading it. Continuous systems are amenable to better process control and optimization. According to Lim et al. (2006), annual production of a 1 m^3 perfusion-based process can be equivalent to that obtained from a 5 m^3 fed-batch process. Perfusion processes can be based on either suspended cells or cells attached to microcarriers. The latter has the obvious advantage that it does not require an elaborate cell retention and recycle system. In the case of anchored cells, once the cells are attached to the microcarrier and merge, they can continue to produce in a protein-free medium for large periods (weeks to months) depending on the cell line.

The major drawbacks of continuous operation are (i) the use of pumps for circulating the cell broths (in suspension systems) that can damage sensitive cells and (ii) elaborate process control for dissolved oxygen, CO_2, pH, and temperature that is more crucial as compared to the relatively flexible fed-batch systems (Catapano et al. 2009).

7B.5   CELL RETENTION TECHNIQUES FOR USE IN CONTINUOUS OPERATION IN SUSPENDED CELL PERFUSION PROCESSES

In contemporary culture technology, there is a perceptible trend for continuous operation with relatively high cell densities. This is achieved by continuously separating the cells from the harvest (product) and recycling them to the reactor. Several excellent reviews on cell retention techniques are available in the literature (Scheirer 1988; Tokashiki and Takamatsu 1993; Woodside et al. 1998; Castilho and Medronho 2002; Kompala and Ozturk 2006). Voisard et al. (2003) have reviewed the techniques reported prior to 2003. Major part of the discussion that follows will refer to this review as well as Nienow (2006). Separation of cells can be achieved on the basis of differences in size, density, electrical charge, and hydrophobicity. Among the various separation techniques available, only two based on (i) size and (ii) density are used industrially. The retention technique/

device should have the following desirable features (Woodside et al. 1998; Castilho and Medronho 2002):

1. It should not have a deleterious effect on the viability of the cells.
2. It should retain live cells in preference to dead cells.
3. It should be possible to clean and sterilize the device "in place" (CIP and SIP).
4. It should provide stable uninterrupted operation.
5. It should have compact yet appropriate design for large-scale operation.

Cell retention is an important characteristic of a device used to retain the cells in the reactor while removing the products continuously. It is defined as

$$R_C = 1 - \left(\frac{C_{\text{Harvest}}}{C_{\text{Bioreactor}}} \right) \qquad (7B.1)$$

The available devices can be classified in two categories: (i) internal or (ii) external to the bioreactor. Each has its merits/drawbacks. In the internal type, the cells are maintained in the same favorable environment that is conducive to growth and expression without significant disruption. The drawback, however, is increased complexity due to additional internals of the retention system inside the bioreactor. These internals generally have a negative impact on the mixing time (Section 7A.8) that is vital to physiological similarity. In contrast, an external retention loop has an inherent flexibility (also see Section 8.3.4). Further, it does not interfere with the internal hydrodynamics of the bioreactor. Unfortunately, the biological species are exposed to unfavorable conditions each time they pass through the external retention system. This can lead to cell death and also change the biological pathways/end products.

7B.5.1   Cell Retention Based on Size: Different Types of Filtration Techniques

7B.5.1.1   Cross-Flow Filtration   The cells are relatively small in size, and therefore, conventional fabric filters are not relevant. Filtration media used for cell recycle are mostly of membrane type such as microfiltration membranes. These membranes have been advantageously used for separating cells from the broth in fermentation processes. However, the variety of cells and the end product in these fermentations are limited. In addition, for example, in ethanol fermentation, requirement for aseptic operation is not as stringent. Fouling or clogging is an inevitable problem with all types of membranes (Mulder 2000). Membranes made by the phase inversion process have a highly intricate pore structure that leads to deposition of particulate matter inside the pores. A typical phase inversion membrane cross section is shown in Figure 7B.3.

Phase inversion membranes belong to the "depth filtration" category. Cells entering the membrane through a wide pore opening tend to plug the membrane at the point of least cross section in the skin. Deposits in these interior pores are extremely difficult to dislodge by conventional methods such as backwashing. In addition, cells trapped in the pores are exposed to an environment, which is entirely

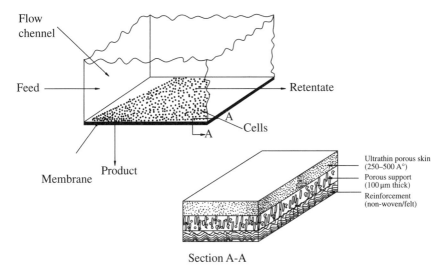

Section A-A

FIGURE 7B.3 Cell retention using cross-flow membrane filtration and typical phase inversion membrane cross-sectional view.

different from that existing in the reactor. This can give rise to undesired metabolic pathways. In contrast to phase inversion membranes, track-etched membranes (http://www.whatman.com/Membranes.aspx: http://www.whatman.com/NucleporeTrack EtchedMembranes.aspx; accessed September 2013) have a narrow pore size distribution. Track-etched membranes are "surface filtration type." They afford high flow rates and are available in materials that have excellent chemical/thermal resistance. Consequently, the extent of fouling for such membranes is expected to be lower than that for the phase inversion type. They are available with both hydrophilic and hydrophobic surfaces. However, literature reports indicate that fouling cannot be entirely prevented. Most of the reported work on membrane filtration of cells has dealt with fermentations involving bacteria and yeast that are less fouling.

   Membrane modules used are classified by the type of flow and membrane geometry. Cross-flow filtration has been widely studied (Fig. 7B.3). In this mode, the broth is pumped parallel to the surface of the membrane. The membrane prevents live/dead cells as well as debris from passing across the skin. The objective is to create a flow past the surface that generates a shear force. This shear decreases the tendency of the cells to deposit on the membrane surface and foul it. It is known that fouling can be reduced by the use of high cross-flow velocities. However, there is an apprehension that the accompanying shear can damage the cells (Section 7B.8). Another important aspect is the possible effects of an external membrane filtration loop on cell viability (Zhang et al. 1993). This can be circumvented by having the filtration module inside the bioreactor similar to silicone rubber membrane-based aeration systems discussed in Section 7B.15.2. Different designs have been investigated to provide higher membrane areas and management of surface fouling. However, most membranes tended to foul in spite of periodic backwashes (Caron et al. 1994). Nucleopore membranes that are supposedly less prone to fouling as mentioned earlier were also found to foul (de la Broise et al. 1991, 1992). Hollow fiber membranes that can provide two orders of magnitude higher surface areas as compared to spiral wound flat sheet membranes were the next obvious choice (Hiller et al. 1993; Zhang et al. 1993; Kyung et al. 1994). However, the problem of fouling persisted. It must also be noted that the best membranes exhibit retention of a range of compounds having different sizes/molecular weight. An additional complicating factor is the change in retention of a "fouled" membrane as compared to a fresh membrane. This is because fouling deposits form a dynamic membrane having a much lower pore size (or molecular weight cut off). Thus, it is difficult to obtain selective retention of cells. A certain portion of high-molecular-weight compounds is also retained (Hiller et al. 1993). This has an adverse effect on product stability. The cost of the membrane system is dictated by sterilization requirements. Thus, the choice for membrane material will favor relatively expensive high-temperature-resistant materials. In conclusion, the available literature indicates that conventional cross-flow microfiltration is yet to mature into a process that can be applied on large scale because of problems listed earlier.

7B.5.1.2 *Controlled Shear Filtration* An alternative to conventional cross-flow membrane filtration is controlled shear filtration or dynamic filtration. In these devices, a relatively high shear is generated at the surface of the filter medium by means of relative motion. Different geometries of the rotor have been used such as concentric cylinder (Rebasmen et al. 1987; Kroner and Nissinen 1988), cone-shaped disc (Vogel and Kroner 1999; Vogel 2002), and plain disc rotor (Castilho and Anspach 2003). The principle involved in this technique is the generation of Taylor vortices.

Taylor vortices (Taylor 1923) were discovered in the annulus of two rotating concentric cylinders. They are a class of flow instabilities that can be advantageously used for wet classification of solid particles in liquid phase. They appear as toroidal vortices placed in a structured pattern, one above the other in the annulus of two concentric rotating cylinders. An investigation by Ohmura et al. (2005) successfully used Taylor vortices to classify and eventually separate a distinct particle phase irrespective of the particle size. Taylor vortices are equally efficient for relatively low density difference between the solid and fluid phases. These vortices can also be formed with the rotation of only one cylinder. However, Couette (1888, 1890) showed that in the case of a single rotating cylinder the phenomenon was observed only with an inner rotating cylinder surrounded by a fixed outer cylinder. A rotating outer/fixed inner cylinder configuration yielded only rotational motion. This type of flow is called Couette flow and it forms the basis of the well-known viscometer. Taylor vortices are formed only above a certain critical speed of rotation of the inner cylinder. The latter type is now known as Taylor–Couette flow, and the device used for segregating the solids as Taylor–Couette device. The structured vortices (Fig. 7B.4) placed one above the other move in opposite directions and have a wavelength that is approximately twice the annular gap between the cylinders.

In a typical vortex flow filter, the motion of the cells between the two cylinders is dictated by settling under applied centrifugal force and transport due to the vertical motion of Taylor vortices. Wereley and Lueptow (1999) simulated the flow field of

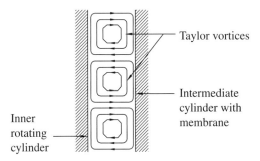

FIGURE 7B.4 Taylor vortices.

the configuration used in devices using Taylor vortices. They showed that neutrally buoyant particles tend to follow the streamlines more than particles that are heavier than the liquid. Particles that are at the outer rim of a vortex are transported inward, whereas particles at the center of a vortex travel toward the outer boundary of the vortex. The particles cannot break away from the vortex limits defined by the gap between the two cylinders and the imposed angular velocity. Consequently, the cells stay within the limits of the vortex at a distance from the porous filter surface. In view of this separation, of the cells from the filter surface, there is a significant reduction in the fouling of the latter. This phenomenon that prevents accumulation of the particles on the filter medium may be termed hydrodynamic hindrance. It is characterized by a dimensionless number, Taylor number:

$$Ta = \left(\frac{\omega r_{\mathrm{I}}(r_{\mathrm{O}} - r_{\mathrm{I}})\rho_{\mathrm{L}}}{\mu_{\mathrm{L}}} \right) \tag{7B.2}$$

The critical speed of rotation of the inner cylinder is expressed in the form of a critical Taylor number Ta_{C}. At this speed, the steady Couette flow is replaced by the vortex-type structures commonly referred to as Taylor vortex pattern. Above Ta_{C}, the centrifugal forces outweigh the viscous forces, thereby causing the instability. Performance of vortex flow filter (VFF) was investigated by Mercille et al. (1994, 2000) and Roth (1997). This device consists of three concentric cylinders. The cell suspension flows in the annular gap between the innermost rotating and fixed intermediate cylinder. The latter supports the filtration media. In the investigation of Mercille et al. (1994), a hydrophilic poly(sulfone) membrane was used. This membrane encountered fouling problems. Further, the perfusion rate did not increase in the same proportion as the surface area of the filter media. Roth (1997) obtained satisfactory performance over an extended period of 3–60 days. Taylor vortices are generated when the annular gap is relatively very small (~2–3 mm). Therefore, maintaining this small gap on large scale will require very careful mechanical design that should include dynamic balancing of the large inner rotating cylinder (Voisard et al. 2003).

Vogel and Kroner (1999) and Castilho and Medronho (2002) developed two separate dynamic filtration devices that use the same principle (Taylor vortices) to

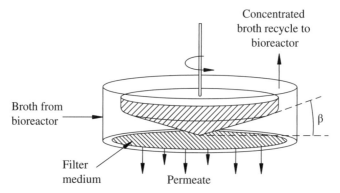

FIGURE 7B.5 Controlled shear membrane filter with conical disc used by Vogel and Kroner (1999). (Reproduced with permission from John Wiley & Sons, Inc. © 1999.)

reduce membrane fouling. In these devices, the shear on the membrane surface and pressure drop through the device can be regulated independently. Vogel and Kroner's design (Fig. 7B.5) consisted of a conical-shaped rotor that rotates above a flat sheet membrane.

A precise shear across the membrane surface can be obtained by adjusting three independent geometric parameters: (i) the distance between the conical rotor and the membrane, (ii) the external angle between the cone and its housing, and (iii) finally the speed at which the rotor revolves can be adjusted to yield a shear rate that is independent of the liquid flow rate (Bouzerar et al. 2000). The technique was demonstrated by Vogel and Kroner (1999) for a batch culture system using a cellulose nitrate membrane of 3 μm rating. It yielded a perfusion rate of 0.024 l/min for 40 min uninterrupted operation without significant increase in the pressure drop across the filter medium. Schmid et al. (1992) used a variant of this technique coupled with a Nucleopore membrane. It has been mentioned earlier that these track-etched membranes are less susceptible to fouling. However, Schmid et al. reported fouling at perfusion rates approximately 25 times lower than those obtained by Vogel and Kroner (1999). The drawbacks of the controlled shear device are (i) relatively very low membrane area per unit volume and (ii) the existence of a rotary part that not only requires maintenance to ensure aseptic conditions but also makes membrane replacement difficult.

7B.5.1.3 Spin Filter   This device (Figs. 7B.6, 7B.7, and 7B.8) was introduced by Himmelfarb et al. (1969).

Its major advantage over Vogel and Kroner's device is the relatively high filtration area that can be made available. This device is available commercially in both versions: installed inside the bioreactor or in an external circuit. However, most of the literature investigations deal with the internally installed device (e.g., van Wezel et al. 1985a). It has been extensively investigated in the literature. The filter medium is located on the inner hollow shaft as against the outer shaft in the case of vortex flow filters. In view of its importance, some key aspects of spin filter performance are discussed in the following.

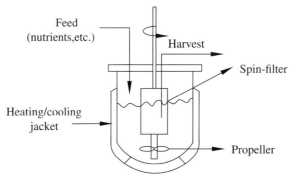

FIGURE 7B.6 Internal spin filter with common drive. (Reproduced from Deo et al. (1996) with permission from John Wiley and Sons. © 1996 American Institute of Chemical Engineers (AIChE).)

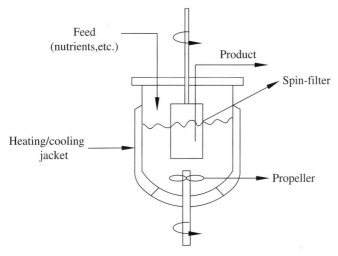

FIGURE 7B.7 Internal spin filter with separate drives. (Reproduced from Vallez-Chetreanu (2006) with permission from the author.)

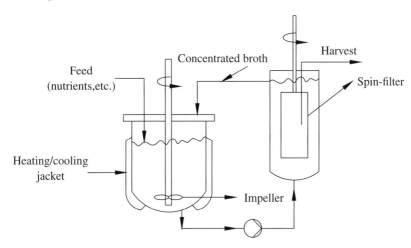

FIGURE 7B.8 External spin filter. (Reproduced from Vallez-Chetreanu (2006) with permission from the author).

Spin Filter Performance   The major factors that affect cell retention by the spin filter are:

1. Effect of filter media

Two types of filter media have been studied: (i) metal (stainless steel) and (ii) polymeric membranes. Irrespective of the filter media used, clogging due to trapped biospecies has been a major problem (Tolbert et al. 1981; Avgerinos et al. 1990; Favre and Thaler 1992; Favre 1993; Deo et al. 1996). Metal screens have been most widely used in spin-filter studies. They are available in a wide range of pore sizes. The range of interest is 10–150 µm since mammalian cells' size is 16–18 µm. Vallez-Chetreanu (2006) has given a comprehensive discussion on the effects of the filter media on spin filter fouling. Stainless steel screens have a positive surface charge, whereas most polymeric materials are neutral. The major advantages of stainless steel screens are (i) availability in wide range of pore sizes, (ii) mechanical strength, and (iii) superior temperature/chemical resistance that allows facile sterilization by both thermal and chemical methods. Polymeric media such as fluoropolymers have also been studied (Avgerinos et al. 1990; Esclade et al. 1991). Cells that need to be retained as well as proteins and DNA carry a negative charge (Stryer 1999). Consequently, they are not strongly adsorbed on polymeric surfaces that are neutral. The hydrophobicity of proteins, however, creates a hydrophobic–hydrophobic attraction with the hydrophobic polymeric surface. In view of this, polymeric membranes that have specially tailored hydrophilic surfaces can be expected to reduce fouling (Lee et al. 1990). Büntemeyer et al. (1994) compared the performance of hydrophilic poly(propylene) (PP) and poly(tetrafluoroethylene) (PTFE) membranes using a hollow fiber module that has a different mechanism of cell retention than a spin filter. It was found that PP membranes afforded higher protein permeability but fouled rapidly as compared to PTFE membranes. The latter evidently supported long-term uninterrupted operation. The pore opening of the filter media is also an important factor. Smaller openings would obviously yield high retention but have a high tendency to foul due to deposition of various biospecies. Siegel et al. (1991) showed that the best performance is obtained when the pore size is approximately equal to the cell size. Vallez-Chetreneau (2006) suggested that the filter opening size should be 20–30% higher than the cell size for achieving longer culture time and better retention. In a significantly different approach, Averginos et al. (1990) attempted to reduce the fouling by means of a change in the morphology of the cells. The large aggregates of CHO they produced apparently did not enter and clog the pores in the filter media. Vallez-Chetreneau (2006) used ultrasonic vibration to dislodge surface-adhering cells. This method yielded a 113% increase in culture time for a static spin filter. The "on time" increased by 75% when the filter was rotated at 2.5 rev/s. It was recommended that a combination of rotation and ultrasonic vibration is the best solution. The life of the piezo element was limited due to reasons not related to clogging. The most probable reason was damage to the element due to continuous use. The author concluded that a combination of periodic (rather than continuous) ultrasonic vibration coupled with (i) larger opening in the filter screen and (ii) reasonably high speed of rotation may give improved results.

2. Cell viability

Internal spin filters have a distinct advantage over other techniques inasmuch as they do not require a pump that could be the cause of cell damage. Most studies with spin filter have brought out a connection between cell viability and filter fouling/plugging. Esclade et al. (1991) found that cell debris and DNA released during operation increased the fouling tendency of the filter due to attachment of the cells. The major reason for cell death arises from the fact that cells trapped in the fouled filter do not enjoy a favorable environment. Vallez-Chetreneau (2006) has summarized the data on fouling and concluded that factors that sustain culture viability (supply of nutrients, etc.) and filter fouling are intimately connected. A vibrant culture was found to decrease fouling.

3. Geometric configuration and operating parameters

The internal spin filter can be mounted on the same shaft as the bioreactor or mounted independently on a separate shaft. The first configuration (Fig. 7B.6) is less complex, whereas the second (Fig. 7B.7), although complex, has the advantage that the surface velocity of the spin filter can be varied independently. An additional feature that has been used by many investigators is the use of a draft tube surrounding the spin filter. The use of such a draft tube has been shown to be advantageous (Sections 3.4.2.4) since it yields a well-directed axial flow. In the present case, this flow is directed parallel to the filter surface and eliminates any inhomogeneity. Effect of tangential velocity of the spin filter surface on cell retentivity was studied by Varecka and Shrier (1987) and Yabannavar et al. (1992, 1994). It was found that the capacity increased significantly with surface velocity. The velocity was varied in the range of 0.28–0.36 m/s. Deo et al. (1996) argued that the role of surface velocity is similar to the cross-flow velocity in conventional membrane filtration. Favre and Thaler (1992) found that velocities exceeding 0.6 m/s reduced attachment of cells to the filter surface and its consequent fouling. Varecka and Scheirer (1987) report an optimum value of 0.28 m/s, whereas Yabannavar et al. (1992, 1994) recommended 0.36 m/s. The cell detachment reported at relatively higher velocities (~0.6 m/s) could prevent rapid fouling. However, these higher rotational speeds pose operational difficulties, particularly mechanical problems associated with vibration of large suspended screens (Vallez-Chetreanu 2006).

There are conflicting literature reports on filter fouling. Favre (1993) suggested that viable cells are preferably attached in comparison with dead cells. Deo et al. (1996) reported an initial fouling of a fresh filter when the rate of filtration exceeded the holding capacity of the filter surface. Extended operation (40–50 days) of the bioreactor caused growth of the cells trapped inside the filter, which led to the second phase of fouling. This study also concluded that fouling of the filter was irreversible. The impeller used in the bioreactor outside the spin filter (Fig. 7B.7) has the important role of keeping the cells in suspension and providing an axial velocity component in the vicinity of the filter surface. Its speed of rotation was found to have an effect on cell retention. Yabannavar et al. (1992) concluded that this impeller should have separate speed control as depicted in Figure 7B.7.

4. Perfusion rate

The perfusion rate is linked with the type of media and the extent of fouling. However, unlike the finding of Mercille et al. (1994) for vortex flow filter, the perfusion rate for spin filter increased in proportion to the area of the filter media. This indicated that scale-up of spin filters could be carried out on filter area basis. Deo et al. (1996) found that the perfusion rate decreased with increasing cell concentration in the range of 1.8–10.8×10^6 cell/cm^3. Further, they found the perfusion rate to be proportional to $(D_{SF} \times \omega)^{0.5}$. The scale-up rule independently suggested by Yabannavar (1994) and Deo et al. is based on the following expression for the perfusion rate:

$$RVD \propto \left(\frac{\pi^3 H_{SF} D_F^{\,3} \omega^2}{R_V \times [C_{Es}]} \right) \qquad (7B.3)$$

Favre (1993) found that his data for constant volume perfusion of two different mouse–mouse hybridoma lines could not fit the classical constant volume filtration equation. He suggested that this could be due to a monolayer type of cell buildup instead of a cake in the conventional filtration. According to Favre, a model principally based on stochastic and steady plugging of the pores in the screen by the cells leading to an equilibrium state can possibly give a rational design procedure for spin filters.

7B.5.2   Separation Based on Body Force Difference

7B.5.2.1 Gravity Settling   The settling rate of cells under gravitational field depends on their size and density difference between the solid and fluid phases. The settling velocity for this case is governed by Stokes' law and is given by

$$V_T = \left(\frac{[\rho_C - \rho_L] g d_P^{\,2}}{18 \mu_L} \right) \qquad (7B.4)$$

Since both d_P and $(\rho_C - \rho_L)$ appear in the numerator of Equation 7B.4, V_T is relatively very low. Voisard et al. (2003) have given the range of settling velocity for cells in different media as 0.01–0.15 m/s. Such low values imply that gravity settlers must provide sufficiently large settling areas. Gravity separators can be both internal and external type. In view of the fact that the cells do not experience any strong shear, their viability, particularly in the internal type, is excellent. However, considering the large settling area required, it may be difficult to accommodate the internal type. The retention is not very high, and therefore, a polishing step is required to obtain a good quality harvest (Vogel and Kroner 1998).

7B.5.2.2 Centrifugal Filtration   Centrifuges have been widely used in the chemical and allied industry. In particular, the pharmaceutical industry extensively uses different designs of centrifuges. Thus, the design and fabrication of centrifuges is well established. The problem of large settling areas in gravity separators is overcome by resorting to application of centrifugal force. In this case, g the acceleration due to gravity in Stokes' equation (Eq. 7B.4) is replaced by the centrifugal force applied. Therefore, the cell settling velocity V_{TCF} is enhanced by increasing the speed of rotation of the centrifuge. The latter can be independently adjusted. The relatively high values of V_{TCF} thus obtained yield rapid settling. Another advantage that centrifuges enjoy arises from the fact that viable cells are heavier than dead cells. Thus, reducing the speed can allow substantial removal of dead cells and debris. Continuous centrifuges have been used for large-scale perfusion systems (Ivory et al. 1995; Johnson et al. 1996). Westfalia Separators AG (Germany) and Kendro Laboratory Products (Switzerland) are among the major specialty centrifuge suppliers. Although the centrifuge is in the external circuit, the cells have a relatively low residence time outside the bioreactor. The presence of a rotary part and possible exposure to shear in the centrifuge (Shirgaonkar et al. 2004), relatively high capital, and maintenance costs, and tendency of mechanical failure of some parts (Johnson et al. 1996) are some of the major drawbacks that centrifuge designers must tackle. Voisard et al. (2003) have given a cogent discussion on the application of centrifuges for cell retention. Special designs that do not have mechanical seals are available (Kendro Centritech®). The absence of a mechanical seal prevents contamination hazard. These designs have a direct channel between the bioreactor and the centrifuge shaft. Further, centrifuges can be used with disposable sterile inserts that have FDA approval. Bracco and Kemp (http://celld.com/documents-adobe/centritech-lab2-anticorps-monoclonaux.pdf; accessed July 2012) have compared the performance of a 0.65 μm pore diameter hollow fiber module and Kendro's proprietary Centritech® system for a myeloma cell line expressing an immunoglobulin G fusion protein (IgG) on two 10 l bioreactors. During a 30-day perfusion trial, the Centritech® system had a 25% higher cell density as compared to the hollow fiber module. The speed of the centrifuge was reduced midway during the trial and dead cells/debris were removed. Results of the trial gave the following results in favor of the centrifuge system: viable cells 74% against 60% and (ii) 3400 mg IgG against 2200mg (~38% more IgG). Several other investigators have used special centrifuge designs for perfusion experiments (Hamamoto et al. 1989; Tokashiki et al. 1990; Takamatsu et al. 1996). The issue of cell viability in the strong centrifugal field needs to be convincingly addressed.

7B.5.2.3 Hydrocyclone This device has been widely used in various industries such as pulp and paper and mineral processing. The separation/classification of particles in a hydrocyclone is also based on centrifugal force differences between particles of varying size/density. Unlike the centrifuge, a hydrocyclone is a static device. The principle of cell separation by hydrocyclone is shown in Figure 7B.9.

The fluid–solid mixture is fed at the top of the hydrocyclone with a high tangential velocity. Larger denser particles gravitate toward the wall since the centrifugal force exceeds the drag force that pulls them toward the central axis. Smaller, lighter particles, on the other hand, rotate along the vortex lines closer to the central axis of the vessel. Thus, larger/heavier particles settle down and are taken out as an "underflow," whereas the smaller/lighter particles go out with the liquid at the top as "overflow." Hydrocyclones have a very small footprint, no moving parts, and hence no maintenance (Svarovsky 1984; Chen et al. 2000; Catapano et al. 2009). Thus, a 25 mm unit can handle flow rates up to 0.5 m³/h. A major drawback is the relatively high

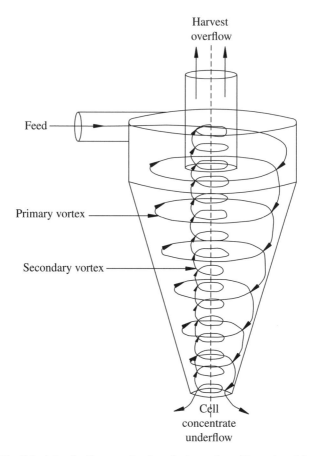

FIGURE 7B.9   Principle of cell separation by a hydrocyclone (Reproduced from US Patent No.6,878,545 B2; date of patent: April 12, 2005. with permission from Professor B. Anspach).

pressure drop (0.1–0.4 MPa) that is likely to be detrimental to shear-sensitive cells. The first reported use of hydrocyclones for separation of mammalian cells is due to Lubberstedt et al. (2000). This study used standard commercial hydrocyclones in the range of 7–10 mm diameter. HeLa cells were separated at pressure drops in the range of 0.1–0.4 MPa. This investigation did not show any decrease in cell viability despite the high pressure drops. This was attributed to relatively very short (0.03–0.1 s) residence time of the cells in the hydrocyclone. Cell retention efficiency of 81% at 0.4 MPa was reported. The cell retention can be further improved by feeding the overflow of the first unit to a second unit in series. The drawback of course is loss of cell viability due to greater exposure to inimical conditions. Jockwer et al. (2001) used computational fluid dynamics (CFD) to develop special designs of hydrocyclones suited for use in retention of shear-sensitive CHO cells. The results showed 94% retention, but the cell viability was found to be affected. At a pressure drop of 0.1 MPa, the extent of dead cells in separated cells ranged from 5 to 12% and the same increased to 7–21% with a small increase in pressure drop to 0.135 MPa. Cells in the overflow exhibited lower viability probably because of a longer residence time and the effect of flow reversal (Fig. 7B.9). With the CFD-based optimal hydrocyclone design, sustained operation over 23 days was possible. The cell viability was in excess of 90%, and with 50% higher cell concentration, the culture yielded three times higher recombinant antibody product. Ahmed (2005) carried out a systematic investigation on the use of hydrocyclones in mammalian cell perfusion cultures. Five different cell lines (BHK 21pSVIL2, HeLa, CHO-3D6, SP-2/0 AG14, and CHO-ATIII) were studied using intermittent application of the hydrocyclone. A double tangential entry hydrocyclone design that yielded high cell separation efficiency was used (Deckwer et al. 2005). The author concluded that cell viability is not as much impacted by the shear as by the formation of two opposing vortices systems (downward for underflow and upward for overflow). For all the cell lines, increase in the pressure brought about a decrease in cell viability. A redeeming feature (similar to centrifuges) was the lower retention of dead cells due to their lower density. Unlike other devices, scale-up of hydrocyclone at constant separation efficiency does not imply increasing its size. On the other hand, smaller sizes yield better results since the particle cut size decreases with decrease in the diameter of the cylindrical portion $d_{PC} \propto (D_{Cylinder})^{1.5}$ (Richardson et al. 2002). For use in large-scale applications, a different approach is required. A bank of hydrocyclones working in parallel as shown in Figure 7B.10 (Svarovsky 1984; Ahmed 2005) may be a good option in this case.

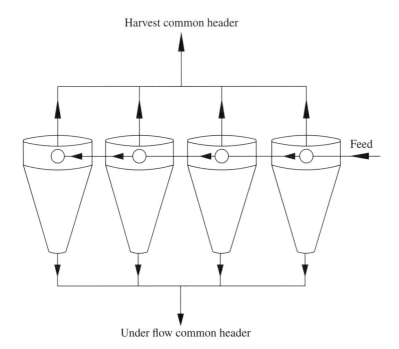

FIGURE 7B.10 Bank of hydrocyclones working in parallel.

7B.5.3 Acoustic Devices

The principle of acoustic particle separation from liquids has been known since the time coagulating effect of ultrasonic waves on liquid emulsions was observed by Sollner and Bondy (1936). Kilburn et al. (1989) were the first to demonstrate the applicability of ultrasonic standing waves for the coagulation and subsequent separation of animal cells and similar microorganisms. Several investigators have subsequently studied various aspects of this principle (Trampler et al. 1994; Doblhoff-Dier et al. 1994; Pui et al. 1995; Hawkes and Coakley 1996; Hawkes et al. 1997; Hawkes et al. 1998a, b; Spengler et al. 2001; Dalm et al. 2004; Dalm 2007). Voisard et al. (2003) have covered some of these studies in their review of various cell retention techniques. Shirgaonkar et al. (2004) have presented an in-depth review on acoustic devices. Literature reports (Gorenflo et al. 2002; Crowley 2004; Dalm et al. 2005; Dalm 2007) indicate successful application of this technique for large-scale (0.2–1 m³/day) continuous cultures.

Hill and Harris (2007) have discussed the mechanism of particle coagulation due to standing acoustic waves. The device generally consists of a flow channel with two walls facing each other. An ultrasonic transducer is glued to one wall while the other wall acts as a reflector. For cell retention applications, transducers generating frequencies greater than 1 MHz (generally 2–3 MHz) are used since lower frequencies are likely to cause cavitation. Cavitation is a phenomenon in which very large energy is dissipated in a small volume due to collapse of bubbles. Evidently, such a phenomenon is extremely harmful to the cells (Shirgaonkar et al. 2004; also see Section 7B.8). Standing acoustic waves that are generated in the channel cause formation of (i) antinodes where the amplitude of pressure is maximum and (ii) nodes where the amplitude is zero. Particles in a fluid flowing through this channel are exposed to a primary force that propels them to the location of minimum acoustic energy. In applications involving cells and other microorganisms suspended in an aqueous medium, the acoustic force acts toward the nodes (Hill and Harris 2007). Figure 7B.11 shows the various stages of particle coagulation under an acoustic standing wave.

Migrating cells form agglomerates at the nodes. These large aggregates have a much higher settling velocity as compared to a single cell. The radiation force responsible for propelling the cells to node locations is a function of (i) the frequency of the transducer, (ii) diameter of the cell, and (iii) its density and compressibility.

7B.5.3.1 Operation of Perfusion System Using Acoustic Device Figure 7B.12 shows a schematic of a perfusion system using an acoustic filter. Detailed description of the device is available at http://www.massetrecovery.com/Pictures10/biosep.pdf (accessed July 2012).

The main parts of the acoustic filter assembly are (i) ultrasonic transducer provided with compressed air cooling facility, (ii) flow channel, (iii) settling

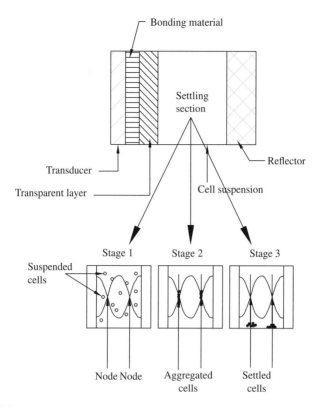

FIGURE 7B.11 Principle of acoustic separation and stages of cell aggregation leading to particle settling. (Reproduced from Hill and Harris (2007) with kind permission from Springer Science + Business Media LLC, New York, USA. © 2007.)

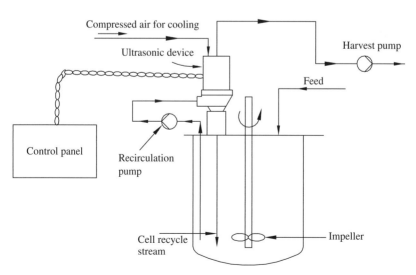

FIGURE 7B.12 Schematic of bioreactor using an acoustic separation device. (Reproduced from USER MANUAL BioSep ADI1015 September 2001 with permission from Applikon Biotechnology B.V. © 2001.)

chamber, (iv) recirculation pump, (v) control panel for adjusting the cycle times, etc. Details of the assembly, its installation/sterilization, operation setting, maintenance, troubleshooting, etc. are described in the user manual referred to earlier.

The procedure described in the aforementioned publication can be summarized as follows:

"The acoustic device and the harvest pump are switched on simultaneously. The acoustic device forms cell aggregates that settle back into the bioreactor and predominantly clear liquid product is removed by the harvest pump. The pump and the acoustic device are simultaneously switched off to increase sedimentation of the cells into the bioreactor. The time between the switching 'on' and 'off' is the duty cycle. Removal of the product and recycle of the cells results in higher cell density in the reactor. The higher cell density requires proportionate increase in the nutrients and hence the nutrient pump is switched on. Continuous addition and removal of harvest, while retaining the cells, creates a continuously perfused bioreactor."

7B.5.3.2   *Effects of Major Operating Variables on the Efficiency of Cell Retention*
Ultrasound Frequency   Most acoustic filters use a relatively narrow frequency range of 2–3 MHz. Therefore, the frequency is considered as a fixed parameter.

Filtration Rate   During the retention "on" cycle, increase in the nutrient supply must be compensated by a higher perfusion rate. This results in a higher upward fluid velocity and consequently a higher drag. Cell aggregates that have a smaller size (and hence lower V_T) are likely to be carried away with the fluid. Further, excessive fluid drag can break the aggregates. Thus, the efficiency of separation decreases with higher flow rate. This issue can be resolved by optimizing the acoustic power input.

Acoustic Power   Gorenflo et al. (2002) have shown that there is an optimum value of the acoustic power up to which the cell retention efficiency increases. Above this optimum value, there is a decrease in the efficiency. This loss of efficiency was attributed to natural convection arising out of heating of the liquid due to higher power dissipation. This factor can be addressed by using a cooling medium to remove the heat generated. Doblhoff-Dier et al. (1994) and Gorenflo et al. (2003) used a two-compartment design (Fig. 7B.13) in which water/air can be used as the cooling medium.

It is understood that the extent of cooling should match the heat generated due to acoustic power dissipation. Thus, the cooling medium flow rate and heat transfer area are important factors. Overcooling was shown to result in a detrimental effect similar to overheating (Gorenflo et al. 2003). Doblhoff-Dier et al. (1994) showed that for power inputs in the range of 0.5 kW/l, air cooling was not sufficient and cooling by water was needed.

Recirculation Rate   The residence time of the cells in the filter needs to be minimized. Once the cells are trapped, they collect in the lower part of the filter

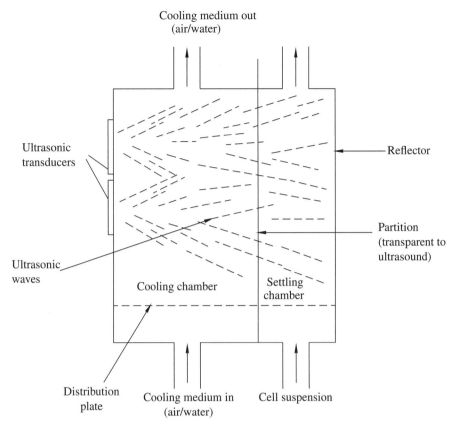

FIGURE 7B.13   Acoustic filter with cooling. (Reproduced from Doblhoff-Dier et al., 1994 with permission from John Wiley and Sons. © 1994, American Institute of Chemical Engineers (AIChE).)

and need to be rapidly recycled into the bioreactor. This is facilitated by a recirculation of the cell suspension into the lower part (Ryll et al. 2000). The recirculation rate should be such that it does not hamper the trapping of cells in the upper part. Experimental results indicate that the optimum recycle rate is in the range of 2–3 times the harvest rate (Gaida et al. 1996; Gorenflo et al. 2002). At relatively higher cell densities, it has been suggested that periodic washing of the lower part can relieve the congestion and increase the retention efficiency. The acoustic chamber is completely flushed and then refilled with a new suspension from the reactor. However, the backwashing frequency needs to be decided with prudence. During backwashing, the acoustic power and harvest pump are switched off. This results in a higher broth volume, which has to be brought back to the steady-state level by increasing the flow rate of the harvest. This has a negative effect on cell retention unless the acoustic power is increased. Increase in power results in an increase in temperature and again a loss of retention efficiency. The sequence starting with turning off the acoustic power and harvest pump results in a discontinuity and a fresh steady-state level takes some time depending on the scale of operation. Gorenflo et al. (2002) demonstrated the favorable effect of backwashing. They suggested an optimum backwash frequency in the range of 12–20/h. Gorenflo et al. (2003) used air backflush instead of recirculation of the cell suspension to flush out the cell aggregate from the acoustic device during the time the acoustic power and harvest pump are switched off (Fig. 7B.12). According to Gorenflo et al. (2003), the filter is partially refilled after being emptied subsequent to backflushing. The advantages claimed are (i) elimination of the cell suspension pump and (ii) a lower residence time of the cells in the separator. However, Gorenflo et al. (2003) also observed that the retention efficiency with air backflush was lower as compared to the conventional recirculation mode. They suggested that optimization of the air backflush technique can yield better results. Backflushing with the medium is another alternative (Merten 2000), but this has to be compensated by higher harvest rates. The latter has a deleterious effect on cell retention as detailed earlier in this section.

The foregoing discussion highlights the need to optimize the various cycles.

Cell Viability   Several investigators have reported that acoustic filters do not have a negative impact on cell viability. Doblhoff-Dier et al. (1994) subjected mammalian cells to ultrasonic radiation. Their study did not find any noticeable loss of cell viability. Similar investigations by Bohm et al. (2000), Wang et al. (2004), and Gherardini et al. (2005) supported the aforementioned findings. Groschl et al. (1998) also did not notice loss of viability at powers as high as 80W in a 50 ml resonator vessel. Pui et al. (1995) have reported the results in terms of a more meaningful intensive quantity and power density. These investigators found that the cells were not damaged at power densities as high as 0.22 kW/l.

Type of Ultrasonic Device   Hill and Harris (2007) have discussed the merits and drawbacks of various materials that can be used for generating the acoustic effect. Lead zirconate titanate type is the most widely used and effective piezo material.

The major advantages of acoustic retention devices are compact, completely sterizable unit, no moving parts, negligible effect on cell viability, etc. (Shirgaonkar et al., 2004; and Hill and Harris, 2007). However, scale-up of these units for large harvest rates is faced with the problem of low retention efficiency. This will imply larger units to accommodate the greater harvest rates and optimization of the various cycles. Shirgaonkar et al.'s (2004) review indicates that there is sufficient confidence in designing units for perfusion rates up to 2 RVD.

Table 7B.4 gives a comparison of the various cell retention devices discussed in this section.

TABLE 7B.4 Comparison of Cell Retention Devices[a]

Type	% retention	Internal/ external	Cell viability	Differentiation between dead and viable cells	SIP/CIP	Maintenance	Capital cost	Feasibility for large scale
Inclined plate separator	Good with proper design	Mostly external	Poor if residence time in the external loop is high	Good	Possible	Very low	Moderate/high	Large area required. Suitable for 0.25–1 m³/day harvest rates
Membrane filtration: sheet/hollow fiber type	Initially good quality; blocked membranes retain macromolecules	Mostly external	Shear damage	Not possible	SIP possible only with expensive high-temperature-resistant polymers. Extensively plugged membrane must be replaced	Backflushing required. Phase inversion-type membranes plug rapidly. Track-etched type has a longer life	Moderate	Membrane area required $\propto (Q_{Harvest})$. Hollow fiber units are more compact
Dynamic membrane filtration	Initially good; blocked membranes retain macromolecules	Mostly external	Shear damage	Not possible	SIP possible only with expensive high-temperature-resistant polymers. Plugging reduced but not eliminated; extensively plugged membrane must be replaced	Rotor needs maintenance	Limited information	Membrane area required $\propto (Q_{Harvest})$. Large area required for high harvest flow rates

(Continued)

TABLE 7B.4 (Continued)

Type	% retention	Internal/ external	Cell viability	Differentiation between dead and viable cells	SIP/CIP	Maintenance	Capital cost	Feasibility for large scale
Vortex flow filters (spin filters)[b]	Good, increases with cell density	Both	Affected by fouling. Cells in plugged filter medium not viable	No information	SIP possible. CIP difficult for extensively plugged screens	Rotor needs maintenance. Extensively plugged screens need special cleaning before reuse	Moderate	Membrane area required $\propto (Q_{Harvest})$. Internal devices need extra reactor volume. Overall good for harvest flows of 0.25 to 1 m³/day
Acoustic filters[b]	Good	Both	Good for small harvest flows and power input. No filter media that can foul	Possible	Both SIP and CIP possible	Minimal except replacement of ultrasonic transducer under continuous operation	Proprietary devices[b]	Overall good for harvest flows of 0.25–1 m³/day
Centrifugal separators*	Good; excellent flexibility	External	Shear damage possible. Cell viability ~90%	Separation of dead cells possible by decreasing the speed	Both SIP and CIP possible	Needs maintenance even if device without mechanical seals is used	Proprietary devices[b]	Very good. Can be used for highest harvest rates >1 m³/day
Hydrocyclones	In the range of 80–90%	External	Shear damage due to vortices interaction	Good	Possible	Robust static device. No special maintenance required	Well established designs from reputed manufacturers[c]	Multiple cyclones in parallel are a feasible alternative. High harvest rates/ small cyclones result in shear damage

[a] Adapted from Voisard et al. (2003) with permission from John Wiley & Sons. © 2003 Wiley Periodicals, Inc.

7B.6 TYPES OF CELLS AND MODES OF GROWTH

To begin the discussion on stirred bioreactors, it is essential to briefly outline some basic aspects in life sciences. Similar to any biological system, cells grow because of a favorable environment and supply of the required nutrients. Four different mechanistic classes identify the different types of growths. Laska and Cooney (2002) have given a succinct description of these modes of growth. According to this classification, the four modes are (i) fission, (ii) budding, (iii), mycelial, and (iv) viral growth. These modes have an important bearing on the choice of a bioreactor, its design in particular, and overall bioprocessing in general. A brief description of the aforementioned four modes and their influence on bioreactor design is discussed as follows (Laska and Cooney 2002):

1. **Fission**: Binary fission is the predominant mode in the growth of bacteria. Bacterial cells are small in size (~ 1 μm). When the conditions are favorable, a bacterium grows slightly in size or length. During this process, a new cell wall grows through the center. Subsequently, two daughter cells are formed. Each of these daughter cells has the same genetic material as the parent cell. Under optimum conditions, the growth occurs in geometric proportion. The rate of growth is thus relatively very high. Typically, the population doubles in 0.5–3 h. Such large growth rates are attractive for industrial applications. However, these must be supported by necessary oxygen supply and removal of metabolic heat as well as inhibitory products when the scale of operation is large. Animal (mammalian) cells are relatively large in size ~ 15–18 μm (Nienow 2006). The rate of growth of these cells is much lower than that of bacteria. Typical doubling times are of the order of one to two days. These large cells lack a protective cell wall. Therefore, they are considered to be sensitive to any force that can disrupt the cell wall, leading to death of the cell. Mammalian cells have a rather limited range of pH in terms of viability. Further, they are also sensitive to osmotic pressures. In addition, mammalian cells require more complex nutrition as compared to bacteria and yeasts.

2. **Budding:** Yeasts growth occurs by division through budding. The doubling times are 1–3 h, which are in the same range as those for bacteria. The budding process yields a mother and daughter cell. These two cells have different growth rates and cell surface characteristics. However, a typical yeast cell population has a broad and time-dependent distribution of ages. This has an important bearing on the formation of the desired product expected from the bioprocess.

3. **Mycelial:** Molds, actinomycetes, and some yeasts are characterized by mycelial growth. Mycelial growth is a process in which elongation and branching of hyphae chain occur. These cultures also exhibit a distribution of ages. Younger cells are located at the hyphae tips. The hyphae combine to yield intertwined cellular strands or mycelia. Increase in population of these mycelia results in increase in the broth viscosity, causing problems of mixing of the broth. The relatively high viscosity also poses problems of process monitoring and

control. The hyphae are also sensitive to fluid shear, which can cause breakdown of hyphae/formation of relatively dense and branched flocs/agglomerates. It is difficult to maintain physiological similarity for these large flocs since they are likely to experience varying microenvironments. These differences can result in differences in cell properties that are important in the context of products derived from the cells. Mass and heat transfer problems are accentuated with increase in viscosity. The shear sensitivity and high viscosity need to be addressed to yield optimum growth conditions.

4. **Viral growth:** Initiation of viral growth occurs with the infection of the host cell. The virus attaches itself to the cell surface and injects viral nucleic acids into the interior of the host cell. The infecting viral genome constructs new viruses from biological molecules synthesized by the host cell under the direction of the former. Rapid replication of viral nucleic acid (more than 500 times) is followed by its encapsulation in coat proteins to constitute a large number of newly born viral particles. Further growth can occur by two methods: (i) in the lytic cycle, the host cells lyse (break) and release new particles capable of infection, and (ii) in the lysogenic cycle, the viral DNA is integrated into the DNA of the host cell. The latter continues to reproduce in the conventional manner.

Bacterial and plant cells have a cell wall followed by a cell membrane, whereas animal cells have only a cell membrane. Plant cell wall is primarily made of cellulose, whereas bacterial cell wall is composed of peptidoglycans. In bacterial and plant cells, the dissolved oxygen has to diffuse through both the cell wall and cell membrane, whereas in mammalian cells, oxygen needs to diffuse through only a thin cell membrane that encloses the cytoplasm. Therefore, it may be argued that the uptake of oxygen by animal cells is faster than the rest on two counts: (i) single barrier for diffusion and (ii) smaller distance over which diffusion must occur. By corollary, it may also be argued that animal cells should grow faster than the rest. This hypothesis is, however, not supported by practically observed doubling rates, which are much lower for mammalian cells than the rest. The earlier logic based on shorter diffusion barrier for oxygen uptake may be used to argue that oxygen uptake is unlikely to be the limiting factor in this case. A complete picture can emerge only from a holistic view comprising among others the gas–liquid, liquid–cell mass transfer steps, and related cell growth parameters.

7B.7   GROWTH PHASES OF CELLS

Since mass transfer and mixing are related to cell growth, a brief discussion on the stages of cell growth is as follows—all cell lines exhibit the following four phases in their growth (Fig. 7B.14):

1. **Lag phase (a–b):** In this phase, the bioactive species acclimatize to the environment constituting the "food" available. The growth rate is therefore relatively low.

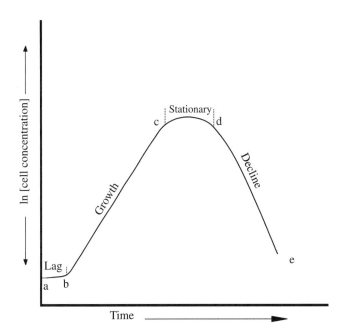

FIGURE 7B.14   Different phases of cell growth.

2. **Log phase (b–c):** After acclimating, the species are ready for growth. They start multiplying relatively fast.

3. **Stationary phase (c–d):** As the population of the species grows, there is a competition among them for the available nutrient supply. The exponential growth of the previous period cannot be sustained under such competition. There is a virtual stoppage of growth that stabilizes the population of the species.

4. **Death phase (d–e):** In all biological processes, growth of the cells is accompanied by formation of toxic species. The accumulation of these species increases the toxic effects. In a batch process, the food supply also diminishes because of consumption during the first three phases. The combined effect of toxicity and malnutrition causes cell death.

In the exponential growth stage, a Monod type of kinetic expression is sufficient since death rate is much smaller than growth rate. The requirements of nutrients (including oxygen) depend on the stage. The largest requirement is in the log phase. Kinetics of cell growth/product formation are very complex. They depend on the cell line (Kromenaker and Srienc 1991; Karkare 2004). Extensive studies on product formation kinetics for hybridomas are available in the literature. Several models have been proposed, which include unstructured and comparatively structured models (Martens 1996). The simple Monod-type approach has also been used, although it may not have general applicability. Any overall kinetic expression should include all the terms that affect growth and death (Karkare 2004). For more information, the reader is referred to Zeng and Bi (2006) and Portner (2009).

7B.8 THE CELL AND ITS VIABILITY IN BIOREACTORS

7B.8.1 Shear Sensitivity

The central figure in cell culture technology is the cell itself. The cell growth discussed earlier depends on the cell and its microenvironment. Although the log phase results in increase in cell population, cell death can occur even in this phase. In particular, mammalian cells have been at the center of much research on the causes of their death for no apparent reasons such as lack of nutrition and accumulation of toxins. As mentioned earlier, mammalian cells have a relatively thin cell wall that encloses their cytoplasm. Therefore, in the initial stages of research using stirred bioreactors, it was assumed that the shear stresses generated by the impeller are the main cause for rupture of the thin mammalian cell wall and consequent death of cells. The fragile nature of animal cells has prompted several investigations on the causes that affect cell viability/death. Recent investigations of Collignon et al. (2010) and Bustamante et al. (2013) were specifically aimed at measuring the shear stress generated by the impeller action. The following discussion considers the viability of animal cells in the two major modes of operation: (i) free cell suspensions and (ii) microcarrier-supported cell systems.

7B.8.1.1 Free Cell Suspension Systems A stirred reactor comprises of the impeller, baffles, and the wall of the reactor. The cells are entrained in a turbulent fluid stream emanating from the impeller. Leist et al. (1990) have indicated among others the following possible causes of damage to cells: (i) impact of collisions of cells with impellers, baffles, and reactor walls, (ii) pressure differences between upstream and downstream faces of the impellers, and (iii) turbulent phenomena. In terms of cell damage in a turbulent fluid from a survey of the literature, the following logical conclusions can be drawn regarding the relative importance of the three possible causes mentioned earlier:

1. **Collisions/impacts**: The relatively very small size and mass of free cells is unlikely to result in cell collisions of high impact strength that can cause damage to the cell wall.
2. **Shear generated by the impeller**: If the biological entity is smaller than Kolmogorov's scale λ_K, then the entity will not be damaged (Kunas and Papoutsakis 1990). This matter has been discussed in some detail by Hua et al. (1993) and later by Nienow (2006). Nienow's analysis indicates that for a commonly used average power input ($\varepsilon_{Avg} \sim 0.25$ W/kg), the derived average value of λ_K is ~45 μm. According to Kolmogorov's theory, the large-scale eddies (size ~ impeller diameter) produced by the impeller decay as they travel away from the impeller. The power input is highest in the immediate vicinity of the impeller. Assuming that ε_I for Rushton turbine is $30 \times \varepsilon_{Avg}$, Nienow estimated that the value of λ_K in the vicinity of the impeller is ~18 μm. Animal cell sizes are in the range of 15–18 μm. Therefore, these rough estimates indicate that turbulent eddies that are nearly the same size as the animal cell are

unlikely to be the source of cell damage (see later discussion on findings of Croughan et al. (1989) and other investigators). It should be noted here that the Rushton turbine generates the highest shear among the various impellers that are available for use.

3. **Energy released in bursting of bubbles:** Starting with Handa et al. (1987) and later Bavarian et al. (1991), Chalmers and Bavarian (1991), Papoutsakis (1991), Kioukia et al. (1992), Trinh et al. (1994), Laska and Cooney (2002), Mollet et al. (2004), and Czermak et al. (2009), the concept of cell damage due to bubble bursting steadily gained acceptance. A computer simulation of the process of bubble rupture was carried out by Garcia-Briones et al. (1994). The results clearly showed that breakup of small bubbles in water (noncoalescing system) dissipated very high levels of energy in a small volume close to the bubble. The extent of energy dissipated was higher than the typical energy dissipation due to impeller-generated turbulence by 3–4 orders of magnitude (Nienow 2006). It was proposed that the fragile mammalian cells are unlikely to survive this shock. This matter is discussed in some detail in the following.

Kioukia et al. (1992) performed cell death experiments on TB/C3 mouse–mouse hybridoma producing anti-IgG monoclonal antibodies under the following conditions: (i) surface aeration at high stirring speeds with and without baffles and (ii) cylindrical columns and spinner flasks with spargers for gas and magnetic stirrers for mixing/ suspension. The dispersion height to diameter ratio was varied in all three types of experiments. The conclusions were:

1. The most important cause of cell death is bubble disengagement. The damage can be reduced by (i) reducing the proportion of the disengagement zone, (ii) reducing the sparging rate and also the extent of small bubbles, and (iii) using Pluronic F68, a surface tension lowering compound that lowers the concentration of cells in the vicinity of the bubbles/in the disengagement zone. This compound belongs to the class of poly(oxyethylene)–poly(polypropylene) block copolymers (Murhammer et al. 1988). Its protective capability increases with dosage up to 0.05 wt %. Typical dosage range is 0.05–0.3 wt %. Lowering of the surface tension also causes a decrease in the energy released during the bubble bursting process.

2. The shear generated by the impeller itself is not responsible for cell death. However, the smaller bubbles produced by the shearing action of the impeller can contribute significantly to cell death. Experiments in unsparged baffled reactors in which oxygen was supplied from headspace did not show any cell damage.

3. In the absence of baffles, there is a tendency to develop a vortex particularly at higher agitation speeds. Starting with a low speed of 3.33 rev/s, the agitation speed was increased to 6.66 rev/s. At the lowest speed with no vortex being formed, the cell viability was unaffected. With increasing speed and a stronger vortex, air is sucked into the liquid. At the highest speed (greater suction of air), cell growth was severely affected. Interestingly, the gas barely came under the influence of the impeller due to the strong vortex motion. This led to the

conclusion that bubble disengagement is the major cause of cell death rather than the trailing vortex system common to stirred reactors in which a gas sparger is used.

4. The conclusions in (2) and (3) were further confirmed for TB/C3 hybridoma by Oh et al. (1989, 1992) who used a high shear Rushton turbine with and without gas sparging. Viability studies in the absence of air sparging showed no noticeable cell damage with a maximum power input of 0.25 W/kg. In contrast, relatively low level of aeration resulted in rapid cell death even at much lower impeller power. These results again prove that impeller-induced shear alone does not result in cell death, whereas the presence of gas bubbles results in significant cell death.

In an extension of the earlier work, Kioukia et al. (1996) cultivated Sf9 insect cells in agitated laboratory bioreactors with and without Pluronic F68 and with and without aeration. Their conclusions can be summarized as follows:

1. Bubble bursts caused by surface-entrained air or sparged air affect the stability of Sf9 cells marginally more than mammalian cells. This marginally higher sensitivity was attributed to the somewhat lower strength of insect cells in comparison with hybridomas, as measured by micromanipulation.

2. Agitation had no effect on the infection rate. Multiplicity of infection had a positive effect on the rate of infection. The rate of infection increased with multiplicity of infection regardless of the species used to induce infection of the cells. This observation was explained using a Brownian diffusion transport mechanism of viral particles to the cell.

3. Use of Pluronic F68 was able to eliminate the damage caused by bursting bubbles. This was independent of the source of the bursting bubbles (whether from surface entrainment or from sparging). Further, the same Pluronic F68 could eliminate cell damage for both the preinfection and postinfection stages. In the preinfection stage, a marginal increase in the growth rate was observed for bioreactors as well as static cultures.

Meier et al. (1999) also showed that in a system with gas sparging, the predominant mode of cell death for animal/insect cells is due to bursting of bubbles. Similar to the findings of Kioukia et al. (1992), cells close to the bubble/trapped in the bubble lamella or adsorbed on the bursting bubble are most likely to die than cells far removed from zones of bursting bubbles. A model for cell attachment to rising bubbles was proposed that could explain various sparging death data available in the literature. According to this model, in the absence of additives that prevent cell attachment, cells are adsorbed on rising bubbles. When these bubbles burst, the adsorbed cells or cells in close proximity to bubbles are exposed to very high levels of energy dissipation and are consequently damaged. Nienow (2006) showed that the bubble bursting phenomenon results in specific energy dissipation rates that are much higher than those associated with impeller-generated shear. The energy dissipation increases with decrease in bubble size. Evidently, sintered/porous spargers that are supposedly good for mass transfer can contribute significantly to cell death.

These observations of Meier et al. (1999) and Nienow (2006) pertaining to the effect of bubble size on cell death have great relevance to design and reactor scale-up, particularly when direct sparging of air/oxygen is used. The sparger used should prevent the formation of small bubbles yet provide adequate mass transfer area. In the presence of either Pluronic F68 or carboxymethyl cellulose, polyvinyl alcohol, polyethylene glycol, and dextran, cell attachment to rising bubbles is reduced to insignificant levels. Nienow (2006) argued that Pluronic F68 causes changes in the hydrophobic bubble surface–cell interactions. Consequently, besides attachment to bubbles, the proximity of the cells to the bubble surface is reduced. Zhang et al. (1992) suggested that the cell membrane bursting tension is possibly increased by the serum and Pluronic F68. Similar observations were made by Ramirez and Mutharasan (1990, 1992). These latter investigators explained the protection mechanism as arising out of a decrease in cell plasma fluidity, which is an indication of strengthening of the cell membrane. Further, as per the model of Meier et al. (1999), cells that are farther from the scene of energy dissipation are not affected. Almost all investigators recommend the use of such surface tension-lowering agents to increase cell viability. This is exactly the reverse of the procedures used in ore dressing. In this case, a "collector" that imparts hydrophobicity to the specific ore to be "won" is used. This hydrophobic nature increases the tendency of the desired part of the ore for attachment to bubbles generated by induced/sparged gas in froth flotation cells (Chapter 9: Gas-Inducing Reactors). This species is then recovered as froth from the flotation cell.

7B.8.1.2 *Microcarrier-Anchored Cells*

Similar to ion exchange resins that carry attached cations/anions, large-scale cultivation of animal cells attached to solid supports for homogeneous production of therapeutic proteins is attracting attention. Microcarriers have the potential to sustain surface-anchored cells at relatively high concentrations. Unfortunately, cells in such systems are more susceptible to damage by hydrodynamic forces as compared to suspended cells. A free cell can overcome the hydrodynamic forces by relaxation that is possible through rotation. Croughan et al. (1989) and Garcia et al. (1994) suggested that the inability of anchored cells for free rotation may be the major cause.

Types of Microcarriers van Wezel (1985b) suggested the use of small spherical beads (diameter ~200–300 μm, specific gravity ~1.02–1.05) for cultivating the cells as attachments (adsorbed) on these solid particles. These particles, referred to as micro-carriers, are suspended in the growth medium. The cells attach to the surface and spread on the microcarrier beads to form a monolayer (van Wezel, 1985b). In a manner analogous to ion exchange resins, both gel-type (nonporous) and macroporous micro-carriers have been used. The gel types are single spheres without any internal pores. The macroporous variety is made of a large number of small beads, which together make a relatively larger sphere (size 500 μm to few millimeters). Bardouille et al. (2001) tested several microcarriers in their investigation of growth and differentiation of permanent and secondary mouse myogenic cell lines. Some of these were cellulose (Asahi microcarrier, Asahi Glass, Tokyo, Japan), polyester (Fibra-Cel; Bibby Sterilin, Staffordshire, Great Britain), DEAE cellulose (DE-53; Whatman, Springfield Mill,

TABLE 7B.5   Typical Physical Properties of Cytodex™ Microcarriers Supplied by General Electric Healthcare[a]

Property	Cytodex™ 1	Cytodex™ 3
Density[b] (g/cm^3)	1.03	1.04
Size[c] d_{50} (μm)	190	175
d_{5-95} (μm)	147–248	141–211
Approx. area[b] (cm^2/g dry weight)	4400	2700
Approx. no of microcarriers/g dry weight	4.3×10^6	3×10^6
Swelling factor[b] (cm^3/g dry weight)	20	15

[a] Reproduced with permission from GE Life Sciences. © 1989–2011 General Electric Company.
[b] In 0.9 % NaCl.
[c] Size is based on diameter at 50% of the volume of a sample of microcarriers (d_{50}) or the range between the diameter at 5 and 95% of the volume of a sample of microcarriers (d_{5-95}). Thus, size is calculated from cumulative volume distributions.

UK), glass (Sigma, Taufkirchen, Germany), plastic (Sigma), and collagen-coated microcarrier beads (Pharmacia, Uppsala, Sweden). The Cytodex™ series available from General Electric Healthcare is another example of microcarriers. Table 7B.5 gives typical properties of microcarriers supplied by General Electric Healthcare.

Cultispher G, S, and GL (Percell Biolytical), Cytopore™, and Cytoline™ are representatives of the macroporous type of carriers. In a manner analogous to macroporous ion exchangers, the internal pore surface is also available for cell attachment. This structure has two advantages: (1) the extensive internal pore area allows increase in cell number density and (2) cells in the internal pores are shielded from damage due to the several types of impacts discussed later or bubble bursting. This advantage is offset by (i) slower rate of attachment of cells on the internal pore surface and (ii) slower kinetics (Karkare 2004). The slower kinetics is because nutrients and oxygen must diffuse (internal diffusion) through a quiescent fluid medium in the pores to reach the cell locations. This additional internal diffusion resistance could be a matter of serious concern. If the internal diffusion resistance curtails supply of oxygen/nutrients, physiological similarity may be lost.

Literature reports on the use of microcarriers using different well-mixed devices for cell growth are available (Bardouille et al. 2001; Malda et al. 2004; Nagel-Heyer et al. 2007). In all these investigations, microcarriers have been shown to be advantageous for producing a large number of cells. Bardouille et al. (2001) have, however, reported aggregation resulting from bridge formation between microcarrier beads. These aggregates are susceptible to mechanical damage effects (collisions/impacts to be precise; see Section 7B.8.1.2) arising out of stirring. Bardouille et al. suggested that a combination of collagen coating and microcarrier geometry such as solid/hollow cylinders is likely to prevent such bridging and hence remove the deleterious effect of "shear." In this connection, it may be emphasized again that macroreticular beads that support cells in the macropores prevent direct exposure of the cells to collisions/impacts. In another study, Sen et al. (2004) used similar microcarrier-based systems to cultivate single cell suspensions of neural stem cells.

Shear Sensitivity of Microcarrier-Anchored Cells   In the case of microcarrier-based cell systems, damage to anchored cells ("shear damage") needs to be explained as due to two distinct effects. These two types are (i) shear damage due to scouring of the microcarrier surface by turbulent eddies and (ii) collision damage:

1. **Damage due to fluid shear:** This can occur when the microcarrier size and eddy size are similar. Croughan et al. (2006) found that in the whole range of eddies characteristic of the turbulent flow, only eddies with a size smaller than 2/3 of microcarrier diameter create microshear rate at cell surface that is sufficient to cause damage. The microcarriers (density slightly higher than culture media) are typically 150–300 μm in size, and eddies smaller than 200 μm are likely to cause damage. Nienow's (2006) estimate of eddy sizes indicates that eddies smaller than 200 μm can be present in large numbers, and therefore, this aspect of damage due to eddies scouring the microcarrier surface cannot be ignored. In this respect, the investigations of Collignon et al. (2010) and Bustamante et al. (2013) deserve specific consideration as discussed later.

2. **Damage due to mechanical impacts:** This term is used to represent cell damage due to various types of collisions of the microcarriers in a stirred bioreactor. These are collisions of microcarriers (i) among themselves and (ii) with the impeller, baffles, and the containing reactor wall. The size and mass of microcarriers are at least one order of magnitude higher than individual suspended cells. It is therefore realistic to assume that the energy involved in microcarrier impacts will be significantly higher. As a result, the various types of impacts cannot be ignored. Damage of cells anchored on microcarriers has some parallels with crystallization where the various collisions referred to earlier result in crystal breakup. Considering the aforementioned mechanisms for cell damage, it can be postulated that cell damage will be dependent on the frequency of collisions and hence, by simple logic, should have the following dependences on operating variables (Croughan et al. 1989; Nienow 1997):

1. Cell death rate $\propto$ microcarrier concentration
2. Cell death rate $\propto$ speed of agitation
3. Cell death rate $\propto$ impeller diameter
4. Cell death rate $\propto$ microcarrier size (particle abrasion is proportional to $(d_p)^6$, Nienow (2006))

Cherry and Papoutsakis (1986, 1988) have shown that these collisions can be an important contributor to cell damage. They defined a turbulence collision severity (TCS), which is a product of the kinetic energy dissipated in collisions and the frequency of collisions, as follows:

$$TCS = \left(\frac{KE_{Interaction} \times \left(\dfrac{f_{Interaction}}{V} \right)}{C_{Beads}} \right) \quad \text{(7B.5)}$$

Cherry and Papoutsakis (1986) and Collignon et al. (2010) have used the following simplified expression for TCS:

$$\text{TCS} = \left(\frac{\left(\frac{P}{V_R} \right) \times \mu}{\rho^2} \right)^{0.75} \left(\frac{\pi^2 \rho_{\text{Bead}} \, d^2_{\text{Bead}} \, \varepsilon_{\text{Bead}}}{72} \right) \tag{7B.6}$$

The above definition of TCS was to found to correlate microcarrier-supported animal culture data of Cherry and Papoutsakis (1986). Using particle imaging velocimetry (PIV), Collignon et al. (2010) measured the macroshear rate from the spatial derivatives of the time-averaged velocities as well as the microshear rate from the relative values of the microcarrier bead size and average Kolmogorov scale computed from the power input (Section 6.3.1.2). These measurements were made for the impellers shown in Figure 7B.15.

The values of TCS were calculated from the PIV and power input data. TCS values showed the following trend: TTP 125 < EE 150 < TTP 50 < A315 125 < 3SB 160 < A310 156 < A315 150. The difference in the TCS values increases with increasing N/N_S. Therefore, for higher N/N_S, TTP125 or EE150 should be preferred if shear damage is the only consideration. In a recent investigation, Bustamante et al. (2013) measured impeller-generated shear and gas–liquid mass transfer simultaneously. These investigators also concluded that the Elephant Ear impellers afforded lower shear rates than the standard disk turbine. It was found that the Elephant Ear impeller in the upflow mode offers not only low shear rates at high speeds of agitation but also higher values of $k_L a$ (Table 7A.2 and Section 7A.10). This observation is important since besides impeller shear, oxygen mass transfer rate and other considerations such as microcarrier suspension and mixing homogeneity also need to be taken into account.

Clark and Hirtenstein (1981) studied the growth of 1 l volume cultures of Vero cells on Cytodex™ 1. They found that for relatively high speeds of agitation, cells would detach from microcarriers, particularly during mitosis. At very low stirring speeds, the growth was poor, possibly due to inadequate suspension of the microcarriers. Croughan et al. (1988) found that there was no noticeable damage to Sephadex-anchored cells at low intensity of agitation. As noted in Section 7A.6.1 for good liquid–solid mass transfer, it is essential to have the solid phase in a state of suspension. Hence, at low impeller speeds, the observation of poor cell growth (Clark and Hirtenstein, 1981) is most likely due to a liquid–microcarrier mass transfer limitation. The optimal stirring speed depends on the type of cell culture and stage of culture cycle, besides the type of vessel/impeller combination, as discussed by Pangarkar et al. (2002) and Collignon et al. (2010) among others.

The "collision/impact caused" cell damage proposition of Cherry and Papoutsakis (1986) is also supported by Nienow (2006). It has been mentioned earlier that particle abrasion is proportional to $(d_p)^6$. This observation indicates that smaller size microcarrier beads should decrease the impact force. Cherry and Papoutsakis (1989a, b)

(a)

(b)

(c)

(d)

(e)

(f)

(g)

(h)

(i)

(j)

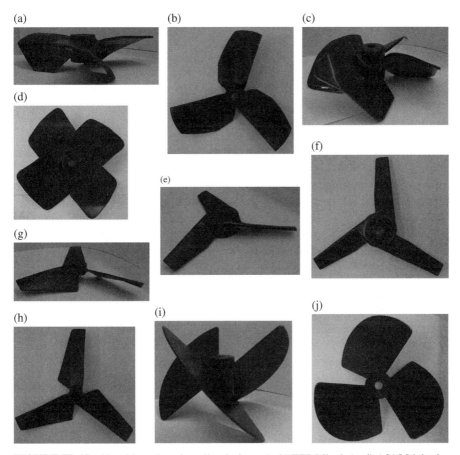

FIGURE 7B.15 Novel low shear impeller designs: (a–b) TTP Mixel, (c–d) A315 Lightnin, (e–f) A310 Lightnin, (g–h) three-streamed-blade VMI-Rayneri, and (i–j) Elephant Ear Applikon. (Reproduced from Collignon et al. (2010) with permission from Elsevier Ltd. © 2014, Elsevier B.V.)

investigated this aspect and found that smaller microcarriers distinctly reduce cell death while simultaneously increasing the cell growth rate. Besides the lower impact damage, smaller microcarrier bead systems provide higher surface area for cell anchorage since surface area/volume $\propto (6/d_p)$. Bardouille et al. (2001) studied a range of microcarriers in their investigation on the yields and myogenic differentiation of murine myogenic cell line C2C12. This study used a range of carriers in stationary cultures. It was found that collagen-coated dextran Cytodex™ 3 gave the best results in terms of both proliferation and differentiation of the cells. This study concluded that it is highly desirable to develop suitable mass culture systems based on biodegradable microcarriers. Such systems can have a great potential for use in the relatively sensitive secondary myogenic stage for restoring skeletal muscle mass in suffering patients.

7B.9 HYDRODYNAMICS

7B.9.1 Mixing in Bioreactors

A bioreactor is a typical three-phase reactor. Chapman et al. (1983a, b, c, d) carried out a systematic investigation on the different mixing processes of importance in stirred three-phase reactors. This was followed by various studies on gas–liquid mixing and solid–liquid mixing (solid suspension), as outlined in Chapter 7A.

Mixing time is a vital parameter affecting the efficiency of a bioreactor. The local concentrations of nutrients, toxins, dissolved oxygen, etc. are all dictated by the mixing process. A low mixing time ensures quick response to any additions (pH equalization, uniform distribution of nutrients, etc.). This results in uniformity of concentrations of all the important ingredients participating in the bioprocess.

Mixing time is a measure of the extent of time over which detectable inhomogeneities are observed (Nienow 1997). It, however, does not reflect on the amplitude of the variations in homogeneities. The sensitivity of a given cell culture system to variations in dissolved oxygen, nutrient concentration, pH, etc. is its own characteristic. Nienow et al.'s (1996) study of CHO 320 and NSO lines in an 8 m^3 stirred reactor is an example of such characteristic behavior. It was found that CHO 320 culture is not affected by dissolved oxygen inhomogeneities or insufficient oxygen supply. On the other hand, NSO cell culture was found to be affected by dissolved oxygen disparities in the reactor and oxygen mass transfer constraints. Several of these factors related to mixing have been known to have a bearing on metabolic pathways (Nielsen and Villadsen 1994) as well as cell viability and expression (Nienow 1990; van't Riet and Tramper 1991).

Mixing time can be measured by several simple techniques that follow change in a certain liquid-phase property resulting from addition of a tracer. The commonly used liquid-phase properties are (i) conductivity, (ii) pH, (iii) decolorization using a starch/iodine–thiosulfate reaction, and (iv) visual decolorization. The starch/iodine–thiosulfate reaction-based decolorization method is particularly suited for multi-impeller high H/T bioreactors in which mixing in individual impeller zones can be followed separately. The tracer method (conductivity or pH change) is a very simple method that has been extensively used (Kushalkar and Pangarkar 1994). The tracer liquid should ideally have similar density and viscosity as the parent liquid in the vessel. The liquid phase for these studies should also include any surface-active agent that is used in the actual system, for example, Pluronic F65. "Cold" experiments without such agents are unlikely to yield reliable estimates of the mixing time since the gas holdup structures with and without surface-active agents are substantially different. Depending on the impeller type, the flow pattern varies as shown in Figure 7A.15. For each flow cycle, the liquid phase passes the probe that is generally located at a fixed position in the vessel. Figure 7B.16 depicts a typical probe response.

A peak in the property that is being measured is observed at completion of each flow loop at the location of the probe. Khang and Levenspiel's (1976) criterion of five circulations (or five peaks in Fig. 7B.16) has been widely used to obtain the mixing time (Nienow 1998). Sardeshpande et al. (2009) measured both mixing time

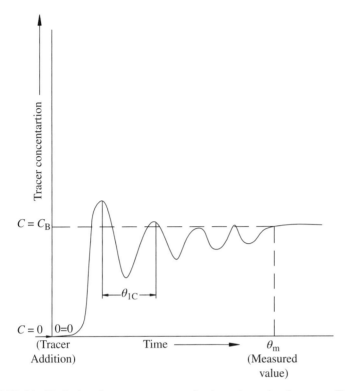

FIGURE 7B.16 Typical probe response to a pulse input in a stirred reactor. (Reproduced from Khang and Levenspiel (1976) with permission from Elsevier. © 1976, Elsevier.)

and circulation loop time in solid–liquid systems. Their results showed that for liquid-phase mixing (and also solid–liquid systems with low solid loadings ~1 wt%), the mixing time ranged from five to seven times the circulation loop time. Edwards et al. (2000) and Nienow (1998) have presented excellent discussions on important aspects of mixing in stirred reactors.

7B.9.1.1 Theories of Liquid-Phase Mixing Literature investigations have used two approaches: (i) a simple method based on the mean flows that uses measurements such as those shown in Figure 7B.16 and (ii) concept of turbulent diffusion that eliminates concentration gradients, thereby causing homogenization. For simplicity, the case of a single impeller ($H/T = 1$) will be considered (Nienow 1998). The second approach is more elegant but relatively complex. The ensuing discussion deals with the first approach (Nienow 1998). The circulation loop time Θ_C can be obtained by a straightforward equation:

$$\Theta_C = \left(\frac{V_B}{Q_{Imp}} \right).$$ Introducing the flow number $F_L = \left(\frac{Q_{Imp}}{ND^3} \right)$ and rearranging give two forms of proportionalities for Θ_M:

Case (i) impellers of same size but different type at equal power input:

$$\Theta_M \propto \left(\frac{N_P^{0.33}}{F_L} \right) \qquad (7B.7)$$

Case (ii) different sizes of same impeller at equal power input:

$$\Theta_M = (D)^{-1.33} \qquad (7B.8)$$

If the equal power input criterion of scale-up (with $D \propto T$) is used,

$$\Theta_M \propto (T)^{0.66} \qquad (7B.9)$$

In a classical reductionist approach to the relatively complex turbulent diffusion model formulation of Ruszkowski (1994), Nienow (1998) deduced the following simple proportionality:

$$\Theta_M \propto \left(\frac{1}{N} \right) \text{ or } N \times \Theta_M = \text{constant} \qquad (7B.10)$$

The above relationship is based on the assumptions that the ratio of local energy dissipation rate and its average value for the entire vessel is the same for different impellers having the same D/T ratio. Supporting evidence for Equation 7B.10 has come from the results of Kramers et al. (1953), Norwood and Metzner (1960), Procházka and Landau (1961), and Holmes et al. (1964). Norwood and Metzner's study clearly showed that for $Re > 10^5$, mixing by turbulent diffusion can be assumed to be predominant such that $N \times \Theta_M = \text{constant}$.

7B.9.1.2 Mixing Time vis-à-vis Mass Transfer and Oxygen Uptake

Nienow (2006) studied the relative importance of mixing time and oxygen uptake rate, (OUR) for CHO 320 and NSO cells. He defined characteristic times for (i) mixing, Θ_M; (ii) mass transfer, Θ_{MT}; and (iii) oxygen uptake, Θ_{OU}. The last two parameters are similar to the characteristic diffusion time for gas–liquid mass transfer τ_D and liquid-phase reaction time τ_R, respectively, as defined in Section 2.6.1.1. When $\Theta_M \gg \Theta_{OU}$, the situation is fraught with significant spatial variation in dissolved oxygen concentration. Similarly, when $\Theta_{MT} \gg \Theta_{OU}$, oxygen starvation can be a serious issue. The mixing and mass transfer problems are particularly accentuated at high cell densities. Modern reactor designs using $\varepsilon_M \geq 0.25$ W/kg can eliminate these problems. Similar to any other biological process, the OUR or the characteristic oxygen uptake time Θ_{OU} is decided by the requirements of a given biological species and the conditions/stage of growth. For instance, Θ_{OU} for animal cell cultures can be less by two orders of magnitude as compared to bacterial systems that are marked by rapid growth (low doubling times) and hence higher Θ_{OU} (Table 7B.6).

The oxygen requirement depends on the cell line and can vary from as low a value of 0.05 to 0.5 mmol/(10^9 cells)/h (Thomas 1990). In general, the cell densities in

TABLE 7B.6 Typical Oxygen Demands and Doubling Times of Biological Species

Biological species	Size[a]	Peak oxygen uptake (mmol/l/h)[b,c]	Doubling time (h)
Bacteria	~1 μm	50–100[b]	0.9–2.1[b]; 0.5–3[a]
		30–70[c]	
Yeast	5–10 μm	25–50[b]	1–3[b]
		30–70[c]	
Plant cells	20–40 μm wide and up to 100 μm long[b]	1–3.5[b]	36–120[b]
		10–30[c]	
Mammalian cells	5–20 μm[a]; ~18 μm[b]	1–5[b]	18–48[a]
		2–10[ce]	14–70[f]

[a] Laska and Cooney (2002).
[b] Junker (2004).
[c] Nienow (2003).
[d] Nienow (2006).
[e] Based on a cell density of 10^{-9} cells/l.
[f] Karkare (2004): Depends on cell line and environmental conditions.

animal cell systems are relatively low as compared to bacterial cultures (Karkare 2004). Considering this observation, it is unlikely that homogeneity with respect to dissolved oxygen concentration is critical for the case of animal cell lines in batch or fed-batch systems. However, perfusion systems with cell recycle are likely to have higher cell densities. Under such conditions, homogeneity with respect to dissolved oxygen concentration will need serious consideration. This argument for oxygen homogeneity does not imply that mixing (Θ_M) is unimportant. Oxygen is only one of the nutrients. Homogeneity with respect to other nutrients/toxins/pH also needs to be considered. In such a case, it may be argued that for relatively slow biological processes, requirements of nutrients/generation of toxic products is also lower. Stretching the argument further, mixing of the broth phase may not be critical, particularly if nutrient addition is through relatively dilute solutions. In practice, however, the use of concentrated solutions is a natural choice to obtain maximum utilization of the reactor capacity. This can result in the "snow ball" effect noticed by Ozturk (1996). This effect can lead to formation of necrosis (Barradas et al. 2012). The problem is further complicated by a peculiar feature common to the majority of cell lines in terms of a requirement of a small amount of dissolved CO_2 at low cell densities (Karkare 2004). Perfusion cultures operating at high cell densities, on the other hand, need a different approach. These can generate larger amounts of CO_2. Such excessive dissolved CO_2 concentrations resulting in a lower pH are detrimental to proliferation of the culture. It will be seen later that stripping of CO_2 can indeed be a problem if the sparged gas is not completely dispersed. Alternatively, oxygen/air supply has to be uniform since this acts as the stripping medium.

The aforementioned discussion indicates the very important role of liquid-phase mixing and the fine balance that is needed for maintaining physiological similarity at the local level in a large-scale industrial reactor without causing excessive damage to the cells.

As explained in Section 7B.7, the overall kinetic expression for the bioprocess must consider several issues. Thus, Θ_{OU} needs to be determined in a manner that is similar to determination of kinetics of a chemical reaction.

The broth mixing time varies with the type of system being handled: (i) liquid alone, (ii) liquid–solid, and finally (iii) gas–liquid–solid.

7B.9.1.3 Single-Phase (Liquid) Mixing

This situation may arise when suspension culture is grown with indirect oxygen supply such as through a membrane (Section 7B.12.2). In free suspension cell bioreactors, the cell is generally not considered as a separate entity in terms of mixing. This is because the cells are relatively very small. Further, the density difference $(\rho_C - \rho_L)$ is less than 100 kg/m³. In view of this, the cells are almost indistinguishable from the broth in which they are suspended. Therefore, as long as the broth is sufficiently mixed, the suspended cells are also considered to be well mixed (uniformity in cell concentration in the reactor). Nienow (1998) has given the following simplified correlation for the mixing time with a single impeller system with $H/T = 1$:

$$\Theta_M = 5.9(T)^{0.66}\left(\frac{D}{T}\right)^{-0.33}(\varepsilon_T)^{-0.33} \tag{7B.11}$$

Equation 7B.11 also indicates that $N\Theta_M$ = constant. Further, since for $H/T = 1$, $\varepsilon_T \propto (N_P N^3 D^5/T^3)$, Equation 7B.11 yields $\Theta_M \propto (N_P)^{-0.33}$ and $\Theta_M \propto (D/T)^{-2}$. These dependences result from the fact that increasing speed of agitation and D/T causes an increase in the pumping rate $Q_L \propto (ND^3)$ and hence leads to a lower mixing time. In addition, higher D/T yields a more even distribution of energy in the vessel (Nienow, 1998). With increasing tank diameter, it is expected that the mixing time should increase. Mersmann et al.'s (1976) experimental measurements showed that $\Theta_M \propto (T)^{0.66}$. Nienow (1998) has confirmed this dependence through a theoretical analysis using the tank diameter as the integral scale of turbulence. An intuitive argument would indicate that a higher power number implies a higher energy dissipation with consequently lower Θ_M. Nienow (1998) showed that $\Theta_M \propto ((N_P^{0.33}ND^3)/Q)$ when impellers of different type but same diameter are used in a vessel of fixed diameter. Alternatively, for a given impeller and fixed vessel diameter but different impeller diameters, $\Theta_M \propto D^{-1.33}$, which again shows that the mixing improves with increase in impeller diameter (or for a fixed T, increase in D/T). For higher H/T ratios, Nienow (1998) indicates that $\Theta_M \propto (H/D)^{2.43}$. This dependence implies that for reactors of $H/T > 1$, mixing time increases significantly and to improve mixing, multiple impellers need to be used.

The correlation given by Equation 7B.11 implies that all impellers with equal D/T yield the same mixing time at equal ε_T. For three-phase systems, ε_T is replaced by ε_{TG}, which includes power input through the gas. This is supported by a satisfactory fit of Equation 7B.11 to mixing time data for a wide variety of impellers that include Prochem Maxflo T, axial flow hydrofoil, equal-sized radial flow Rushton turbine, and Scaba 6SRGT impellers (Nienow 1998). These impellers afford different flow patterns. For example, radial flow impellers have a longer liquid circulation loop stretching horizontally first and then splitting (upward and downward) at baffles/wall. Contrary to this, axial/mixed flow impellers have a shorter liquid circulation loop reaching the bottom where major concentration gradients in solid and nutrient concentration/pH are likely to exist. Further, ε_{TG} used in Equation 7B.11 is itself a function of the type and size of the impeller and includes the gas velocity. In spite of these differences, the mixing times are correlated satisfactorily by a single correlation. This has led Nienow (1998) to argue that the turbulence theory rather than the concept of mixing by bulk flow is appropriate for correlating mixing time data. In Chapter 7A, it has been established that correlations based on simple power input result in geometry-/system-dependent variations. It should be mentioned here that although Equation 7B.11 has been tested for different impellers, the variation in D/T and C/T has not been as wide as for the correlations based on relative suspension/dispersion in Chapter 7A. Therefore, Equation 7B.11 should be used with caution for a widely differing system geometry (type of impeller, C/T, D/T, etc.).

7B.9.1.4 Mixing in Two-Phase (Gas–Liquid) Systems

7B.9.1.4 Mixing in Two-Phase (Gas–Liquid) Systems Mixing time in two-phase stirred reactors is decided by the relative values of the gas flow rate and the impeller speed. An additional factor is the type of impeller. The introduction of a gas can have a dramatic effect on the power drawn by the impeller. This matter has been discussed in Section 7A.3. Figure 7A.6 shows that there is a severe drop in power drawn by downflow impellers. The radial flow disc turbine shows an intermediate power drop, whereas the upflow pitched turbine shows noticeable drop only at relatively very high gas rates. Scaba 6SRGT yields the most stable performance, closely followed by the upflow pitched turbine. Nienow and Bujalski (2004) have highlighted the versatility of various upflow impellers. Some of the advantages mentioned by these investigators that are relevant to bioreactors are as follows: (i) they can operate with a minimal drop in power drawn or torque from the transitional regime up to the turbulent regime. This indicates stable operation over a wide range of power inputs. (ii) In the case of tall ($H/T > 1$) vessels, liquid-phase mixing time can be reduced to less than half of that afforded by radial flow-type impellers. This is a major advantage in animal cell systems sensitive to nutrient concentration gradients. (iii) Suspension of solids can also be achieved at much lower power inputs than the radial or wide blade downflow hydrofoils, some of which are shown in Figure 7B.15. This advantage persists even at relatively high aeration rates. (iv) As shown by Bhattacharya and Kresta (2004), upflow impellers can allow surface addition and eliminate problems associated with subsurface additions. (v) They act as highly effective foam breakers when located near the top surface. It will be shown in Section 7B.10 that for typical operating conditions for animal cell culture reactors, the aeration rates are very low (typically 0.005–0.01 vvm; Nienow et al. 1996). Therefore, the impact of aeration on the overall hydrodynamics including mixing time is significant only for the downflow impellers. Consequently, Equation 7B.11 can be used for a pilot estimate of the mixing time for two-phase systems. Equation 7B.11 has been derived from data in the turbulent regime. Therefore, its applicability should be ascertained by estimating the impeller Reynolds number Re_I, $[ND^2\rho_L/\mu_L]$ and confirming that the same is in the turbulent regime. In most suspended cell culture systems, Pluronic F68 or a similar surface-active agent is added to protect the cells from bursting bubbles. It is well known that such compounds cause a considerable decrease in the surface tension of the media used (Dey 1998 *cf* Nienow 2006) and thereby cause changes in the gas holdup structure. Therefore, Equation 7B.11 should be used with caution. However, Nienow (1998) and Nienow and Bujalski (2002) suggest that the mixing time for single liquid phase and gas–liquid system is the same if (i) the power input per unit mass is same and (ii) the impeller is not flooded in the case of the two-phase (gas–liquid) system. The use of Equation 7B.11 with ε_{TG} in place of ε_T was claimed to yield a fairly reliable estimate of Θ_M.

Hari-Prajitno et al. (1998) studied liquid-phase mixing for multiple hydrofoil impellers (APV B-2 with angle of pitch 30° and 45°) in both up- and downflow modes A-315 and Rushton turbines in two phases (gas–liquid systems). They also found that the upflow variety yields a stable operation under aeration. Further, the mixing time in both single and multiple systems were 40 and 20% shorter than Rushton turbine and downflow A-315, respectively, under otherwise identical conditions. Equation 7B.11 was found to correlate the mixing time data satisfactorily. Hari-Prajitno et al. (1998) have recommended optimum multi-impeller configuration as well as location of feed point that yields the best homogenization results.

7B.9.1.5 Mixing Time in Two-Phase (Solid–Liquid) System

Mixing in solid–liquid stirred reactors is of relevance for microcarrier-based cell cultivation systems with very low aeration rates (negligible effect of aeration) or systems using aeration through a membrane. Typical microcarrier loading used in cell culture systems is as follows: batch, 5 g/l and, perfusion, >20 g/l (Ibrahim and Nienow 2004). Liquid mixing in solid–liquid stirred reactors has been extensively studied, and a large number of correlations are available for predicting the mixing time (Edwards et al. 2000; Sardeshpande et al. 2009). Most studies have used solids that were relatively heavier ($\Delta\rho = \rho_S - \rho_L > 1000$ kg/m³). In animal cell culture systems, the microcarriers have much lower density with $\Delta\rho \sim 30$–50 kg/m³. In general, it has been found that the mixing time is higher for solid–liquid systems as compared to liquid alone (Michelletti et al. 2003; Sardeshpande et al. 2009). Further, most investigators have found that the behavior of mixing time is related to solid suspension. The process of solid suspension has been discussed in Section 7A.7.1. As the solids are progressively suspended, the dimensionless mixing time ($N \times \Theta_M$) increases and attains a maximum. Further increase in the agitation speed is accompanied by a steady decrease in $N \times \Theta_M$ till a constant value that is approximately the same as for liquid alone is reached (Bujalski et al. 1999). This behavior is common to both dense ($\rho_S > \rho_L$) and light (Kuzmanic et al. 2008) particles. The higher value of Θ_M in the presence of solid is because a certain portion of the power introduced is spent on solid motion till gradually all the solids are suspended. There are two conditions: $N = N_S$ and $N = N_{CS}$ (Section 7A.7.1). A constant value of Θ_M is achieved at $N = N_{CS}$. (It should be noted that stirred reactors are commonly operated at $N = N_S$ rather $N = N_{CS}$.) For particles heavier than the liquid, there is only one maximum in $N \times \Theta_M$ versus N (Fig. 7B.17)

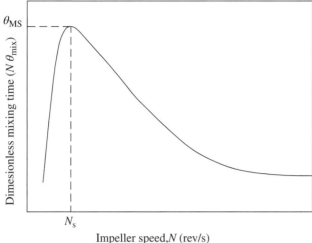

FIGURE 7B.17   Typical plot of dimensionless mixing time versus impeller speed for particles heavier than the medium. (Adapted from Sardeshpande et al. (2009) with permission from American Chemical Society. © 2009, American Chemical Society.)

plot, and this corresponds to N_S (Bujalski et al. 1999). This maximum becomes sharper as the solid loading increases.

Ibrahim and Nienow (2004) have reported solid suspension data for Cytodex™ 3 microcarriers with three different impellers: Chemineer HE-3 hydrofoils, single as well as dual Ekato InterMIG impellers, and six-blade PTD. The "S" values in Zwietering's correlations (Zwietering 1958) have been listed for these impellers for different types of tank bottom shapes. It was found that the N_S values for the small density difference as in the present case were relatively low to the extent that impeller Reynolds number was below that required for a fully turbulent flow. The corresponding power input required for just suspension was $\sim 1 \times 10^{-3}$ kW/m^3. In this case, solid suspension is unlikely to be an important factor. This type of system may therefore be treated as that corresponding to mixing in liquid phase alone and Equation 7B.11 can be used. It must be pointed out that other correlations developed for higher density particles show a significant effect on mixing time for solid–liquid system mainly because a substantial portion of the energy is used in suspending the solids.

7B.9.1.6   Mixing Time in Three-Phase (Gas–Liquid–Solid) System

There are no reported experimental data on gas–liquid–solid systems that are relevant to animal cell culture using microcarrier beads as support for the cells. As mentioned earlier in the case of gas–liquid and solid–liquid systems, because of (i) the low density difference ($\Delta\rho \sim 30$–50 kg/m^3) and (ii) low aeration rates, there is insignificant impact of introduction of the solid and gas phases. Further, if an upflow impeller is used, the power drop due to aeration is less than 10% at low aeration rates (Fig. 7A.6). Therefore, Equation 7B.11 can be used in this case also.

7B.9.1.7   General Comments on Liquid-Phase Mixing   Liquid-phase mixing in a three-phase stirred bioreactor is a complex phenomenon. For a given impeller–vessel combination, a large number of variables have an important bearing on the mixing time. These are (i) geometric variables such as the sparger type and distance between the sparger and impeller; (ii) solid-phase properties such as diameter, density, and solid loading; (iii) liquid-phase properties such as viscosity, surface tension, and presence of surfactants; (iv) operating variables, that is, gas velocity; and (v) the location of feed point. Therefore, prediction of the mixing time is a formidable task. It is practically impossible to arrive at a unique correlation satisfying various design and system parameters. Under these circumstances, it is best to simulate the **actual** system in a reasonable sized stirred reactor (diameter ~ 0.5 m) using the proposed impeller type and other geometric proportions (*C/T*, *D/T*, type and size of sparger, etc.) to obtain mixing time data. These results should give a rough estimate of the mixing time in the large vessel although it is known that mixing time increases with vessel size. Some thumb rules have been proposed in terms of the impeller tip speed. Czermak et al. (2009) suggest a tip speed between 1 and 2 m/s. Ma et al. (2006) indicate that 2 m/s is sufficient to give good mixing. As mentioned earlier in this chapter as well as Chapter 7A, such thumb rules must be used with great caution.

Mixing is more critical in high-density cell cultures. Ozturk (1996) has elaborated on the problems in such cultures. The viscosity of the medium can increase by a factor of 2–3 at cell densities of the order of 5×10^7 cells/ml. Mixing times in such cases could be in the range of several minutes, particularly for large industrial reactors. Aqueous alkali is generally added to counter the decrease in pH. Ozturk has brought out the need for a well-defined strategy for pH control, which includes addition of the base in the discharge stream at the eye of the impeller and an on–off control so that sufficient time is allowed for the pH probe to respond to an addition. Proper management of the aeration system and alteration in cell metabolism can allow minimization of the need for base addition (Ozturk 1996).

7B.9.1.8 Nutrient Addition and Mixing   The importance of physiological similarity and role of mixing in bioreactors has been already emphasized. To achieve physiological similarity in a larger-scale reactor, it is required that mixing of the contents of the reactor should be as good as in the small scale. It has been mentioned (Eq. 7B.9) that the mixing time increases with the size of the reactor. To meet the requirements of physiological similarity in different locations of the reactor, it is desirable to have good macromixing. It is now a generally accepted practice (Baldyga and Bourne 1999) that addition of reactants/nutrients should be in the region of highest local specific energy dissipation rate ε_{Max}. The strength and volumetric flow of the discharge stream in this region is such that it can disperse any material added to the reactor relatively quickly than any other position in the reactor. For a bioreactor, in view of the importance of maintaining uniform conditions *vis-à-vis* nutrients, pH, etc., it is extremely important to follow this practice. Such a practice entails subsurface addition with attendant choking/blocking of the feed pipe. Assirelli et al. (2005) suggested the use of a feed pipe that revolves with the impeller. For instance, they used the location for maximum energy dissipation in disc turbine-agitated vessel determined from the work of Assirelli et al. (2002). This location is at $1.04 \times (D/2)$ from the top of the blade tip (in the trailing vortex of the impeller blade). However, this is at the expense of considerable complexity. Bhattacharya and Kresta (2004) proposed a novel method of surface addition. Using the upflow type of impeller, the feed point is located at a position that corresponds to the radius at which the upflow from the impeller hits the surface. In this manner, the feed is swept by the flow from the impeller even though its location is at the surface where the energy dissipation is lowest. These investigators showed that the proposed technique could yield better results. However, this method was found to be sensitive to the radial position of the feed pipe and the submergence of the impeller in the liquid. Nienow and Bujalski (2004) suggest that the performance can be improved further through the use of an upflow hydrofoil ostensibly because of a better focused upflow of the hydrofoil.

Nutritional Requirements of Animal Cells   Mammalian cells have a very intricate nutrition requirement. Karkare (2004), Burgener and Butler (2006), Eibl et al. (2009), and Sharma et al. (2011) have presented exhaustive discussions on this issue. The medium used should have amino acids, vitamins, salts, and, in some cases, protein supplements/antibiotics. Glucose is generally added to the medium as the primary carbohydrate. In the initial stages of animal cell culture, fetal calf serum was widely used. Serum is the supernatant of clotted blood and has most of the nutritional components required. Calf serum's greatest advantage is its high embryonic growth factor content. The actual serum content as a percentage of the medium varies from 1 to 20 vol. %. The drawbacks of such serum are (i) its variable nature/composition; (ii) possibility of viral contamination that creates difficulties in process validation; (iii) high cost, which can be as high as 85% of the total medium cost; (iv) lack of regular supply; (v) high protein content that can create difficulties in product separation; and (vi) health risk arising from undefined agents (Section 7B.2; Karkare 2004; Burgener and Butler 2006).

The chapter by Burgener and Butler (2006) on Media Development gives a very good summary of the various media available for industrial use.

In view of the aforementioned drawbacks, efforts have been made to develop suitable serum-free media (Section 7B.2). Such media are available for commonly used cell lines such as CHO and hybridomas. Unfortunately, poor understanding of growth factors and hormone requirements has retarded the development of effective synthetic formulations. The final aim would be cultivation in a serum- and protein-free formulation that avoids the risks and allows facile product separation (Butler and Meneses-Acosta 2012; Rodrigues et al. 2013).

7B.10   GAS DISPERSION

7B.10.1   Importance of Gas Dispersion

In the case of a culture in which oxygen is provided by aeration, it is necessary to operate the reactor above the speed at which the gas is completely dispersed. This allows utilization of the entire vessel volume for gas–liquid contacting. Gas dispersion is important in the case of bioreactors because incomplete dispersion has serious implications on the bioprocess. Regions that are deficient in oxygen can exhibit a completely different physiological behavior. Two situations are possible in this regard: (i) the section of the bioreactor below the sparger is starved of oxygen and hence follows a significantly different metabolic pathway and (ii) inadequate stripping of carbon dioxide because of absence of stripping gas. This causes a buildup of dissolved CO_2. The importance of complete gas dispersion is further explained in a semiquantitative treatment in the following.

Figure 7B.18 depicts the concentration profiles for oxygen absorption and CO_2 desorption. Although CO_2 is relatively more soluble in aqueous media than oxygen, it is still not so highly soluble that the gas-side mass transfer can contribute significantly to the overall resistance. The contribution of gas-side resistance is approximately 10–15%, and therefore, ignoring it results in a decrease in the assumed driving force and hence a small increase in the predicted $k_L a$ value. At steady state (equilibrium), the rates of various steps involved in desorption of CO_2 being equal,

$$[R]_{CO_2} = K_{SL}\left\{[CO_2]_{Cell} - [CO_2]_{Bulk\ liquid}\right\} = k_L a\left\{[CO_2]_{Bulk\ liquid} - [CO_2 *]\right\}$$

Applying Henry's law, $[CO_2 *] = (H)_{CO_2} \times (p_I)_{CO_2}$

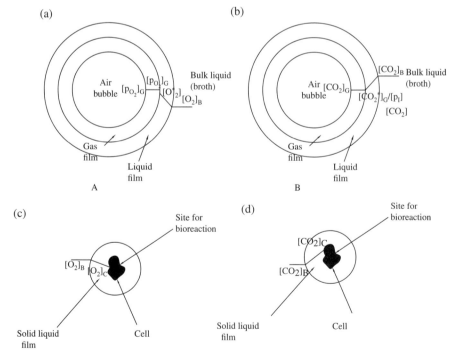

FIGURE 7B.18 Concentration profile for transfer of oxygen across (A) gas–liquid film and (C) cell–liquid film. Concentration profiles for transfer of CO_2 across (B) liquid–gas film and (D) cell–liquid film. (Adapted from Nienow (2003) with permission from John Wiley & Sons, Inc. © 1969.)

$(p_I)_{CO_2}$ is the partial pressure of CO_2 at the gas–liquid interface. For $N > N_{CD}$, $(p_I)_{CO_2}$ can be obtained from the flow rates of CO_2, Q_{CO_2}, and the stripping gas (air), Q_{Air}, as follows:

$$(p_I)_{CO_2} = \left(\frac{Q_{CO_2}}{Q_{CO_2} + Q_{Air}} \right) \tag{7B.12}$$

For the case when $Q_{Air} \gg Q_{CO_2}$,

$$(p_I)_{CO_2} = \left(\frac{Q_{CO_2}}{Q_{Air}} \right) \tag{7B.13}$$

A lower $(p_I)_{CO_2}$ results in lower $[CO_2^*]$ or a higher driving force for desorption of CO_2. From Equation 7B.13, for a given metabolic process rate (rate of formation of CO_2 or Q_{CO_2}), $(p_I)_{CO_2}$ decreases with increase in stripping air flow rate. However, as explained later in this section, the maximum allowable sparging rate is ~0.1 vvm (Ozturk, 1996). For a given rate of aeration, stirred reactor configuration, and operating conditions, the value of $k_L a$ is fixed based on the relative dispersion criterion outlined in Section 7A.5.3 and used for estimating $k_L a$ from Equation 7B.14 in Section 7A.6:

$$k_L a = 3.35 \times \left(\frac{N}{N_{CD}} \right)^{1.464} (V_G) \tag{7A.18}$$

At $N = N_{CD}$, the values of Q_{CO_2} and Q_{Air} are spatially uniform. Thus, operating at $N = N_{CD}$, $(p_I)_{CO_2}$ or dissolved CO_2 concentration is also expected to be spatially uniform.

7B.10.2   Effect of Dissolved Carbon Dioxide on Bioprocess Rate

Carbon dioxide is produced in the process of cell metabolism. It can be both harmful and useful depending on the type and stage of the bioprocess. In general, most cell lines need a small quantity of dissolved CO_2, particularly when the cell density is low. This situation is reversed at high cell densities (Karkare 2004). The solubility of carbon dioxide is approximately 25 times that of oxygen. Thus, CO_2 dissolution is thermodynamically favored. Dissolution of carbon dioxide in aqueous media results in formation of carbonic acid. A direct consequence is a decrease in the pH of the broth. To counter this change in pH, a base (sodium carbonate or hydroxide) is added. Such additions result in increase in the osmolality and greater potential for cell damage. The effects of dissolved CO_2 and osmolality have not been decoupled. The majority of cell cultures show good tolerance to osmolality over the range of 270–330 mOsm/kg. The effect depends on the cell type. For example, Kimura and Mille (1997) and deZengotita et al. (1998) found that CHO cells exhibit lower inhibition than hybridomas. The same authors also found that lowering of the osmolality

TABLE 7B.7 Environmental parameters for mammalian cell cultivation

Parameter	Range	Typical value
Broth pH	6.6–7.6[a]; (7–7.5)[b]	7[a]
Temperature (°C)	33–39[a]; (36–38)[c]	37[a]
Dissolved oxygen (Pa)	0.7–40[a,d]	10[a]
Osmolality (mOsm/kg)[e]	280–360[a,e]	300[a]
Dissolved CO_2 (Pa)	0.9–20[a]	7[a]
Tolerable shear rate (s^{-1})	0–3000[a]	1500[a]
Production cycle	10–30 days[c]	

[a] Karkare (2004).
[b] Junker (2004).
[c] Nienow (2006).
[d] Malda et al. (2004) indicate that the specific rate of oxygen consumption of adult articular chondrocytes in microcarrier bioreactor culture is approximately constant over an oxygen tension range of 4–21%. (Oxygen tension is another method of expressing partial pressure of oxygen. It is defined as the pressure that oxygen in a mixture of gases would exert if it were present alone. In classical chemical engineering, it is referred to as "partial pressure of oxygen.")
[e] Milliosmolar or milliosmole (mOsm): 1 Osmole is equal to 1 mol of solute divided by the number of ions formed per molecule of the dissolved solute.

partially reduced the deleterious effect of dissolved CO_2. The optimum levels of dissolved CO_2 in terms of partial pressure of CO_2 are in the range of 40–60 mm Hg (deZengotita et al. 1998).

Table 7B.7 gives the typical environmental parameters for mammalian cell cultivation.

Chu and Robinson (2001) and Meier (2005) and Nienow (2006) indicate that high levels of dissolved CO_2 and associated higher osmolality are the main hurdles in scale-up. Indeed, as shown later, much higher dissolved CO_2 levels can be reached in large-scale reactors that do not have (i) adequate means of stripping the CO_2 formed and also (ii) poor gas dispersion.

Incomplete gas dispersion could also result in poor mixing of the liquid phase and localized high concentration of dissolved CO_2 since there is a very low degree of turbulence in the region below the gas sparger. There are several literature studies that prove the deleterious effect of high dissolved CO_2 concentrations. In a 3-day batch study on recombinant CHO cells, Drapeau et al. (1990) found that under otherwise identical conditions (constant dissolved oxygen, pH, and other relevant variables), there was a dramatic decrease in cellular growth and specific productivity with increasing concentration of dissolved CO_2. The resulting high CO_2 concentrations caused local cell lysis and poor productivity. Gray et al.'s (1996) study on CHO cells with adequate oxygen supply showed that poor CO_2 removal curtailed cell growth. In the case of a bioreactor for TGFß, Garnier et al. (1996) have shown that improper aeration caused carbon dioxide accumulation that inhibited protein production. Mostafa and Gu (2003) studied scale-up of protein X (a therapeutic glycoprotein) culture from 1.5 to 1000 l. The two sets of cultures used exactly identical conditions. It was found that there was a 40% loss of productivity in the 1000 l reactor as compared to the bench scale. An important difference was that in the large-scale reactor, the aeration rate (in terms of vessel volume per minute (vvm)) had to be lowered to 86% of that in the bench scale due to foaming. This small reduction in vvm caused a dramatic 250% increase in dissolved CO_2. An obvious conclusion is that the reduced aeration rate resulted in poor stripping of CO_2 and higher dissolved CO_2 concentration in the 1000 l reactor. This was based on observations that the resulting lower pH affects pH-dependent enzymatic reactions (McQueen and Bailey 1990; deZengotita et al. 2002). Using a combination of an open pipe air sparger and an antifoam agent, Mostafa and Gu (2003) could maintain the dissolved CO_2 at the lower level required. In the absence of specific information, it is difficult to calculate the N_{CD} values in the bench and large-scale reactors. Ozturk (1996) has provided conclusive evidence on requirement of adequate stripping of CO_2. A silicone membrane aeration system was used to supply oxygen in a 200 l reactor for cultivation of hybridoma cells. A significant imbalance between dissolution of oxygen and stripping of CO_2 resulted in accumulation of CO_2. This is a clear evidence of the effect of stripping on accumulation of CO_2. These findings show that adequate stripping of CO_2 should be one of the factors that must be addressed during scale-up. Adequate dispersion of air/oxygen should be an answer to this problem. Additionally, sweeping of the headspace to remove accumulated CO_2 was also found to curtail base addition.

Gas dispersion in stirred reactors has been exhaustively discussed in Sections 7A.3 and 7A.4 for two- and three-phase systems, respectively. Therefore, this section will be restricted to the specific requirements of stirred reactors for animal cell culture.

7B.10.3 Factors That Affect Gas Dispersion

Gas dispersion in bioreactors is a process that has a complex dependence on various factors. Firstly, the gas flow rate has an important bearing on the hydrodynamics of two-/three-phase reactors. A measure of aeration rate that has been conventionally used to represent the air flow rate in stirred bioreactors is vvm. Chapman et al. (1983b) have shown that the use of vvm eliminates the scale-up effect since vvm is the reciprocal of the residence time of the gas phase (V_B/Q_G). Schluter and Deckwer's (1992) work on relatively large-diameter (up to 3 m) vessels also supports the concept of vvm for scale-up of stirred tank reactors. In cell culture systems, the aeration rate is decided by the oxygen requirement of the culture. In terms of vvm, the oxygen requirements of various biological species are as follows: (i) bacterial and mycelial, 0.5–1.5, and (ii) animal cells, ~0.01 (Oh et al. 1989). The low oxygen requirement of animal cell lines is a consequence of their relatively slow growth. This is evinced from their doubling times, which are more than 20 times higher than bacterial cells. The peak oxygen requirement (Table 7B.6) of animal cell cultures is also much lower than that of bacterial cells. As a result, animal cell lines require lower aeration rates at least for the current low cell density batch/fed-batch cultures. Perfusion systems with high cell density may require higher aeration rates that may be deleterious to cell viability. Ozturk (1996), in his review, has indicated that the maximum sparging rate that is limited by cell lysis and foaming factors is ~0.1 vvm.

The second important factor that affects gas dispersion is the presence/absence of surface-active agents in the medium. The average bubble size is decided by the surface tension of the medium and presence of surface-active substances as discussed by Walter and Blanch (1986). The details are available in Section 7A.5.1.2. The former has a bearing on the critical speed for gas dispersion N_{CD}. Surface tension lowering agents such as Pluronic F68 widely used in mammalian cell culture tend to decrease the average bubbles size and hence result in a lower N_{CD}. However, these small bubbles possess greater potential for cell damage (in the absence of Pluronic F68). Another interacting parameter is the broth viscosity. Different broths/systems may have varying viscosities. Higher viscosity results in larger bubble sizes and consequently higher N_{CD}. Proper selection of sparger/impeller system is important for minimizing cell damage and foaming and maximizing the mass transfer rates. The use of Pluronic F68 increases the foaming tendency of the broth. Prevalence of foam has been found to cause loss of cells and product and to introduce the hazard of contamination (Vardar-Sukan 1998). Antifoaming agents can be used to prevent excessive foaming. The antifoaming agent must be compatible with the bioprocess requirements. Junker (2007) has presented a critical review of the factors that cause foaming and methods for detection and prevention of the same with or without an antifoaming agent. It is difficult to predict the combined effect of Pluronic F68 and an antifoaming agent on N_{CD}. Estimation of N_{CD} is discussed in Section 7B.10.4.

7B.10.4   Estimation of N_{CD}

Correlations for N_{CD} are given in Section 7A.5.4. N_{CD} generally decreases with increase in impeller diameter but increases with increase in gas flow rate and reactor diameter for a fixed impeller diameter. The low aeration rate requirements in batch/fed-batch animal cell systems imply that N_{CD} is also low. Low impeller speeds have a further advantage that surface aeration, which can be harmful to cell viability, is also eliminated (Chapman et al. 1980). Correlations for N_{CD} for standard impellers such as Rushton turbines and mixed flow impellers (pitched blade turbines, both up- and downflow types) are available in the literature, as discussed in Section 7A.10. It has been noted that correlations for N_{CD} for the low shear/low mechanical impact severity impellers (Lightnin, Elephant Ear Appelkon, Mixel TTP, etc.) studied by Collignon et al. (2010) are not available. There is an urgent need for studies on gas holdup and dispersion to prove the utility of these impellers in three-phase systems. However, the studies of Nienow and his group (Bujalski et al. 1990; Bujalski 1987; Bujalski et al. 1988; Bujalski et al. 1990; Nienow and Bujalski 2004; Simmons et al. 2007) and also of Fishwick et al. (2005) have proven that the 45° upflow pitched blade turbine is a very versatile impeller for gas dispersion and solid suspension. Reliable data/correlations are available for this impeller. Therefore, in the present state of knowledge, it should be considered as a good alternative. It must, however, be noted that these correlations are unlikely to give satisfactory prediction in the presence of surface-active substances common to animal cell culture systems. Ideally, it will be desirable to measure N_{CD} using the real system composition as discussed in Section 7B.9.1.7.

7B.11   SOLID SUSPENSION

For cell growth on microcarriers, it is important that the microcarrier receives adequate supply of the nutrient. This implies effective mass transfer to the microcarriers from the bulk liquid medium. The sequential mass transfer steps have been shown in Figure 7B.18. For adequate mass transfer from medium to microcarrier, the latter have to be kept in suspension as explained in Section 7A.7. Thus, there is an additional minimum agitation intensity consideration, namely that for solid suspension. Current commercial microcarriers are slightly denser ($\rho_S \sim 1040$ kg/m^3) than culture media. Pangarkar et al. (2002) have given a comprehensive review of the relevant literature on solid suspension in both solid–liquid and gas–liquid–solid systems up to 2002. Collignon et al. (2010) have covered the post-2002 literature on solid suspension for solid–liquid systems (without gas sparging). The aforementioned reviews indicate that abundant literature is available on the critical speed for solid suspension in both two- (N_S) and three-phase (N_{SG}) systems at least for the conventional impellers such as the disc and pitched blade turbines. However, N_{SG} is likely to change due to the presence of surface-active agents. In this event, similar to N_{CD}, it is advisable to determine N_{SG} using the real system composition as discussed in Section 7B.9.1.7.

7B.11.1   Two-Phase (Solid–Liquid) Systems

The critical speed for solid suspension in a stirred reactor is a strong function of the type of impeller and other important geometric parameters (C/T, D/T, etc.; Table 7A.3). Thus, the Rushton turbine is much less energy efficient for solid suspension than a mixed flow pitched blade turbine. The downflow blade turbine is preferred in solid–liquid (two-phase) systems, whereas the upflow is preferred in gas–liquid–solid systems. There is an optimum configuration for minimum power input for solid suspension. For the conventional impellers (various types of turbines), Pangarkar et al. (2002) have presented a discussion on the optimum geometry with respect to solid suspension and solid–liquid mass transfer. They have pointed out that operation of the impeller just above N_S gives optimum solid–liquid (liquid–cell wall in the case of cell culture bioreactors) mass transfer. Speeds of agitation higher than N_S do not result in concomitant increase in liquid–solid mass transfer. On the other hand, the power input P increases rapidly since $P \propto N^3$. In the present application, besides higher operating cost, unresolved doubts about shear sensitivity of cells (due to shear/impact/mechanical damage, etc.) also dictate that unnecessary higher power inputs should be avoided. Ibrahim and Nienow (2004) carried out a systematic study on the suspension of microcarriers in vessels with different geometries of the base. The microcarrier used was hydrated Pharmacia Cytodex 3® (size 140–210 µm and density 1040 kg/m^3) at ~20 g/l loading in phosphate buffer saline solution. A fully baffled vessel (four baffles of width $= 0.1 \times T$) having diameter of 0.29 m and $H/T = 1$ with different bottom shapes and impellers was used in this study. It was found that the HE3 impellers at $0.35 < D/T < 0.45$ and $C/T = 0.25$ with all the three base shapes were able to suspend the microcarriers at an average specific power input of $\varepsilon_{Avg} = 0.5$ kW/m^3. However, there was a significant time lag for achieving full suspension. To eliminate this time lag, it was suggested that the operation should begin with much higher power input than that for just suspension ($N > N_S$). After suspension is achieved and cell attachment begins, the agitation speed can be reduced to the just-suspended condition. In practice, this speed should be 10–20% higher than N_S. Nienow (2006) has indicated that even at a relatively low ε_V of 0.5 kW/m^3, maximum value ε_I encountered at the critical suspension speed could be >>1 kW/m^3. Such high values can cause mechanical damage at least in the impeller discharge zone. In this respect, low shear/low mechanical impact severity impellers used by Collignon et al. (2010) are superior alternatives.

7B.11.2   Three-Phase (Gas–Liquid–Solid) Systems

In Section 7A.4, it has been explained that in a three-phase (gas–liquid–solid) system, the critical speed for suspension of the solids N_{SG} depends on the type of impeller and the gas flow rate (Chapman et al. 1983a). The behavior of various impellers has been elaborated in Section 7A.7.1.3. Aeration rates are relatively low in the case of bioreactors for animal cell culture. In addition, the microcarrier supports used have a marginally higher density than water, and probably, these could be neutrally buoyant in denser broths. Therefore, it is expected that the difference between N_{SG} and N_S is unlikely to be substantial. Further, if the six-blade upflow pitched blade turbine is used, the discussion in Section 7A.7.1.3 suggests that relatively stable operation can be achieved. An optimum configuration that is consistent with solid suspension and gas dispersion suggested by Pangarkar et al. (2002) is recommended. The operating power input should be chosen such that it is slightly higher than N_{SG} yet does not cause significant damage to the cells because of mechanical impact. A major problem with sparged systems using microcarriers is the tendency of the microcarriers to float into the foam. It has been suggested that this drawback can be overcome by aeration of the liquid in a cage that is either rotated or vibrated (Karkare 2004). The constant relative motion of the cage prevents clogging that can occur due to bridges particularly in a perfusion-type process. This reactor has some similarities with the Carberry spinning basket reactor (Carberry 1964). A number of commercial reactors in this category are available (Karkare 2004). Additional advantages of these reactors include protection of the anchored cells from shear and well-directed flow patterns that yield relatively better mass transfer. This variant cannot be classified as a stirred reactor in the strict sense since there is no impeller used.

Correlations for N_{SG} for conventional impellers (six-blade disc turbine, up- and downflow pitched turbines, Scaba 6SRGT, Chemineer HE3 downflow, A315 Lightnin' upflow, propeller) are given in Table 7A.4. The relative values of N_{SG} and N_{CD} depend on the type of impeller used. In the reactor design examples in Section 7A.10, estimates using standard correlations show that for both the six-blade pitched turbines, $N_{SG} > N_{CD}$. As a safe design practice, the higher of these two values should be used as the basis for operating speed. Therefore, correlations for both N_{CD} and N_{SG} in stirred reactors need to be developed for real systems of appropriate surface tension (with additives) and viscosity. These studies should cover not only the relatively new impeller types (Collignon et al. 2010; Bustamante et al. 2013) but also the six-blade upflow pitched turbine found to be equally good by Simmons et al. (2007), as discussed in Section 7B.9.1.7.

7B.12   MASS TRANSFER

7B.12.1   Fractional Gas Holdup (ε_G)

Fractional gas holdup is an indication of the effective interfacial area in any gas–liquid system. Detailed discussion on fractional gas holdup in stirred reactors is available in Section 7A.5. Analogous to other parameters, the gas holdup is also a function of the operating parameters (superficial gas velocity, type of impeller and its size/position in the reactor, etc.) and system properties (liquid-phase viscosity, surface tension, solid density and loading, presence of surfactant, etc.). As discussed in Section 7A.5, Yawalkar et al. (2002a) have been able to obtain a unique correlation for the gas holdup using the concept of relative dispersion N/N_{CD}:

$$\varepsilon_G = 0.104 \left(\frac{N}{N_{CD}} \right)^{0.62} (vvm)^{0.64} \tag{7B.14}$$

A modified version of Equation 7B.14 is

$$\varepsilon_G = 0.122 \left(\frac{N}{N_{CD}} \right)^{0.64} (vvm)^{0.69} (T)^{0.22} \left(\frac{D}{T} \right)^{0.14} \tag{7B.15}$$

This correlation was found to give a better fit of data covering a broad range of system properties, impeller types/size, and geometric configurations. Both Equations 7B.14 and 7B.15 require information on N_{CD}. In the case of stirred bioreactors for both suspension and microcarrier-anchored culture, it is desirable to use low shear/low mechanical impact severity impellers (Collignon et al. 2010; Bustamante et al. 2013; Fig. 7B.15). For these new impellers, correlations for N_{CD} are not available; but as mentioned earlier, the upflow pitched blade turbine with six blades has been found to be equally good in terms of gas dispersion and solid suspension and the same can be recommended. The value of N_{CD} to be used in Equation 7B.15 should be established for the actual system with, for example, Pluronic F68 and any antifoaming agent that may be used (Section 7B.9.1.7).

7B.12.2   Gas–Liquid Mass Transfer

Figure 7B.18 shows the various mass transfer steps involved in a bioreactor.

Yawalkar et al. (2002b) used the concept of relative dispersion to unify all gas–liquid mass transfer data available in the literature for stirred reactors. This matter has been discussed in great detail in Section 7A.5. The correlation proposed was

$$k_L\underline{a} = 3.35 \left(\frac{N}{N_{CD}} \right)^{1.464} (V_G) \tag{7B.18}$$

It is evident that the above correlation also requires knowledge of N_{CD}. Therefore, studies for establishing reliable correlations for N_{CD} for the new impellers deserve immediate attention.

Several investigators have shown that addition of Pluronic F68 causes significant reduction in $k_L a$ (Nienow and Lavery 1987; Kawase et al. 1992; Garcia-Ochoa and Gomez 2005). Garcia-Ochoa and Gomez (2009) suggest that the decrease in $k_L a$ could be 40–80%. It has been explained earlier that Pluronic F68 also affects the bubble size (Section 7A.5.1b and Walter and Blanch 1986) and hence N_{CD}. N_{CD} accounts for all the changes in physical properties such as surface tension and viscosity. In addition, the system geometries (C/T, D/T) are also included in N_{CD}. Consequently, as long as the appropriate/representative value of N_{CD} is used, Equation 7B.16 should yield a reasonably reliable estimate of $k_L a$. In terms of energy efficiency, Sardeing et al.'s (2004) data also support the argument of Nienow's group (Section 7B.10) that under otherwise identical conditions, upflow impellers afford better mass transfer (Section 7A.6).

Currently, cell densities ~5 × 10⁶ cells/cm³ are being achieved (Nienow 2006). There is a growing interest in operating at cell densities higher than these. One of the limitations on high cell density is the oxygen transfer rate afforded by the type of aeration device used. Ozturk (1996) used literature data to obtain broad ranges of cell densities achievable with different types of aeration devices (Fig. 7B.19).

It is evident from Figure 7B.19 that for achieving the highest densities (>10⁷ cells/ ml), surface aeration is totally inadequate. Membrane aeration can eliminate the problem of cell death due to bubble bursting (Lehmann et al. 1988; Luttman et al. 1994). Silicone rubber membrane-based aeration can increase the oxygen aeration rate to a level ~5 × 10⁷ cells/ml. Unfortunately, the membrane length required in terms of the membrane area for oxygen supply at a given flux ((kg)$_{oxygen}$/m²s) imposes limi-

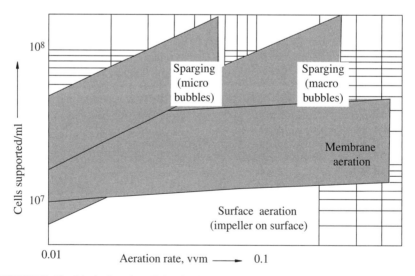

FIGURE 7B.19 Limitations in cell density based on oxygen delivery in different aeration systems. (Reproduced from Ozturk (1996) with kind permission from Springer. © 1996, Springer.)

tations on the maximum reactor size, which is ~200 l. Further, desorption of CO_2 is seriously affected due to inadequate stripping as elaborated in Section 7B.10.2. Moreira et al. (1995) studied the efficacy of three methods of supplying the required oxygen: surface aeration, supply through membrane, and gas sparging. They found that the membrane-based system could yield higher oxygen transfer rates than sparging, particularly for small reactors (the drawbacks of this system are discussed in Section 7B.15.2). Systems using macrobubble sparging also yielded nearly same cell densities. Sparging has its own problems of cell viability, foaming, etc. The use of Pluronic F68 can decrease cell lysis due to bubble bursting, but it is detrimental to gas–liquid mass transfer.

Figure 7B.19 indicates that the oxygen requirements for the highest cell densities can be met with sparging of air using appropriately designed pipe or ring spargers.

Ring-type spargers with larger holes (>1 mm) produce bubbles in the range of 6–8 mm. Sparger design is discussed in Section 10.7.5.1. Evidently, these spargers with larger holes can suppress cell death and foaming but yield lower gas–liquid mass transfer area and hence lower volumetric mass transfer coefficient $k_L a$. At this stage, it needs to be pointed out that most literature studies refer to an overall effect of small/large bubbles on mass transfer rates. The correct expression for the effect of the size of bubbles is through the gas holdup or the effective gas–liquid mass transfer area, a rather than true k_L. Considering that cell damage is more due to bursting of bubbles and not due to impeller shear (Section 7B.8), higher intensities of agitation can be used to increase $k_L a$. Nienow (2006) suggests that a variable speed drive may be used such that at the beginning for low cell densities, the speed of agitation is maintained at a low level and then increased. Bustamante et al. (2013) also suggest a lower initial operating speed, which is followed by higher speeds as the cell density increases. There are two reasons for this strategy: (i) at low cell densities typical of the start of the growth phase, lower operating speeds afford low impeller shear while meeting the relatively low oxygen demand and (ii) at higher cell densities, the oxygen demand is higher and a higher operating speed is required. This strategy will require a detailed model of growth/oxygen demand that should include the changes in physical properties (viscosity in particular). Alternatively, a high absorption rate of oxygen can also be realized by using enriched air. However, since the respiration coefficient for animal cells is unity, use of enriched air is fraught with higher dissolved CO_2/pH control/higher osmolality. In the event that enriched air is used, it must be ensured that the aeration rate (vvm) is maintained at the same level as for normal air since decrease in aeration rate results in lower $k_L a$ (Eq. 7B.16) and lower rate of stripping of CO_2 (Mostafa and Gu 2003; Benz 2011). Therefore, this alternative must be weighed carefully.

7B.12.3 Liquid–Cell Mass Transfer

There is a singular lack of information on this aspect of mass transfer in bioreactors. Figure 7B.18 shows that oxygen dissolved in the bulk liquid must diffuse across the liquid–cell film before it is taken up by the cell. Although there is no literature information on liquid–cell mass transfer in stirred bioreactors, some conclusions can be drawn based on analogy from mass transfer in stirred reactors for chemical reactions (Chapter 7A). In the case of cell culture reactors bioreactors, three major situations can be identified:

1. Mass transfer to freely suspended cells

These cells are relatively very small (~20 μm) and have density only slightly higher than the broth. Therefore, most investigators have implicitly assumed that the suspension of these cells under normal conditions of agitation (adequate power input) is undeniable. For these relatively very small cells, the characteristic Reynolds number is likely to be <1 (Nienow 2006). Under these conditions, the Ranz and Marshall equation (Eq. 6.4 in Chapter 6) predicts that the particle Sherwood number is given by the limiting value:

$$Sh_C = \frac{K_{SL}d_C}{D_A} = 2 \qquad (7B.17)$$

Knowledge of d_C and D_A allows calculation of K_{SL}. The suspension viscosities are similar to that of water (Nienow, 2006), which gives $D_A \sim 2 \times 10^{-9}$ m²/s. This yields an estimate of K_{SL} around 2.5×10^{-4} m/s, which is approximately the same as that for small particles at the state of just suspension ($N = N_S$) in a conventional stirred reactor for chemical systems (Section 7A.7).

2. Mass transfer to cells anchored on microcarriers

Microcarriers have a relatively large size (150–300 μm). For this situation, the value of K_{SL} can be calculated by the universal correlation proposed by Pangarkar et al. (2002) (Eq. 7A.21):

Single impeller system:

$$K_{SL} = 1.8 \times 10^{-3} \left(\frac{N}{N_{SG}} \right) (Sc)^{-0.53} \qquad (7A.24)$$

Multi-impeller system (Dutta and Pangarkar 1996):

$$K_{SL} = 1.02 \times 10^{-3} \left(\frac{N}{N_{SG}} \right)^{0.97} (Sc)^{-0.45} \qquad (7B.18)$$

Both these correlations require a knowledge of the minimum speed for just suspension of the solids. Details of estimation of the critical speed for solid suspension are discussed earlier in Section 7B.11 as well as Chapter 7A.

3. Oxygen uptake by the cell resulting from biological reaction of oxygen with the cell

The gas–liquid and liquid–cell mass transfer steps are in series. The last step in the growth/differentiation of the cell is the biological uptake of oxygen by the cell. This may be considered equivalent to a reaction on a catalytic surface. This step has not received as much attention in the literature as the gas–liquid mass transfer step. Malda et al. (2004) in their investigation of oxygen tension (~partial pressure) on adult artic-ular chondrocytes in microcarrier bioreactor culture did not find any significant difference in specific rate of oxygen consumption over an oxygen tension range of 4–21%. This is in contrast to the suggested Monod-type dependence of oxygen con-sumption by chondrocytes (Haselgrove et al. 1993). Malda et al. (2004) proposed that this may be due to a relatively low (<4%) Monod constant. The low Monod constant may result in a situation such that the effect of oxygen tension in the range of 4–21% used by Haselgrove et al. (1993) provides more than adequate oxygen for the processes occurring in the cell. This observation is difficult to generalize for other cell lines. In the particular case of Malda et al., the observed lack of dependence on oxygen partial pressure could be interpreted as a result of a (i) relatively slow liquid–cell mass transfer step (ii) low mass transfer rate through the cell wall, and (iii) slow uptake of oxygen by the cells as compared to the rate of the gas–liquid mass transfer step, par-ticularly under the conditions used. The overall effect in this situation is such that the rate of gas–liquid mass transfer is sufficiently large (even at 4% oxygen) in comparison with the rates of other steps occurring in series. Consequently, the gas–liquid mass transfer rate or oxygen tension does not influence the overall process. From a chemical engineering point of view, it is necessary to study the effect of oxygen partial pressure under carefully defined hydrodynamic conditions for individual cell lines. Such data will reveal the importance of the gas–liquid mass transfer step. The liquid–cell mass transfer and cell wall permeation coefficient can be estimated as discussed in Section 7B.15. The intrinsic specific oxygen consumption rate of the biological pro-cess is a characteristic of the cell line and this needs to be addressed independently. Unfortunately, most chemical engineers engaged in research on bioreactors have given importance *only* to the gas–liquid mass transfer step. This is a serious lacuna since there may be situations in which liquid–cell mass transfer or uptake of oxygen by the cell may be important. As a result, the corresponding mass transfer coefficients need to be established by careful studies. Malda et al. (2004) also suggest that the uptake of oxygen by the cells needs a more detailed investigation. Obviously, consid-erable additional information is required so that the rate-controlling step is clearly defined. At this stage, it is important that chemical engineers try to include the last two steps in their research in collaboration with molecular biologists/biochemists so that a rational bioreactor design based on the rate-controlling step (Section 7A.10: Worked Example) can be performed. Such a study should be an integral part of any research and development program for cell culture process.

7B.13   FOAMING IN CELL CULTURE SYSTEMS: EFFECTS ON HYDRODYNAMICS AND MASS TRANSFER

As a rule, pure liquids do not foam when contacted with a gas. A foaming liquid gener-ally prevents local thinning of the liquid film surrounding a bubble. On the other hand, a nonfoaming liquid allows thinning and thereby coalescence of neighboring bubbles

(Vardar-Sukan 1998). Aeration results in a froth that is thermodynamically unstable. The froth begins to drain trapped liquid as soon as it is formed. Depending on the presence/absence of surface-active substances, the froth can, respectively, yield a foam that has a very low liquid holdup or collapse, yielding separate gas and liquid phases. Surface-active agents affect the behavior of gas–liquid interfaces. Surfactants and anti-foaming agents are two types of surface-active compounds that are common. Both reduce equilibrium surface tension but have different effects. The former obstructs coalescence, thereby producing a lower average bubble size. Antifoams, on the other hand, promote coalescence that disrupts stable foams (Kawase and Moo-Young 1990; Vardar-Sukan 1998; Morao et al. 1999; Karakashev and Grozdanova 2012). As explained earlier, animal cell culture medium has a complex composition that has the potential to produce stable foams when aerated. For batch and fed-batch systems, foaming creates problems in the functioning of the reactor itself. In the case of perfusion cultures, the foam may carry over into the cell recycle system and become a hindrance in the separation and recycle of the cells. Therefore, an antifoaming agent is invariably used.

For the range of concentrations used (<50 ppm), antifoaming agents have a negative effect on $k_L a$. The precise nature of the effect can be better understood separately through the effects on the effective interfacial area, a, and the true mass transfer coefficient, k_L (Morao et al. 1999). Two opposing effects decide the absolute value of a: (1) reduction in equilibrium surface tension causes a lower mean bubble diameter $\{a \propto [\sigma]^{-0.6}\}$ (Calderbank 1958), while (2) the antifoaming action results in coalescence. Several investigators (Akita and Yoshida 1974; Koide et al. 1985 and Kawase; Moo-Young 1990) have reported that in the case of a sparged reactor, the above opposing effects counterbalance each other with the result that a is not affected by the presence of an antifoaming agent. For stirred reactors, Yagi and Yoshida (1974) found that there is a significant effect of antifoaming agents on a. On the other hand, Morao et al. (1999) found that the effect of the antifoaming agent on the gas holdup, ε_G ($a \propto \varepsilon_G$), varied with the type of antifoaming agent used. For silicone and soybean oils, ε_G decreased, whereas for poly(propylene glycol), PPG, there was an increase in ε_G. It was argued that for soybean and silicone oils the coalescing effect of the antifoaming agent dominated over the effect that causes lowering of surface tension. For PPG, this argument was reversed. However, the measured $k_L a$ values were lower for all antifoaming agents, regardless of coalescing (tap water) and noncoalescing (aqueous sodium sulfite) systems for antifoam concentrations between 2 and 50 ppm. Similar results were also reported earlier by Benedek and Heideger (1971), Kimura and Nakao (1986), Lavery and Nienow (1987), and Liu et al. (1994). Most investigators suggested that the major contribution to reduction in $k_L a$ arises due reduction in k_L. The latter was attributed to coverage of bubbles by the antifoaming agent and consequent reduction in internal circulation in the bubble (Kawase and Moo-Young 1990). These investigators also found that the reduction in $k_L a$ was lower in the case of stirred reactors as compared to bubble columns. This is probably due to the shearing action of the impeller in the former. An interesting observation was the increase in $k_L a$ above a critical concentration (>100 ppm) of the antifoaming agent (Lavery and Nienow 1987; Rols et al. 1990; Vardar-Sukan 1992; Liu et al. 1994). As the concentration of the antifoaming agent increases, three conflicting effects simultaneously occur: surface tension decreases, bubble coalescence increases, and finally bubble surface rigidity increases. Morao et al. (1999) argued that the last two negative effects attain a limiting value as the concentration of the antifoaming agent reaches a certain critical value. Beyond this value, the decrease in surface tension predominates and $k_L a$ increases because of increase in a. Rols et al. (1990) proposed a facilitated mass transfer mode in the case of some antifoaming agents such as soybean oil. This was attributed to a higher solubility of oxygen in the antifoaming agent, which is not necessarily true for other antifoaming agents.

From the available data, it can be concluded that for the low concentration range, most antifoaming agents cause a substantial decrease in $k_L a$. The exact decrease depends on the type and concentration of the antifoaming agent used. For example, the reduction expressed in terms of the ratio ($[k_L a]_{AF}/[k_L a]_{NF}$) ranges from 0.8 to 0.3 for a silicone antifoaming agent in 0.2 M sodium sulfate solution over a relatively small concentration range of 0.2–10 ppm (Morao et al. 1999).

Summary of Hydrodynamic and Mass Transfer Effects   The discussion earlier brings out the conflicting requirements of broth homogeneity, oxygen transfer, CO_2 stripping/osmolality/pH, and cell viability. As pointed out by Nienow (2006), this conundrum needs to be addressed through detailed systematic studies. Cold studies (without cells) preferably on large scale (at least a reactor diameter of 0.5 m) and in presence of representative concentrations of surface-active agent (Pluronic F68 or equivalent as well as antifoaming agent) could be a beginning in this direction.

7B.14   HEAT TRANSFER IN STIRRED BIOREACTORS

Uptake of oxygen by animal cell cultures is relatively slow. It is therefore anticipated that the associated heat effect will be low. Van't Riet and Tramper (1991) and Nienow (2006) suggest the following equation for estimating the heat released:

$$\Delta H = 4.6 \times 10^5 \times OUR^* \, (W/m^3) \tag{7B.19}$$

$$OUR^* : \left(\frac{mol}{m^3 s} \right)$$

Safe engineering practice requires design for the worst conditions. Therefore, the OUR in Equation 7B.19 to be used is the highest value/peak OUR for animal cells listed in Table 7B.6: 5 mmol/l/h. Based on this maximum OUR, the heat effect is

$$\Delta H = 630 \, W/m^3 \, \text{or} \, 730 \, kcal/m^3/h$$

Cooney et al. (1969), Junker (2004), and Benz (2011) suggest that the heat transfer duty can be estimated as 110 kcal/mol of oxygen uptake. Using this value, the heat evolved is 550 kcal/m³/h, which is 25% lower than the value obtained from

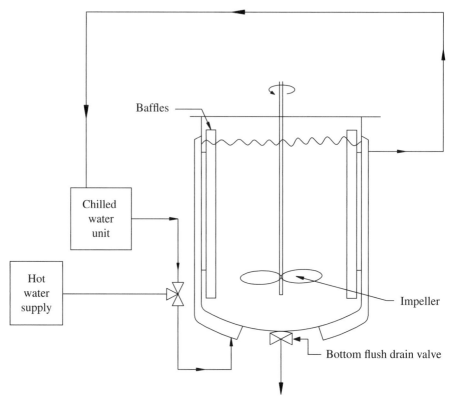

Bioreactor with cooling/heating jacket

FIGURE 7B.20 Bioreactor with cooling jacket. (Reproduced from Benz (2011) with permission from American Institute of Chemical Engineers. © 2011, American Institute of Chemical Engineers (AIChE).)

Equation 7B.19 of Van't Riet and Tramper (1991) and Nienow (2006). In addition to this direct heat load arising from the biological process, heat is generated due to dissipation of impeller power and expansion of air. If the entering air is relatively dry, it can absorb some part of the total heat load as it equilibrates with the media and evaporates water during the equilibration (Benz 2011). The total heat load estimated earlier is significantly lower than that for a chemical reaction. However, the optimum temperature (33–39°C) range for an animal cell process is relatively very narrow (Table 7B.7). If water available at ~32°C from a cooling tower is to be used as cooling medium, the temperature driving force realized (3–5°C) is too low for safe continuous operation. Benz (2011) has given correlations for bottom jackets, sidewall jacket, internal helical cooling coils, and vertical tube baffles. Simple calculations indicate that with a typical overall heat transfer coefficient of 250 W/m² K, the maximum heat duty of 630 W/m² can be met with a jacket (Fig. 7B.20). However, the margin of error is small with cooling tower water particularly during periods of high humidity and also malfunctioning of the cooling

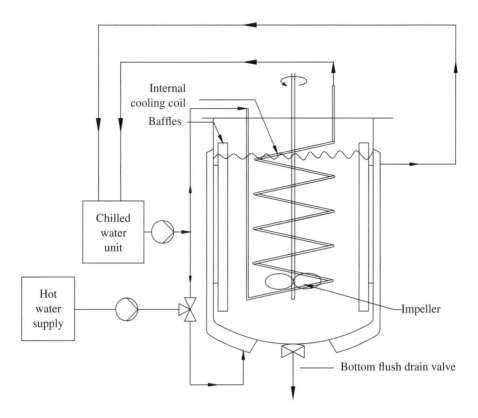

Bioreactor with jacket and internal cooling coil

FIGURE 7B.21 Bioreactor with jacket and internal cooling coil. (Adapted from Benz (2011) with permission from American Institute of Chemical Engineers. © 2011, American Institute of Chemical Engineers (AIChE).)

tower. The problem is aggravated in the case of insect cells for which the operating temperature window is further narrowed (between 25 and 30°C; Nienow 2006). In either case, it is prudent to increase the temperature driving force by resorting to use of chilled water. The available alternatives in terms of heat transfer surface are shown in Figures 7B.20, 7B.21, 7B.22, and 7B.23.

The drawbacks associated with internal cooling coil (Fig. 7B.21) have been discussed in Sections 7A2, 7A.8, and 7A.10.8. Vertical tube bundle (Fig. 7B.22) serving as baffles has the drawback that biological growth fouls the coils in the bundle and it is also difficult to clean the bundle in place. The external heat exchanger loop (Fig. 7B.23) is fraught with the problem of exposure of the cells to inimical environment such as deprivation of oxygen and thermal shock and hence possible cell damage/loss of productivity. Therefore, cooling/heating jackets that do not interfere with the hydrodynamics or bioprocess should be preferred particularly for animal cells that have low heat removal demand. A jacket using chilled water should be the most appropriate solution.

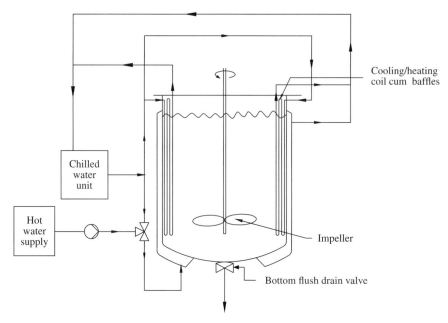

FIGURE 7B.22 Bioreactor with cooling jacket and cooling coil cum baffles. (Adapted from Benz (2011) with permission from American Institute of Chemical Engineers. © 2011, American Institute of Chemical Engineers (AIChE).)

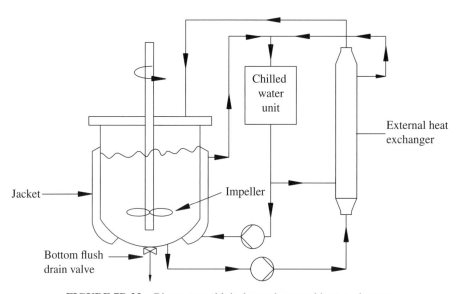

FIGURE 7B.23 Bioreactor with jacket and external heat exchanger.

7B.15   WORKED CELL CULTURE REACTOR DESIGN EXAMPLE

7B.15.1   Conventional Batch Stirred Reactor with Air Sparging for Microcarrier-Supported Cells: A Simple Design Methodology for Discerning the Rate-Controlling Step

Duty   It is required to design a stirred reactor using microcarrier-supported animal cells. The working capacity of the reactor (volume of broth) is 20 m^3. The dispersion height to diameter ratio = 1. Disengagement volume equivalent to 1/3 of the total volume is to be provided above the broth level. Therefore, total reactor volume is 30 m^3 and vessel diameter is 3 m. Impeller to be used is six-blade upflow pitched turbine with $D/T = C/T = 0.33$.

Microcarrier Properties   Density = 1040 kg/m^3, particle size = 200 μm, cell wall thickness ~ 1 μm, solid loading = 5 g/l (Section 7B.9.1.5 and Nienow and Bujalski 2004), vvm = 0.008 (Nienow (1996) has suggested that aeration rates as low as 0.005 vvm can be sufficient for animal cell lines), $V_G = (\text{vvm}/60) \times (V_B/A_C) = 3.77 \times 10^{-4}$ m/s, and $Q_G = (\text{vvm}/60) \times V_B = 2.66 \times 10^{-3}$ m^3/s, diffusivity of oxygen in water at 37°C ~ 2 × 10^{-9} m^2/s, and viscosity of the broth ~1 × 10^{-3} Pa s. Thus, Schmidt number = 500. The microcarrier used is a gel type with no internal pores. The microcarrier surface area per cubic meter of the broth is 144 m^2/m^3. Diffusion coefficient of oxygen in the cell wall is assumed to be 1 × 10^{-11} m^2/s and cell diameter = 20 μm. Surface area of a spherical cell = 1.25 × 10^{-9} m^2. Assuming complete coverage of the microcarrier support, the cell density is 1.14 × 10^5 cells/ml of broth. This value is lower than 5 × 10^6 indicated by Nienow (2006) because of the low microcarrier loading used. The cell density can be significantly increased through (i) higher microcarrier loading and (ii) the use of macroreticular (microporous) microcarriers.

Reactor Configuration   Reactor diameter = 3 m. Impeller diameter ($D/T = C/T = 0.33$) is 1 m.

Impeller to be used is six-blade upflow pitched turbine (6 PTU).

Calculation of Critical Stirring Speeds N_{CD} for 6 PTU:

$$N_{CD} = \sqrt[11]{8.26 \times 10^{-6} \left(\frac{Q_G}{D^3}\right)\left(\frac{g}{D}\right)^5 \left(\frac{T_0}{T}\right)^{2.5}} \tag{7B.20}$$

(Value of T_0 for use in this equation given by Bujalski et al. (1988) is 1 m.)

$$N_{CD} = 0.43 \text{ rev/s}$$

Calculation of N_{SG}

1. Calculation of N_S: Equation for estimation of N_S is (Section 7A.8.3)

$$N_S = 6.9 \, v^{0.1} \left[g \, (\rho_s - \rho_E)/s\rho_L\right]^{0.45} X^{0.13} \, dp^{0.2} \, D^{-0.85} \, (\text{Bujalski 1987})$$

N_S: 0.2 rev/s

$$N_{SG} = N_S \left(1 + \text{vvm}\right)^{0.11} = 0.20 \left(1 + 0.008\right)^{0.11} \quad \left(\text{Bujalski et al. 1990}\right)$$

N_{SG}: 0.2 rev/s

The dependence of the relative values of N_{CD} and N_{SG} has been discussed in Sections 7A.7.1.3 and 7B.11.2. In the present case, N_{CD} has been found to be higher than N_{SG}. However, the difference in the critical speeds for solid suspension in two- and three-phase systems, ΔN_{SG}, is practically zero. This behavior has been explained in Sections 7A.7.1.3 and 7B.11.2. The difference is further lower for two reasons: (i) the low density difference ($\Delta \rho \sim 30\text{--}50 \text{ kg/m}^3$) between the microcarrier support and the broth and (ii) relatively very low vvm of 0.008. As per conventional design practice, the higher among N_{CD} and N_{SG} (0.43 rev/s) is selected. Further, considering the fact that correlations do not yield a precise estimate, a safety margin should be incorporated by operating at a reasonably higher speed than the estimated N_{CD}. Thus, the operating speed of agitation is given by $N_{\text{Operating}}$ $= 1.2 \times N_{CD} = 0.51$ rev/s. P_G is assumed to be nearly equal to P_O since V_G is very low, and for upflow pitched blade turbine, introduction of gas has negligible effect on P_G (Fig. 7A.6):

Power required, $P_G = N_P \left\{\rho_L \left[(N_{\text{operating}})^3 (D)^5 = 1.7 \times 1000 \times 0.13 \times (1)^5 = 0.225 \text{ kW}\right.\right.$

$(N_P = 1.7, \text{Ibrahim and Nienow 1995})$

The working reactor volume is 20 m³. Therefore, $\varepsilon_{\text{Avg}} = 225/20{,}000 \sim 1.1 \times 10^{-2} \text{ W/kg}$. This value is sufficiently low for use in animal cell bioreactors. The low power requirement can be attributed to the relatively very low vvm used. The low vvm in turn is a result of the much lower oxygen utilization rate (lower peak OUR/higher doubling times as compared to bacteria/yeast in Table 7B.6 in such systems).

Mass Transfer Coefficients

1. Section 7A.6:

$$k_L \underline{a} = 3.35 \left[\frac{N}{N_{CD}}\right]^{1.464} \quad V_G = 1.64 \times 10^{-3} \text{ s}^{-1} \tag{7B.21}$$

2. Section 7A.7.2:

$$K_{SL} = 1.8 \times 10^{-3} \frac{N}{N_{SG}} Sc^{-0.53} = 1.7 \times 10^{-4} \text{ m/s} \tag{7B.22}$$

$$K_{SL} \left(\frac{A_P}{V_B}\right) = 1.7 \times 10^{-4} \times 144 \sim 0.024 \text{ s}^{-1} \tag{7B.23}$$

3. Permeation coefficient in the cell wall, $k'_P = D_A/\delta_{CW}$

$$k'_P = (1\times10^{-11}/1\times10^{-6}) = 1\times10^{-5}\,\text{m/s} \qquad (7B.24)$$

For a spherical cell, area/volume $= 6/d_C = 3 \times 10^5\,\text{m}^2/\text{m}^3$

$$k_P = k'_P\left(\frac{A_C}{V_C}\right) = 1\times10^{-5} \times 3\times10^5 = 3\,/\,\text{s} \qquad (7B.25)$$

The very high rate of permeation of oxygen in the 20 µm diameter cell having a very thin (1 µm) wall can be explained in a manner similar to that in Section 3.4.2.4 Spray Column Reactor (Li et al. 2013). It may be noted that this is despite a relatively very low value assigned to the diffusion coefficient of oxygen in the cell wall.

Comparison of the Three Mass Transfer Resistances   The resistances of the various steps estimated through the reciprocals of the transfer coefficients are as follows:

1. Resistance of gas–liquid mass transfer: $1/k_L a = 610\,\text{s}$

2. Resistance of liquid–cell wall mass transfer: $1\bigg/\left[K_{SL}\left(\dfrac{A_P}{V_D}\right)\right] = 1/0.024 = 41.7\,\text{s}$

3. Resistance for permeation across the cell wall $= 1/k'_P\left(\dfrac{A_C}{V_C}\right) = 0.33\,\text{s}$

4. Total resistance ~ 651 s

5. Percentage of resistance due to gas-liquid mass transfer $= \left(\dfrac{610}{652}\right)100 \sim 93.55\%$

6. Percentage of resistance due to liquid-cell wall mass transfer $= \left(\dfrac{41.7}{652}\right)\times 100 \sim 6.4\%$

7. Percentage of resistance due to permeation across the cell wall $= \left(\dfrac{0.33}{652}\right)\times 100 \sim 0.05\%$

Conclusion on Controlling Step   The aforementioned estimates reveal that among the diffusional steps the major resistance to the overall oxygen transfer process to the cell is contributed by the gas–liquid mass transfer step. This supports the extensive attention that has been granted to this step by various groups worldwide.

It should be mentioned that in this analysis only the diffusional resistances have been considered. The kinetics of uptake of nutrients and oxygen by the cell, its growth, and formation of products is a bioprocess. This factor has not been included in the earlier estimation of resistances. As explained in Section 7B.7, kinetic description of the bioprocess is very complex. It depends on a variety of factors that include nutrient supply/limitation/simultaneous changes in medium concentration (nutrients and metabolites) and varies with cell line. It is, therefore, not possible to make a general comparison similar to the rates of the physical (diffusional) processes.

7B.15.2 Reactor Using Membrane-Based Oxygen Transfer

It has been indicated earlier that one of the alternatives to eliminate cell damage due to bursting bubbles is membrane-based oxygen transfer. Hollow fiber membranes can give large permeation area per unit volume (Pangarkar and Ray 2014). Since it is desirable to use a compact device, hollow fiber membrane module is the preferred choice. Dense membranes made from a highly permeable polymer such as poly(dimethyl siloxane), popularly known as silicone rubber, are preferred (Kuhlmann 1987; Ozturk 1996). Microporous membranes made of PTFE (Mano et al. 1990; Schneider et al. 1995) and polypropylene (Lehmann et al. 1988) have also been suggested. In the case of the microporous variety, the pressure on the membrane side must be maintained below the bubble point of the membrane to prevent breakthrough of the gas phase. In such a system, the oxygen permeates across the dense membrane and is taken up by the broth. This method also prevents formation of foam and facilitates precise control of oxygen concentration consistent with the requirement of the specific bioprocess (Luttman et al. 1994). Ozturk (1996) has shown that for achieving higher cell densities, relatively higher oxygen transfer rates are required. Fenge and Lullau (2006) have cited a quantitative estimate of oxygen consumption rates. The range specified for different cell lines such as CHO, BHK, hybridoma cells, insect cells, and hematopoietic cells is 0.02–0.6 $\mu mol/10^6$ cells/h. This demand can be met for 10^6 cells/ml if the oxygen mass transfer coefficient $k_L a$ is in the range of 1.1×10^{-4} to $1.1 \times 10^{-3}/s^1$. From these approximate estimates, Fenge and Lullau (2006) concluded that for the anticipated high cell densities, compact membrane-based system may not be feasible even when pure oxygen is used. The analysis in Section 7B.15.1 indicates that $k_L a$ values in the above range indicated by Fenge and Lullau (2006) can indeed be assured with conventional coarse spargers even at a relatively low vvm of 0.008.

Apprehensions regarding deposition of cell aggregates and microcarriers on the membrane surface also persist. More importantly, the accumulation of CO_2 formed is a serious issue (Nienow 2006). The consequences of high dissolved CO_2 concentrations have been detailed earlier (Sections 7B.9.1.5, 7B.10.1, 7B.10.2, and 7B.12.2). Respiration coefficient for animal cells is approximately unity. Thus, CO_2 formation is substantial. The issue is further compounded by the relatively high solubility of CO_2. This implies that the system should be able to strip out large amounts of CO_2. This argument also indicates that the use of membrane-based oxygen transfer is not practical. On the other hand, direct sparging of air with a low vvm in combination with a suitable surfactant (Pluronic F68) can give the desired result without much complication (Nienow 2006).

7B.15.3 Heat Transfer Area Required

The subject of heat transfer requirements in animal cell cultures has been discussed in Section 7B.14.1. The amount of heat released by the bioreaction based on the maximum OUR has been estimated as $\Delta H = 630$ W/m^3. Additional heat load due to impeller agitation and air expansion is assumed as 10% of ΔH. Consequently, the total heat load Q_H is ~700 W/m^3. A dimpled jacket will be used. Figure 7B.24 is the block diagram for the purpose of heat transfer design.

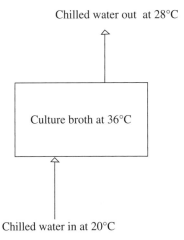

Chilled water out at 28°C

Culture broth at 36°C

Chilled water in at 20°C

FIGURE 7B.24   Block diagram for heat transfer design.

Chilled water available at 20°C serves as the cooling medium. The reactor is to be operated isothermally at 36°C. Assuming an overall heat transfer coefficient of 250 W/m²/K,

$$Q_H \times V_B = U \times A_{\text{Jacket}} \times \left(\Delta T_K \right)_{\text{lm}} \, A_{\text{Jacket}} \sim 5 \, \text{m}^2$$

The external cylindrical area of the vessel $= \pi \times T \times H \sim 28$ m². This is five times more than the heat transfer area prescribed for the duty, and therefore, a safe and stable operation is assured. Indeed, cooling tower water available at 32°C and leaving the jacket at 34°C can also be used since it requires approximately 20 m² of heat transfer area in the jacket. However, (i) seasonal variations in humidity, (ii) possibility of malfunctioning of the cooling tower, and (iii) the narrow temperature window for viability of animal cells dictate that chilled water should be used.

7B.16 SPECIAL ASPECTS OF STIRRED BIOREACTOR DESIGN

The special features of a stirred bioreactor have been emphasized at various stages in this chapter. The main difference between stirred chemical reactors and stirred biore-actors for animal cell culture technology is that the latter deals with living species. These living species are very sensitive to contamination, availability of nutrients, dissolved oxygen, composition of the medium, shear forces, etc. A chemical compound in a chemical reaction will not be totally damaged if one of the reactants taking part in the reaction is not available. Living cells, on the other hand, need a constant supply of the various nutrients and assiduous temperature control. In the absence of such nutrients/temperature control, the viability of living cells is seriously affected and can even lead to complete cell damage. Considering these aspects, the mechanical design of a reactor for cell culture is significantly different from that of a chemical reactor. Major part of the matter in the following is derived from Bartow and Spark (2004) and Junker (2004) who have presented excellent discussion of these aspects. Interested readers are referred to this literature for more details including P&ID, control systems, etc, which are not presented here.

7B.16.1   The Reactor Vessel

This is generally a jacketed vessel made of SS316L. This material is cost-effective in terms of corrosion resistance for solutions commonly used in cell culture technology. Welded joints are preferred over flanged joints. The reactor internal wall should be smooth and free of crevices/sharp corners that are difficult to clean. All internal surfaces including welds need to be polished to minimum 25 microinch surface average to prevent the possibility of adherence of cells that are difficult to clean/dislodge. For extreme situations, electropolishing is advised. The H/T ratio of the vessel varies between 1 and 2. However, for effective mixing, $H/T = 1$ is generally preferred. Mechanical design of the vessel is for a duty ranging from 0.4 to 0.5 MPa abs to full vacuum. It is common practice to choose a jacket that strengthens the vessel. Dimpled jackets that are inexpensive and serve both heating and cooling cycles are favored. The drain valve must be flush with the dish end. Self-draining valves with polished interior parts should be preferred. Sanitary diaphragm designs introduced over a period of time are the preferred choice. Other types (ball valves) have been known to be the cause of batch failures. The latter have been traced to contamination in crevices in the valve that could not be sanitized/sterilized to the required level. The drain valve must have "CIP" and "SIP" ports. Pipe fittings such as tees and elbows should be of the "long tangent" type, which prevents sharp corners where dead liquid can form pools due to lack of dynamic liquid. Further, these fittings should be suitable for welding by "orbital welding" technique. Considering the fact that rough pipe/fitting surfaces can also harbor contaminants, it is common practice to mechanically polish all internal surfaces in contact with the broth to very exacting standards. Electropolishing of these fittings to 20 microinches Ra (average roughness ~0.5 μm) is also recommended for very strict compliance.

7B.16.2   Sterilizing System

All additions such as the inoculums, nutrients, pH control solutions, or gases (carbon dioxide/nitrogen wherever required) and the source of oxygen must be completely sterile. Air as the source of oxygen has to be continuously fed. This must be filtered through a suitable hydrophobic, fully sterilizable filter to remove all types of bacteria/viruses. Steam at 0.4–0.5 MPa abs is generally used as the sterilizing medium.

7B.16.3   Measurement Probes

Probes for dissolved oxygen, carbon dioxide/pH, temperature, etc. are essential components that are linked to the control system. These parts are not easy to sterilize and therefore could lead to contamination from a previous batch. Utmost care needs to be taken in this regard. Therefore, bare minimum/absolutely essential measurement probes are advised.

7B.16.4 Agitator Seals

One major drawback of the stirred reactor is that it has a moving part that must be maintained in "fail-safe" conditions at all times during the operation. This has been highlighted in the Multiphase Reactor Selection Guide in Chapter 3, Table 3.3. For a stirred cell culture reactor, the agitator seal is a source of contamination and hence should be such that it completely isolates the contents of the reactor from the surroundings. Double mechanical seals similar to those used in the pumps for the venturi loop reactor (Chapter 8) are widely used. The seal is generally lubricated with pure steam or sterile condensate. This arrangement maintains a positive pressure in the seal that prevents any adventitious leakage out of the reactor. The agitator shaft is generally made of two pieces. Double mechanical seals with tungsten carbide faces are favored in view of the ease of sterilization and maintenance in repeated cycles without loss of sealing ability. Top mounted as well as bottom mounted agitator systems can be used. The latter has the advantage that it allows better utilization of the reactor volume. On the other hand, it has the drawback that both the shaft and mechanical seal remain submerged in the broth at all times. Magnetic drives are an attractive proposition for small reactors/low power inputs.

7B.16.5 Gasket and O-Ring Materials

Since steam sterilization (SIP) is a frequent incidence, all gaskets and O-ring materials must be able to withstand the applied temperature. Junker (2003) suggests ethylene–propylene monomer (EPDM) or fluoroelastomers such as Viton® or Teflon®-EPDM sandwich gaskets as most suitable for the range of temperatures used. Reuse of gaskets may require special care in terms of possible faults developed during operation or reactor opening. In such a case, it is prudent to use a low-cost, "use-and-discard"-type elastomer.

7B.16.6 Vent Gas System

Bioreactors are operated at a positive pressure to avoid accidental ingress from the surroundings. The vent gas arrangement generally has a back pressure valve to accomplish this objective. Biological processes are classified in terms of biological safety levels (BSL) depending on the organism in use. Situations in which the organism is a safety hazard (type BSL-1) require careful containment of the contents of the reactor. The containment can be effected by the same type of filters that are used for sterilization of feed air. These can be used in the vent line to prevent accidental release of the organism into the surrounding. For BSL-2-type organisms, a secondary containment of all rotary parts, sample ports, etc. is suggested (Bartow and Spark 2004).

7B.16.7 Cell Retention Systems in Perfusion Culture

Various cell retention systems for cell retention and recycle in perfusion culture have been discussed in Section 7B.5. Most perfusion cultures are operated over a relatively long period from 2 weeks to more than 1 month. If the same cell retention system is to be used in future campaigns, it must also adhere to the SIP/CIP requirements. The summary given in Table 7B.4 indicates that among the various techniques, centrifugal and acoustic separators have a definite advantage over the other types. Spin filters and membrane filtration systems have a strong tendency to clog (dead cells, bridged cell groups, debris, etc.). It is practically impossible to clean (in place) these devices to the level required. Centrifuges and acoustic devices are amenable to both CIP and SIP.

7B.17 CONCLUDING REMARKS

Design of stirred bioreactor is fraught with many unknown parameters and could even classify as a chemical engineer's nightmare. Several constraints are to be satisfied simultaneously and these include very small operating windows. Considerable progress has been made in understanding the causes of damage to cells, in particular the fragile animal cells. The hydrodynamic effects are, however, not adequately studied in real systems and much needs to be investigated in this area. Particularly, reliable correlations for mixing time, N_{CD} and N_{SG}, need to be developed for real-life systems containing surfactants, free cells, or cells supported on microcarrier media. With respect to mass transfer, diffusion of oxygen across the cell wall is an uncharted area. Finally, the uptake of oxygen by the cell also needs to be understood from a chemical engineer's point of view to have a better understanding of the rate-controlling step in the overall process of cell growth and expression of products. Chemical engineers and molecular biologists should collaborate in such efforts for a better understanding of the various phenomena.

NOMENCLATURE

a:	Effective gas–liquid interfacial area (m^2/m^3)
A_{Jacket}:	Heat transfer area of the jacket (m^2)
A_M:	Effective membrane area for permeation ($\pi \times d_o \times L$) (m^2)
A_p:	Total particle surface area (m^2)
$[A*]$:	Interfacial concentration of solute (kmol/m^3)
C:	Clearance of the impeller from the vessel bottom (m)
$C_{Bioreactor}$:	Cell concentration in the bioreactor outside side spin filter (wt/volume) or (number of cells/ml)
$[CO_2]_{Bulk\ liquid}$:	Concentration of dissolved CO_2 in bulk liquid (kmol/m^3)
$[CO_2*]$:	Interfacial concentration of CO_2 (kmol/m^3)
$C_{Harvest}$:	Cell concentration in the harvest (wt/volume) or (number of cells/ml)
$[C]_{iO2}$ or $[C*]_{O2}$:	Concentration of oxygen at the gas–liquid interface (kmol/m^3)
$[C]_{BO2}$:	Concentration of oxygen in the bulk liquid phase (kmol/m^3)
$[CO_2]_{Cell}$:	Concentration of CO_2 in the cell (kmol/m^3)

$C_{\text{Bioreactor}}$:   Cell concentration in the bioreactor (wt/volume) or (number of cells/ml)

$[C]_{\text{Cell}}$:   Concentration of cells (number of cells/ml)

D: Impeller diameter (m)

D_A or D_{O2}:   Diffusivity of gaseous species A or oxygen in the liquid phase (m²/s)

d_C: Representative linear dimension of a cell (m or μm)

D_{Cylinder}:   Diameter of the cylindrical portion of a hydrocyclone (m)

d_O: Outside diameter of hollow fiber membrane (m)

d_p: Particle diameter (m)

d_{PC}:   Particle cut diameter of a hydrocyclone or membrane retention device (m)

D_{SF}:   Diameter of spin filter rotor (m)

F_L: Flow number, $[Q_{\text{Imp}}/ND^3]$ (—)

g: Acceleration due to gravity, 9.81 (m/s²)

H: Height of dispersion or broth in the stirred reactor (m)

H_{SF}:   Height of spin filter over which filtration occurs (m)

k_p: Volumetric permeation coefficient across the cell wall (1/s)

k'_p: True permeation coefficient across the cell wall (m/s)

k_L: True gas–liquid mass transfer coefficient (m/s)

$k_L a$:   Volumetric gas–liquid mass transfer coefficient (1/s)

$[k_L a]_{\text{AF}}$:   Volumetric gas–liquid mass transfer coefficient in the presence of antifoaming agent (1/s)

$[k_L a]_{\text{NF}}$:   Volumetric gas–liquid mass transfer coefficient for nonfoaming system (1/s)

K_p: Permeation coefficient (m/s)

K_{SL}:   True cell–liquid or microcarrier–liquid mass transfer coefficient (m/s)

L: Total length of hollow fiber membrane (m)

MAb: Monoclonal antibodies

N: Speed of agitation (rev/s)

N_{CD}:   Critical speed for gas dispersion (rev/s)

N_p: Power number, $[P/N^3D^5\rho_L]$ (—)

N_s: Critical speed for solid suspension in solid–liquid system (rev/s)

N_{SF}:   Speed of rotation of spin filter (rev/s)

N_{SG}:   Critical speed for suspension of solids in the presence of gas (rev/s)

P_G: Power input in the presence of gas (W)

p_i: Partial pressure of oxygen at the interface (atm or N/m² or MPa)

$(p_G)_{\text{CO}_2}$:   Partial pressure of CO_2 in the bulk gas phase (atm or N/m² or MPa)

$(p_I)_{\text{CO}_2}$:   Partial pressure of CO_2 at the gas–liquid interface (atm or N/m² or MPa)

P_O: Power input in the absence of gas (W)

Q_{Air}:   Flow rate of air (m³/s)

Q_{CO_2}:   Flow rate or rate of formation of CO_2 (m³/s)

Q_G: Volumetric gas flow rate (m³/s)

Q_H: Total heat to be removed from the bioreactor (W/m³)

$Q_{Harvest}$:   Harvest flow rate from cell retention device (m³/s)

Q_{Imp}:   Volumetric fluid discharge just off the impeller blade (m³/s)

RVD, vvd: Volume of medium/vessel volume per day as a measure of perfusion rate in spin filter (m³/m³ day)

R_C: Cell retention (%)

$[R]_{CO_2}$:   Rate of desorption of CO_2 (mol/m³s)

Re_I: Impeller Reynolds number, $[ND^2 \rho_L/\mu_L]$ (—)

r_I: Radius of inner rotating cylinder of spin filter (m)

r_O: Radius of outer rotating cylinder of spin filter (m)

R_V: Reactor volume in spin filter retention system (m³)

Sc: Schmidt number, $[\mu/\rho D_A]$ or $[\mu/\rho D_{O2}]$ (—)

Sh_C: Sherwood number based on cell dimension, $(K_{SL}d_C/D_A)$ (—)

T: Diameter of the bioreactor (m)

Ta: Taylor number, $\left(\dfrac{\omega r_I (r_O - r_I) \rho_L}{\mu_L}\right)$ (—)

Ta_C: Critical Taylor number (—)

T_K: Temperature (K)

$(\Delta T_K)_{lm}$:   Log mean temperature difference (K)

V_B: Volume of the broth (m³)

V_D: Dispersion volume (m³)

V_G: Superficial gas velocity based on empty vessel cross-sectional area (m/s)

V_T: Terminal settling velocity solid particle (m/s)

V_{TCF}:   Settling velocity under a centrifugal field (m/s)

vvm: Vessel volume per minute as a measure of aeration rate (m³/min)

Greek Letters

δ_{CW}:   Cell wall thickness (μm)

δ_M: Thickness of the selective skin layer of membrane (μm)

ε_{Avg}:   Average power dissipation per unit mass (W/kg)

ε_G: Fractional gas holdup (—)

ε_I: Power dissipation at the tip of the impeller (W/kg)

ε_M: Power dissipation per unit mass (W/kg)

ε_{Max}:   Maximum power dissipation per unit mass (W/kg)

ε_{TG}:   Total power dissipation per unit mass including that due to gas sparging (W/kg)

ε_V: Power input per unit volume kW/m^3
Θ_C: Circulation time (s)
Θ_M: Characteristic mixing time (s)
Θ_{MT}: Characteristic mass transfer time (s)
Θ_{OU}: Characteristic oxygen uptake time (s)
λ_K: Kolmogorov's scale (m or μm)
μ_L: Liquid/broth viscosity (Pa s)
ρ_L, ρ_B: Liquid or broth density (kg/m^3)
ρ_{MB}: Density of microcarrier beads (kg/m^3)
ρ_P or ρ_C: Density of the solid particles or cells (kg/m^3)
σ: Surface tension of the broth (N/m)
τ_D: Characteristic diffusion time for gas–liquid mass transfer (s)
τ_{Rbulk}: Characteristic bulk liquid-phase reaction time (s)
ω: Tangential speed of spin filter (2 × π × rev/s) (1/s)

Subscripts

B: Broth
C: Refers to cell
G: Gas phase
L: Liquid phase
P, S: Particle, solid phase

REFERENCES

Ahmed EAE. (2005) Application of hydrocyclone for cell separation in mammalian cell perfusion cultures. Ph.D. Thesis, Tech. Univ. Braunschweig, Germany.

Akita K, Yoshida F. (1974) Bubble size, interfacial area, and liquid phase mass transfer coefficients in bubble columns. Ind. Eng. Chem. Proc. Des. Dev., 13:84–88.

Allison DW, Borgschulte T, Bausch CL, Bahr SM, Kaple MW, Kayser KJ. (2007) Deciphering the mechanisms of therapeutic protein production. Chem. Eng. Prog., 103(10):48–52.

Assirelli M, Bujalski W, Eaglesham A, Nienow AW. (2002) Study of micromixing in a stirred tank using a Rushton turbine: comparison of feed position and other mixing devices. Trans. IChemE., Part A 80:855–863.

Assirelli M. (2004) Micromixing studies in turbulent stirred baffled and unbaffled vessels agitated by a Rushton turbine: an experimental study. Ph.D. Thesis, The University of Birmingham, UK. cf: Assirelli et al. (2005).

Assirelli M, Bujalski W, Eaglesham A, Nienow AW. (2005) Intensifying micromixing in a semi-batch reactor using a Rushton turbine. Chem. Eng. Sci., 60:2333–2339.

Avgerinos GC, Drapeau D, Socolow JS, Mao J, Hsiao K, Broeze RJ. (1990) Spin filter perfusion system for high density cell culture: production of recombinant urinary type plasminogen activator in CHO cells. Nat. Biotechnol., 8(1):54–58.

Baldyga J, Bourne JR. (1999) Turbulent mixing and chemical reactions. John Wiley & Sons, Chichester, UK.

Bardouille C, Lehmann J, Heimann P, Jockusch H. (2001) Growth and differentiation of permanent and secondary mouse myogenic cell lines on microcarriers. Appl. Microbiol. Biotechnol., 55:556–562.

Barradas OP, Jandt U, Phan LDM, Villanueva ME, Schaletzky M, Rath A, Freund S, Reichl U, Skerhutt E, Scholz S, Noll T, Sandig V, Pörtner R, Zeng A-P. (2012) Evaluation of criteria for bioreactor comparison and operation standardization for mammalian cell culture. Eng. Life Sci., 12(5):1–11.

Bartow M, Spark E (2004) Bioreactor design for mammalian cell cultures. Chem. Eng., Jan. 4:49–54.

Bavarian F, Fan LS, Chalmers JJ. (1991) Microscopic visualization of insect cell-bubble interactions. I: Rising bubbles, air–medium interface, and the foam layer. Biotechnol. Bioeng., 7:140–150.

Benedek A, Heideger WJ. (1971) Effect of additives on mass transfer in turbine agitation. Biotechnol. Bioeng., 13:663–684.

Benz GT (2011) Bioreactor design for chemical engineers. Chem. Eng. Prog., 107:21–26.

Bhattacharya S, Kresta SM. (2004) Surface feed with minimum by-product formation for competitive reactions. Chem. Eng. Res. Des., 82:1153–1160.

Bliem R, Katinger H. (1988) Scale up engineering in animal cell technology: Part II. Trends Biotechnol., 6:224–230.

Bohm H, Anthony P, Davey MR, Briarty LG, Power JB, Lowe KC, Benes E, Groschl M. (2000) Viability of plant cell suspensions exposed to homogeneous ultrasonic fields of different energy density and wave type. Ultrasonics, 38(1–8):629–632.

Bouzerar R, Jaffrin MY, Ding L, Paulier P. (2000) Influence of geometry and angular velocity on performance of a rotating disc filter. AICHEJ, 46:257–265.

Broise de la D, Noiseux M, Lemieux R. (1991) Long-term perfusion culture of hybridoma: a "grow or die" cell cycle system. Biotechnol. Bioeng., 38:781–787.

Broise de la D, Noiseux M, Massie B, Lemieux R. (1992) Hybridoma perfusion systems: a comparison study. Biotechnol. Bioeng., 40:25–32.

Bujalski W. (1987) Three phase mixing: studies of geometry, viscosity and scale. Ph.D. thesis, University of Birmingham cf: Bujalski et al. (1990).

Bujalski W, Konno M, Nienow AW. (1988) Scale up of 45° pitched blade agitators for gas dispersion and solid suspension. In Proceedings of Sixth European Mixing Conference, Italy, 389–398. AIDIC-BHRA Fluid Engineering Center, Cranfield, UK.

Bujalski W, Nienow AW, Huoxing L. (1990) The use of upward pumping 45° pitched blade turbine impellers in three-phase reactors. Chem. Eng. Sci., 45:415–421.

Bujalski W, Takenaka K, Paolini S, Jahoda M, Paglianti A, Takahashi K, Nienow AW, Etchells AW. (1999) Suspension and liquid homogenization in high solids concentration stirred chemical reactors. Chem. Eng. Res. Des., 77:241–247.

Butler M, Meneses-Acosta A. (2012) Recent advances in technology supporting biopharmaceutical production from mammalian cells. Appl. Microbiol. Biotechnol., 96(4):885–894.

Boraston R, Thompson PW, Garland S, Birch JR. (1984) Growth and oxygen requirements of antibody producing mouse hybridoma cells in suspension culture. Develop. Biol. Standard, 55:103–111.

Büntemeyer H, Bohme C, Lehmann J. (1994) Evaluation of membranes for use in online cell–separation during mammalian–cell perfusion processes. Cytotechnology, 15(1–3): 243–251.

Burgener A, Butler M. (2006) Medium Development. In: Ozturk SS, Hu WS, editors. Cell culture technology for pharmaceutical and cell-based therapies. Taylor & Francis, New York, USA. p 41–80.

Bustamante MCC, Cerri MO, Badino AC. (2013) Comparison between average shear rates in conventional bioreactor with Rushton and Elephant ear impellers. Chem. Eng. Sci., 90:92–100.

Calderbank PH. (1958) Physical rate processes in industrial fermentation: I—the interfacial area in gas-liquid contact with mechanical agitation. Trans. Inst. Chem. Eng., UK, 36:443–463.

Capstick PB, Telling RC, Chapman WG, Stewart DL (1962) Growth of cloned strain of hamster kidney cells in suspended cultures and their susceptibility to the virus of foot and mouth disease. Nature, 195:1163–1164.

Carberry JJ. (1964) Designing catalytic reactors. Ind. Eng. Chem., 56:39–46.

Caron AW, Tom RL, Kamen AA, Massie B. (1994) Baculovirus expression system scale up by perfusion of high-density Sf-9 cell-cultures. Biotechnol. Bioeng., 43:881–891.

Carrel A. (1923) A method for the physiological study of tissues in vitro. J. Exp. Med., 38:407–418. This article can be downloaded free from the archives at www.jem.org.

Carrel A, Rivers TM (1927) La fabrication du vaccin in vitro. CRSoc Biologique., 96:848–850.

Carinhas N, Oliveira R, Alves PM, Carrondo MJT, Teixeira AP. (2012) Systems biotechnology of animal cells: the road to prediction. Trends Biotechnol., 30(7):377–385.

Castilho LR, Medronho RA. (2002) Cell retention devices for suspended cell perfusion cultures. Adv. Biochem. Eng. Biotechnol., 74:129–169.

Castilho LR, Anspach FB (2003) CFD–aided design of a dynamic filter for mammalian cell separation. Biotechnol. Bioeng., 83:514–524.

Catapano G, Czermak P, Eibl R, Eibl D, Portner R (2009) Bioreactor design and scale-up in cell and tissue reaction engineering. In: Eibl R, Eibl D, Portner R, Catapano G, Czermak P, editors. Cell and tissue reaction engineering. Springer-Verlag, Berlin, Heidelberg, Germany. p 83–171.

Chalmers JJ, Bavarian F. (1991) Microscopic visualization of insect cell-bubble interactions II: the bubble film and bubble rupture. Biotechnol. Prog., 7:151–158.

Chapman CM, Nienow AW and Middleton JC. (1980) Surface aeration in a small agitated and sparged vessel. Biotechnol. Bioeng., 22:981–994.

Chapman CM, Nienow AW, Cooke M, Middleton JC. (1983a) Particle-gas-liquid mixing in stirred vessels. Part I—Particle-liquid motion. Chem. Eng. Res. Des., 61(2):71–81.

Chapman CM, Nienow AW, Cooke M, Middleton JC. (1983b) Particle–gas–liquid mixing in stirred vessels. Part II— Gas–liquid mixing. Chem. Eng. Res. Des., 61(2):82–95.

Chapman CM, Nienow AW, Cooke M, Middleton JC (1983c) Particle–gas–liquid mixing in stirred vessels. Part III—Three-phase mixing. Chem. Eng. Res. Des., 61(3):167–181.

Chapman CM, Nienow AW, Cooke M, Middleton JC. (1983d) Particle–gas–liquid mixing in stirred vessels. Part IV—Mass transfer and final conclusions. Chem. Eng. Res. Des., 61(3):182–185.

Chen W, Zydek N, Parma F. (2000) Evaluation of application of hydrocyclone models for practical applications. Chem. Eng. J., 80:295–303.

Chen S, Oh SKW. (2010) Human embryonic stem cell expansion and differentiation for clinical applications. Available online on: www.aiche.org/SBE/Publications/Articles.aspx; http://www.lw20.com/201205191054769.html cf: Cortes-Caminero M. (2010) The engineering of stem cells. SBE supplement: Stem cell Eng. Chem. Eng. Prog., Nov.:34.

Chen AKL, Reuveny S, Oh SKW. (2013) Application of human mesenchymal and pluripotent stem cell microcarrier cultures in cellular therapy: achievements and future direction. Biotechnol. Adv., 31(7):1032–1046.

Cherry RS, Papoutsakis ET. (1986) Hydrodynamic effects on cells in agitated tissue culture reactors. Bioprocess. Eng., 1:29–41.

Cherry RS, Papoutsakis ET (1988) Physical mechanisms of cell damage in microcarrier cell culture bioreactors. Biotechnol. Bioeng., 32:1001–1014.

Cherry RS, Papoutsakis ET. (1989a) Modeling of contact-inhibited animal cell growth on flat surfaces and spheres. Biotechnol. Bioeng., 33:300–305.

Cherry RS, Papoutsakis ET. (1989b) Growth and death rates of bovine embryonic kidney cells in turbulent microcarrier bioreactors. Bioprocess. Eng., 4:81–89.

Chu L, Robinson DK (2001) Industrial choices for protein production by large-scale cell culture. Curr. Opin. Biotechnol., 12:180–187.

Clark JM, Hirtenstein MD. (1981) Optimizing culture conditions for the production of animal cells in microcarrier culture. Ann. NY Acad. Sci., 369:33–46.

Couette MM. (1888) Sur un novel appareil pour l'etude dy frottement des fluides. CR Acad. Sci., 107(Aug 06):388–390.

Couette MM. (1890) Études sur le frottement des liquides, Ann. Chim. Phys., 6, Ser. 21:433–510.

Collignon M-L, Delafosse A, Crine M, Toye D. (2010) Axial impeller selection for anchorage dependent animal cell culture in stirred bioreactors: methodology based on the impeller comparison at just-suspended speed of rotation. Chem. Eng. Sci., 65:5929–5941.

Cooney CL, Wang DIC, Mateles RI. (1969) Measurement of heat evolution and correlation with oxygen consumption during microbial growth. Biotechnol. Bioeng., 11:269–281.

Cortes-Caminero M (2010) The engineering of stem cells. Chem. Eng. Prog., 106(11): 34.

Croughan MS, Hamel JF, Wang DIC. (1988) Effects of microcarrier concentration in animal cell culture. Biotechnol. Bioeng., 32:975–982.

Croughan MS, Sayre ES, Wang DIC. (1989) Viscous reduction of turbulence damage in animal-cell culture. Biotechnol. Bioeng., 33:862–872.

Croughan MS, Hamel JF, Wang DIC. (2006) Hydrodynamic effects on animal cells grown in microcarrier cultures. Biotechnol. Bioeng., 95:295–305.

Crowley J. (2004) Using Sound waves for cGMP manufacturing of a fusion protein with mammalian cells. Bioprocess. Int., 2(3):46–50.

Czermak P, Portner R, Brix A. (2009) Special engineering aspects. In: Eibl R, Eibl D, Portner R, Catapano G, Czermak P, editors. Cell and tissue reaction engineering. Springer-Verlag, Berlin, Heidelberg, Germany. p 83–171.

Dalm MCF, Cuijten SMR, van Grunsven WMJ, Tramper J, Martens DE. (2004) Effect of feed and bleed rate on hybridoma cells in an acoustic perfusion bioreactor: Part I. Cell density, viability, and cell-cycle distribution. Biotechnol. Bioeng., 88:547–557.

Dalm MCF, Jansen M, Keijzer TMP, van Grunsven WMJ, Oudshoorn A, Tramper J, Martens DE. (2005) Stable hybridoma cultivation in a pilot-scale acoustic perfusion system:

long-term process performance and effect of recirculation rate. Biotechnol. Bioeng., 91:894–900.

Dalm MCF. (2007) Acoustic perfusion processes for hybridoma cultures: viability, cell cycle and metabolic analysis. Ph.D. Thesis, University of Wacheningen, the Netherlands.

Deckwer W-D, Medrinho RA, Anspach FB, Lubberstedt M. (2005) Method for separating viable cells from cell suspensions. U.S. Patent 6,878,545 B2, WO 2001/85902.

Deo YM, Mahadevan MD, Fuchs R. (1996) Practical considerations in operation and scale-up of spin filter based bioreactors for monoclonal antibody production. Biotechnol. Prog., 12:57–64.

Dey D. (1998) Cell–bubble interactions during bubble disengagement in aerated bioreactors. Ph.D. Thesis, The University of Birmingham, UK. *cf:* Nienow (2006).

deZengotita VM, Kimura R, Miller WM. (1998) Effects of CO_2 and osmolality on hybridoma cells: growth, metabolism and monoclonal antibody production. Cytotechnology, 28:213–227.

deZengotita VM, Schmelzer AE, Miller WM. (2002) Characterization of hybridoma cell response to elevated dissolved CO_2 and osmolality: intracellular pH, cell size, apoptosis, and metabolism. Biotechnol. Bioeng., 77:369–380.

Doblhoff-Dier O, Gaida T, Katinger H, Burger W, Groschl M, Benes E. (1994) A novel ultra-sonic resonance field device for the retention of animal-cells. Biotechnol. Prog., 10:428–432.

Drapeau D, Luan Y, Whitford JC, Lavin D, Adamson R. (1990) Cell culture scale up in stirred tank bioreactor. SIM Annual Meeting, Orlando, Florida, 1990.

Dutta NN, Pangarkar VG. (1996) Particle-liquid mass transfer in multi-impeller agitated three phase reactors. Chem. Eng. Commun., 146:65–84.

Edwards MF, Baker MR, Godfrey JC. (2000) Mixing of liquids in stirred tanks. Chapter 8. In: Harnby N, Edwards MF, Nienow AW, editors. Mixing in the Process Industries. 2nd ed. Butterworth-Heinemann, Oxford, UK. p 322–363.

Eibl D, Eibl R, Portner R. (2009) Mammalian cell culture technology: an emerging field. In: Eibl R, Eibl D, Portner R, Catapano G, Czermak P, editors. Cell and tissue reaction engineering. Springer-Verlag Berlin Heidelberg, Germany, p 3–12.

Eibl R, Eibl D. (2009) Bioreactors for mammalian cells: general overview. In: Eibl R, Eibl D, Portner R, Catapano G, Czermak P, editors. Cell and tissue reaction engineering. Springer-Verlag, Berlin Heidelberg, Germany, p 55–82.

Enfors S-O, Jahic M, Rozkov A, Xu B, Hecker M, Jurgen B, Kruger E, Schweder T, Hamer G, O'Beirne D, Noisommit-Rizzi N, Reuss M, Boone L, Hewitt C, McFarlane C, Nienow AW, Kovacs T, Tragardh C, Fuchs L, Revstedt J, Friberg PC, Hjertager B, Blomsten G, Skogman H, Hjort S, Hoeks F, Lin H.-Y, Neubauer MP, van der Lans R, Luyben K, Vrabel P, Manelius A. (2001) Physiological responses to mixing in large scale bioreactors. J. Biotechnol., 85:175–185.

Esclade LRJ, Carrel S, Péringer P. (1991) Influence of the screen material on the fouling of the spin filters. Biotechnol. Bioeng. 38:159–168.

Favre E, Thaler T. (1992) An engineering analysis of rotating sieves for hybridoma cell retention in stirred tank bioreactors. Cytotechnology, 9:11–19.

Favre E. (1993) Constant flow-rate filtration of hybridoma cells suspensions. J. Chem. Technol. Biotechnol., 58:107–112.

Fenge C, Lüllau E. (2006) Cell culture bioreactors. In: Ozturk SS, Hu WS, editors. Cell culture technology for pharmaceutical and cell-based therapies. Taylor & Francis, New York, USA. p 155–224.

Fernandes TG, Rodrigues CAV, Diogo MM, Cabral JMS. (2014) Stem cell bioprocessing for regenerative medicine. J. Chem. Technol. Biotechnol., 89:34–47. doi: 10.1002/jctb.4189.

Fishwick RP, Winterbottom JM, Parker DJ, Fan XF and Stitt EH. (2005) Hydrodynamic measurements of up- and down-pumping pitched-blade turbines in gassed, agitated vessels, using positron emission particle tracking. Ind. Eng. Chem. Res., 44:6371–6380.

Fleischaker RJ, Sinskey A. (1981) Oxygen demand and supply in cell culture. Eur. J. Appl. Microbiol. Biotechnol., 12:193–197.

Fox R. (2008) FLUENT flow lab collaborations -Taylor-Couette flow. http://www.ansys.com/products/flowlab/collaborations/pdfs/taylor_couette_flow.pdf. Accessed on January 2008.

Frank RG. (2007) Regulation of follow-on biologics. N. Engl. J. Med., 357:841–843.

Furey J. (2002) Scale-up of a cell culture perfusion process. Gen. Eng. News, 22:62–63.

Gaida T, Doblhoff-Dier O, Strutzenberger K, Katinger H, Burger W, Groschl M, Handl B, Benes E. (1996) Selective retention of viable cells in ultrasonic resonance field devices. Biotechnol. Prog., 12:73–76.

Garcia-Briones MA, Brodkey RS, Chalmers JJ. (1994) Computer simulations of the rupture of a gas bubble at a gas–liquid interface and its implications in animal cell damage. Chem. Eng. Sci., 49:2301–2320.

Garcia-Ochoa F, Gomez E. (2005) Prediction of gas–liquid mass transfer in sparged stirred tank bioreactors. Biotechnol. Bioeng., 92:761–772.

Garcia-Ochoa F, Gomez E. (2009) Bioreactor scale up and oxygen transfer rates in microbial systems: an overview. Biotechnol. Adv., 27:153–176.

Garnier A, Voyer R, Rosanne T, Perret S, Jardin B, Kamen A. (1996) Dissolved carbon dioxide accumulation in large scale and high cell density production of TGFβ receptor with baculovirus infected Sf-9 cells. Cytotechnology, 22:53–63.

Gherardini L, Cousins CM, Hawkes JJ, Spengler J, Radel S, Lawler H, Devcic-Kuhar B, Groschl M. (2005) A new immobilisation method to arrange particles in a gel matrix by ultrasound standing waves. Ultrasound Med. Biol., 31(2):261–272.

Gonzalez-Valdez J, Aguilar-Yanez JM, Benavides J, Rito-Palomares M. (2013) DNA based vaccines offer improved vaccination supply for the developing world. J. Chem. Technol. Biotechnol. (wileyonlinelibrary.com). doi: 10.1002/jctb.4046.

Gorenflo VM, Smith L, Dedinsky B, Persson B, Piret JM. (2002) Scale-up and optimization of an acoustic filter for 200 L/day perfusion of a CHO cell culture. Biotechnol. Bioeng., 80:438–444.

Gorenflo VM, Angepat S, Bowen BD, Piret JM. (2003) Optimization of an acoustic cell filter with a novel air-backflush system. Biotechnol. Prog., 19:30–36.

Griffiths JB (1992) Animal cell culture processes-batch or continuous? J. Biotechnol. 22: 21–30.

Griffiths B (2007) The development of animal cell products: history and overview. In: Stacey G, Davis J, editors. Medicines from animal cell culture. John Wiley & Sons, Ltd, Chichester, UK. p 567–587. doi:10.1002/9780470723791.c.

Gray DR, Chen S, Howarth W, Inlow D, Maiorella RL. (1996) CO_2 in large-scale and high density CHO cell perfusion culture. Cytotechnology, 22:65–78.

Gröschl M, Burger W, Handl B. (1998) Ultrasonic separation of suspended particles—Part III: Application in Biotechnology. Acustica, 84:815–822.

Hamamoto K, Ishimaru K, Tokashiki M. (1989) Perfusion culture of hybridoma cells using a centrifuge to separate cells from culture mixture. J. Ferment. Bioeng. 67:190–194.

Handa A, Emery AN, Spier PE. (1987) On the evaluation of gas–liquid interfacial effects on hybridoma viability in bubble column bioreactors. Dev. Biol. Stand., 66:241–253.

Hansen S, Leslie RGQ. (2006) TGN1412: scrutinizing preclinical trials of antibody-based medicines. Nature, 441:282.

Hari-Prajitno D, Mishra VP, Takenaka K, Bujalski W, Nienow AW, McKemmie J. (1998) Gas–liquid mixing studies with multiple up- and down-pumping hydrofoil impellers: power characteristics and mixing time. Can. J. Chem. Eng., 76:1056–1068.

Harrison R. (1907) Observations on the living developing nerve fiber. Anat. Rec. 1,116–128; Also see: Proc. Soc. Exp. Med, N.Y.:140–143.

Harrison R. (1910) The outgrowth of the nerve fiber as a mode of protoplasmic movement. J. Exp. Zool., 9:787–846.

Haselgrove JC, Shapiro IM, Silverstone SF. (1993) Computer modeling of the oxygen supply and demand of cells of the avian growth cartilage. Am. J. Physiol., 265:C497–C506.

Hawkes JJ, Coakley WT. (1996) A continuous flow ultrasonic cell filtering method. Enzyme Microb. Technol., 19(1):57–62.

Hawkes JJ, Limaye MS, Coakley WT. (1997) Filtration of bacteria and yeast by ultrasound-enhanced sedimentation. J. Appl. Microbiol., 82(1):39–47.

Hawkes JJ, Barrow D, Cefai J, Coakley WT. (1998a) A laminar flow expansion chamber facilitating downstream manipulation of particles concentrated using an ultrasonic standing wave. Ultrasonics, 36(8):901–903.

Hawkes JJ, Barrow D, Coakley WT. (1998b) Microparticle manipulation in millimetre scale ultrasonic standing wave chambers. Ultrasonics, 36(9):925–931.

Henry CM. (2000) The pharmaceutical industry. Chem. Eng. News., 78(43):85–100.

Hill M, Harris N (2007) Ultrasonic particle manipulation. In: Hardt S, Schönfeld F, editors. Microfluidic Technologies for Miniaturized Analysis Systems. Springer Science + Business Media LLC, New York, USA. p 357–383.

Hiller GW, Clark DS, Blanch HW. (1993) Cell retention-chemostat studies of hybridoma cells—analysis of hybridoma growth and metabolism in continuous suspension culture on serum-free medium. Biotechnol. Bioeng., 42:185–195.

Himmelfarb P, Thayer PS, Martin HE. (1969) Spin filter culture: the propagation of mammalian cells in suspension. Science, 164:555–557.

Hoeks FWJMM, Khan M, Guter D, Willems M, Osman JJ, Mommers R, Wayte J. (2004) Industrial applications of mixing and mass transfer studies. Alvin W. Nienow's Day; Stirred, Not Shaken Symposium, The University of Birmingham, Birmingham, UK.

Holmes DB, Voncken RM, Dekker JA. (1964) Fluid flow in turbine-stirred, baffled tanks— I: Circulation time. Chem. Eng. Sci., 19:201–208.

Horwitz EM, Prockop DJ, Fitzpatrick LA, Koo WWK, Gordon PL, Neel M, Sussman M, Orchard P, Marx JC, Pyeritz RE, Brenner MK. (1999) Transplantability and therapeutic effects of bone marrow-derived mesenchymal cells in children with osteogenesis imperfecta. Nat. Med., 5:309–313.

Horwitz EM, Gordon PL, Koo WK, Marx JC, Neel MD, McNall RY, Muul L, Hofmann T. (2002) Isolated allogeneic bone marrow-derived mesenchymal cells engraft and stimulate growth in children with osteogenesis imperfecta: implications for cell therapy of bone. Proc. Natl. Acad. Sci. USA, 99:8932–8937.

Hou JJC, Codamo J, Pilbrough W, Hughes B, Gray PB, Munro TP. (2011) New frontiers in cell line development: challenges for biosimilars. J. Chem. Technol. Biotechnol., 86:895–904.

Hua J, Erickson LE, Yiin T-Y, Glasgow LA. (1993) A review of the effects of shear and interfacial phenomena on cell viability. Crit. Rev. Biotechnol., 13(4):305–328.

Hubbard B. (2009) Future Directions in purification technology. Second International Conference on Accelerating Biopharmaceutical Development. March, 9–12, 2009, Coronado, CA, USA.

Ibrahim S, Nienow AW. (1995) Power curves and flow patterns for a range of impellers in Newtonian fluids: 40 < Re < 5 × 105. Trans. IChemE, 73, Part A:485–491.

Ibrahim S, Nienow AW. (2004) Suspension of microcarriers for cell culture with axial flow impellers. Trans. IChemE, Part A, Chem. Eng. Res. Des., 82(A9):1082–1088.

Ivory CF, Gilmartin M, Gobie WA, McDonald CA, Zollars RL. (1995) A hybrid centrifuge rotor for continuous bioprocessing. Biotechnol. Prog., 11(1):21–32.

Jaworski Z, Dyster KN, Nienow AW. (2001) The effect of size, location and pumping direction of pitched blade turbine impellers on flow patterns: LDA measurements and CFD predictions. Trans. IChemE, Pt A, Chem. Eng. Res. Des., 79:887–894.

Jockwer A, Medronho RA, Wagner R, Anspach FB, Deckwer W-D. (2001) The use of hydrocyclones for mammalian cell retention in perfusion bioreactors. In: Lindner-Olsson E, Chatzissavidou N, Lullau E, editors. Animal cell technology: from target to market. Kluwer Academic Pubs., Dordrecht, the Netherlands. p 301–305.

Johnson M, Lanthier S, Massie B, Lefebvre G, Kamen AA. (1996) Use of the centritech lab centrifuge for perfusion culture of hybridoma cells in protein-free medium. Biotechnol. Prog., 12(6):855–864.

Johnston MD, Christofinis G, Ball GD, Fantes KH, Finter NB. (1979) A culture system for producing large amounts of human lymphoblastoid interferon. Dev. Biol. Stand. 42:189–192.

Junker B. (2004) Fermentation. In: Kirk-Othmer, Encyclopedia of chemical technology. Vol. 11. John Wiley & Sons, Inc., New York, USA. p 1–55.

Junker B. (2007, Jul–Aug) Foam and its mitigation in fermentation systems. Biotechnol. Prog., 23(4):767–784.

Kadouri A, Spier RE. (1997) Some myths and messages concerning the batch and continuous culture of animal cells. Cytotechnology, 24:89–98.

Karakashev SI, Grozdanova MV. (2012) Foams and antifoams. Adv. Colloid Interface Sci., 176:1–17.

Karkare SB. (2004) Cell culture technology. In: Kirk-Othmer, Encyclopedia of chemical technology. Vol. 5. John Wiley & Sons Inc., New York, USA. p 345–360.

Kawase Y, Moo-Young M. (1990) The effect of antifoam agents on mass transfer in bioreactors. Bioprocess. Eng., 5:169–173.

Kawase Y, Halard B, Moo-Young M. (1992) Liquid-phase mass transfer coefficients in bioreactors. Biotechnol. Bioeng., 39:1133–1140.

Kayser O, Warzecha H. (2012) Pharmaceutical biotechnology and industrial applications— learning lessons from molecular biology. In: Kayser O, Warzecha H, editors. Pharmaceutical biotechnology: drug discovery and clinical applications. 2nd ed. Wiley-VCH Verlag GmbH & Co. KGaA, Weiheim, Germany. p 1–13.

Khang SJ, Levenspiel O. (1976) New scale up and design criteria for stirrer agitated batch mixing vessels. Chem. Eng. Sci., 31:569–577.

Kilburn DG, Clarke DJ, Coakley WT, Bardsley DW. (1989) Enhanced sedimentation of mammalian cells following acoustic aggregation. Biotechnol. Bioeng., 34:559–562.

Kimura M, Nakao K. (1986) Measurements of oxygen transfer coefficient using glucose oxidase reaction. J. Chem. Eng. Jpn., 31:786–794.

Kimura R, Miller WM. (1997) Glycosylation of CHO-derived recombinant tPA produced under elevated pCO_2. Biotechnol. Prog., 73:311–317.

Kioukia N, Nienow AW, Emery AN, Al-Rubeai M. (1992) The impact of fluid mechanics on the biological performance of free suspension animal cell cultures: further studies. Trans. Inst. Chem. Eng. UK, 70:143–148.

Kioukia N, Nienow AW, Al-Rubeai M, Emery AN. (1996) Influence of agitation and sparging on the growth rate and infection of insect cells in bioreactors and a comparison with hybridoma culture. Biotechnol. Prog., 12:779–785.

Koide K, Yamazoe S, Harada S. (1985) Effects of surface active substances on gas holdup and gas liquid mass transfer in bubble column. J. Chem. Eng. Jpn., 18:287–292.

Kompala DS, Ozturk SS. (2006) Optimization of high cell density perfusion bioreactors. In: Ozturk SS, Hu WS, editors. Cell culture technology for pharmaceutical and cell-based therapies. Taylor & Francis, New York, USA. p 387–416.

Kramers H, Baars GM, Knoll WH. (1953) A comparative study on the rate of mixing in stirred tanks. Chem. Eng. Sci., 2:35–42.

Kromenaker SJ, Srienc F. (1991) Cell-cycle dependent protein accumulation by producer and non-producer murine hybridoma cell lines: a population analysis. Biotechnol. Bioeng., 38:665–677.

Kroner KH, Nissinen V. (1988) Dynamic filtration of microbial suspensions using an axially rotating filter. J. Membrane Sci., 36:85–100.

Kuhlmann W. (1987) Optimization of a membrane oxygenation system for cell culture in stirred tank reactors. Dev. Biol. Stand., 66:263–268.

Kunas KT, Papoutsakis ET. (1990) Damage mechanisms of suspended animal cells in agitated bioreactors with and without bubble entrainment. Biotechnol. Bioeng., 36:476–483.

Kushalkar KB, Pangarkar VG. (1994) Particle-liquid mass transfer in a bubble column with a draft tube. Chem. Eng. Sci., 49:139–144.

Kuzmanic N, Zanetic R, Akrap M. (2008) Impact of floating suspended solids on the homogenization of the liquid phase in dual-impeller agitated vessel. Chem Eng Process.: Process Intensification, 47:663–669.

Kyung Y-S, Peshwa MV, Gryte DM, Hu W-S. (1994) High density culture of mammalian cells with dynamic perfusion based on on-line oxygen uptake rate measurements. Cytotechnology, 14:183–190.

Langheinrich C, Nienow AW, Eddleston T, Stevenson NC, Emery AN, Clayton TM, Slater NKH. (2002) Oxygen transfer in stirred bioreactors under animal cell culture conditions. Food Bioproduct. Proc. (Trans I Chem.E, Part C), 80:39–44.

Lapinskas E. (2010) Scaling up research to commercial manufacturing. Chem. Eng. Prog., 106(11):44–55.

Laska ME, Cooney CL. (2002) Bioreactors, continuous stirred-tank reactors. In Flickinger MC, Drew SW, editors. Encyclopedia of Bioprocess Technology, John Wiley & Sons, Inc., New York, USA. p 353–371.

Lavery M, Nienow AW. (1987) Oxygen transfer in animal cell culture medium Biotechnol. Bioeng., 30:368–373.

Lee JH, Kopeckova P, Kopecek J, Andrade JD. (1990) Surface properties of copolymers of alkyl methacrylates with methoxy (polyethylene oxide) methacrylates and their application as protein-resistant coatings. Biomaterials, 11:455–464.

Lehmann J, Vorlop J, Büntemeyer H. (1988) Bubble-free reactors and their development for continuous culture with cell recycle. In: Spier RE, Griffiths JB, editors. Animal cell biotechnology. Vol. 3. Academic Press, New York, USA. p 221–237.

Leist CH, Meyer HP, Fiechter A. (1990) Potential problems of animal cells in suspension culture. J. Biotechnol., 15:1–46.

Li M, Niu F, Zuo X, Metelski PD, Busch DH, Subramaniam B. (2013) A spray reactor concept for catalytic oxidation of *p*-xylene to produce high-purity terephthalic acid. Chem. Eng. Sci., 104:93–102.

Lim AC, Washbrook J, Titchener-Hooker NJ, Farid SS. (2006) A computer-aided approach to compare the production economics of fed-batch and perfusion culture under uncertainty. Biotechnol. Bioeng., 93:687–697.

Lindbergh CA. (1939) A culture flask for the circulation of a large quantity of fluid medium. J. Exp. Med., 70:231–238. Available from archives at www.jem.org

Liu H, Chiung WC, Wang YC. (1994) Effect of lard oil, olive oil and castor oil on oxygen transfer in an agitated fermenter. Biotechnol. Tech., 8(1):17–20.

Lubberstedt M, Anspach FB, Deckwer W-D. (2000) Abtrennung tierischer zellen mit hydrozyklonnen. Chem. Eng. Technol., 72:1089–1090 (*cf* Ahmed, 2005).

Lubiniecki AS, Lupker JH. (1994) Purified protein products of rDNA technology expressed in animal cell culture. Biologicals, 22(2):161–169.

Lubiniecki AS. (1998) Historical reflections on cell culture engineering. Cytotechnology, 28:139–145.

Luttman R, Florek P, Preil W. (1994) Silicone-tubing aerated bioreactors for somatic embryo production. Plant Cell Tissue Organ Cult., 39(2):157–170.

Ma N, Mollet M, Chalmers JJ. (2006) Aeration, mixing and hydrodynamics in bioreactors. In: Ozturk SS, Hu WS, editors. Cell culture technology for pharmaceutical and cell-based therapies. Taylor & Francis, New York, USA. p 387–416.

Malda J, van den Brink P, Meeuwse P, Grojec M, Martens DE, Tramper J, Riesle J, van Blitterswijk CA. (2004) Effect of oxygen tension on adult articular chondrocytes in microcarrier bioreactor culture. Tissue Eng., 10:987–994.

Mano T, Kimura T, Iijima S, Takashiki K, Takeuchi H, Kobayashi T. (1990) Comparison of oxygen supply methods for cultures of shear-stress sensitive organisms including animal cell culture. J. Chem. Technol. Biotechnol., 47:259–271.

Marshall E. (2006a) Violent reaction to monoclonal antibody therapy remains a mystery. Science, 311:1688–1689.

Marshall E. (2006b) Lessons from a failed drug trial. Science, 313:901.

Martens DG. (1996) Growth and death of animal cells in bioreactors. Ph.D. Thesis, Univ. of Wageningen, the Netherlands.

McFarlane CM, Nienow AW. (1996) Studies of high solidity ratio hydrofoil impellers for aerated bioreactors. 4. Comparison of impeller types. Biotechnol. Prog., 12(1):9–15.

McQueen A, Bailey JE. (1990) Effect of ammonia ion and extracellular pH on hybridoma cell metabolism and antibody production. Biotechnol. Bioeng., 35:1067–1077.

Meier SJ, Hatton TA, Wang DIC. (1999) Cell death from bursting bubbles: role of cell attachment to rising bubbles in sparged reactors. Biotechnol. Bioeng., 62:468–478.

Meier SJ. (2005) Cell culture scale-up: mixing, mass transfer, and use of appropriate scale-down models. Biochem. Eng. XIV. Harrison Hot Springs, Canada.

Mercille S, Johnson M, Lemieux R, Massie B. (1994) Filtration-based perfusion of hybridoma cultures in protein-free medium-reduction of membrane fouling by medium supplementation with DNAse-I. Biotechnol. Bioeng., 43:833–846.

Mercille S, Johnson M, Lanthier S, Kamen AA, Massie B. (2000) Understanding factors that limit the productivity of suspension-based perfusion cultures operated at high medium renewal rates. Biotechnol. Bioeng., 67:435–450.

Mersmann A, Einenkel W-D, Kappel M. (1976) Design and scale-up of agitated vessels. Int. Chem. Eng., 16:590–603.

Merten O-W. (2000) Constructive improvement of the ultrasonic separation device ADI 1015. Cytotechnology, 34(1–2):175–179.

Meyer RA, editor (2013) Stem cells: from biology to therapy. Vols 1 and 2. Wiley-Blackwell, Hoboken, NJ, USA.

Michelletti M, Nikiforaki L, Lee KC, Yianneskis M. (2003) Particle concentration and mixing characteristics of moderate-to-dense solid-liquid suspensions. Ind. Eng. Chem. Res., 42:6236–6249.

Middleton JC. (2000) Gas–liquid dispersion and mixing. In: Harnby N, Edwards MF, Nienow AW, editors. Mixing in the process industries. 2nd ed. Butterworth-Heinemann, Oxford, UK. p. 322–363.

Mollet M, Ma N, Zhao Y, Brodkey RS, Taticek K, Chalmers JJ. (2004) Bioprocess equipment: characterization of energy dissipation rate and its potential to damage cells. Biotechnol. Prog., 20:1437–1448.

Monk J. (2009) Exploring new paradigms for bioprocessing. Chem. Eng. Prog., 105(7): 19–22.

Monk J. (2010) Stem cell engineering. Chem. Eng. Prog., 106(11):35–36.

Morao A, Maia CI, Fonseca MMR, Vasconcelos JMT, Alves SS. (1999) Effect of antifoam addition on gas-liquid mass transfer in stirred fermenters. Bioprocess Eng., 20:165–172.

Moreira JL, Cruz PE, Santana PC, Feliciano AS, Lehmann J, Carrondo MJT. (1995) Influence of power input and aeration method on mass transfer in a laboratory animal cell culture vessel. J. Chem. Technol. Biotechnol., 62:118–131.

Mostafa SS, Gu X. (2003) Strategies for improved dCO_2 removal in large-scale fed-batch cultures. Biotechnol. Prog., 19(1):45–51.

Mulder MHV. (2000) Basic principles of membrane technology. Kluwer Academic Pub. Dordrecht, the Netherlands.

Murhammer DW, Goochee CF. (1988) Scale up of insect cell cultures: protective effects of Pluronic F-68. Nat. Biotechnol., 6(12):1411–1418.

Murphy MB, Blaski D, Buchanan RM, Tasciotti E. (2010) Engineering better ways to heal broken bones. Chem. Eng. Prog., 106(11):37–43.

Nagel-Heyer S, Leist Ch, Lünse S, Goepfert C, Pörtner R. (2007) From biopsy to cartilage-carrier constructs by using microcarrier cultures as sub-process. Cell Technol. Cell Prod., 3:367–369.

Nielsen JH, Villadsen J. (1994) Bioreaction engineering. Plenum Press, New York, USA.

Nienow AW, Lavery M. (1987) Oxygen transfer in animal cell culture medium. Biotechnol. Bioeng. 30:368–373.

Nienow AW. (1990) Agitators for mycelial fermentations. Trends Biotechnol., 8:224–233.

Nienow AW, Langheinrich C, Stevenson NC, Emery AN, Clayton TM, Slater NKH. (1996) Homogenization and oxygen transfer rates in large agitated and sparged animal cell bioreactors: some implications for growth and production. Cytotechnology. 22:87–94.

Nienow AW. (1997) On impeller circulation and mixing effectiveness in the turbulent flow regime. Chem. Eng. Sci., 52:2557–2565.

Nienow AW. (1998) Hydrodynamics of stirred bioreactors. Appl. Mech. Rev., 51(1):3–32.

Nienow AW. (2000) The suspension of solid particles. In: Harnby N, Edwards MF, Nienow AW, editors. Mixing in the process industries. 2nd ed. Butterworth-Heinemann, Oxford, UK. p 364–393.

Nienow AW, Bujalski W. (2004) Recent studies on agitated three-phase (gas–solid–liquid) systems in the turbulent regime. Trans. IChemE, 80, Part A:832–838.

Nienow AW. (2003) Aeration biotechnology. In Kirk & Othmer, Encyclopedia of chemical technology. Vol.1. John Wiley & Sons Inc., , New York, USA. p 730–747.

Nienow AW, Bujalski W. (2004) The versatility of up-pumping hydrofoil agitators. Trans. IChemE, Part A, Chem. Eng. Res. Des., 82(A9):1073–1081.

Nienow AW. (2006) Reactor engineering in large scale animal cell culture. Cytotechnology, 50:9–33.

Norwood KW, Metzner AB. (1960) Flow patterns and mixing rates in agitated vessels. AICHEJ, 6:432–437.

Oh SKW, Nienow AW, Al-Rubeai M, Emery AN. (1989) The effect of agitation intensity with and without continuous sparging on the growth and antibody production of hybridomas cells. J. Biotechnol., 12:45–62.

Oh SKW, Nienow AW, Al-Rubeai M, Emery AN. (1992) Further studies of the culture of mouse hybridomas in an agitated reactor with and without continuous sparging. J. Biotechnol., 22:245–270.

Ohmura N, Suemasu T, Asamura Y. (2005) Particle classification in Taylor vortex flow with an axial flow. J. Phys., 14:64–71.

Olmer R, Lange A, Selzer S, Kasper C, Haverich A, Martin U, Zweigerdt R. (2012) Suspension culture of human pluripotent stem cells in controlled, stirred bioreactors. Tissue Eng., Part C-Methods, 18:772–784.

Ozturk SS. (1996) Engineering challenges in high density culture systems. Cytotechnology, 22:3–16.

Owens OVH, Gey MK, Gey GO. (1953) A new method for the cultivation of mammalian cells suspended in agitated fluid medium (Abstract). Proc. Am. Assoc. Cancer Res., 1:41.

Pangarkar VG, Yawalkar AA, Sharma MM, Beenackers AACM. (2002) Particle–liquid mass transfer coefficient in two/three-phase stirred tank reactors. Ind. Eng. Chem. Res., 41: 4141–4167.

Pangarkar VG, Ray SK (2014) Pervaporation: theory, practice and applications in the chemical and allied industries. In: Pabby AK, Rizvi SSH, Sastre AM, editors. Handbook of membrane separations: chemical, pharmaceutical, food and biotechnological applications. 2nd ed. Marcel-Dekker (Taylor & Francis): to be published March, 2015.

Papoutsakis ET. (1991) Fluid-mechanical damage of animal cells in bioreactors. Trends Biotechnol., 9:427–437.

Porter JR, Ruckh TT, Popat KC. (2009) Bone tissue engineering: a review in bone biomimetics and drug delivery strategies. Biotechnol. Prog., 25:1539–1560.

Portner R. (2009) Characteristics of mammalian cells and requirements for cultivation. In: Eibl R, Eibl D, Portner R, Catapano G, Czermak P, editors. Cell and tissue reaction engineering. Springer-Verlag, Berlin, Heidelberg, Germany. p 13–54.

Procházka J, Landau J. (1961) Homogenization of miscible liquids by rotary impellers. Collect. Czech. Chem. Commun., 26:2961–2974.

Projan SJ, Gill D, Lu Z, Herrmann SH. (2004) Small molecules for small minds? The case for biologic pharmaceuticals. Expert Opin Biol Ther, 4(8):1345–1350.

Pui PWS, Trampler F, Sonderhoff SA, Groeschl M, Kilburn DG, Piret JM. (1995) Batch and semicontinuous aggregation and sedimentation of hybridoma cells by acoustic-resonance fields. Biotechnol. Prog., 11:146–152.

Ramirez OT, Mutharasan R. (1990) The role of the plasma membrane fluidity on the shear sensitivity of hybridomas grown under hydrodynamic stress. Biotechnol. Bioeng., 36: 911–920.

Ramirez OT, Mutharasan R. (1992) Effect of serum on the plasma membrane fluidity of hybridomas: an insight into its shear protective mechanisms. Biotechnol. Prog., 8:40–50.

Rebasmen E, Goldinger W, Scheirer W, Merten O-W, Palfi GE. (1987) Use of a dynamic filtration method for separation of animal cells. Dev. Biol. Stand., 66:273–277.

Richardson RF, Harker JH, Backhurst JR (2002) Chapter 1: Particulate solids. In: Coulson and Richardson's Chemical engineering. Vol. 2, 5th ed., Particle Technology and Separation Processes. Butterworth-Heinemann (An imprint of Elsevier Science), Oxford, UK. p 1–94.

Rodrigues, ME, Costa, AR, Henriques M, Cunnah P, Melton DW, Azeredo J, Oliveira R. (2013) Advances and drawbacks of the adaptation to serum-free culture of CHO-K1 cells for monoclonal antibody production. Appl. Biochem. Biotechnol., 169(4):1279–1291.

Rols JL, Condoret JS, Fonade C, Goma G. (1990) Mechanism of enhanced oxygen transfer in fermentation using emulsified oxygen vectors. Biotechnol. Bioeng., 35:427–435.

Roth G, Smith CE, Schoofs GM, Montgomery TJ, Ayala JL, Horwitz JI. (1997) Using an external vortex flow filtration device for perfusion cell culture. Biopharm, 10:30–35.

Rowley JA. (2010) Developing cell therapy biomanufacturing processes. Chem. Eng. Prog., 106(10):50–55.

Ruszkowski S. (1994) Rational method for measuring blending performance and comparison of different impeller types. In: Proceedings of Eighth European Mixing Conf., IChemE, Rugby, UK. p 283–291.

Ryll T, Dutina G, Reyes A, Gunson J, Krummen L, Etcheverry T. (2000) Performance of small-scale CHO perfusion cultures using an acoustic cell filtration device for cell retention: characterization of separation efficiency and impact of perfusion on product quality. Biotechnol. Bioeng., 69:440–449.

Samanen J. (2013) Similarities and differences in the discovery and use of biopharmaceuticals and small molecule chemo-therapeutics. In: Ganellin CR, Roy Jefferis R, Roberts S, editors. Introduction to biological and small molecule drug research and development. Elsevier. p 161–200.

Sardeing R, Aubin J, Poux M, Xuereb C. (2004) Gas–liquid mass transfer: influence of sparger location. Trans. IChemE, Part A, Chem. Eng. Res. Des., 82(A9):1161–1168.

Sardeshpande MV, Sagi AR, Juvekar VA, Ranade VV. (2009) Solid suspension and liquid phase mixing in solid–liquid stirred tanks. Ind. Eng. Chem. Res., 48:9713–9722.

Scheirer W. (1988) High-density growth of animal cells within cell retention fermenters equipped with membranes. Anim. Cell Biotechnol., 3:263–281.

Schluter V, Deckwer W-D. (1992) Gas/liquid mass transfer in stirred vessels. Chem. Eng. Sci., 47:2357–2362.

Schmid G, Wilke CR, Blanch HW. (1992) Continuous hybridoma suspension cultures with and without cell retention: kinetics of growth, metabolism and product formation. J. Biotechnol., 22:31–40.

Schneider M, Reymod F, Marison IW, von Stockar U. (1995) Bubble-free oxygenation by means of hydrophobic porous membranes. Enzyme Microb. Technol., 17: 839–847.

Sekhon BS. (2010) Biopharmaceuticals: an overview. Thai. J. Pharm. Sci., 34:1–19.

Sen A, Kallos MS, Behie LA. (2004) New tissue dissociation protocol for scaled-up production of neural stem cells in suspension bioreactors. Tissue Eng., 10:904–913.

Sharma S, Raju R, Sui S, Hu W-S. (2011) Stem cell culture engineering-process scale up and beyond. Biotechnol. J., 6:1317–1329.

Sheffi J (2009) Generic biologic drugs: what are they and will they save billions of dollars? http://getbetterhealth.com/generic-biologic-drugs-what-are-they-and-will-they-save-billions-of-dollars/2009.06.29. Accessed on August 7, 2014.

Shirgaonkar IZ, Lanthier S, Kamen A. (2004) Acoustic cell filter: a proven cell retention technology for perfusion of animal cell cultures. Biotechnol. Adv., 22:433–444.

Siegel U, Fenge C, Fraune E. (1991) Spin filter for continuous perfusion of suspension cells. In: Murakami H, Shirahata S, Tachibana H, editors. Animal cell technology: basic and applied aspects. Kluwer Academic Publishers, Fukuoka, Japan. p 434–436.

Simmons MJH, Zhu H, Bujalski W, Hewitt CJ, Nienow AW. (2007) Mixing in a model bioreactor using agitators with a high solidity ratio and deep blades. Chem. Eng. Res. Des., 85:551–559.

Singec I, Jandial R, Crain A, Nikkhah G, Snyder EY. (2007) The leading edge of stem cell therapeutics. Ann. Rev. Med., 58:313–328.

Söllner K, Bondy C. (1936) The mechanism of coagulation by ultrasonic waves. Trans. Faraday Soc., 32:616–623.

Spengler JF, Jekel M, Christensen KT, Adrian RJ, Hawkes JJ, Coakley WT. (2001) Observation of yeast cell movement and aggregation in a small-scale MHz-ultrasonic standing wave field. Bioseparation, 9:329–341.

Stebbings R, Findlay L, Edwards C, Eastwood D, Bird C, North D, Mistry Y, Dilger P, Liefooghe E, Cludts I, Fox B, Tarrant G, Robinson J, Meager T, Dolman C, Thorpe SJ, Bristow A, Wadhwa M, Thorpe R, Poole S. (2007) Cytokine storm in the phase 1 trial of monoclonal antibody TGN1412: better understanding the causes to improve preclinical testing of immunotherapeutic. J. Immunol., 179:3325–3331.

Steiner D, Khaner H, Cohen M, Even-Ram S, Gil Y, Itsykson P, Turetsky T, Idelson M, Aizenman E, Ram R, Berman-Zaken Y, Reubinoff B. (2010) Derivation, propagation and controlled differentiation of human embryonic stem cells in suspension. Nat. Biotechnol., 28(4):361–364.

Stryer L. (1999) Biochemistry. New York: W. H. Freeman & Co. p 1064.

Suntharalingam G, Perry MR, Ward S, Brett SJ, Castello-Cortes A, Brunner MD, Panoskaltsis N. (2006) Cytokine storm in a phase 1 trial of the anti-CD28 monoclonal antibody TGN1412. New Engl. J. Med., 355:1018–1028.

Svarovsky L. (1984) Hydrocyclones. Holt, Reinhart and Winston Ltd. London, UK.

Takamatsu H, Hamamoto K, Ishimaru K, Yokoyama S, Tokashiki M. (1996) Large-scale perfusion culture process for suspended mammalian cells that uses a centrifuge with multiple settling zones. Appl. Microbiol. Biotechnol., 45:454–457.

Taylor G. (1923) Stability of a viscous liquid contained between two rotating cylinders, Phil. Trans. R. Soc. Lond.: Series A 223:289–343.

Thomas JN. (1990) Mammalian cell physiology. In: Lubiniecki AS, editor. Large-scale mammalian cell culture technology. Marcel Dekker, New York, USA. p 93–145.

Tokashiki M, Arai T, Hamamoto K, Ishimaru K. (1990) High density culture of hybridoma cells using a perfusion culture vessel with an external centrifuge. Cytotechnology, 3:239–244.

Tokashiki M, Takamatsu H. (1993) Perfusion culture apparatus for suspended mammalian cells. Cytotechnology, 13:149–159.

Tolbert WR, Feder J, Kimes RC. (1981) Large scale rotating filter perfusion system for high density growth of mammalian suspension culture. In Vitro, 17:885–890.

Tramper J. (1995) Oxygen gradients in animal cell bioreactors. In: Beuvery EC, Griffiths JB, Zeijlemaker WP, editors. Animal cell technology: developments towards the 21st century. Kluwer Academic Publishers, the Netherlands. p 883–891.

Trampler F, Sonderhoff SA, Pui PWS, Kilburn DG, Piret JM. (1994) Acoustic cell filter for high-density perfusion culture of hybridoma cells. Biotechnology, 12:281–284.

Trinh K, Garcia-Briones MA, Hink FH, Chalmers JJ. (1994) Quantification of damage to suspended insect cells as a result of bubble rupture. Biotechnol. Bioeng., 43:37–45.

Vallez-chetreanu F. (2006) Characterization of the mechanism of action of spin-filters for animal cell perfusion cultures. Thèse no 3488 (2006), École Polytechnique Fédérale de Lausanne, Lausanne, France.

van Wezel AL, Steenis G. (1978) Production of an inactivated rabies vaccine in primary dog kidney cells. Dev. Biol. Stand., 40:69–75.

van Wezel AL, van der Velden-de-Groot CAM, van Herwaarden JAM. (1980) The production of inactivated polio vaccine on serially cultivated kidney cells from captive-bred monkeys. Dev. Biol. Stand., 46:151–158.

van Wezel AL, van der Velden-de-Groot CAM, de Haan HH, van den Heuven N, Schasfoort R. (1985) Large scale animal cell cultivation for production of cellular biologicals. Dev. Biol. Stand., 60:229–236.

van Wezel AL. (1985) Monolayer growth systems: homogeneous unit processes. In: Spier RE, Griffiths JB, editors. Animal cell biotechnology. Academic Press Inc., London, UK. p 266–281.

Van't Riet K, Tramper J. (1991) Basic bioreactor design. Marcel Dekker, New York, USA.

Vardar-Sukan F. (1992) Foaming and its control in bioprocesses. In: Vardar-Sukan F, Sukan SS, editors. Recent advances in biotechnology. NATO ASI series. Vol. 210, Springer-Science+Business Media, B.V., Dordrecht, The Netherlands. p 113–146.

Vardar-Sukan F. (1998) Foaming: consequences, prevention and destruction. Biotechnol. Adv., 16:913–948.

Varecka R, Scheirer W. (1987) Use of a rotating wire cage for retention of animal cells in a perfusion fermenter. Dev. Biol. Stand., 66:269–272.

Varley J, Birch J. (1999) Reactor design for large scale suspension animal cell culture. Cytotechnology, 29:177–205.

Vogel JH, Kroner KH. (1999) Controlled shear filtration—a novel technique for animal-cell separation. Biotechnol. Bioeng., 63:663–674.

Vogel JH, Anspach FB, Kroner KH, Piret JM, Haynes CA. (2002) Controlled shear affinity filtration (CSAF): a new technology for integration of cell separation and protein isolation from mammalian cell cultures. Biotechnol. Bioeng., 78:805–813.

Voisard D, Meuly F, Ruffieux PA, Baer G, Kadouri A. (2003) Potential of cell retention techniques for large scale high-density perfusion culture of suspended mammalian cells. Biotechnol. Bioeng., 82:751–765.

Walsh G. (2003) Biopharmaceuticals: Biochemistry and Biotechnology. 2nd ed. John Wiley & Sons Ltd, Chichester, UK.

Walsh G. (2004) Second-generation biopharmaceuticals. Eur. J. Pharma. Biopharm., 58:185–196.

Walter JF, Blanch HW. (1986) Bubble break-up in gas–liquid bioreactors: break-up in turbulent flows. Chem. Eng. J., 32 (1), B7–B17.

Wang ZW, Grabenstetter P, Feke DL, Belovich JM. (2004) Retention and viability characteristics of mammalian cells in an acoustically driven polymer mesh. Biotechnol. Prog., 20(1):384–387.

Wayte J, Boraston R, Bland H, Varley J, Brown M. (1997) pH Effects on growth and productivity of cell lines producing monoclonal antibodies: control in large scale fermenters. Gen. Eng. Biotechnol. 17(2–3):125–132.

Wereley ST, Lueptow RM. (1999) Inertial particle motion in a Taylor Couette rotating filter. Phys. Fluids, 11:325–333.

Werner R, Walz F, Noe W, Konrad A. (1992) Safety and economic aspects of continuous mammalian cell culture. J. Biotechnol., 22:51–68.

Whitford WG. (2006) Fed-batch mammalian cell culture in bioproduction. Bioprocess Int., 4(4):30–40.

Woodside SM, Bowen BD, Piret JM. (1998) Mammalian cell retention devices for stirred perfusion bioreactors. Cytotechnology, 28:163–175.

Yabannavar VM, Singh V, Connelly NV. (1992) Mammalian cell retention in a spin filter perfusion bioreactor. Biotechnol. Bioeng., 40:925–933.

Yabannavar VM, Singh V, Connelly NV. (1994) Scale up of spin filter perfusion bioreactor for mammalian cell retention. Biotechnol. Bioeng., 43:159–164.

Yang ST, Basu S. (2006) Animal cell culture. In: Lee S, editor. Encyclopedia of chemical processing. Vol. 1. Taylor & Francis/Marcel-Dekker, New York, USA. p 67–79.

Yawalkar AA, Beenackers AACM, Pangarkar VG. (2002a) Gas holdup in stirred tank reactors. Can. J. Chem. Eng., 80:158–166.

Yawalkar AA, Heesink ABM, Versteeg GF, Pangarkar VG. (2002b) Gas–liquid mass transfer coefficient in stirred tank reactor. Can. J. Chem. Eng., 80:840–848.

Yagi H, Yoshida F. (1974) Oxygen absorption in fermenters—effects of surfactants, antifoaming agents and sterilized cells. J. Ferment. Technol., 52:905–916.

Zeng A-P, Bi J-X. (2006) Cell culture kinetics and modelling. In: Ozturk SS, Hu WS, editors. Cell culture technology for pharmaceutical and cell-based therapies. Taylor & Francis, New York, USA. p 299–348.

Zhang Z, Al-Rubeai M, Thomas CR. (1992) Effect of Pluronic-F68 on the mechanical properties of mammalian cells. Enzyme Microb. Technol., 14:980–983.

Zhang S, Handacorrigan A, Spier RE. (1993) A comparison of oxygenation methods for high-density perfusion cultures of animal-cells. Biotechnol. Bioeng., 41:685–692.

Zhou HJ. (2007) Biologics in the pipeline: large molecules with high hopes or bigger risks? Clin. Pharmacol., 47:550–552.

Zwietering ThN. (1958) Suspending of solid particles in liquids by agitators. Chem. Eng. Sci., 8:244–253.

8

VENTURI LOOP REACTOR

8.1 INTRODUCTION

The venturi loop reactor is ideally suited for fast reactions involving a pure expensive gas-phase reactant and with simultaneous requirements of relatively high reaction pressure (>2 MPa) and heat removal ($\Delta H_R > 100$ kJ/mol) (Chapter 3 and Table 3.3). In the following sections, a brief overview of the areas of application and advantages of the venturi loop reactor is presented.

8.2 APPLICATION AREAS FOR THE VENTURI LOOP REACTOR

The venturi loop reactor belongs to the type in which the solid catalyst is suspended in a gas–liquid dispersion. Its most common application is in the case of a dead-end system using an expensive pure gas (Table 3.3). In addition, these applications involve relatively high pressure and temperature reactions. Due to the use of high pressure, the vapor pressure of the liquid-phase reactant amounts to only a small fraction of the total pressure. Thus, for such cases, even considering the vapor pressure of the liquid-phase reactant, the gas-side resistance for the main reacting gaseous species can be neglected. Therefore, in all the subsequent discussion, it is presumed that the gas–liquid mass transfer is entirely controlled by the resistance offered by the liquid film. The principal factors governing the overall process are (i) true mass transfer coefficient, k_L; (ii) interfacial area, a; (iii) the solid–liquid mass transfer coefficient, K_{SL}, along with the catalyst surface area (A_p/V_D); and (iv) the intrinsic kinetics of the

Design of Multiphase Reactors, First Edition. Vishwas Govind Pangarkar.
© 2015 John Wiley & Sons, Inc. Published 2015 by John Wiley & Sons, Inc.

reaction. The theory of mass transfer accompanied by a chemical reaction has been discussed earlier in Chapter 2. For a general p–qth-order reaction:

$$A + \psi\, B \rightarrow \text{products}$$

The Hatta number that is indicative of the relative rates of reaction in the gas–liquid film and the rate of diffusion across the liquid film is defined as

$$\text{Ha} = \frac{\left(\sqrt{\dfrac{2}{p+1} D_A k_{p-q} [B_B]^q [A^*]^{p-1}} \right)}{k_L} \tag{2.4}$$

Three-phase catalytic reactions are the focus of discussion in the present chapter. Further, in most applications, the catalysts used are expensive. Therefore, effective utilization and sustenance of the catalyst activity are important aspects in the economics. This initial discussion is based on the value of the Hatta number defined by Equation 2.4. The reactor utilization is related to its ability to achieve the intrinsic enhancement of the reaction rate by the catalyst used. As discussed in Section 2.2, the intrinsic kinetics of the catalyzed reaction are unaffected by the type of multiphase reactor. Therefore, the reactor utilization needs to be *reinterpreted* keeping in view the large changes in both k_L and a that are possible with a venturi loop reactor.

8.2.1   Two Phase (Gas–Liquid Reactions)

8.2.1.1   Regime A: Ha<0.3 and $k_L a \gg \varepsilon_L k_2 B_B$ (Eq. 2.6)—This condition defines a very slow reaction. The mass transfer from gas to liquid is unimportant as the reaction is kinetically controlled. Size of the reactor is therefore decided solely by the kinetics of the reaction. Alternatively, it can be argued that all mass transfer limitations have been eliminated and the reaction is achieving the *maximum intrinsic* rate. This observation is very relevant to the venturi loop reactor since it can achieve relatively very high values of $k_L a$ with the result that $k_L a \gg \varepsilon_L k_2 B_B$ and maximum intrinsic rate is attained.

8.2.1.2   Regime B: $0.3 < \text{Ha} < 1.0$ and $k_L a \ll \varepsilon_L k_2 B_B$ —In this case, the gas–liquid mass transfer rate decides the rate of absorption of the gas. As an example in a stirred tank reactor with relatively low value of $k_L a$, the inequality given by Equation 2.7 is likely to hold, and the reactor will operate in a mass transfer-controlled regime. The solved reactor design problem in Section 7A.10 supports this observation.

However, if this same reaction is carried out in a venturi loop reactor, the order of magnitude large values of $k_L a$ in the venturi section is likely to result in a situation such that Equation 2.6 is valid. Consequently, the mass transfer limitation can be eliminated when the venturi loop reactor replaces the stirred tank type. Thus, the reaction can achieve the *maximum intrinsic* rate or operate at the *maximum* possible capacity. This matter has been briefly discussed in Section 3.4.2.4 for liquid-phase oxidation of substituted benzenes. The solved reactor design problem in Section 8.13 shows that this is indeed the case for catalytic hydrogenation of aniline to cyclohexylamine.

The industrially important example of oxidation of p-xylene to terephthalic acid has been discussed in Section 3.4.2.3. Most investigators have found that for this case Ha < 1. However, it is not clear whether the situation corresponds to Regime A or B. Nonetheless, there is general agreement that nonuniformity of oxygen holdup and/or inadequate gas–liquid mass transfer leads to formation of side products. Li et al. (2013) have shown that a spray column is very effective in near-total elimination of by-products. Li et al.'s arguments (Section 3.4.2.4) are equally applicable to the venturi loop system that offers high mass transfer coefficients. The major advantage of the latter over spray column is a relatively well-established scale-up procedure for large-scale applications.

8.2.1.3   Regime C: Ha > 3.0—This situation corresponds to a fast reaction occurring in the liquid film. In many situations involving fast reactions, gas–liquid mixing assumes great importance. Rapid mixing, uniform gas holdup, and catalyst concentration coupled with near-isothermal conditions achieved in the case of a venturi loop reactor are advantageous in realizing high selectivity (Dierendonck et al. 1988). Several additional features such as removal of reaction by-product in the external gas circuit facilitate further improvement in the yield (Table 8.1).

8.2.2   Three-Phase (Gas–Liquid–Solid-Catalyzed) Reactions

We now consider three-phase (gas–liquid–solid) reactions that constitute a major class of application of the venturi loop reactor. Figure 2.2 depicts the concentration profiles for this class of reactions. The overall rate of the gas–liquid–solid-catalyzed reaction for a first-order reaction is given by Equation 2.16:

$$R_V = k_L a([A^*] - [A_B]) = \left(K_{SL} \left(\frac{A_P}{V_D} \right) \right)([A_B] - [A_S]) = k_R[A_S] \qquad (2.16)$$

The relative resistances of the three steps involved are (i) gas–liquid mass transfer, $(1/k_L a)$; (ii) solid–liquid mass transfer, $(1/K_{SL}(A_P/V_D))$; and (iii) surface reaction that is first order with respect to the gas and zero order with respect to the liquid-phase reactant, $(1/k_R)$. As argued in Section 8.2.1.1, **Regime A**, to achieve the *maximum intrinsic* reaction rate, the mass transfer resistances need to be eliminated. Continuing this argument further, it may be stated that a venturi loop reactor that affords relatively very high values of both $k_L a$ and K_{SL} should ensure such a situation. A major advantage of the high rates of mass transfer of the gaseous solute (in comparison to the rate of surface reaction) is that the catalyst is always saturated with the gaseous solute. It is well known that noble metals can form ligands with organic compounds particularly in hydrogen-deficient environments. In the author's experience, many hydrogenation catalysts suffer from an "ill-defined" deactivation in stirred reactors. A careful analysis of the organic phase and material balance on the active metal shows that the so-called deactivation is actually a loss of the active metal by the ligand formation discussed earlier caused by deficiency in supply of hydrogen. Most users have to either replenish the catalyst after every batch or use fresh catalyst for each new batch. The used catalyst is sent to the catalyst suppliers for

TABLE 8.1   Industrially Important Reactions Carried Out in Venturi Loop Reactors with Relative Merits Over Stirred Reactors

Reaction	Results and remarks	Reference
Hydrogenation of α-methyl styrene to cumene	Very high yield of the order of 90%	Greenwood (1986)
Hydrogenation of aniline to cyclohexylamine amine	Enhanced yield of cyclohexylamine Yield increased from 81% in conventional stirred reactor to 93.5% (please refer to Section 8.13, Worked Examples for Design of Venturi Loop Reactor: Hydrogenation of Aniline to Cyclohexylamine)	Greenwood (1986)
Hydrogenation of glucose to sorbitol	Relatively short time of ~4 h at a lower pressure of 2.5 MPa. Unreacted sugar as low as 0.03%	Chaudhari and Mills (2004)
Hydrogenation of p-nitro benzoic acid to p-amino benzoic acid	Catalyst consumption reduced by 80% as compared to conventional stirred reactor Batch time simultaneously reduced by 100% without any change in product quality.	Leuteritz (1973), Malone (1980), Greenwood (1986), Malone and Merten (1992), Concordia (1990)
Hydrogenation of o- and p- nitro toluene to corresponding toluidines	Very rapid hydrogenation in a system without solvent. Reaction almost complete in ~1.3 h	Greenwood (1986), Malone and Merten (1992)
Hydrogenation of dinitro toluene to toluenediamine	Rapid hydrogenation. Excellent alternative to very high-pressure (20–30 MPa) vapor-phase hydrogenation or cascade of three stirred reactors using a solvent for controlling temperature	Duveen (1998)
Hydrogenation of bisphenol A	Substantial enhancement in the rate of hydrogenation	Malone (1980)
Hydrogenation of m-nitro benzofluoride to m-amino benzofluoride	Reduction in catalyst consumption by ~25% as compared to conventional stirred tank reactor. Hydrogenation batch time reduced to 2–3 h. Splitting of fluorine also reduced to 10% of that in a stirred reactor under otherwise similar conditions.	Greenwood (1986), Malone and Merten (1992)
Hydrogenation of fatty nitriles to fatty amines	Significant reduction in batch time with change over to venturi loop reactor. Single unit can produce primary as well as secondary amines. Example: hydrogenation of acrylonitrile to primary amine and then to diamine. Rapid heating/cooling. Fast removal of water produced in the reaction by	Greenwood (1986), Duveen (1998), Opre (2009)

Reaction	Results and remarks	Reference
	means of a gas flow loop (condenser (Fig. 8.2) or absorber (Fig. 8.3) depending on requirements) increases the equilibrium yield in reversible reactions. Catalyst loading reduced by a factor of 2 as compared to conventional stirred tank reactor	
Hydrogenation of fatty acid methyl esters to alcohols	Pressures as high as 20–30 MPa possible as against <2 MPa for stirred reactors	Duveen (1998)
Hydrogenation of 2-chloro nitro anisole to 2-chloro anisidine	Significant (33%) reduction in batch time. Simultaneous increase in yield from ~91% to ~98%. Solvent requirement reduced by 10%. Exceptional reduction in catalyst loading by a factor of 50	Greenwood (1986)
Reductive N-alkylation of aromatic amines	Reduction in reaction time by a factor of 2. Efficient heat transfer allows almost isothermal operation, thereby preventing hydrogenation of the aromatic ring and controlling the reduction of the nitro group	Malone and Merten (1992)
Hydrogenation of 2,5 and 3,4-dichloro nitrobenzene	Substantial reduction in batch time	Malone (1980)
Hydrogenations of o-,p-nitro chlorobenzene	Substantial reduction in both reaction time and severity of reaction conditions	Malone (1980)
Hydrogenation of aliphatic nitro compound (nitro glycol)	Batch time lowered by a factor of 4. Yield increased from 72% to 96.5%. Pressure lowered from 7.5 to 3.0 MPa	Greenwood (1986)
Hydrogenation of aldehydes	Pressure reduced to 3–9 MPa from 10–15 MPa as compared to conventional stirred tank reactor. Reaction time reduced by a factor of 3	Greenwood (1986)
Reductive amination of an aldehyde	Catalyst loading and reaction time reduced by a factor 2 and 2.5, respectively, as compared to conventional stirred tank reactor	Greenwood (1986)
Reductive alkylation of an aldehyde with a ketone	Reduction in reaction time by 25%, catalyst loading reduced by a factor of 4, improvement in yield from 93% to 98%	Greenwood (1986)

(Continued)

TABLE 8.1   (Continued)

Reaction	Results and remarks	Reference
Ring saturation of alkyl benzene	Batch time lowered by a factor of 2; catalyst loading reduced by factor of 2	Greenwood (1986)
Heterocyclic ring saturation	Solventless system increases productivity as compared to conventional stirred tank reactor. Reaction time reduced by 80%; yield increased from 83 to 94%	Greenwood (1986)
Selective chlorination of phenol to 2,4-dichlorophenol	Appropriate design of venturi loop reactor reduces chlorination batch time by a factor of 5–7 from 20 to 40 h required in a conventional stirred tank reactor. Highly efficient heat transfer in the external loop affords nearly isothermal operation at 45–50 °C, resulting in a high yield (more than 90%) of 2,4-dichloro phenol while suppressing formation of 2,6 dichloro phenol. Efficient removal of reaction by-product (hydrogen chloride) in the external gas circuit (Fig. 8.3) improves the yield	Leuteritz et al. (1976), Opre (2009)
Carbonylation of methanol to methyl formate using sodium methoxide as homogeneous catalyst	Almost complete conversion of methanol in less than 1 h	Leuteritz et al. (1976), Malone (1980)
Carbonylation of dimethyl amine using sodium methoxide as homogeneous catalyst	Substantially higher yields and shorter batch times as compared to a conventional stirred reactor under otherwise similar conditions	Leuteritz et al. (1976), Malone (1980)
Ethoxylations and propoxylations	Safe and almost isothermal operation. Residual epoxide in the product is less than 50 ppm. Much lower dioxane formation decreases the hazard. Relatively very high ratios of epoxide to base reactant are possible	Yang et al. (1996), Geser and Malone (2000), Di Serio et al. (2005)
Oxidations with pure oxygen	Safe operation outside the explosive limits through the use of nitrogen purge in the headspace. Dead-end operation results in practically no vent gas and hence relatively very low operating cost	Duveen (1998). Also refer to Section 3.4.2.4, **Reactor Selection**, for oxidation of *p*-xylene to terephthalic acid

Reaction	Results and remarks	Reference
Aerobic treatment of industrial wastewaters	High rate of oxygen transfer, well-directed circulation, lower power requirement, absence of moving parts	Lubbecke et al. (1995), Bloor et al. (1995), Petruccioli et al. (2002), Gourich et al. (2005, 2007, 2008), Farizoglu and Keskinler (2007)
Pretreatment of exhausts containing volatile organic compounds to ensure stable operation of biofilters	venturi loop reactor using aqueous surfactant solution yields stable operation of biofilters despite concentration excursions of the volatile organic compounds	Park et al. (2008)

"recovery/reactivation." Leuteritz et al. (1976) and Greenwood (1986) have also reported such losses. According to Greenwood (1986), the noble metal loss can be as high as 40%. These observations confirm the advantage afforded by the venturi loop reactor. This matter is discussed later in Section 8.9. The earlier discussion clearly brings out the superiority of the venturi loop reactor over the conventional stirred tank reactor. Buss AG recommends venturi loop reactor for the following types of reactions:

1. Hydrogenation of double bonds, acid chlorides, aromatic nitro compounds, carbonyl groups, partial reduction of triple bonds, etc.
2. Aminating hydrogenation
3. Carbonylation
4. Saturation of aromatic ring
5. Alkylation
6. Hydrogenation of vegetable oils
7. Ethoxylation/propoxylation of long-chain alcohols

Several reported pilot plant data on the performance of this reactor are given in Table 8.1.

8.3   ADVANTAGES OF THE VENTURI LOOP REACTOR: A DETAILED COMPARISON

In the following sections, a more detailed semiquantitative discussion of the specific advantages is presented.

8.3.1   Relatively Very High Mass Transfer Rates

The high power input in a relatively small volume of the ejector results in very high levels of turbulence in this section. The mixing shock phenomenon discussed later in Section 8.6 disperses the gas as very fine bubbles in the liquid. This leads to very high ε_G and a. Leuteritz (1973) and Leuteritz et al. (1976) have indicated that the size of gas bubbles generated is in the range of 0.5–1 mm with corresponding interfacial areas from 800 to 2500/m. Besides this, the very high levels of turbulence and a strong shear field yield relatively very high values of the true gas–liquid mass transfer coefficient, k_L. The overall result is relatively high values of the volumetric mass transfer coefficient $k_L\underline{a}$ averaged over the entire assembly comprising the ejector and the holding vessel (Nardin and Cramers 1996). It has been conclusively shown by Cramers et al. (1992b) and Havelka et al. (1997 *cf* Dierendonck et al. 1988) that the venturi section affords $k_L\underline{a}$ values (~10/s) that are at least 2 orders of magnitude higher than a conventional stirred tank reactor. The strong shear field particularly in the venturi section also ensures high values of the solid–liquid mass transfer coefficient K_{SL}. In comparison, the mass transfer coefficients in the holding vessel are relatively lower because of lower energy dissipation per unit volume and certain extent of coalescence of the bubbles. The overall (venturi + holding vessel) values depend on the relative volumes of the vessel and the venturi section. Dierendonck et al. (1998) have reported overall values of $k_L\underline{a}$ for venturi loop system in the range of 0.2–1.5/s. Such high $k_L\underline{a}$ and $(K_{SL}A_P/V_D)$ values ensure that both the liquid and the solid catalysts are essentially saturated with the reactive gas species (Stefoglo et al. 1999). A comparison of the above values of the mass transfer coefficients with a conventional stirred reactor (Section 7A.10) or its variants (Section 9.5) reveals the shortcomings of the latter and also explains the observed leaching of noble metal catalysts as soluble complexes in stirred reactors.

8.3.2   Lower Reaction Pressure

In a majority of the cases, a mass transfer limitation is eliminated through increase in the driving force for gas–liquid mass transfer. The first alternative is to increase dissolved gas concentration, [A^*]. Assuming that Henry's law applies, which is true for most sparingly soluble gaseous solutes, the solubility [A^*] is given by

$$[A^*] = H_A \times p_A$$

When pure solute (e.g., hydrogen) is used and also when the liquid-phase reactant has a marginal contribution to the total pressure:

$$p_A = P_{total}$$

Increase in total pressure implies a higher capital cost of the reactor and is therefore not the optimum solution. As mentioned earlier, the gas–liquid mass transfer and solid–liquid mass transfer coefficients offered by a venturi loop reactor are at least an order of magnitude higher than their nearest competitor, the conventional stirred tank reactor or gas-inducing reactor. Therefore, if the gas–liquid mass transfer coefficient can be so increased by a factor of 10 through the use of a venturi loop reactor, then the sole reason for excessive reactor pressure to achieve the required higher value of mass transfer rate vanishes. In conclusion, the venturi loop reactor can allow operation at much lower pressures as compared to the stirred reactor or its variants.

8.3.3   Well-Mixed Liquid Phase

The high circulation rate through the liquid nozzle and the intense mixing in the ejector and subsequently in the holding vessel result in a well-mixed liquid phase. This also ensures high degree of homogeneity in concentration and temperature. The circulation number defined as the ratio of volumetric liquid flow rate to the reactor volume is in the range of 60–100/h depending on the process requirements (Dierendonck et al. 1988). Leuteritz et al. (1976) and Bhutada (1989) have reported that the liquid phase may be considered as fully back mixed.

A majority of applications of venturi loop systems (Table 8.1) pertain to situations in which the gas-phase reactant is a pure gas. Further, the operating pressures are also high (~1 MPa and above); therefore, partial pressure of the liquid under saturation conditions is also negligible. As a result, back mixing of the gas phase is inconsequential.

8.3.4   Efficient Temperature Control

In the case of a stirred tank reactor with a jacket, the heat transfer area ($\pi H_{HV} T$) or (πT^2) for $H_{HV}/T_{HV} = 1$ is decided by the diameter and height of the tank. For larger reactor sizes, the jacket may not be able to extract the heat of reaction. Use of external heat exchanger loop/internal coils to augment the heat transfer can be considered to cater to excessive heat loads. The external heat exchanger increases the operating cost besides the capital cost. The internal coil alternative has been shown to be counterproductive in Section 7A.8. On the other hand, the external heat exchanger is an integral part of the venturi loop reactor. Its surface area can be chosen independent of the venturi and holding vessel volume. Further, this external heat exchanger does not interfere with the mixing and mass transfer within the reactor and yet provides a heat transfer coefficient that is three times higher than that for internal coils (Leuteritz et al. 1976). Thus, relatively very large heats of reaction can be removed in the external heat exchanger loop without any detrimental effects on the hydrodynamics of the venturi and holding vessel. A combination of adequate heat transfer and large circulation rates results in nearly isothermal (±1 °C) operation of the entire system, avoids hot spots within the reactor, and greatly improves the selectivity.

8.3.5   Efficient Solid Suspension and Well-Mixed Solid (Catalyst) Phase

Above a certain liquid circulation rate (or power input), the venturi loop reactor maintains the solid catalyst in complete suspension without any sedimentation problems. Further, as shown by Bhutada and Pangarkar (1989), above this power input, the solid concentration is uniform in both the axial and radial directions. This clearly shows that as long as a certain minimum power input is maintained, the catalyst concentration is independent of the location in the venturi loop reactor. In sharp contrast to this, there is substantial variation of solid concentration in conventional stirred (or sparged) reactors operated at the just-suspended condition (Section 7A.7.1). This characteristic behavior helps in attaining uniform rates throughout the reactor volume, avoids hot spots, and enhances selectivity.

8.3.6   Suitability for Dead-End System

Ejectors are self-priming devices. They induce the unreacted gas and redisperse it. This occurs without any external device since the unreacted gas is discharged at a higher pressure than the suction pressure. Thus, a separate recycle compressor is not necessary. This is a significant advantage inasmuch as process economics dictates that recycle of the expensive unconverted gas is obligatory. The suction chamber is connected to a gas supply/storage, and a pressure control device supplies only a quantity of gas equal to that consumed by the reaction occurring in the reactor. On the other hand, in the case of a conventional stirred reactor, such an automatic recycle is not possible. The unreacted gas must be compressed to the reactor pressure through an external recycle compressor. Conventional gas compressors can cause contamination due to entrained oil mist in the gas. The alternative oil-free hydrogen recycle compressors are expensive. The overall capital and operating cost of a conventional stirred reactor is therefore relatively high as compared to a venturi loop system.

8.3.7   Excellent Draining/Cleaning Features

There are no internal parts such as baffles/spargers/internal cooling/heating coils in the venturi loop reactor. This facilitates quick and efficient draining of the product with a bottom flush valve at the end of a batch. If the same reactor is to be used for another product, the same simple construction allows rapid and efficient cleaning. In view of this, the venturi loop reactor can be readily adapted to multipurpose, campaign-based manufacture of high-value chemicals.

8.3.8   Easy Scale-Up

Scale-up factors as high as 500 have been claimed without any problem (Greenwood 1986; Dierendonck et al. 1988 and Table 2.1). The criterion that has been suggested is constancy of power input per unit volume (valid if the gas and the liquid are fully back mixed) or circulation number defined as $Q_L/V_D = n$ (/h) between 60 and 120 (/h) Dierendonck et al. 1988).

Table 8.2 gives a semiquantitative comparison of venturi loop reactor with stirred tank reactor.

8.4   THE EJECTOR-BASED LIQUID JET VENTURI LOOP REACTOR

The venturi loop reactor was introduced by Buss ChemTech, Switzerland in 1950. (http://www.buss-ct.com/e/reaction_technology/loop-reactor.php?navanchor=2110031; http://www.buss-ct.com/e/company/publications/reaction_technology/BCT_Loop_Reactor_Technology_2009-02.pdf. Accessed Nov., 2011). The company has optimized more than 800 reactions for its customers. Several articles have been published by the company on its technology, which will be referred to frequently in the following sections.

TABLE 8.2 Comparison of Venturi Loop Reactor with Stirred Tank Reactor

Parameter	Unit	Stirred tank reactor and variants including gas-inducing reactor	Venturi loop reactor
Operating pressure	MPa	Maximum up to three (limited by sealing of the shaft)	Up to 20[c]
Maximum operating temperature	°C	350	320
Maximum temperature fluctuation	°C	10–20 (<2 if external heat exchanger is used)[a]	1–1.5
Maximum viscosity	Pa s	40×10^{-3}	40×10^{-3}
Catalyst loading required	Weight %	≈ 5	< 2. In most cases, the catalyst loading is substantially less (~10–50% lower) than that for stirred reactors
Power input	kW/m³	2; ~10 if external heat exchanger is used	3–10
Gas–liquid mass transfer coefficient, $k_L a$	/s	2–8×10^{-2}	~10 for the venturi section). Overall ~1–2
Particle–liquid mass transfer coefficient K_{SL}[b]	m/s	4–8×10^{-5} m/s (Section 7A.10)	No information available in the literature. However, in view of the extremely high level of turbulence in the venturi section, it is safe to assume that K_{SL} is as high as k_L
Catalyst loss	%	Could be as high as 50–70%	Substantially lower/ negligible loss due to high H_2 coverage of the catalyst (Leuteritz et al., 1976; Greenwood, 1986)
Scale-up factor (please see Table 2.1)	–	$>10,000$[d] ~ 1000[e]	500[f] ~ 200[g]

Source: Adapted from Dierendonck et al. (1988) with permission from BHRA, The Fluid Engineering Centre, Bedford, England.

[a] See Sections 7A.10.

[b] Depends also on particle shape.

[c] Duveen (1998).

[d] Homogeneous reaction: Vogel (2005).

[e] Heterogeneous two-/three-phase reaction: Chapter 7A.

[f] Based on constant power dissipation per unit total volume in the ejector (Dierendonck et al., 1988; 1998).

[g] Author's experience.

8.4.1 Operational Features

The liquid-driven jet ejector is the most important part of a venturi loop reactor. It is a simple device that finds great acceptance as an alternative to the steam jet ejector and mechanical vacuum pumps for extracting gases (gas pump)/creating moderate vacuum. The geometry of this section has a very important bearing on the rate of induction of the gas per unit power input and subsequent mass transfer in the venturi. The design features of this part have been widely studied (Section 8.6), and there is substantial level of confidence in designing it, at least for creating vacuum. A high velocity jet of the liquid-phase reactant containing the catalyst is fed to the nozzle in the ejector section of the venturi. This section is located at the top of the holding vessel in the system known as Buss Loop Reactor (BLR®). The high pressure of the liquid in the nozzle is converted into a high kinetic energy jet that induces the flow of the gas phase into the suction chamber of the ejector. The latter is connected to the gas storage. Since the volume of the venturi section is relatively small and the liquid circulation rate is high, there is a very large power input per unit volume of the venturi section given by $(\Delta P_{Noz} Q_L)/V_{Ej}$. The consequence of such highly concentrated power input is a very high level of turbulence within the venturi section. This ensures high shear rates, very small bubble size, and hence relatively very high mass transfer rates.

The ejector-based reactor can be constructed in various configurations. Firstly, the jet can be either upflow or downflow. In this book, only the downflow type is considered. In this type, the venturi loop reactor can be with or without a draft tube. In case there is a draft tube, there is an additional internal circulation loop within the reactor. An outer circulation loop already exists because the liquid + catalyst slurry is continuously recycled by the external pump through the heat exchanger and the primary nozzle.

8.4.2 Components and Their Functions

8.4.2.1 The Conventional Venturi Loop Reactor/BLR®   Figure 8.1 shows a schematic of the venturi loop reactor. The unit has a self-priming ejector located at the top of the reactor. As described earlier, the motive liquid sucks the gas into the suction chamber. The gas is simultaneously mixed and compressed due to the transfer of momentum from the primary fluid. Depending on the relative rates of the primary and sucked fluid, different regimes of flow are observed in the diffuser as discussed later. The flow issuing out of the diffuser into the holding vessel is a two-phase jet comprising of the gas and the liquid + catalyst slurry. The unreacted gas eventually rises out of the dispersion in the holding vessel at a pressure greater than that in the suction chamber and is therefore automatically recycled. The liquid (or slurry, when solid catalyst is used) flows out from the bottom and is pumped firstly through a heat exchanger to remove (or add) heat and then back to the nozzle, thus completing the external loop. Cross-flow slurry filtration, product withdrawal, or feed addition can be performed in this external loop. Optional absorber/condenser can also be provided in this external loop to remove undesired by-products (Figs. 8.2 and 8.3; Duveen 1998; Opre 2009). The most crucial parts of this unit are the ejector (comprising of the nozzle for the primary fluid, suction chamber, diffuser, mixing tube, and diverging cone) and the recirculation pump. The pump delivers a slurry at high flow rates and heads and can present some design/operational problems. Pumps made out of

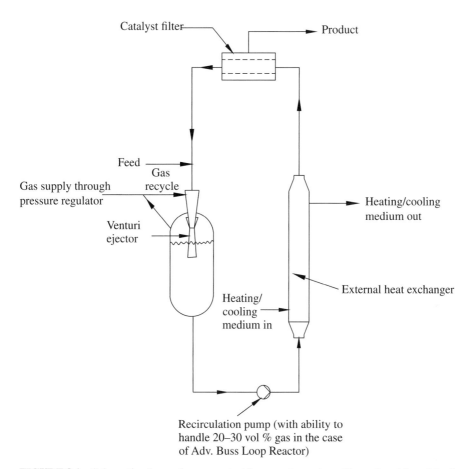

FIGURE 8.1 Schematic of a continuous venturi loop reactor system. (Reproduced from http://www.buss-ct.com/e/reaction_technology/Continuous_Reactions.php?navanchor=2110004 with permission from BUSS ChemTech AG.)

Cr–Ni–Mo stainless steel variety are suitable for most duties. The pump can have a single, double, or triple mechanical seal. The seal is protected from ingress of solid particles by maintaining the sealing liquid on the nonreaction side at a higher pressure than within the pump (Greenwood 1986). Alternatively, a pump with magnetic drive can also be used. Deflection of the short pump shaft is relatively less as compared to that of the long impeller shaft in a stirred vessel. As a result, downtime caused due to seal replacement is also much lower in the loop reactor (Greenwood 1986; Baier 2001). Improved pump impeller and housing allow circulation of slurries with gas holdup as high as 30% without any problem. These proprietary pump designs can handle solid loadings as high as 8 wt % without abrasion problems and are rated for use up to pressures as high as 20 MPa (Duveen 1998; BCT Loop® Reactor Technology (2009–2002) http://www.buss-ct.com/e/company/publications/reaction_technology/BCT_Loop_Reactor_Technology_2009-02.pdf. Accessed Nov 2011). For continuous operation, a cross-flow catalyst filter is incorporated in the external circuit.

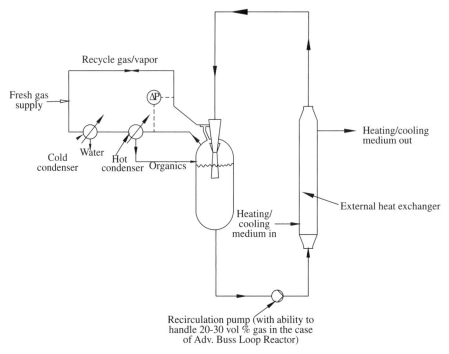

FIGURE 8.2 Continuous removal of a condensable volatile by-product. (Reproduced from: http://www.buss-ct.com/e/reaction_technology/In_Situ_Removal_of_By_Products.php?navid=34 with permission from BUSS ChemTech AG.)

Nardin and Cramers (1996) and Opre (2009) have described several examples in which the addition of a suitable external separation loop can be advantageously used to remove a by-product of the reaction. A condensable vapor-like water is easily eliminated in the external loop condenser(s) (Fig. 8.2).

Similarly, an absorber can be incorporated in the secondary external loop to remove a gaseous by-product that may affect the conversion of the reactant (Fig. 8.3).

The manufacture of pure monochloroacetic acid (MCAA) through the hydrogenation of dichloroacetic acid with *in situ* removal of by-product hydrogen chloride and the alkylation of polyether alcohols with *in situ* removal of methanol have also been described (http://www.buss-ct.com/e/reaction_technology/In_Situ_Removal_of_By_Products.php?navid=34). In both cases, removal of a product of the reaction eliminates the thermodynamic limitation. As a result, the reaction time can be substantially lowered (~50%) with simultaneous improvement in the yield and purity of the main product.

8.4.2.2   The Advanced Buss Loop Reactor (ABLR)® This reactor has been described in detail by Baier (2001). This unit includes two major modifications of the previous version termed BLR®. A shorter venturi without a diffuser or draft tube has replaced the relatively long prior design. In this form, the combination of primary

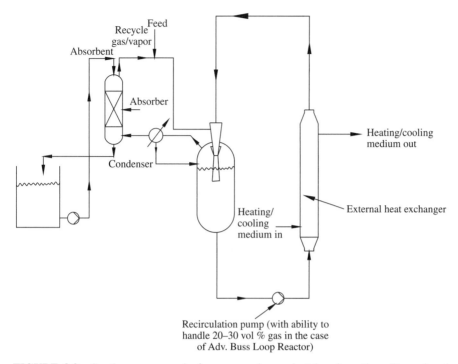

FIGURE 8.3 Continuous removal of a gaseous by-product by absorption. (Reproduced from: http://www.kresta-industries.com/index.php/de/industrialengineering-de/reaction-technology/in-situ-removal-of-by-products with permission from BUSS ChemTech AG.)

nozzle, suction chamber, and venturi section acts as a section in which the various phases are intimately mixed. This section is referred to as "reaction mixer." Baier (2001) has given a schematic of the Advanced Buss Loop Reactor (ABLR®) (also see Advantages of ABLR with external gas circuit: http://www.dechema.de/index.php?id= 116220&tagung=9&file=7717&site=achema2009&lang=en&path=1%252 52C79095). The three major parts of the ABLR® are (i) the "reaction mixer," (ii) the holding or reaction vessel, and (iii) the recirculation loop, which now has a special type of pump described in Section 8.4.2.1. Unlike the conventional BLR, the mass transfer contribution of the "reaction mixer" in this advanced version is reportedly not very high due to its smaller volume as compared to the BLR® (Baier 2001). This suggests that the ejector volume is lower than 1–2% of the holding vessel volume in its predecessor. The liquid issuing out of the venturi preserves a major portion of its original momentum. From the description available (Baier 2001), the outlet of the venturi delivers a "plunging jet" rather than the customary two-phase mixture being delivered directly below the liquid surface in the BLR. More recent pump designs capable of handling a mixture comprising up to 50% gas without cavitation are used (Baier 2001). The reaction continues in the external loop since all the three participants, catalyst, the liquid-phase reactant, and gas, continue to be in intimate contact at the temperature and pressure of the reaction. The operation of the ABLR® is otherwise similar to its predecessor.

According to Baier (2001), the ABLR® yields the same mass transfer performance (defined as oxygen transfer rate obtained by cobalt-catalyzed oxidation of sodium sulfite) at half the power input as compared to the conventional BLR.

8.5   THE EJECTOR–DIFFUSER SYSTEM AND ITS COMPONENTS

Typical ejector-based gas–liquid nozzle-diffuser system used in a venturi loop reactor is shown in Figure 8.1. The suction chamber is followed by a diffuser section. The latter comprises of an entry region that is sequentially followed by a mixing tube (optional) and a diverging cone that is typically venturi shaped. The mixing tube as the name suggests causes mixing of the primary and entrained fluids. This mixing can be alternatively termed as transfer of momentum between the slow moving entrained fluid and faster primary fluid. Following the mixing tube, the mixture of primary and entrained fluids enters the diverging cone. The latter has a low angle of divergence to prevent boundary layer separation/eddy formation. This section of the diffuser assembly is meant to recover the pressure by converting kinetic energy into pressure energy. It also partially helps in continuing the rapid mass transfer experienced in the mixing tube.

Figure 8.4 shows details of a diffuser without a parallel throat (mixing tube length=0). Figure 8.5 shows an ejector assembly with bell shaped entry, mixing tube and pressure recovery diverging section.

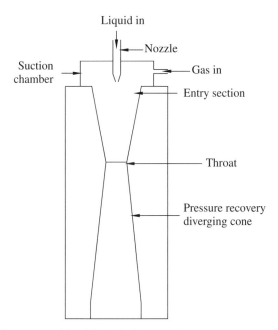

FIGURE 8.4   Ejector assembly with conical entry and pressure recovery section but without mixing tube. (Reproduced from Bhutada and Pangarkar (1987) with permission from Taylor & Francis Ltd. Copyright © 1987 Taylor & Francis Ltd. http://www.tandfonline.com/loi/gcec20.)

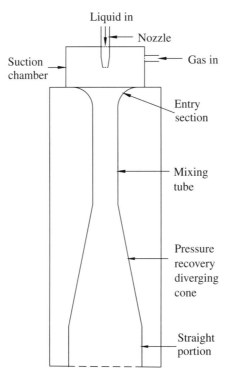

FIGURE 8.5 Ejector assembly with bell-shaped entry, mixing tube, and pressure recovery diverging section. (Reproduced from Bhutada and Pangarkar (1987) with permission from Taylor & Francis Ltd. Copyright © 1987 Taylor & Francis Ltd. http://www.tandfonline.com/loi/gcec20.)

8.6 HYDRODYNAMICS OF LIQUID JET EJECTOR

To achieve an optimal design of this crucial part, it is necessary to understand the hydrodynamics of the ejector. A simple description of the overall phenomenon is through the conservation of energy as implied in Bernoulli's theorem. Several investigators have studied the performance characteristics of ejectors used for either pumping of a gas or as a gas–liquid contacting device (Flugel 1939 *cf*: Engel 1963; Folsom 1948; Engel 1963; Bonnington 1956, 1964; Berman and Efimochkin 1964; Davies et al. 1967a, b; Witte 1965, 1969; Cairns and Na 1969; Bhat et al. 1972; Hill 1973; Cunningham 1974; Cunningham and Dopkin 1974; Biswas and Mitra 1981; Schugerl 1982; Ben Brahim et al. 1984; Henzler 1981, 1983; Gamisans et al. 2004; Kandakure et al. 2005; Mandal et al. 2005a, b; Kim et al. 2007; Liao 2008; Yadav and Patwardhan 2008; Li and Li 2011; Utomo et al. 2008). Bonnington and King (1976) and Sun and Eames (1995) have presented comprehensive reviews on various aspects of ejector design for the corresponding periods. A typical horizontal flow liquid jet ejector assembly is shown in Figure 8.6.

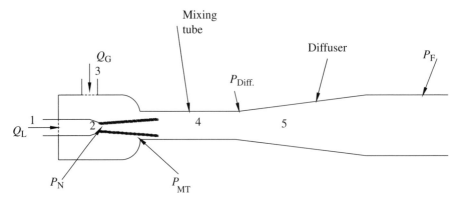

Figure 8.6 Geometry used by Neve (1988). (Reproduced from Neve (1988) with permission from Elsevier. Copyright © 1988 Published by Elsevier Inc.)
Numerals in the figure refer to locations in Figure 2 of Ben Brahim et al. (1984).

Referring to Figure 8.6, the efficiency of an ejector-based gas pump was defined (Neve 1988) as follows:

$$\eta_{GP} = \left(P_{MT} \left(\frac{Q_G}{Q_L} \right) \right) \ln \left(P_{Diff} / P_{MT} / (P_N - P_{Diff}) \right) \tag{8.1}$$

For applications of the ejector as a contacting device, Bhutada and Pangarkar (1987) defined the ejector efficiency as follows:

$$\eta'_I = \left(\frac{M_G}{M_L} \right) \left(\frac{\Delta P_G}{\Delta P_L} \right) \tag{8.2}$$

Since the term ΔP_G is not relevant in the application of the ejector as a gas–liquid contacting device, Equation 8.2 was further simplified as

$$\eta'_I = \left(\frac{M_G}{M_L \times \Delta P_L} \right) \tag{8.3}$$

Since $M_L \times \Delta P_L$ is equal to power supplied through the primary fluid, Equation 8.3 gives the mass rate of gas induction per unit power input. This variant of the definition of ejector efficiency is a reflection of the contacting efficiency of the ejector since for a given pressure–temperature condition, M_G or Q_G (mass or volumetric gas induction rate) decides the gas holdup and effective gas–liquid interfacial area. In addition, the power input term $M_L \times \Delta P_L$ indicates the extent of turbulence in the contactor and consequently the gas–liquid mass transfer coefficient $k_L a$.

The devices used by the initial investigators had relatively very low (~10%) efficiency likely due to their very low rates of induction (or low Q_G / Q_L). Witte (1962, 1965, 1969) pointed out that the large gas–liquid interfacial area provided by liquid-propelled gas ejector is a major advantage in their use as gas–liquid contactor. Witte (1965) described the remarkable phenomenon of "mixing shock" initially observed by von Pawell Ramingen as early as 1936. This phenomenon converts an initial jet

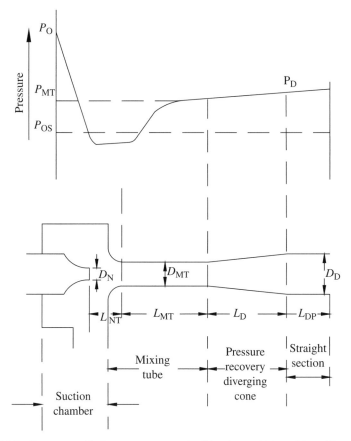

FIGURE 8.7 Pressure profile from entrance to exit of the ejector. (Adapted from Henzler (1983) with permission from John Wiley and Sons Inc. Copyright © 1983 John Wiley and Sons Inc.)

flow regime to froth or homogeneous bubble flow regime. It is observed when a pressure buildup is forced on the ejector by, for instance, throttling the valve on the outlet of the ejector. The differential pressure across the ejector is generally referred to as "back pressure." Figure 8.7 shows a typical pressure profile across the ejector.

It can be seen that there is a steep increase in pressure as the fluid mixture travels across the mixing tube. This is due to the mixing shock that is characterized by very fine bubbles dispersed in a continuous liquid phase. It is also the source of the relatively high gas–liquid effective interfacial area noted by Witte (1965). Witte (1965) carried out a systematic analysis of the effects of major geometric and operating variables on the performance of liquid–gas ejectors. The objective of his study was to derive optimum geometric parameters for achieving a high efficiency. It was observed that the efficiency could be increased from the low initial value of approximately 10–25%. This efficiency could be additionally enhanced to a maximum of 45% by a minor change of recirculation of the liquid coming out of the gas–liquid separator. Witte (1965) further showed that the mixing shock results in relatively high compression ratios ranging from 9 to 30 depending on the ejector geometry. In Figure 8.7, the mixing tube is followed by a pressure recovery divergent cone in which the pressure recovery continues. The advantages of this part are discussed later in Section 8.7.1.7. Witte's (1969) analysis of the mixing shock phenomenon used both isothermal and adiabatic approaches. The mechanism for gas entrainment was assumed to be similar to that due to entrapment of gas on impact of a liquid on a free surface (Engel 1966, 1967). The model predictions of the compression ratio and gas entrainment rate exhibited satisfactory agreement with the experimental results. As mentioned earlier, the compression ratio for the gas is strictly not relevant to the use of liquid ejectors in venturi loop reactors as compared to steam ejectors. Cunningham (1974) and Cunningham and Dopkin (1974) subsequently made significant contributions. Cunningham's (1974) work dealt with the performance of gas pumps, whereas Cunningham and Dopkin (1974) studied the length of breakup of liquid jets. When the flow of the gas–liquid mixture from the diffuser is throttled, a resistance is created. In the case of the venturi loop reactor, the back pressure could arise from a static head on the diffuser when it is immersed in the dispersion in the holding tank (Fig. 8.1; Bhutada and Pangarkar 1987). This resistance has a pronounced impact on the gas phase, whereas the high-energy liquid phase is unlikely to suffer any serious effect due to this relatively small increase in the resistance. Variation of the back pressure by regulating the pressure at the outlet of the diffuser can facilitate the location of the mixing shock in the diffuser. According to Witte (1965, 1969), prior to the mixing shock, there is a slip between the two phases. It was implicitly assumed that beyond the mixing shock this slip vanishes. However, Radhakrishnan et al. (1986) and Neve (1991) have given evidence of dissimilar velocities of the two phases (imperfect mixing) particularly when the secondary to primary fluid volumetric flow ratio is high. The extent of mixing is an important aspect in modeling of two-phase flow using the computational fluid dynamics (CFD) approach (Kandakure et al. 2005).

8.6.1   Flow Regimes

8.6.1.1   *Flow Regimes in the Ejector*   Various flow regimes have been observed depending on whether the flow in the ejector is horizontal or vertical (downward).

Horizontal Mode   Four different regimes of flow have been observed in this mode of liquid jet gas ejectors. The flow regime prevailing in the diffuser is a strong function of the following operating variables: (i) flow rates of the primary and secondary fluids and (ii) pressure differential across the ejector. Bhat et al. (1972) have observed coaxial flow, homogeneous bubbly flow, and stratified flow regimes in the diffuser in increasing order of the pressure differential across the diffuser. The flow behavior of multi-jet nozzle ejectors was studied by Biswas and Mitra (1981) in the horizontal configuration. This study also reported the effect of back pressure. The mass ratio of induced gas and primary fluid (mass ratio, M_R) was plotted as a function of the fractional energy utilization, F, which is related to the pressure at the gas outlet.

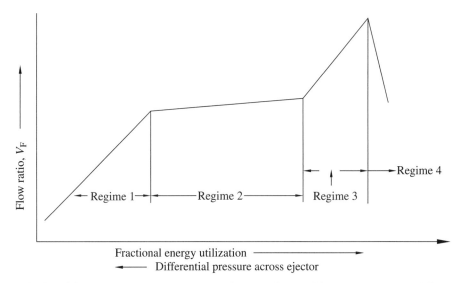

FIGURE 8.8 Flow regimes in horizontal ejectors with multi-jet nozzles. (Adapted from Biswas and Mitra (1981) with permission from John Wiley and Sons. Copyright © 1981 Canadian Society for Chemical Engineering.)

Figure 8.8 shows the aforementioned four regimes as a function of F: at relatively low values of F, the gas is dispersed in the form of fine bubbles, and the rate of induction increases rapidly (Regime 1). As the gas outlet pressure is decreased, a certain state is reached at which the gas induction rate is almost independent of the decrease in outlet gas pressure (Regime 2). The third distinct regime is observed with further decrease in outlet gas pressure. In this regime, a distinct increase in rate of gas induction is observed concomitant with formation of gas slugs (Regime 3). With further increase such that the suction pressure and outlet gas–liquid separator pressure are equal, the energy is completely utilized. This stage is accompanied by a sharp drop in the rate of gas induction (Regime 4).

Vertical Downflow Mode In the case of vertical downflow, only the coaxial and homogeneous bubbly flow regimes have been reported (Ben Brahim et al. 1984; Bhutada and Pangarkar 1987; Panchal et al. 1991).

The different types of flow regimes possible in the venturi section of the venturi loop reactor are:

1. For low gas to liquid flow ratio, homogeneous bubble flow with very small bubbles dispersed in the continuous liquid phase is observed. This type of bubbly flow (also referred to as emulsion) regime covers the entire diffuser from its throat to the outlet.

2. At higher gas to liquid flow ratios, two separate regimes are observed in the different sections of the ejector: (1) a jet flow of liquid surrounded by a coaxial gas phase prevails in the throat and mixing tube followed by (2) bubbly flow in the diverging section.

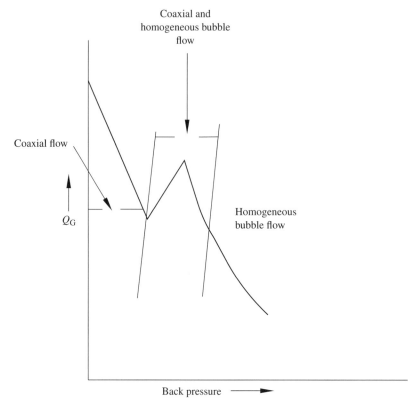

FIGURE 8.9   Flow regime map for vertical downflow ejector. (Reproduced from Ben Brahim et al. (1984) with permission Elsevier. Copyright © 1984 Published by Elsevier Ltd.)

The flow regime changes with respect to the back pressure on the diffuser outlet observed by Ben Brahim et al. (1984) are shown in Figure 8.9.

The back pressure on the suction chamber is decided by (i) the extent to which the diffuser is dipped in the dispersion in the holding tank and (ii) the density of this dispersion at the top of the holding tank. Elgozali et al. (2002) have observed that there is a critical liquid flow rate (or kinetic energy of the primary fluid), which is required to overcome the existing back pressure on the inlet of the secondary fluid. According to Bernoulli's theorem also, a threshold kinetic energy must be supplied to create a negative pressure that balances this back pressure. In addition to the "dip," the back pressure on the diffuser can be further increased by throttling the valve in the gas outlet line from the holding tank (Fig. 8.1). Bhutada and Pangarkar (1987) found that the flow regime is very sensitive to the "dip" of the diffuser in the dispersion in the holding vessel. At constant liquid rate, a change in the "dip" resulted in a shift of the flow regime from one to the other depending on the actual gas to liquid flow rates ratio (Bhutada and Pangarkar 1987; Dirix and van der Wiele 1990; Panchal et al. 1991). For the case of a typical venturi loop reactor system (Fig. 8.1), Bhutada and

Pangarkar (1987) investigated the regimes at varying back pressure, using converging–diverging type of diffusers (Figs. 8.4 and 8.5) and a straight tube diffuser. The "dip" was maintained around 0.1 m of dispersion in their investigation. For the converging–diverging diffusers, they observed a homogeneous bubbly flow regime at the end of the diffuser. As the back pressure was increased by throttling the valve in the gas outlet line of the holding tank, this homogeneous bubbly flow rose into the diffuser and finally covered the entire diffuser. The coaxial flow regime observed by Ben Brahim et al. (1984) was not observed for these diffusers by Bhutada and Pangarkar (1987), presumably because the diffuser was always under a certain back pressure determined by the "dip" of the diffuser. For the straight tube diffuser with the same "dip" of 0.1 m of dispersion, the entire diffuser always exhibited the homogeneous bubbly flow regime. A small increase in the back pressure caused the homogeneous bubbly mixture to spill into suction chamber. For the horizontal arrangement also, Witte (1965) observed a similar phenomenon. It was found that when the pressure at the diffuser outlet was increased, at a certain back pressure, the mixing shock was blown in the reverse direction back into the suction chamber. Additional increase in the back pressure resulted in flooding of the suction chamber followed by backup of the liquid in the gas inlet line, which is exactly similar to the observation of Bhutada and Pangarkar (1987) for the straight tube diffuser in the vertical downflow mode.

Cramers et al. (1992b) have given a flow regime map (Fig. 8.10) for the pilot venturi loop reactor used in their work.

They found that as the liquid flow rate is increased, the flow ratio Q_G/Q_L at which the transition from bubbly to jet flow regime occurs decreases. Dirix and van der Wiele (1990) observed the prevalence of bubble flow regime at low gas to liquid flow rates ($Q_G/Q_L < 1.3$). These investigators observed the jet flow regime for the range $1.3 < Q_G/Q_L < 3$. Otake et al. (1981) also observed a similar transition in nearly the same range: $1 < Q_G/Q_L < 2$. Ben Brahim et al. (1984) studied the variation of the regime for a vertical downflow ejector. With increasing back pressure, the following regimes were observed: (i) coaxial flow over entire ejector, (ii) coaxial flow near suction

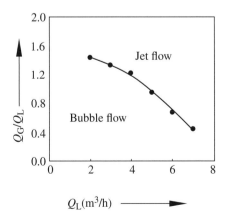

FIGURE 8.10 Flow regime map for venturi section. (Reproduced from Cramers et al. (1992b) with permission from Elsevier. Copyright © 1992 Published by Elsevier Ltd.)

chamber and homogeneous bubble flow toward diffuser end, and (iii) finally homogeneous bubble flow regime over the entire ejector. For low-viscosity primary fluids similar to water, the changes in the regimes were distinctly defined, one followed by the next. However, as the viscosity of the primary fluid increased the band gap (coaxial flow near suction chamber and homogeneous bubble flow toward diffuser end) the width decreased. Finally, for the highest-viscosity liquid used ($\mu_L = 84$ mPa s), the intermediate regime vanished and coaxial flow directly changed over to homogeneous bubble flow regime at a certain back pressure. At constant back pressure, Bhutada and Pangarkar (1987) and Panchal (1990) observed a sudden decrease in the rate of entrainment at a particular value of the power input. Visual observations indicated that at this condition the diffuser exhibited a change over to the homogeneous bubble flow regime. Beyond this power input, the entrainment rate continued to increase with increase in power input. In a manner analogous to Cramers et al. (1992b), Gourich et al. (2005, 2007, 2008) concluded that the transition from jet flow to homogeneous bubble flow can be predicted from a plot of Q_G versus Q_L. They found that at the transition point this plot shows a sharp change in the slope. They further noted that the transition occurs at the maximum in Q_G/Q_L in a plot of Q_G/Q_L versus Q_L. These investigators reported a regime transition at Q_G/Q_L in the range of 0.6–0.8 depending on the value of T/H_{TV}. The transition point was also found to depend on the tank diameter and the back pressure on the outlet of the diffuser. It must, however, be pointed out that in all of their investigations, Gourich et al. did not use the standard venturi loop system employed by Bhutada and Pangarkar (1987), Panchal et al. (1991), or Cramers et al. (1992b), as is evident from the description of the equipment used.

The ejector geometric parameters, such as the area ratio, A_R; mixing tube length; diffuser angle and length; etc. also play an important role in deciding the flow behavior as determined by the volume ratio, Q_G/Q_L. For a given ejector design, Q_G depends on the value of Q_L. In the standard venturi loop reactor, the "dip" of the diffuser itself creates a back pressure and yields the homogeneous bubble flow regime. Bhutada and Pangarkar (1987) used an ultra low pressure drop device at the outlet of their system for measuring the gas induction rate. The induction rates so obtained are representative of the intrinsic induction rates unimpeded by any flow restriction other than the "dip." The ABLR described by Baier (2001) appears to use a diffuser discharging the mixture of primary and secondary fluids well above the top of the dispersion ("dip"=0) in the holding vessel (Section 8.4.2.2).

8.6.1.2   Flow Regimes in the Main Holding Vessel   The momentum of the two-phase jet issuing out of the diffuser is decided mainly by the momentum of the liquid phase. The latter in turn is decided by $M_L \times \Delta P_{Noz}$. The power input per unit volume of the liquid in the reactor is ($M_L \times \Delta P_{Noz}/V_L$). This is an important parameter deciding the type of flow regime in the holding vessel. At relatively low P_{Noz}/V_L, the momentum of this two-phase jet is low. As a consequence, the expanding two-phase jet flow is arrested at a distance well above the bottom of the holding vessel. Further, with the expansion of the jet being incomplete, it does not pervade the entire cross section of the holding vessel but exists in a small zone around the central core of the holding vessel. Therefore, the entire volume of the holding vessel (cross section and height from bottom to the diffuser) is not utilized for mass transfer. The gas phase in the arrested jet disengages and flows upward due to its higher buoyancy. The radial and axial penetration of the two-phase jet increases as P_{Noz}/V_L increases. At a certain P_{Noz}/V_L $[P_{Noz}/V_L]_{Crit}$, the two-phase jet reaches the bottom and the entire holding vessel behaves as a multiphase contactor. The core of the holding vessel acts as a downflow bubble column, whereas the outer annulus is similar to the conventional bubble column. The $[P_{Noz}/V_L]_{Crit}$ value is important for solid suspension in the main vessel as discussed later in some detail in Section 8.8.

Most of the studies referred to in the previous discussion used two-phase (gas–liquid) systems. The considerations are substantially similar when the liquid jet ejector is to be used as a three-phase catalytic reactor, particularly because the catalyst loading commonly used in a venturi loop system is relatively low. In terms of mass transfer, besides gas–liquid mass transfer, solid–liquid mass transfer step assumes great importance. As discussed in Chapters 7A and 7B, factors relating to dispersion of the gas phase, suspension of solids, and concentration profile of the catalyst phase need to be addressed in the case of a three-phase reactor.

8.6.2 Prediction of Rate of Gas Induction

8.6.2.1 Macroscopic Approaches Based on Energy Balance In the initial stages, different investigators alluded the induction to drag of the primary fluid on the entrained fluid, surface roughness of the jet (swirl pieces deliberately incorporated in the jet to increase its roughness or sharp-edged orifices), etc. The "jet envelope mechanism" has been supported by van de Sande and Smith (1973, 1975), Evans and Jameson (1991), Cramers et al. (1992b), Bin (1993), and Baier (2001). This mechanism relies on the fact that turbulent jets are inherently unstable. Instabilities in the surface of the jet give rise to a gas envelope surrounding the core of the jet. Gas trapped in these instabilities is sucked by the primary fluid jet. Evidently, according to this mechanism, a disturbance that causes surface instabilities such as the previously mentioned swirl piece in the primary nozzle should enhance the rate of gas induction. The literature evidence with respect to this "instability"-promoted gas induction is discussed later in Sections 8.7.1.1 and 8.7.2.1

Despite the widespread use of ejectors for "moving" gases, there is very little supporting basis or theory that can lend credence to a precise microscopic mechanism of gas induction. In the absence of powerful computing devices, the early investigators resorted to prediction of the performance of ejectors using the macroscopic approach based on an energy balance. Most of these investigations focused on the use of ejectors as liquid jet gas pumps (Folsom 1948; Bonnington 1956, 1964; Witte 1965; Davies et al. 1967; Witte 1969; Cairns and Na 1969; Cunningham 1974; Cunningham and Dopkin 1974; Radhakrishnan et al. 1986; Neve 1988). The objective was to obtain optimum geometry/design parameters with the pumping efficiency as the important performance indicator. These initial theoretical treatments of ejectors used macroscopic considerations such as Bernoulli's principle to derive performance criteria for ejectors. The principles first laid down by them were subsequently improved upon by many workers, in particular by Engel (1963), Mueller (1964), Kar and Reddy (1968), and Cunningham (1974). These theoretical analyses were one-dimensional (1D) flow models based on the momentum and energy exchange equations. Various energy losses due to friction, mixing, etc. in the nozzle, suction chamber, entry section, mixing tube, and the diverging section were introduced to account for nonideality and then experimentally evaluated. The involvement of fluid mechanics was understandably kept to a minimum because there was very little understanding of the precise nature of the entrainment and mixing phenomena, flow patterns, etc. One of the basic difficulties in the use of momentum and energy balance theory for ejectors involving two-phase systems (liquid–gas) is the determination of the properties of the fluid mixture or in other words providing an "equation of state" for the mixture. In such cases, the simplest option is to assume that the fluids are homogeneously mixed in the mixing tube and hence the average volume-weighted physical properties can be used.

The "mixing shock" phenomenon originally observed by Von Pawell-Rammingen (1936) has been discussed earlier in Section 8.6. Witte (1965, 1969) went on to elaborate on this observation. It was found that a resistance in the outlet gas line could cause significant changes as observed in transparent equipment. At a certain "back pressure" on the diffuser outlet, the flow changed from separated gas and liquid flow streams to a "gas in liquid" dispersion. Variation of the "back pressure" resulted in a change in the position of the mixing shock in the mixing tube. Witte (1965) observed some similarities of such a mixing shock in an ejector with the plane shock wave in gas dynamics. However, some salient differences between the two were also noted. The most important differences are as follows: (i) due to the relatively very high heat capacity of the liquid phase, there is a negligible change in the temperature over the mixing shock zone. (ii) Prior to the mixing shock, there is a significant difference in the velocities of the gas and liquid phases. Therefore, unlike the plane shock wave in gas dynamics, neither is the mixture homogeneous nor is the flow supersonic. The gas velocity prior to the mixing shock can be substantially different from its liquid counterpart. (iii) The zone over which this mixing shock occurs is relatively broad than that in gas dynamics. Witte (1969) presented a mathematical analysis of the problem with a model for the pressure and entropy changes occurring before, within, and after the shock. Such macroscopic approaches based on momentum balance have been used for both the horizontal and vertical downflow modes. The vertical downflow mode is relevant to venturi loop reactor. A number of investigators have used the momentum balance to obtain empirical correlations for the induction rates (Engel 1963; Mueller 1964; Kar and Reddy 1968; Cairns and Na 1969; Bhat et al. 1972; Cunningham 1974). These models are more relevant to the use of an ejector as a "gas pump." They relate the overall loss coefficient to the pressure recovery and estimate the gas induction rate. Similar to other momentum balance approaches, these suffer from the following drawbacks:

1. Their applicability is restricted to the specific geometric configuration used.
2. Since an averaging of physical properties is used, such models are further restricted to the post-"mixing shock" homogeneous bubble flow regime.

In the following, the semiempirical approach of Ben Brahim et al. (1984) is used for illustrating application of the macroscopic approach. This approach represents the effects of various parameters in the form of dimensionless groups and is based on the principle of dynamic similarity (Section 5.2.2.1). It suffers from the drawback that their applicability is restricted to the specific geometric configuration used. Figure 8.11 shows the geometry considered and the different elements of the ejector used. The momentum balances are applied to individual sections separately inasmuch as these elements perform different functions. Figure 8.11 shows the various zones. The notations for the equations that follow are also shown in the same.

For zone 0–1 secondary fluid entrainment, neglecting potential energy difference in comparison with other forms of energy involved:

$$(P_1 - P_0) = \left(\frac{1}{2}\right) \rho_S (V_{S1}^2 - V_{S0}^2) - E_{1-2} \tag{8.4}$$

For zones 2–3, besides neglecting the potential energy component, the following assumptions were made: (i) the secondary fluid is an ideal gas and the primary fluid is incompressible; (ii) at point 2, the flow is in the homogeneous bubble flow regime, and the flow properties can be represented by weighted mixture properties:

$$V_P (P_3 - P_2) + nRT \ln \left(\frac{P_3}{P_2}\right) = \left(\frac{1}{2}\right)(M_S + M_P)V_{Mix2}^2 - E_{2-3} \tag{8.5}$$

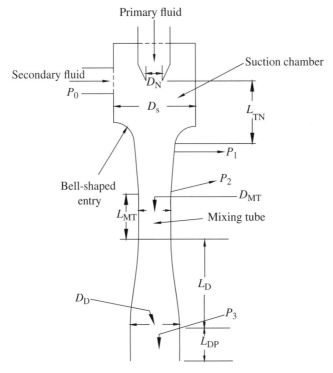

FIGURE 8.11 Vertical downflow ejector assembly (Adapted from Ben Brahim et al. (1984) with permission from Elsevier. Copyright © 1984 Published by Elsevier Ltd.)

Equation 8.5 can be simplified to

$$(P_3 - P_2) = \left(\frac{1}{2}\right)(\rho_{Mix2} V_P^2) - E_{Mix2-3} \tag{8.6}$$

where $= \rho_{Mix2} = \left(\dfrac{M_S + M_P}{V_{Mix2} + V_{S2}(1+\varepsilon)}\right)$; $(E_{2-3}) = \left(\dfrac{E_{Mix2-3}}{V_{Mix2} + V_{S2}(1+\varepsilon)}\right)$; and E_{1-2} and E_{2-3} represent the energy lost in the various processes such as gas–liquid mixing and expansion.

A macroscopic balance for sections 1–2 that include the force exerted by the primary and secondary fluids on the entry section wall is

$$(P_2 a_2) - (P_1 a_1) = \left(M_S V_{S1} + M_P V_{PNt}\right) - \left(M_S + M_P\right) V_{Mix2} - F_1 \tag{8.7}$$

Combining Equations 8.4, 8.6, and 8.7 yields the following equation:

$$\left(P_3 - P_0\left(\frac{a_1}{a_2}\right)\right) = \frac{1}{2}\frac{a_1 \rho_S \left(V_{S0}^2 - V_{S1}^2\right)}{a_2} + \frac{1}{2}\rho_{Mix2} V_{Mix2}^2 +$$
$$\frac{1}{a_2}\left[\left(M_S V_{S1} + M_P V_{PNt}\right) - \left(M_S + M_P\right) V_{Mix2}\right] - \left(E_2 + E_1\frac{a_1}{a_2} + \frac{F_1}{a_2}\right) \tag{8.8}$$

At this stage, Ben Brahim et al. (1984) combined the energy losses, E_{i-j}, and represented them as function of the kinetic energy of the primary fluid jet at the tip of the nozzle: $\dfrac{1}{2}K_{overall}M_{PNt}^2$. Further, the following dimensionless terms were introduced to simplify Equation 8.8:

$$A_{2-N} = \frac{a_2}{a_N}; \ M_R = \frac{M_S}{M_P}; \ A_{1-N} = \frac{a_1}{a_N}; \ \rho_{P-S} = \frac{\rho_P}{\rho_S}; \ \rho_{Mix2} = \left(\frac{M_S\left(M_S + M_P\right)}{\left(\dfrac{M_S}{\rho_S}\right)(1+\varepsilon_S) + \dfrac{M_P}{\rho_P}}\right);$$

$$V_{S0} = \frac{M_S}{\rho_S a_0}; V_{PNt} = \frac{M_P A_{2-N}}{\rho_P a_2}; \text{ and } A_{2-0} = \frac{a_2}{a_0} \text{ and }$$

$$\beta = \frac{P_3 - P_0(a_1/a_2)}{\dfrac{1}{2}\rho_P V_{PNt}^2} \tag{8.9}$$

Using the above dimensionless groups, Equation 8.8 was simplified to yield

$$M_R^2 \rho_{P-S}\left[A_{2-0}^2 \frac{A_{1-N}}{A_{2-N}} + \frac{A_{2-N}\left(A_{1-N} - 2\right)}{\left(A_{1-N} - 1\right)^2}\right] +$$
$$2A_{2-N} + \frac{1+M_R}{M_R \rho_{P-S}(1+\varepsilon_S)}\left[\left(M_R \rho_{P-S}(1+\varepsilon_S) + 1\right)\right] + \left(K_{Overall} + \beta\right)A_{2-N}^2 = 0 \tag{8.10}$$

Equation 8.10 is a simplified form using the assumption that the potential energy difference can be neglected. This assumption is valid for the industrial version of the venturi loop reactor in which the potential energy difference between the nozzle entry and exit is a small fraction of the total energy. On the other hand, in the experimental arrangement used by Ben Brahim et al. (1984), the total height of the ejector and extended parallel diffuser was significant (~1 m), and therefore, the potential energy factors were not ignored in their equation corresponding to Equation 8.10.

Ben Brahim et al. (1984) investigated the performance of a vertical downflow ejector using three different liquids—water, monoethylene glycol, and Tellus oil—as the primary fluids and air as the secondary fluid. The primary fluid viscosity was varied from ~1 to 84×10^{-3} mPa s, and surface tension from 3.2×10^{-2} to 7.3×10^{-2} N/m. Using the experimental data collected, they obtained the values of K_{Overall} as a function of β. It was found that K_{Overall} could be correlated uniquely with β by a simple straight line relationship: $K_{\text{Overall}} = -(m\beta) + \text{constant}$. This correlation fitted the data of Acharjee et al. (1975), which were incidentally for the vertical upflow mode. This agreement indicates that Ben Brahim et al. also could have neglected the potential energy variation in deriving Equation 8.10.

Dimensional Analysis: The flow phenomena in an ejector are complex, and hence, it is not possible to predict the flow rate of the secondary fluid *a priori*. In conventional chemical engineering, such phenomena are treated by a dimensional analysis of the dependent and independent variables based on the Buckingham π theorem. In the present case, the dependent variable is the rate of entrainment of the secondary fluid, and the independent variables are the physical properties, ejector configuration, and operating parameters. The latter are defined mainly by the flow rate of the primary fluid. Ben Brahim et al. (1984) gave the following separate dependences for the primary and secondary fluids:

$$M_{\text{P}} = f_1 \left(V_{\text{PNt}}, \rho_{\text{P}}, \mu_{\text{P}}, \sigma_{\text{P}}, D_{\text{N}}, D_{\text{T}}, g \right) \tag{8.11}$$

$$M_{\text{S}} = f_2 \left(V_{\text{S0}}, \rho_{\text{S}}, \mu_{\text{S}}, D_0, \Delta P \right) \tag{8.12}$$

Solving the above functional dependences in a manner analogous to Davies et al. (1967b), Ben Brahim et al. gave the following equation in the form of the dimensionless groups involved:

$$\frac{M_{\text{S}}}{M_{\text{P}}} = M_{\text{R}} = \text{Constant} \left(\frac{D_{\text{N}} V_{\text{PNt}} \rho_{\text{P}}}{\mu_{\text{P}}} \right)^a \left(\frac{g \mu_{\text{P}}^4}{\rho_{\text{P}} \sigma_{\text{P}}^3} \right)^b \left(A_{\text{TN}} \right)^c \left(\frac{D_0 V_{\text{S0}} \rho_{\text{S}}}{\mu_{\text{S}}} \right)^x \left(\frac{\Delta P}{\rho_{\text{S}} V_{\text{S0}}^2} \right)^y \tag{8.13}$$

The primary and secondary fluid Reynolds numbers are interrelated and hence only one needs to be retained. The last term in Equation 8.13 is the secondary fluid Euler number, E_{US}. Ben Brahim et al. found that the latter also bears a linear relationship with the primary fluid Reynolds number. Therefore, M_{OP} was retained as the only representative dimensionless group for the primary fluid properties in Equation 8.13. This simplification yielded the following form:

$$(M_R) = \text{Constant} \left(\frac{\Delta P}{\rho_S V_{S0}^2} \right)^a \left(\frac{g \mu_P^4}{\rho_P \sigma_P^3} \right)^b (A_R)^c \qquad (8.14)$$

Ben Brahim et al. (1984) correlated their data on the lines of Equation 8.14. However, since they used only one nozzle (constant $A_{TN} = 4$), only the first and second terms in Equation 8.14 were retained. The correlation obtained was

$$(M_R) = 43.86 \times 10^{-3} \left(\frac{\Delta P}{\rho_S V_{S0}^2} \right)^{-0.38} \left(\frac{g \mu_P^4}{\rho_P \sigma_P^3} \right)^{-0.01} \quad \text{For } A_R = 4 \qquad (8.15)$$

Bhat et al. (1972) and Davies et al. (1963) studied the gas entrainment characteristics of gas–liquid ejectors in the horizontal mode. The physical properties of the primary fluid were again represented by the Morton number $M_{OP} = \left(g \mu_P^4 / \rho_P \sigma_P^3 \right)$. The range of physical properties of the liquid phase covered in the investigation of Davies et al. was the same as that in the investigation of Ben Brahim et al. (1984). The viscosity range covered by Bhat et al. was relatively narrow. The dependences of mass ratio on the physical properties group observed by Davies et al. and Bhat et al. were $M_R \propto \left(g \mu_P^4 / \rho_P \sigma_P^3 \right)^{-0.04}$ and $M_R \propto \left(g \mu_P^4 / \rho_P \sigma_P^3 \right)^{-0.02}$, respectively. All the three investigations show a relatively weak dependence of the mass ratio on the physical properties. Acharjee et al. (1975) observed a stronger negative $\left[M_R \propto \left(g \mu_P^4 / \rho_P \sigma_P^3 \right)^{-0.30} \right]$ dependence as compared to most other investigators. It may be argued that the weak dependence observed by Ben Brahim et al., Davies et al., and Bhat et al. is realistic since the very strong shear field in the ejector implies a predominance of inertial forces. Table 8.3 summarizes the correlations available in the literature for predicting the mass ratio.

8.6.2.2 Computational Approaches Based on Fluid Mechanics   The problem of describing mixing of the fluid phases was recognized as early as 1948 by Folsom. In his review of the early theoretical approaches for understanding the mixing process, Folsom concluded that the use of Tollmien streamline approach was reasonably satisfactory (Goff and Coogan 1942). In the relatively common applications involving operation of jet pumps under an adverse pressure gradient, the mixing process required use of a mixing efficiency parameter/mixing loss coefficient. These models could not elucidate the gas–liquid efficiency parameter that is so important in the application of ejectors as mass transfer devices. This had to be determined empirically from experimental data. The investigation of Goff and Coogan (1942) modeled an air ejector by approximating the flow with Tollmien streamlines. Prandtl's mixing length theory was used to obtain generalized velocity distribution for fluids of substantially different densities such as air–water. A majority of the theoretical treatment up to the 1970s used the 1D models. Keenan and Neumann (1950) were the first to present a comprehensive experimental and theoretical investigation on ejectors. Their analyses used the

TABLE 8.3 Correlations for Mass Ratio Available in the Literature for Gas–Liquid Ejectors in the Downflow Mode

Investigator(s)	System used	Range of physical properties covered	Geometrical parameters covered	Correlation proposed	Comments
Ben Brahim et al. (1984)	Air–water/monoethylene glycol/high-viscosity oil.	$1 < \mu_P < 84$ mPa s; $3.2 \times 10^{-2} < \sigma_P < 7.3 \times 10^{-2}$ N/m	$D_N = 1.25 \times 10^{-3}$; $\left(\dfrac{L_{MT}}{D_{MT}}\right) = 7$; $A_R = 4$; $\left(\dfrac{L_D}{D_{MT}}\right) = 36.8$. Diverging diffuser followed by an extended parallel tube of 1 m length	$M_R = 43.86 \times 10^{-3} \left(\dfrac{\Delta P_G}{\rho_S V_{SEnt}^{2}}\right)^{-0.38} \left(\dfrac{g \mu_P^{4}}{\rho_P \sigma_P^{3}}\right)^{-0.01}$	No variation in important ejector geometric parameters. M_R very weak function of physical properties.
Bhutada and Pangarkar (1987, 1988)	Air–water, air–aqueous carboxymethyl cellulose.	$5.3 \times 10^{-2} < \sigma_P < 7.3 \times 10^{-2}$ N/m; $\tau = k \times (\gamma')^n$ $9.8 \times 10^{-3} < k < 8.2 \times 10^{-1}$ Pa (s)^n; $0.58 < n < 0.98$; $\gamma' = 5000 \times V_G$	Single-orifice nozzles of $D_N = 0.005, 0.008, 0.01,$ and 0.012 m. Several diffusers having bell-shaped, conical, and straight entry regions with diameters varying from 0.016 to 0.05 m, and different mixing tube diameters and lengths were used. Diffuser combinations denoted as A, B, C, D, and E. Holding vessel diameter, $T_{HV} = 0.3$ m and $H/T_{HV} = 3.33$.	Diffuser A: $M_R = 5.58 \times 10^{-4} (A_R)^{0.22} \left(\dfrac{\Delta P_G}{\rho_P V_{PNT}^{2}}\right)^{-0.167}$ Diffuser B: $M_R = 0.23 \times 10^{-4} (A_R)^{0.07} \left(\dfrac{\Delta P_G}{\rho_P V_{PNT}^{2}}\right)^{-0.187}$ Diffuser C: $M_R = 9.67 \times 10^{-4} (A_R)^{0.1} \left(\dfrac{\Delta P_G}{\rho_P V_{PNT}^{2}}\right)^{-0.135}$ Diffuser D: $M_R = 6.77 \times 10^{-4} (A_R)^{0.13} \left(\dfrac{\Delta P_G}{\rho_P V_{PNT}^{2}}\right)^{-0.20}$ Draft tubes with $\left(\dfrac{A_{DT}}{A_{HV}}\right)$ of 2.25 and 9 used	Diffuser E gave the highest induction rate. Q_G increases with decrease in mixing tube length. No effect of presence of draft tube on Q_G. Pressure recovery diverging cone helps in improving Q_G. Optimum value of A_R found to be between 2.5 and 4. Practically no effect of viscosity on M_R.

(Continued)

TABLE 8.3 (Continued)

Investigator (s)	System used	Range of physical properties covered	Geometrical parameters covered	Correlation proposed	Comments
Dutta and Raghavan (1987)	$Air + CO_2$–aqueous sodium carbonate–bicarbonate buffer. $Air + CO_2$–aqueous NaOH. Tricresyl phosphate used as antifoaming agent.	Not appreciable.	Two nozzles of 0.0045 and 0.0065 m. Two holding vessels: (i) $T_{HV} = 0.15$ m and $H/T_{HV} = 5$; (ii) $T_{HV} = 0.3$ m and $H/T_{HV} = 4$. Diffusers of design similar to C and E (Bhutada and Pangarkar, 1987).	$M_R = 2.4 \times 10^{-3} \left(\dfrac{\Delta P_G}{\rho_P V_{PNT}^2} \right)^{-0.82}$ $\left(\dfrac{g \mu_P^4}{\rho_P \sigma_P^3} \right)^{-0.01}$	At low power inputs, extra CO_2 added. Results strictly not for self-aspirating conditions.
Bhutada (1989)	Air–water	No variation.	Diffuser E with multiorifice nozzles. Each orifice of 0.005 m dia. Pitch varied from 1.5 to $2.5 \times D_{H}$.	$M_R = 7.37 \times 10^{-4} (A_R)^{0.44}$ $\left(\dfrac{\Delta P_G}{\rho_P V_{PNT}^2} \right)^{-0.117}$	Multiorifice nozzle yields higher value of rate of gas induction as compared to a single-orifice nozzle having the same area as the combined hole area for a given multiorifice nozzle. Pitch of $2 \times D_H$ found to yield the highest values of gas induction rate.

Reference	System	Property variation	Geometric variables	Correlation	Remarks
(1991)			Bhutada and Pangarkar (1987) with multiorifice nozzles. Orifices diameter varied from 0.003 to 0.006 m. Pitch: $1\times D_H$ to $2\times D_H$. Other details the same as Bhutada and Pangarkar (1987).	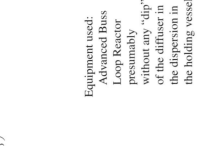$\left(\dfrac{\Delta P_G}{\rho_P V_{PNT}^2 N_O}\right)^{-0.117}$	nozzle the same as Bhutada (1989).
Baier (2001)	Noncoalescing: air–demineralized water. Coalescing: air–0.25 M aqueous sodium sulfate. Air–aqueous glucose (40–60 wt %)	$1 < \mu_P < 98$ mPa s;	D_N: 0.0062 and 0.008 m (with and without swirl devices); $2.25 < A_R < 3.74$; $10 < \dfrac{L_{MT}}{D_{MT}} < 31$; $T_{HV} = 0.3$ m and $H_{HV}/T_{HV} = 2.22$	$\left(\dfrac{Q_G}{Q_L}\right) = 0.8\left[\left(\dfrac{D_{MT}}{D_N}\right)^2 - 1\right]$	Equipment used: Advanced Buss Loop Reactor presumably without any "dip" of the diffuser in the dispersion in the holding vessel.
Mandal et al. (2005b)	Air–aqueous carboxymethyl cellulose	$\tau = k\times(\gamma)^n$, $2.18\times10^{-3} < k < 6.92\times10^{-1}$ Pa (s)^n; $0.85 < n < 0.94$; $\mu_{Eff} = 8^{n-1} V_L^{\,n-1} T_{HV}^{\,1-n} k$	D_N, 0.005, 0.006, 0.007, 0.008 m; D_T, 0.019 m; $5.65 < A_R < 14.5$; $\dfrac{L_{MT}}{D_{MT}} = 9.7$; $T_{HV} = 0.05$ m and $H/T_{HV} = 37.25$.	$\left(\dfrac{Q_G}{Q_L}\right) = 1.63\times10^{-4}\left(\dfrac{\rho_P T_{HV} V_P}{\mu_P}\right)^{0.824} \times \left(\dfrac{g\mu_P^4}{\rho_P \sigma_P^3}\right)^{-0.234}\left(\dfrac{A_N}{A_{HV}}\right)^{0.061}\left(\dfrac{H_{CL}}{T_{HV}}\right)^{-1.733}$	Geometry different from conventional venturi loop reactor.
Gourich et al. (2005)	Air–water	No variation	Four orifices of dia. 0.01 m; convergent section length of 0.096 m, angle of convergence of 20°; $L_{MT}\sim0$; diffuser length of 0.225 m, angle of divergence of 7.6°; $2 < H/T_{HV} < 4$.	$Q_L = 0.043 Q_G^{0.49}$	$4.5\times10^{-4} < Q_L < 13.5\times10^{-4}$ m³/s. Geometry not similar to venturi loop system.

constant pressure and constant area mixing models, which later on came to be known as **1D models**. The constant pressure model assumes that the pressures of the primary and secondary fluid at the exit of the nozzle are same. Therefore, the fluids enter the mixing tube at equal pressure (Huang et al. 1999). These 1D models formed the basis of many subsequent investigations. A number of experimental investigations were conducted using such constant area mixing tubes to empirically derive the wall friction losses and obtain optimum geometries for attaining high pumping efficiencies (Gosline and O'Brien 1934; Mueller 1964; Hansen and Kinnavy 1965a, b; Hansen and Na 1966). Helmbold et al.'s (1954) experimental investigation brought out the advantages of jet pumps having mixing tubes with variable cross section. They showed that depending on the inlet conditions, these jet pumps having convergent–divergent mixing tubes could yield increase in pressure ranging as widely as from 60% higher to 40% lower in comparison with the pressure recovery in conventional mixing tubes having constant area. The analysis of pressure profile along the mixing tube in such variable area jet pumps calls for a two-dimensional (2D) flow model. With the advent of machine computing, an explicit change in the modeling procedure was noted. Sanger (1971) provided a FORTRAN program for designing jet pumps. Nilavalagan et al. (1988) pointed out that the 1D models cannot correctly describe the important aspect of mixing. On the other hand, the subsequent 2D models appear to be adequate for describing the mixing process and also yield realistic criteria for optimizing the geometric parameters of the ejector system. Hedges and Hill (1972) and Hill (1973) pioneered the use of such 2D models for mixing tubes with both constant and varying diameters. The flow pattern in these devices is relatively complex. In the vicinity of the tip of the primary liquid nozzle, the jet is made up of a core of the liquid phase. In this region, the original velocity distribution of the jet is modified to yield a bell-shaped distribution that is a distinguishing feature of the fully developed jet. Further downstream, the jet continues to widen and entrain the surrounding fluid. The rate of entrainment is a function of the momentum of the jet. At low entrainment rates, the entrainment process of the secondary fluid is complete by the time the rim of the jet reaches the wall of the mixing tube. Further downstream, the flow is dominated by the boundary layer on the wall of the mixing tube. At this stage, the bell-shaped profile is converted to the conventional profile in a tube. At sufficiently high rates, entrainment of the secondary fluid is complete even before the rim of the jet reaches the mixing tube wall. Beyond this point, a recirculating eddy pattern is produced in which a flow reversal occurs at the wall. In this zone, the eddies formed suck the secondary fluid from upstream positions and release it to downstream locations. The flow assumes a forward direction once it passes this position of eddy formation. The second type of flow is relevant to the operating region of jet pumps and also their design. Hill (1973) used this physical picture to obtain the pressure and velocity profiles. These were subsequently used to predict the rate of entrainment for an ejector using air as both primary and secondary fluid. The study of Nilavalagan et al. (1988) also dealt with a similar air–air ejector and focused on the mixing process in a mixing tube having constant area since major part of the energy is consumed in the mixing process and in overcoming frictional resistances. The 2D axisymmetric models used the theory of confined jets. The flow field was divided in four possible distinct regions: potential core, shear layer, eddy recirculation (in case it is present), and finally the developed flow zone. Nilavalagan et al. used an algorithm based on the Patankar–Spalding finite difference technique to obtain the velocity and pressure profiles in each case. Both Hill (1973) and Nilavalagan et al. (1988) have pointed out the importance of the Craya–Curtet number, C_t (Aloqaily et al. 2007). The Craya–Curtet number was originally introduced for the case of a single burner in limekilns. It is used to describe confined single jets that are symmetric. It is a measure of the relative importance of the primary jet momentum in determining the entrainment and flow in the mixing tube. In the case of jet pumps, the range of C_t is $0.2 < C_t < 1.5$. When C_t is approximately less than 0.8, the aforementioned recirculation is present. For the higher range of $C_t > 2$, the boundary layer at the wall is the principal deciding factor. Unfortunately, most 2D models reported so far deal with a situation in which the same fluid is used as primary and secondary fluid, for example, air, whereas in gas–liquid contacting, two different phases with widely different properties (density and viscosity) are encountered.

The advent of CFD has resulted in significant improvements. CFD has been conveniently harnessed by several investigators to understand hydrodynamics of ejectors for a variety of situations. The lack of basic theory for gas induction in ejectors has been a strong motivation for the application of CFD to investigate the hydrodynamic aspects of liquid ejectors. The major advantage of this approach is that it can quantify the effects of different geometric combinations constituting an ejector on its performance characteristics. Such approaches have been described by Neve (1993), Kandakure et al. (2005), Yadav and Patwardhan (2008), Kim et al. (2007), Utomo et al. (2008), and Li and Li (2011). The major approaches used for numerical computation of multiphase flows are (i) Euler–Lagrangian and (ii) Euler–Euler approaches. Kandakure (2006) has discussed these approaches for application to multiphase systems. A brief discussion of the same is given in the following sections.

The physical picture used in the CFD approach considers the primary fluid phase as a continuum. The time-averaged Navier–Stokes equations are then solved for this continuum. The secondary phase (dispersed phase) is treated by following it (drops/bubbles or solid particles) through the flow field determined in the previous step. Mass, momentum, and energy exchange can occur between the two phases involved.

Euler–Lagrangian Approach   In this approach, the continuous phase is treated as a continuum, and time-averaged Navier–Stokes equations are solved to obtain the flow profile in the whole domain. Effect of the dispersed phase is obtained by tracking a sufficiently large number of particles, bubbles, or droplets through the calculated flow field. Momentum, mass, and energy exchanges between the dispersed and continuous phases are allowed. The flow path of individual particles is established at specified intervals. Equation 8.16, which is essentially a force balance on the particle, is used for finding the profile of the secondary phase. A major premise in this approach is that the dispersed phase volume fraction is relatively low such as in combustion of liquid/solid fuels and spray dryers. In the present application, dispersed phase holdup can be of the order of 10–20% (Bhutada and Pangarkar 1987; Cramers et al. 1992a), and hence, this approach is not suitable, particularly in the desired range of operating conditions. This limitation is because the Euler–Lagrangian approach does not take into account complex interactions among particles of dispersed phase when present in large concentration (Fluent documentation 2006):

$$\frac{dV_S}{dt} = F_D\left(V_P - V_S\right) + g\frac{\left(\rho_S - \rho_P\right)}{\rho_P} + F_X \tag{8.16}$$

where

F_D: drag force experienced by the dispersed phase
F_X: other forces, that is, lift, virtual mass, etc.
V_S and V_P: velocities of secondary and primary phases, respectively
g: acceleration due to gravity

In view of the aforementioned restrictions imposed, this approach has not found favor with investigations on ejector-based systems for gas–liquid contacting (Kandakure et al. 2005).

Euler–Euler Approach   In the Euler–Euler approach, the different phases are treated mathematically as interpenetrating continua (Fluent documentation 2006). The contents of different phases are represented by their respective volume fractions. These volume fractions are assumed to be continuous functions of space and time and their sum is equal to unity. Conservation equations given in the following (Eqs. 8.17 and 8.18) for each phase are derived to obtain a set of equations, which have similar structure for all phases. These equations are closed by providing constitutive relations that are obtained from empirical information for drag, lift, virtual mass, etc. (Fluent documentation 2006).

The continuity equation is as follows:

$$\frac{\partial}{\partial t}\left(\varepsilon_P\rho_P\right) + \nabla\cdot\left(\varepsilon_P\rho_P\vec{V}_P\right) = \sum_{s=1}^{n}\left(m_{SP} - m_{PS}\right) + S_P \tag{8.17}$$

m_{SP}:   mass transferred from secondary to primary phase
m_{PS}:   mass transferred from primary to secondary phase
S_P: source term for mass generation or consumption for phase P

The momentum equation for z direction is:

$$\frac{\partial}{\partial t}\left(\varepsilon_P\rho_P\vec{V}_P\right) + \nabla\left(\varepsilon_P\rho_P\vec{V}_P\vec{V}_P\right) = -\varepsilon_P\nabla P + \nabla\cdot\overline{\overline{\tau}}_P + \varepsilon_P\rho_P\vec{g} + \sum_{s=1}^{n}\left(\bar{R}_{PS} + m_{SP}\vec{V}_{SP} - m_{PS}\vec{V}_{PS}\right)$$
$$+ \left(\vec{F}_P + \vec{F}_{\text{lift,P}} + \vec{F}_{\text{Vm,p}}\right) \tag{8.18}$$

For $m_{SP} > 0$, $V_{SP} = V_S$.
For $m_{PS} > 0$, $V_{PS} = V_P$.
ε_P: volume fraction of phase P
∇P: pressure drop in the system

$\bar{\bar{\tau}}_p$:   stress tensor (includes viscous as well as effects due to turbulence)

$\vec{R}_{PS}$:   interphase forces (includes forces due to friction, pressure, cohesion, etc.)

$\vec{V}_{SP}$:   slip velocity—velocity difference between secondary phase 1 and primary phase

$\vec{V}_{PS}$:   slip velocity—velocity difference between primary and other secondary phases

$\vec{F}_P$:   external body force

$\vec{F}_{Lift,P}$:   lift force

$\vec{F}_{Vm,P}$:   force due to virtual mass

$\vec{F}_{Vm,p}$:   force due to virtual mass

The Euler–Euler approach essentially has two broad classifications (FLUENT documentation 2006):

Homogeneous phase model: In this approach, all phases are clubbed into a single pseudophase. All the properties of the pseudophase are weighted averages of the corresponding property of each phase. All the phases share the same temperature, pressure, and velocity. Conservation equations are solved for this pseudophase to get the flow profile in the domain.

Separated phase flow model: In this approach (also denoted as *n*-fluid model, where *n* represents the number of phases present in the system), all the phases have their own distinct temperature, pressure, and velocity fields. Therefore, conservation equations for all the phases are solved simultaneously. This approach also takes into account the interaction among different phases through constitutive equations for drag, lift, virtual mass, etc.

The solution of detailed *n*-fluid model for complex geometry and physics is challenging even with contemporary high-power computing machines. Further, presence of a large number of constitutive relations makes this model dependent on the accuracy of these relations. Therefore, a number of assumptions must be made to simplify the *n*-fluid model depending on the complexity of the physical picture adopted.

The most popular simplified version of *n*-fluid model is the mixture model. In the mixture model, different phases share the same temperature and pressure profiles. However, the velocity field is different for each phase. In this model, conservation equations are solved for the mixture, and the velocity difference between phases is handled through the concept of slip velocity, which is the difference between the velocities of two phases. In the commercially available software FLUENT, this slip velocity is found through the algebraic formulation proposed by Manninen et al. (1996). The basic assumption of the algebraic slip mixture model is that a local equilibrium between the phases should be reached over short spatial length scale:

$$\vec{V}_{SP} = \frac{\tau_S}{f_{Drag}} \frac{(\rho_S - \rho_M)}{\rho_S} \vec{a} \tag{8.19}$$

τ_S: secondary phase relaxation time $= \dfrac{\rho_S d_S^2}{18\mu_P}$

f_{Drag}: drag force experienced by secondary phase

ρ_M: density of the mixture

$\vec{a}$: acceleration of the particle that is calculated from the mixture flow field

Kandakure et al. (2005) used FLUENT 6.1 to solve the governing equations for the mixture model. The model used for quantitative description of the turbulent flow was the standard κ–ε model. The built-in slip velocity of Manninen et al. (1996) used in FLUENT is relevant to flows generated by forces arising out of buoyancy of the dispersed phase such as in bubble columns and tray columns. In the case of cocurrent downflow venturi loop system, the flow is driven by a very high kinetic energy liquid and is in a direction opposite to the buoyancy of the secondary (gas) phase. To overcome this limitation, Kandakure et al. (2005) assigned the slip velocity a value that is some fraction of the primary fluid velocity in the axial direction. This procedure was validated by using different values of the slip velocity and comparing the entrainment rates obtained from the mixture model with the experimental data of Bhutada and Pangarkar (1987). The flow simulation covered the same range of L_{MT}/D_T from 0 to 8, as used in the experimental study of Bhutada and Pangarkar. This method is analogous to fitting of the model predictions to experimental data to evaluate the actual slip and is similar to those described in Section 8.6.1.1. Figure 8.12 reproduced from

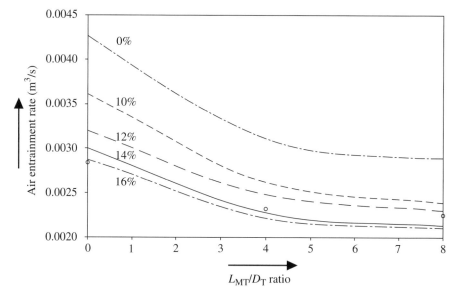

FIGURE 8.12 Comparison of entrainment rate predicted from computational fluid dynamics modeling with the data of Bhutada and Pangarkar (1987) **o**: Experimental data of Bhutada and Pangarkar (1987). (Reproduced from Kandakure et al., 2005 with permission from Elsevier. Copyright © 2005 Elsevier Limited. All rights reserved.)

Kandakure et al. (2005) compares the calculated entrainment rates at different fractions of slip velocity with the experimental data of Bhutada and Pangarkar.

Kandakure et al. varied the slip from 0 to 16%. At 0% slip, the primary and secondary fluid velocities are equal. This implies very high degree of transfer of momentum from primary to secondary fluid such that both phases have identical velocity. Consequently, a higher secondary fluid velocity results in higher entrainment rates. This case of slip velocity = 0 yielded entrainment rates that were higher by 40–50% than the experimental values. The difference is relatively very wide and leads to the conclusion that the phenomenon of complete mixing suggested by Witte (1965, 1969) is strictly not valid. This problem was realized as early as 1942 by Goff and Coogan who defined a mixing efficiency parameter and similarly by Radhakrishnan et al. (1986) and Neve (1991). The "slip" could not be ignored, particularly at high secondary to primary fluid flow ratios. With finite momentum transfer from primary to the secondary fluid (finite slip), the secondary fluid velocity/entrainment rates decrease. Eventually, it was found that a slip between 12 and 14% resulted in good agreement of model predictions with the experimental data of Bhutada and Pangarkar (1987). For all other simulations, Kandakure et al. chose the same slip equal to 13% of the local primary fluid velocity. Diffusers A, B, and C of Bhutada and Pangarkar were considered for model predictions. These diffusers have substantially different geometries, and therefore, the applicability of a unique value of slip (13%) for all the three diffusers is perplexing. The different mixing tube lengths used in the three diffusers are bound to yield different extents of momentum transfer, and therefore, fixing the slip arbitrarily at some percentage of the local primary fluid velocity because of agreement with experimental data cannot be justified. The fact that Kandakure et al.'s prediction are not universal can be seen from their observation that "when the air flow rate was plotted against the nozzle pressure drop or the pressure drop between the air inlet and the throat entry, all the data could not be collapsed onto a single curve. This could also be a reason for the differences in the exponents in the correlations of previous authors...." It is evident that a pseudo-unique empirical parameter similar to the 13% slip used by Kandakure et al. could not be obtained in cases involving correlations based on dimensionless groups. In these dimensionless groups-based correlations, different empirical constants were required for different diffusers. Considering this fact and the riddle of "13% slip velocity satisfying all geometries" but subsequently not able to provide a unique correlation for all diffusers indicates that the CFD approach can yield a qualitative picture of the pressure and velocity profiles. However, in this complex case of a gas–liquid ejector, it falls short of providing quantitative predictions. If such quantitative predictions are to be made, then probably a "geometry-related parameter" (e.g., different slip velocities for different geometries) needs to be defined. Understandably, this means fitting the CFD predictions to experimental data for individual diffusers. This is a classic case of a multiphase system that eludes fundamental analysis, which can yield *a priori* predictions. In such cases, it is very tempting to resort to simple correlations involving dimensionless groups obtained from regression of available experimental data as the best equivalent alternative. In the latter case, it is clearly understood that the correlation validity is for the range of geometries and operating conditions covered in the experimental work. This is similar to the inclusion of a "geometry-related parameter" suggested previously for quantitative predictions from CFD.

Yadav and Patwardhan (2008) and Yuan et al. (2011) also used a similar approach. Yadav and Patwardhan (2008) used the same percentage of slip velocity as Kandakure et al. for their simulation of vertical downflow air–water ejector. Yuan et al. (2011), who studied the horizontal mode, found that a 12% slip gave good agreement with experimental data, which is again almost the same as Kandakure et al. and Yadav and Patwardhan (2008). This is another perplexing observation. Kim et al. (2007) used a horizontal air–water ejector as a distributor at the bottom of a bubble column. These investigators, however, used the formulation of Manninen et al. (1996) for the slip velocity. Their model predictions also were shown to agree with the experimental data of Bhutada and Pangarkar (1987). One striking observation of Kim et al. is the formation of a vortex just before the two-phase flow enters the diffuser. This behavior was also noted as early as 1973 (Hill 1973). It has an important bearing on the rate of entrainment since any energy loss in the eddies must reflect on the rate of entrainment. In the following sections, the results of CFD simulations of Yadav and Patwardhan (2008) and their comparisons with the data of Bhutada and Pangarkar (1987) are briefly discussed.

Effect of Geometric Parameters

1. The effect of the free area ratio, A_{SN}, $\left(D_S^2 - D_N^2 / D_N^2 \right)$, was understandably found to be related to the projection ratio (PR). For low values of the area ratio, the entrainment rate was relatively low, and it increased with increase in the area ratio to attain a maximum value. The maximum values in entrainment rates were attained at A_{SN} of 8 for PR $= 2.5$ and 6.6 for PR $= 5$ and 14.5 (Yadav and Patwardhan 2008). Above the corresponding optimum area ratio, a marginal decrease in the entrainment rate was observed. For free area ratios $A_{SN} > 16.4$, there was no change in the entrainment rate. These findings are consistent with the fact that a smaller suction chamber size can result in pressure losses due to radial flows and eddy circulation. Bhutada (1989) has also indicated that the secondary fluid inlet and suction chamber should have a sufficiently large area.

2. The simulations of Yadav and Patwardhan yielded an optimum value of $L_{NT}/D_T = 5$. The velocity profiles for low values of L_{NT}/D_T indicate the existence of radial velocity components that cause recirculation and hence pressure loss of the secondary fluid. Figure 8.13 reproduced from Kim et al. clearly shows the circulatory flows.

3. Simulations for different convergence angles (2.5–90°) showed that the entrainment rate increased with an increase in the convergence angle and exhibited a maximum at 10°. This observation is also related to the formation of circulatory flows (eddies) resulting in drag and consequent lower pressure driving force for the secondary fluid.

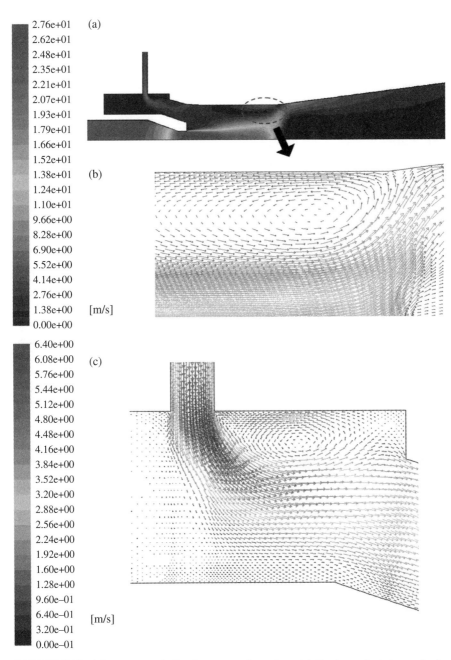

FIGURE 8.13 Flow patterns indicating vortex formation. (a) Contour of mixture velocity magnitude, (b) mixture velocity vector of mixing chamber, and (c) air velocity vector of suction section. The geometric parameters used were as follows: D_N, 3×10^{-3} m; L_{MT}, 9×10^{-3} m, $(L_{MT}/D_T) = 3$; L_{Diff}, 30×10^{-3} m; static head over ejector, 0.618 m water column; and Q_P, 1.8×10^{-4} m^3/s (Reproduced from Kim et al. (2007) with permission from Elsevier. Copyright © 2007 Elsevier Ltd. All rights reserved.)

The CFD analysis of Yadav and Patwardhan yielded the following optimum suction chamber proportions: free area ratio, $\left(D_S^2 - D_N^2 / D_N^2 \right) = 6.6$; PR, $L_{NT}/D_T = 5$; and a convergence angle between $5°$ and $15°$. The radial component practically vanishes for $L_{NT}/D_T = 5$, and hence, any further increase in L_{NT}/D_T does not lead to improvement in entrainment rates.

8.7   DESIGN OF VENTURI LOOP REACTOR

The ejector is at the heart of the venturi loop system. Therefore, a discussion on the design of the latter needs to include the information available on design of ejectors, the related hydrodynamics, and mass transfer parameters. This section reviews the same and makes recommendations regarding configuration of the venturi section in relation to the entire system.

There is considerable information available in the literature on the design of ejectors (steam jet ejectors, water jet pumps, air injectors, etc.) supported by extensive experimental data. Most of this information deals with its use as an evacuator and the focus is on ejector optimization for maximizing the gas pumping efficiency. The major advantage of the venturi loop reactor is its relatively very high mass transfer coefficient due to the excellent gas–liquid contact achieved in the ejector section. Therefore, the ejector section needs careful consideration to achieve this aim. The major mass transfer parameter is the volumetric liquid side mass transfer coefficient, $k_L a$. The variables that decide $k_L a$ are (i) the effective gas–liquid interfacial area, a, that is related to the gas holdup, ε_G. The gas induction rate and the shear field generated in the ejector determine the value of ε_G and, consequently, the value of a. (ii) the true liquid side mass transfer coefficient, k_L. The mass ratio of the secondary to primary fluid in turn decides both k_L and $\underline{a}$. For the venturi loop reactor the volumetric induction efficiency parameter η_{IP}, Q_G/P_{Noz} is more relevant. This definition has a built in energy efficiency since it is defined in terms of the power input through the primary nozzle required to induce a given volume of the gas under a specified "dip" or back pressure. In the following sections, the reported design information on the various constituents of an ejector that affect Q_G and $k_L a$ is reviewed. The focus will be on arriving at the most energy efficient ejector design from a mass transfer perspective. Besides the effects of design and operating parameters on the mass transfer factors, the effects of system physical properties will also be discussed.

8.7.1   Mass Ratio of Secondary to Primary Fluid

Several investigators have studied this aspect. A variety of geometric parameters and flow directions (horizontal, upflow, and downflow) have been investigated. Table 8.3 summarizes results of these investigations for the downflow configuration that is relevant to the present application.

Table 8.4 gives a summary of the literature on volumetric efficiency parameter in terms of the total power input, η_{IP}, as (Q_G/P_{Noz}). Also included is the correlation derived from the data of Panchal et al. (1991).

TABLE 8.4 Correlations for induction Efficiency Parameter Available in the Literature for Gas–liquid Ejectors in the Downflow Mode

Investigator (s)	System used	Range of physical properties covered	Geometrical parameters covered	Correlation proposed
Bhutada and Pangarkar (1987)	Air–water, air–aqueous carboxymethyl cellulose	$5.3 \times 10^{-2} < \sigma_p < 7.3 \times 10^{-2}$ N/m; $\tau = k \times (\gamma)^n$ $9.8 \times 10^{-3} < k <$ 8.2×10^{-1} Pa (s)^n; $0.58 < n < 0.98$; $\gamma = 5000 \times V_G$	Single-orifice nozzles of $D_N = 0.005$, 0.008, 0.01, and 0.012 m. Several diffusers having bell-shaped, conical, and straight entry regions with diameters varying from 0.016 to 0.05 m, and different mixing tube diameters and lengths were used. Diffuser combinations denoted as A, B, C, D, and E. Holding vessel diameter, $T_{HV} = 0.3$ m and $H/T_{HV} = 3.33$	[a]Diffuser A (8 mm dia. nozzle): $\eta_{IP} = 4.63 \times 10^{-3} P_{Noz}^{-0.47}$ $(A_R = 4)$ [a]Diffuser B (8 mm dia. nozzle): $\eta_{IP} = 4.53 \times 10^{-3} P_{Noz}^{-0.57}$ $(A_R = 4)$ [a]Diffuser C (8 and 10 mm dia. nozzles) $\eta_{IP} = 4.06 \times 10^{-3} P_{Noz}^{-0.70}$ $(A_R = 4$ and 2.25) [a]Diffuser C (5 mm dia. nozzle) : $\eta_{IP} = 2.13 \times 10^{-3} P_{Noz}^{-0.66}$ $(A_R = 10.25)$ [a]Diffuser E (8 and 10 mm dia. nozzles): $\eta_{IP} = 5.49 \times 10^{-3} P_{Noz}^{-0.58}$ [a]Diffuser E (5 mm dia. nozzle): $\eta_{IP} = 6.31 \times 10^{-3} P_{Noz}^{-0.1}$
Panchal et al. (1991)	Air–water	No variation	Diffusers E and C of Bhutada and Pangarkar (1987) with multiorifice nozzles. Each orifice of 0.005 m diameter. Orifice diameter varied from 0.003 to 0.006 m. Pitch: $1 \times D_H$ to $2 \times D_H$. Other details the same as Bhutada and Pangarkar (1987)	[a]Diffuser C (multiorifice nozzle: 4 Nos. 5 mm diameter holes on pitch of 7.5 mm; pitch $= 1.5 \times D_H$): $\eta_{IP} = 4.94 \times 10^{-3} P_{Noz}^{-0.64}$ [a]Diffuser C (multiorifice nozzle: 4 Nos. 5 mm diameter holes on pitch of 10 mm; pitch $= 2 \times D_H$): $\eta_{IP} = 7.1 \times 10^{-3} P_{Noz}^{-0.53}$

[a]From regression of published data.

The ejector characteristics are determined by the following three important geometric parameters: (i) the PR defined as L_{NT}/D_T, (ii) suction chamber free area ratio $\left(D_S^2 - D_N^2 / D_N^2 \right)$, and (iii) the angle of convergence from the end of the straight part of the suction chamber to the point of entry in the diffuser. The effects of the various parts of the ejector on the mass ratio and induction efficiency parameter are discussed in the following.

8.7.1.1 *Effect of Primary Fluid Nozzle Design*

The primary nozzle, which delivers the slurry to the venturi chamber, has an important role in any ejector device. The rate at which the gas is aspirated depends on the flow rate of the primary liquid and its pressure at the nozzle outlet. Different investigators have used varied designs of these nozzles that included sharp-edged nozzles, nozzles with swirl pieces, and multi-jet nozzles (Bhutada 1989; Cramers et al. 1992a, 1993; Panchal et al. 1991; Cramers et al. 2001; Baier 2001; Yadav and Patwardhan 2008). Optimization of the nozzle geometry with respect to the suction chamber, dimensions of the diffuser, etc. is important from an overall high induction and mass transfer efficiency viewpoint.

Bonnington (1964) suggested that to prevent energy loss, the primary fluid nozzle should have a smooth decrease in its cross section accomplished through appropriate tapering. Henzler (1983) suggested a nozzle having 20° convergence angle and aspect ratio (ratio of the length of the tapered portion of the nozzle to the pipe diameter as shown in Fig. 8.14) of 0.5 for liquids in a liquid jet ejector. Burgess and Molloy (1973) studied the stability of the jet for the case of plunging jet reactors. They found that the nozzle aspect ratio has an important bearing on the gas induction rate. Increase in the aspect ratio increases the roughness of the jet surface. An aspect ratio of zero gave the smoothest jet. The study of Burgess and Molloy (1973) showed that smooth jets gave lower induction rates but higher effective interfacial area as compared to rough jets. Variation of the aspect ratio is a convenient method for getting the desired surface roughness. Other important nozzle parameters are (Grandchamp et al. 2012) the angle of contraction and the contraction ratio (Fig. 8.14).

Depending on the liquid viscosity, the angle of convergence can vary from 15° to 100°. Davies and Jackson (1985) in their study of turbulent jets for cutting and mining work made a potential flow analysis that brought out the importance of the boundary layer and pressure drop gradient across the nozzle. The algorithm provided is a convenient means for selecting a nozzle geometry that produces the thinnest boundary layers with least separation near the nozzle exit. Davies and Jackson suggested that the acceleration created in the nozzle should be gradual, culminating into a high pressure gradient at the exit. The final contraction piece suggested by Davies and Jackson has an elliptical shape that is significantly different from that shown in Figure 8.14.

Panchal (1990) studied gas induction characteristics of multi-jet nozzles with smooth and rough edges. It was found that depending on the power input and ejector design, nozzles with rough edges gave 10–20% higher gas induction rates. Assuming

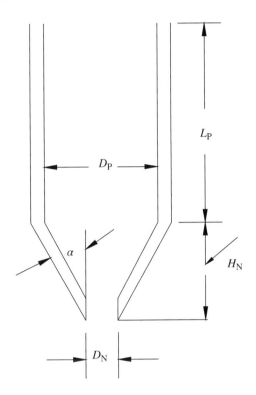

$$\text{Angle of contraction, } \alpha; \text{ contraction ratio,} \left(\frac{D_\text{P}}{D_\text{N}}\right)^2; \text{ aspect ratio,} \left(\frac{H_\text{N}}{D_\text{P}}\right)$$

FIGURE 8.14   Major geometric parameters for the primary fluid nozzle.

that the conventional definition applies to the individual orifices, these nozzles had relatively very low aspect ratios (0.06–0.13). Panchal et al. (1991) also showed that optimized multiorifice nozzles gave 20–30% higher gas induction rates as compared to a single-orifice nozzle having the same cross-sectional area under otherwise identical conditions. Multiorifice nozzles occlude the entrained gas between the jets. However, when the orifice spacing is too low, the individual jets merge, whereas for too large a spacing, the enhancement was reduced. It was found that an orifice pitch of $2 \times D_\text{H}$ gave the highest enhancement in gas induction rate. The improved enhancement, however, must be viewed in the context of decrease in the mass transfer coefficient as discussed later. Henzler (1981), Cramers et al. (1993), Havelka et al. (1997), Baier (2001), and Elgozali et al. (2002) found that introduction of a deliberate disturbance in the nozzle by means of a swirl pieces resulted in increase in the rate of gas induction. Evidently, besides the jet momentum, the nature of the jet surface also affects the gas induction rates. The last two studies showed that the introduction of a swirl device stabilized the location of the mixing shock zone, whereas in its absence, the location of this zone was a strong function of the nozzle velocity. Havelka et al. found that an optimized combination of nozzle and swirl piece yielded gas induction rates that were 50–80% higher than comparable nozzle geometry but without a swirl piece. Elgozali et al. (2002) reported a 200% enhancement in the induction rate for a nozzle using a swirl piece. Baier (2001) carried out a systematic study of the effect of a swirl device on various aspects of an ejector. The swirl piece converts the axial velocity into a tangential component. In this case, the jet spreads faster and its momentum is distributed over a wider cross-sectional area. It is also evident that the conversion from axial to tangential component consumes a certain part of the jet momentum. The swirl body number is defined as the ratio of the tangential and axial momentum components (Duquenne et al. 1993). Havelka et al. (1997) studied the gas induction characteristics of nozzles containing different types of swirl pieces. They varied the swirl body number, S_{WB}, from 0 (no swirl piece) to ~0.3. Three distinct zones were noted (Figure 10 of Havelka et al. 1997): (1) for $0.06 < S_{WB} < 0.12$, there was a sharp increase in the gas induction rate; (2) for $0.12 < S_{WB} < 0.2$, induction rate remained comparatively constant; and (3) for $S_{WB} > 0.2$, there was a decline in the induction rate. At very low S_{WB}, the jet behaved similar to a single jet without swirl piece, whereas at very high S_{WB}, energy dissipation due to jet rotation on the throat surface was too high to yield the required driving force for gas suction. It was found that at equal S_{WB} the induction rates were similar. Both Havelka et al. (1997) and Baier (2001), however, indicated that S_{WB} values calculated from swirl piece geometry (as assumed by Havelka et al. 1997) could be significantly different from the actual value obtained from measurements of fluid velocities at the exit of the nozzle.

Zahradnik and Rylek (1991) suggested that the pressure drop across the nozzle, ΔP_N, should be in the range of 0.5–0.8 MPa and not more than 1 MPa in any case. This allows the calculation of the liquid velocity through the nozzle from the following equation:

$$\Delta P_N = \zeta U_N^2 \rho_L / 2 \qquad (8.20)$$

where ζ is the nozzle discharge coefficient. For $0.5 < D_N/D_P < 0.8$, Zahradnik and Rylek suggested that the nozzle discharge coefficient $\zeta \sim 1$ for the above range. The corresponding allowable liquid velocity at the exit of the nozzle for the range of ΔP_N 0.5 to 0.8 MPa is 20–50 m/s. Bhutada (1989) and Dirix and van der Wiele, however, found that for the nozzles used in their study, ΔP_N values as low 0.3 and 0.2 MPa gave smooth, coherent jets.

8.7.1.2 Effect of the PR: L_{NT}/D_T   Several investigators have studied the effect of PR on the gas induction rate for different configurations. Table 8.5 gives a summary of the available literature dealing with the effect of PR.

The optimum values of PR reported in Table 8.5 must be treated with prudence since the same are related to the geometry of the entry section. Yadav and Patwardhan (2008) carried out a CFD study for a circular suction chamber followed by a conical section. The model calculations showed that for low PR the secondary fluid develops

TABLE 8.5   Studies On Effect of PR and Optimum PR

Reference	Range of PR covered	Optimum PR reported
Kroll (1947)	0.5–5	Not reported
Davies et al. (1967a)	Not reported	1.9
Sanger (1969)	0–3	1.0
Bhat et al. (1972)	9	Not reported
Cunningham and Dopkin (1974)	1–7	3
Acharjee et al. (1975)		2
Biswas et al. (1975)	Not reported	1.9
Henzler (1983)	Literature data reviewed	$0.4 < PR < 0.9$
Bhutada and Pangarkar (1987)	1.5–4	No appreciable change in gas induction rate over the range of PR covered. Recommended PR = 3
Huang et al. (1999), Rusly (2005)		1.5
Hammoud (2006)	1.25, 1.5, 1.75 (Water–water pump)	Upflow: optimum PR = 1.25 for maximum jet pump efficiency at $\Delta P_N = 0.15\,MPa$ Downflow: optimum PR = 1 for maximum suction to delivery pressure ratio of 0.93 at $\Delta P_N = 0.15\,MPa$

a significant radial velocity component, which leads to pressure loss. Consequently, the gas induction rate is lower. With increase in PR, the radial component decreases, and finally at PR > 5, it is negligible and a constant gas induction rate is obtained. The logic behind the definition of PR as L_{NT}/D_T is difficult to explain since it is not a dimensionless quantity that relates one linear parameter to another in the same direction. On the other hand, it may be argued that the absolute value of L_{NT} is more important than PR because the former decides the position of the nozzle tip with respect to the throat at which it starts mixing with the secondary fluid. For a given nozzle velocity, L_{NT} also represents the contact time between the two fluids in the suction chamber. This may affect the rate of entrainment of the secondary fluid. L_{NT} also decides the extent of expansion/coherence of the jet before it enters the mixing throat/tube. For $L_{NT} = 0$, the suction zone is extremely limited. With increasing values of L_{NT}, the free suction zone area increases, and further, jet expansion as well as it roughness adds to the entrainment process. However, for too large a value of l_{NT}, breakup of the jet/loss of momentum before it enters the mixing tube may have detrimental result. Henzler (1983) has indicated that a premature breakup of the jet can generate radial flow and consequent pressure loss. This was also concluded by Yadav and Patwardhan (2008) from their CFD analysis. Although the definition of PR based on the throat diameter is not appropriate, most authors have used the same and reported optimum values for the conditions used. Donald and Singer (1959) studied the entrainment of fluids into a tube by a jet of the same fluid. They measured the angle of divergence of the jet for different nozzle sizes and fluids. Based on this measured angle of divergence and the distance between the jet and diffuser tube, they developed equations for predicting entrainment rates. Although this approach has not found much application, it underlines the importance of the absolute distance between the nozzle exit and the mixing tube.

8.7.1.3 Suction Chamber Geometry

The first interaction between the primary and secondary fluids takes place in the suction chamber. Unfortunately, most experimental investigations have not given due importance to this important part of the ejector. Qualitatively, several investigators (Mellanby 1928; Watson 1933 *cf.* Kroll 1947; Bonnington 1964; Zahradnik et al. 1982; Henzler 1983; Bhutada 1989) have indicated that the suction chamber should be large enough with respect to the nozzle diameter. This is necessary to prevent undesired pressure losses due to stagnation/circulation of the secondary fluid. In a venturi loop reactor, the secondary fluid usually enters horizontally. It subsequently changes direction to almost vertically downward as it flows across the suction chamber of a standard venturi loop system. Any loss of pressure during this change of direction should result in lower entrainment rates. It has been suggested that the suction chamber design should be such that it prevents sudden cross-sectional changes for the secondary fluid. Vogel (1956) suggested that elliptical suction chambers can meet this requirement. Yadav and Patwardhan (2008) used CFD to study the flow behavior in the suction chamber. They varied the geometry of the suction chamber covering a range of PR and the dimensionless suction chamber free area ratio, A_{SN}, $\left(D_S^2 - D_N^2 / D_N^2 \right)$. Although not mentioned, the simulation was done for PR = 2.5, 5, and 14.5 by changing the length between the nozzle tip and throat, L_{NT}, for a constant nozzle diameter $D_N = 7.91 \times 10^{-3}$. The angle of converging section with respect to the vertical was 10°. Their model calculations showed three broad regions of variation of the induction rate with A_{SN}: (i) beginning with very low values of A_{SN} and for a given value of PR, the induction rate showed a rapid initial increase and reached a maximum value at a certain A_{SN}; (ii) beyond this value of A_{SN}, the induction rate decreased but not as sharply as in case (i); and (iii) for $A_{SN} > 16.4$, there was no further change in the induction rate. The maximum in the induction rate was obtained at $A_{SN} = 8$ for PR of 2.5. The optimum A_{SN} for the other PR (PR = 5 and 14.5) studied was 6.6. Further, the variation in induction rates was identical for PR values of 5 and 14.5. For the specific geometry studied, the absolute free area in the suction chamber increased with increase in A_{SN} for PR = 5 and 14.5. For PR = 2.5 and $A_{SN} \geq 10.8$, the tip of the nozzle was inside the converging section of the suction chamber. Therefore, for this situation, the absolute free area in the suction chamber remained constant despite increase in A_{SN}. This observation further supports the importance of the absolute value of L_{NT} mentioned in Section 8.7.1.1. The simulations showed that for low A_{SN} the axial velocities were relatively high, resulting in loss of pressure. With increasing A_{SN}, the axial component decreases. Consequently, it results in a lower pressure loss and higher induction rates. This corresponds to the broad region (i) mentioned previously. In region (ii), a radial flow was observed in a substantial part of the suction chamber, which caused reduction in the induction rate. Region (iii) corresponded to a situation ($A_{SN} \geq 16.4$) in which both the radial and axial components were relatively low. Thus, there was no further increase in the induction rate.

8.7.1.4 Design of Entry Region to the Diffuser
Various geometries of the entry sections to the diffuser have been reported in the literature. However, no specific design criteria have been evolved. Basic principles of fluid mechanics suggest that the change in the direction of the secondary fluid from originally horizontal to vertically downward should be as smooth as possible. Sharp contours can cause radial flow. The resultant boundary layer separation introduces significant pressure loss and is deleterious in this regard. Therefore, a parallel coaxial entry of the two streams is suggested (Kroll 1948). Henzler (1983) suggested a gradually converging bell-shaped entry to the mixing tube prior to the diffuser. Such a bell-shaped contour was used by Bhutada and Pangarkar (1987). However, a conical entry shape was also found to be equally good in their study. The same authors suggest an optimum value of 40 for the ratio of diffuser entry area (shown in Figure 8.11) to nozzle cross-sectional area A_{EN}.

8.7.1.5 Throat Area to Nozzle Area (A_R) Ratio
Kastner and Spooner (1950), Engel (1963), Kar and Reddy (1968), Henzler (1983), and Bhutada and Pangarkar (1987) have investigated ejector performance for varying A_R values (Figs. 8.4, 8.5, 8.6 and 8.11). There is general agreement in all these studies that the optimum value of $A_R = 4$. The CFD analysis of Kandakure et al. (2005) showed that for low values of A_R, there is recirculation of the entrained gas in the suction chamber. This phenomenon is also observed in spray towers for gas absorption, wherein the relatively high nozzle momentum of the sprayed liquid causes severe gas recirculation. The recirculation is severe to the extent that the gas phase in this application is assumed to be completely back mixed (Section 3.3.4). In the present case, a high nozzle diameter further causes a reduction in the annular area available for flow of the gas leading to recirculation in the conical portion of the suction chamber. The overall effect is a reduction in the entrainment rate. It is easy to visualize choking of the mixing tube/throat when the nozzle diameter is nearly the same as the former. For higher A_R (smaller nozzle diameters/lower nozzle momentum), as expected, the simulation results showed a relatively mild recirculation restricted mainly to the area near the gas inlet. These CFD simulations agreed with the optimum value of $A_R = 4$ obtained in the experimental study of Bhutada and Pangarkar (1987). This concurrence could also be because the basis of the simulations itself was on matching the predictions with the experimental data of Bhutada and Pangarkar. Notwithstanding this empiricism in Kandakure et al. (2005) (and also Yadav and Patwardhan, 2008), there is overwhelming evidence in favor of an optimum $A_R = 4$ (Henzler 1983).

For a venturi loop reactor as a gas–liquid contactor, the important parameter is the gas holdup, ε_G. Bhutada and Pangarkar (1987) have shown that ε_G is independent of A_R and is solely decided by the value of $[P_{Noz}/V_L]$ for a given type of diffuser. The

CFD simulations of Kandakure et al. (2005) also showed that for $A_R \geq 4$ there is very little change in the major ejector performance parameters.

8.7.1.6 Effect of Mixing Tube Length The function of the mixing tube is to transfer the momentum from the primary to the secondary fluid. This process results in consumption of energy and buildup of pressure in the mixing tube. Kandakure et al. (2005) calculated the pressure at the exit of the mixing tube for $L_{MT}/D_T = 0$, 4, and 8 (Fig. 8.11). Their results showed a steady increase in the pressure at the exit of the mixing tube with increasing L_{MT}/D_T. The increasing pressure obviously causes a decrease in the driving force for gas induction. Bhutada and Pangarkar (1987) studied the effect of the mixing tube length on the rate of entrainment. Expectedly, they found that the highest entrainment rate was obtained at $L_{MT}/D_T = 0$. For nozzles with swirl pieces, Havelka et al. (1997) also found that the gas induction rate was maximum for a mixing tube length = 0. Neve's (1993) CFD analysis indicated that the optimum value of L_{MT}/D_T is in the range of 3.

8.7.1.7 Effect of Pressure Recovery Diverging Section The pressure recovery diverging section succeeds the mixing tube and is necessary for reconverting the kinetic energy into pressure energy by gradual deceleration. Thus, the fluid leaving the diffuser has a higher pressure than at the inlet to the suction chamber. The self-aspirating nature of the venturi loop system is derived from this phenomenon, which automatically recycles the relatively high-pressure gas coming out from the diffuser back into the lower-pressure zone at the suction inlet. Bhutada and Pangarkar measured entrainment rates for different diffusers with and without a pressure recovery diverging section. Their results clearly brought out the beneficial effect of the pressure recovery section on Q_G. Q_G was higher by ~20% in the presence of the diverging section under otherwise identical conditions. It is known that diverging flows have a tendency to create recirculation patterns arising out of boundary layer separation when there is a sudden increase in the flow area. Such separation results in eddy formation and loss of momentum. A relatively small angle of divergence should suppress boundary layer separation. Kroll (1948), Engel (1963), Cunningham (1974), and Henzler (1983) have suggested an angle of 5°. Kar and Reddy (1968) suggest that this angle can be between 2.5° and 7.5°. Higher divergence angles were found to be detrimental. Henzler (1983) recommended that the length of the diverging cone should be such that the pressure recovery exceeds the frictional loss suffered in the mixing tube. He suggested an optimum length $= 8 \times D_T$. Henzler further suggested that the addition of a straight cylindrical pipe of diameter equal to that at the exit of the diverging cone and length of $3 < L_{DP}/D_D < 6$ is beneficial for further pressure recovery. It must be pointed out that in the optimum design of an ejector for the venturi loop reactor, high-pressure recovery is not a major requirement. Nonetheless, besides the increase in Q_G provided by the diverging cone, the pressure recovery also helps in facile recycle of the unreacted gas.

8.7.1.8 Effect of Physical Properties of the Primary Fluid Most investigators have reported relatively mild negative dependence of Q_G on the viscosity and surface tension of the primary fluid. Cunningham (1957) and Henzler (1983) have reported a decrease in Q_G with decrease in the primary fluid Reynolds number caused by higher viscosity. According to Henzler, the effect is negligible for $Re_T > 2 \times 10^4$. Ben Brahim et al. (1984) covered a very wide range of viscosities and surface tension (Table 8.3). The effects of these physical properties were lumped into Morton number for the primary fluid, $\left(g\mu_P^4 / \rho_P \sigma_P^3 \right)$. The exponent on the Morton number varied from -0.01 (Ben Brahim et al.) to -0.234 (Mandal et al. 2005b). The higher negative exponent in Mandal et al.'s (2005b) correlation could also be due to inclusion of μ_P and ρ_P in another dimensionless number $(\rho_P T_{HV} V_P/\mu_P)$ not considered by Ben Brahim et al.

8.7.1.9 Effect of the Presence of a Draft Tube

Bhutada and Pangarkar (1987) and Panchal et al. (1991) measured rates of gas entrainment with and without draft tubes of different diameters with respect to the holding vessel. They did not find any effect of the draft tube on Q_G.

8.7.2 Gas Holdup

A summary of the literature investigations on gas holdup in venturi loop systems and their results is given Table 8.6. It is evident that although gases other than air have been used, there is a singular lack of studies for organic liquids. Since a majority of the applications of the venturi loop reactor are in reactions involving organic compounds (Table 8.1), there is an urgent need for such systems.

Besides factors that affect the induction rate, the other major factors that affect the gas holdup in venturi loop system are the interaction between the liquid jet and induced gas in the mixing tube and the diverging cone. The effects of the geometries of the individual components of a venturi loop system are discussed in the following.

8.7.2.1 Effect of the Primary Fluid Nozzle Design

Effects of several types of primary nozzles on the gas holdup have been studied in the literature. The variables investigated are (i) nozzle aspect and contraction ratio, (ii) roughness of the orifice, (iii) nozzles with swirl piece, and (iv) multiorifice nozzles. For nozzle geometry variables (i)–(iii), it was found that although there was an increase in Q_G, the gas holdup ε_G showed the opposite trend. The only exception to this was the case of multiorifice nozzles.

As discussed earlier, a higher aspect ratio yields a jet having a rough surface. The latter yields higher rate of induction due to the occlusion of induced gas in surface voids. However, this same property causes the jet to break up more easily. For jets with greater roughness and higher L_{NT}, the jet may break before it enters the mixing tube/throat. In such a case, most of the momentum in the jet is lost by impact with boundary walls. Therefore, the shear force available for gas dispersion in the mixing tube is lower. As a consequence, any feature that causes jet roughness and eventual breakup of the jet yields lower gas holdup. Baier (2001) carried out a systematic study of the effect of swirl piece design on ε_G. He could not discern any clear trend on the effect of presence of a swirl piece. For multiorifice nozzles, Bhutada (1989)

TABLE 8.6 Correlations For Gas Holdup in Gas–liquid Ejectors in the Downflow Mode Available in the Literature

Investigator (s)	System used	Range of physical properties covered	Geometrical parameters covered	Correlation proposed	Comments
Bhutada and Pangarkar (1987)	Air–water; air–aqueous carboxymethyl cellulose	$5.3 \times 10^{-2} < \sigma_P < 7.3 \times 10^{-2}$ N/m; $\tau = k \times (\gamma)^n$ 9.8×10^{-3}; $0.58 < n < 0.98$; $\gamma = 5000 \times V_G$	Single-orifice nozzles of $D_N = 0.005, 0.008, 0.01,$ and 0.012 m. Several diffusers having bell-shaped, conical, and straight entry regions with diameters varying from 0.016 to 0.05 m and different mixing tube diameters and lengths were used. Diffuser combinations denoted as A, B, C, D, and E. Holding vessel diameter $T_{HV} = 0.3$ m and $H/T_{HV} = 3.33.$	Diffuser: $A: \varepsilon_G = 2.16 \times 10^{-2} \left(\dfrac{P_{Noz}}{V_L} \right)^{0.77}$ $B: \varepsilon_G = 2.0 \times 10^{-2} \left(\dfrac{P_{Noz}}{V_L} \right)^{0.88}$ $C: \varepsilon_G = 3.2 \times 10^{-2} \left(\dfrac{P_{Noz}}{V_L} \right)^{0.72}$ $D: \varepsilon_G = 2.8 \times 10^{-2} \left(\dfrac{P_{Noz}}{V_L} \right)^{0.72}$ $E: \varepsilon_G = 3.0 \times 10^{-2} \left(\dfrac{P_{Noz}}{V_L} \right)^{0.73}$	Negligible effect of viscosity and surface tension.
Bhutada and Pangarkar (1988)	Air–water/ Newtonian carboxymethyl cellulose. Glass beads (150, 250, and 450 μm diameter; solid loading 0.5 to 5 wt %).	$0.85 \times 10^{-4} < \mu_P < 3.5 \times 10^{-4}$ Pa s; $\sigma_P = 7.2 \times 10^{-2}$	Diffusers C and E (Table 8.3)	Diffuser: $C: \varepsilon_G = 3.2 \times 10^{-2} \left(\dfrac{P_{Noz}}{V_L} \right)^{0.72}$ $E: \varepsilon_G = 3.0 \times 10^{-2} \left(\dfrac{P_{Noz}}{V_L} \right)^{0.73}$	No change in ε_G due to presence of solids.

Reference	System	Variation	Details	Equations	Remarks
Dutta and Raghavan (1987)	$Air + CO_2$–aqueous sodium carbonate–bicarbonate buffer. $Air + CO_2$–aqueous NaOH. Tricresyl phosphate used as antifoaming agent.	Not appreciable	Two nozzles of 0.0045 and 0.0065 m. Two holding vessels (i) $T_{HV} = 0.15$ m and $H/T_{HV} = 5$ and (ii) $T_{HV} = 0.3$ m and $H/T_{HV} = 4$. Diffusers of design similar to C and E (Bhutada and Pangarkar, 1987).	Diffuser: $C: \varepsilon_G = 4.0 \times 10^{-2} \left(\dfrac{P_{Noz}}{V_L} \right)^{0.27}$ $E: \varepsilon_G = 5.0 \times 10^{-2} \left(\dfrac{P_{Noz}}{V_L} \right)^{0.298}$	Overall ε_G higher for diffuser E. Diffuser C gives much better dispersion in the ejector but relatively poor in the holding vessel.
Panchal (1990)	Air–water	No variation.	Same as Panchal et al. (1991). Two draft tubes of 0.092 and 0.168 m diameter and $\left(\dfrac{A_{DFT}}{A_{VH}} \right) = 0.094$ and 0.31 used with diffusers C and E of Bhutada and Pangarkar (1987).	10 mm multiorifice nozzle, 4 holes of 5 mm at pitch of $2 \times D_{H}$, $\left(\dfrac{A_{DT}}{A_{VH}} \right) = 0.092$: Diffuser C: $\left(\varepsilon_G \right)_{DT} = 0.317 \left(\dfrac{P_{Noz}}{V_L} \right)^{1.78}$ $\left(\varepsilon_G \right)_{AN} = 0.2 \left(\dfrac{P_{Noz}}{V_L} \right)^{1.7}$ $\left(\varepsilon_G \right)_{AVG} = 0.245 \left(\dfrac{P_{Noz}}{V_L} \right)^{1.65}$ Diffuser E: $\left(\varepsilon_G \right)_{DT} = 11.23 \left(\dfrac{P_{Noz}}{V_L} \right)^{0.38}$ $\left(\varepsilon_G \right)_{AN} = 9.1 \times \left(\dfrac{P_{Noz}}{V_L} \right)^{0.31}$ $\left(\varepsilon_G \right)_{AVG} = 9.39 \times \left(\dfrac{P_{Noz}}{V_L} \right)^{0.31}$	Gas holdup is much higher in the draft tube as compared to the holding vessel. Average gas holdup higher by ~80% as compared to that without draft tube. 0.092 m diameter draft tube gives 200–300% higher holdup at low (~5 kW/m³), but at high (>10 kW/m³) gas holdups are similar.

(Continued)

TABLE 8.6 (Continued)

Investigator (s)	System used	Range of physical properties covered	Geometrical parameters covered	Correlation proposed	Comments
Panchal et al. (1991)	Air–water	No variation	Diffusers E and C with multiorifice and single-orifice nozzles of total area equal to 0.01 m dia. nozzle. Each orifice of 0.005 m dia. Orifices diameter varied from 0.003 to 0.006 m. Pitch: $1 \times D_N$ to $2 \times D_N$. Draft tube of 0.091 m dia. Coaxial with the diffuser. Other details same as Bhutada and Pangarkar (1987).	Diffuser C with draft tube: $$\varepsilon_G = 6.4 \times 10^{-2} \left(\frac{P_{Noz}}{V_L} \right)^{0.47}$$ Diffuser E with draft tube: $$\varepsilon_G = 6.04 \times 10^{-2} \left(\frac{P_{Noz}}{V_L} \right)^{0.61}$$ Diffuser E without draft tube: $$\varepsilon_G = 8.33 \times 10^{-2} \left(\frac{P_{Noz}}{V_L} \right)^{0.42}$$	At high power inputs, $(\varepsilon_G)_{MON} > (\varepsilon_G)_{SON}$. At low power inputs, ε_G nearly same. Overall, ε_G higher by ~80 in the presence of draft tube.
Cramers et al. (1992a)	Water–Air/He/Ar/ CO_2/SF_6	$0.18 < \rho_G < 6.18\,kg/m^3$; $0.02 < V_G < 0.1\,m/s$	D_{NT}: 0.009 m, optimum design as per Henzler (1981); $H_{HV}/T_{HV} = 5$	$$\varepsilon_G = 7.7 \times V_G \times \left(\frac{\rho_G}{\rho_L} \right)^{0.11}$$	ε_G weak function of gas density. Minor improvements in ε_G at high pressure.

Reference	System	Parameter variation	Geometry	Correlation	Remarks
Tinge and Casado (2002)	Nitrogen–water/activated carbon ($D_P \sim 20\,\mu m$; loading: 0–4.8 wt %)	Total pressure varied from 0.2 to 1.5 MPa; $2.3 < \rho_G < 17\,kg/m^3$; activated carbon Norit SX4 ($d_{P50\%} = 20.4\,\mu m$), loading: up to 4.8 wt %	Ejector dimensions similar to that of Cramers et al. (1992a), $H_{HV}/T_{HV} = 5$	$\varepsilon_G = 0.873 \times V_G^{0.55} \times \rho_G^{0.17}$	ε_G increases at low solid loading but decreases at high solid loadings. Minor positive effect of total pressure on ε_G
Gourich et al. (2005)	Air–water	No variation	Four orifices of dia. 0.01 m; suction chamber length of 0.096 m, angle of convergence of 20°; $L_{MT} \sim 0$; diffuser length of 0.225 m, angle of divergence of 7.6°; $2 < H/T_{HV} < 4$	$\varepsilon_G = 0.11 \times Q_G^{0.55}$ $4.5\times10^{-4} < Q_L < 13.5\times10^{-4}\,m/s$: $\varepsilon_G = 0.0121(Q_L)^{0.264}$ $13.5\times10^{-4} < Q_L < 27.8\times10^{-4}\,m/s$: $\varepsilon_G = 0.0012(Q_L)^{1.21}$	Nonstandard system

and Panchal et al. (1991) found that for optimized multiorifice nozzle (orifice pitch $= 2 \times D_H$), both gas induction rate and holdup improved significantly. Further, Panchal (1990) showed that multiorifice nozzles with rough edges did not suffer from the drawback mentioned previously. Multiorifice nozzles with rough surface yielded both higher Q_G and ε_G (+20%) as compared to a single-orifice nozzle having the same total area as the combined area of the multiorifice nozzle.

8.7.2.2 Effect of the PR   There is no specific study of the effect of PR on the gas holdup. However, it may be argued that for a given value of Q_G, the gas holdup is mainly decided by the shearing action of the primary liquid phase in the mixing tube/ throat. PR, which relates to dimensions outside the mixing tube/throat, is unlikely to have any effect on the gas holdup. Bhutada and Pangarkar (1987) have reported gas holdups for a variety of geometries for a constant PR of 3. For the estimation of the gas holdup, the correlations for respective geometries may be obtained from these correlations of Bhutada and Pangarkar for PR $= 3$.

8.7.2.3 Effect of Entry Region to the Diffuser   In this case also, there is no specific study of the effect of the entry region on the gas holdup. However, following the logic in Section 8.7.1.4, it may be argued that the optimum A_{EN} value of 40 for diffuser C suggested by Bhutada and Pangarkar (1987) may be adhered to in calculating the entry area for a given nozzle diameter. Similarly, for the estimation of the gas holdup, the correlations for respective geometries given by Bhutada and Pangarkar may be used.

8.7.2.4 Effect of Throat Area to Nozzle Area (A_R) Ratio   Several investigators (Engel 1963; Henzler 1983) suggested an optimum area ratio of 4 for obtaining higher efficiencies of a liquid jet gas pump. Bhutada and Pangarkar (1987) varied A_R from 2.54 to 10 in their experimental investigation of the effect of A_R on the gas holdup. It was found that for a given diffuser, the gas holdup is independent of the value of A_R. Thus, estimate of the gas holdup may be obtained from the respective correlations of Bhutada and Pangarkar.

8.7.2.5 Effect of Mixing Tube Length   During the mixing process, a high level of shear creates a very fine dispersion of the induced gas phase. The resulting very small (~ 1 mm) bubbles afford relatively very high effective gas–liquid mass transfer area ($40,000$–$70,000$ m^2/m^3), as shown by Cramers et al. (1992b). Ideally, the mixing tube length should be just sufficient so that the momentum transfer is complete and the mixture issuing out is homogeneous with respect to momentum of the two phases. A shorter length leads to incomplete momentum transfer, whereas an excessive length leads to additional pressure drop. The effect of the mixing tube depends on two geometrical parameters: (i) diameter and (ii) length. Ratio of the mixing tube area and nozzle area denoted as A_R fixes the diameter of the mixing tube for a given nozzle diameter. Considerable information on mixing tube geometry is available. The optimum mixing tube length and diameters obtained by various workers have been reviewed by Henzler (1983). Reinterpreting the earlier data in terms of ejector loss coefficients and efficiencies, Henzler (1983) suggested that for $A_R \geq 2.5$ the optimum throat length is eight times the throat or mixing tube diameter. Bonnington (1964) also suggested an optimum mixing tube length of $8 \times D_{MT}$ to allow adequate momentum transfer. Henzler has further indicated that for multiorifice nozzles the mixing tube length required for ensuring complete mixing of the primary and secondary phases is lower than that for single-orifice nozzles. This may be related to greater loss of momentum at the larger total surface area of such nozzles.

Bhutada and Pangarkar (1987) used mixing tubes having lengths of 0 (no parallel throat), 4, and $8 \times D_T$ with smooth single-orifice nozzles. The data obtained reflect the distinctly different trends for Q_G and ε_G. It was found that reduction in the mixing tube length, L_{MT}, resulted in an increase in Q_G, whereas increase in L_{MT} resulted in decrease in Q_G. Generally, the gas holdup has a positive dependence on the gas flow rate. However, despite the variations in Q_G, there was no change in ε_G. The explanation given is as follows: increase in the mixing tube length yields lower Q_G. Two factors counter the decrease in Q_G to yield almost identical ε_G: (1) for a fixed Q_L, a lower Q_G implies lower Q_G/Q_L or greater momentum of the liquid for a given gas mass flow rate, and (2) a longer mixing tube allows greater momentum transfer between the two phases. These two factors result in a greater shearing action and lower average bubble size. Evidently, the lower Q_G is counterbalanced by a lower average bubble diameter with the result that there is negligible change in ε_G. The situation is reversed with decrease in L_{MT}. Thus, although there is an increase in Q_G, the higher Q_G/Q_L and lower extent of shear result in a higher average bubble diameter. Consequently, despite the greater number of bubbles due to a higher Q_G, the total gas holdup remains unaffected.

8.7.2.6 Effect of Pressure Recovery Diverging Section

The pressure recovery diverging cone is not a very crucial component of ejectors used in venturi loop systems. However, since it yields better Q_G, Bhutada and Pangarkar (1987) studied the effect of addition of a pressure recovery diverging section. The length of this diverging cone was fixed at $8 \times D_T$ and angle of divergence was $5°$. It was found that the use of this additional part gave approximately 20 % higher gas holdup values. It can therefore be concluded that inclusion of this part yields improvement in overall performance.

8.7.2.7 Effect of Physical Properties of the Primary and Secondary Fluids

The only available investigation on the effect of the physical properties of the primary liquid on ε_G is that of Bhutada and Pangarkar (1988). Newtonian solutions of carboxymethyl cellulose were used to vary the viscosity of the liquid by a factor of 4. Over this range, there was practically no change in the gas holdup. As regards the secondary fluid (gas in this case), Cramers et al. (1992a) and Tinge and Casado (2002) have observed a minor effect of the density of the gas. This effect was studied separately by increasing the (i) molecular weight of the gas (Cramers et al. 1992a) and (ii) total pressure (Tinge and Casado 2002). In the case of three-phase sparged reactors, it is known that increase in the total system pressure has a stabilizing effect on gas bubbles (Deckwer 1992; Lau et al. 2004) through a hindrance to the coalescence of the bubbles. Thus, higher pressures result in smaller bubble diameters, higher gas hold, and consequently higher effective area for mass transfer. Apparently, this effect of pressure does not apply to venturi loop system since the effect of gas density is not substantial (Cramers et al. 1992a; Tinge and Casado 2002). The most likely reason for this relatively small effect is the strong shear field produced in the venturi loop system, which dominates its hydrodynamics. Evidently, liquid and gas physical properties that affect coalescence/bubble breakup have only a minor effect. The correlations of Ben Brahim et al. (1984) and Dutta and Raghavan (1987) in Table 8.3 for gas induction rate also bear testimony to this observation.

8.7.2.8 Effect of the Presence of a Draft Tube

Bhutada (1989), Panchal (1990), and Panchal et al. (1991) studied the effect of a draft tube on the gas holdup. Bhutada's (1989) study involved two draft tubes with $[A_{DT}/A_{HV}] = 0.44$ and 0.11 and single-orifice nozzles. It was found that for the larger draft tube, the two-phase jet was arrested in the draft tube even at relatively high $[P_{Noz}/V_L]$ of 0.5 kW/m³. As a result, the holdup was approximately 25% lower than that without the draft tube. On the other hand, for the smaller draft tube, the velocity of the two-phase jet was sufficiently high at $[P_{Noz}/V_L] > 0.25$ kW/m³. Above this power input, the two-phase jet issued out of the draft tube and reached the bottom of the holding vessel. At the bottom of the vessel, the gas component reversed and occupied the entire column. The overall holdup was 25–60% higher than that for the same diffuser without the draft tube under otherwise similar conditions.

Panchal (1990) measured the gas holdup separately in (i) the draft tube, (ii) the annulus between the draft tube and the holding vessel, and (iii) the average holdup for the entire venturi loop system. These studies using multiorifice nozzles also revealed the beneficial effect of the draft tube. However, this benefit was again realized only above a certain critical power input, $[P_{Noz}/V_L]_{Crit}$. As explained earlier, at power inputs higher than this critical value, the two-phase jet has sufficient momentum to overcome the inertia of the gas–liquid dispersion in the draft tube and issue out into the annulus. Figure 8.15 depicts the progress of the two-phase jet with increasing power input.

For diffuser C, the gas holdup in the draft tube was at least 200% higher than the average value for the entire system. This difference was lower for diffuser E. Behr et al. (2008 cf Behr et al. 2009) have also reported almost 200% higher gas holdup in the presence of a draft tube.

8.7.2.9 Effect of Presence of Solids:

The venturi loop reactor is predominantly used for three-phase solid-catalyzed reactions. The effect of the solid phase on the gas holdup has been extensively studied for three-phase sparged reactors. However, very limited information is available for the venturi loop system. The results of Bhutada and Pangarkar (1989) for spherical glass beads do not show any effect of the presence of the solids used on the gas holdup. Tinge and Casado (2002) observed an initial increase (up to 0.3 wt % solid content) in the gas holdup followed by a decline at higher solid loadings up to 4.8 wt %. The difference between the results of these two investigations can be attributed to the nature of the solids used. As discussed

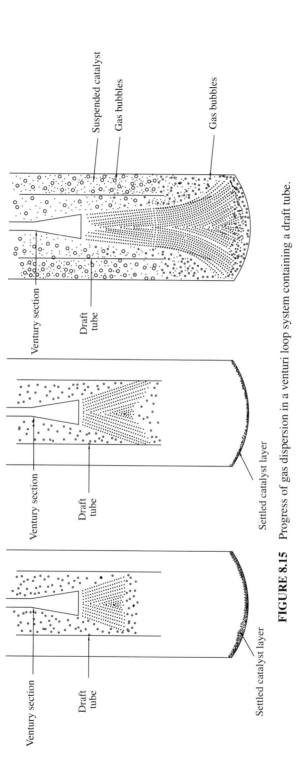

FIGURE 8.15 Progress of gas dispersion in a venturi loop system containing a draft tube.

earlier in Sections 8.6.2.1, 8.7.2.2, and 8.7.2.7, the strong shear field in the venturi section is the main source of the high gas holdup offered by the venturi loop system. It is, therefore, not surprising that other factors such as presence of solid do not have a major effect on ε_G.

8.7.3   Gas–Liquid Mass Transfer: Mass Transfer Coefficient ($k_L \underline{a}$) and Effective Interfacial Area (a)

The venturi loop system comprises of two distinct parts: (1) the ejector and (2) the holding vessel. Inasmuch as these have vastly differing hydrodynamics, their mass transfer behavior needs separate discussion.

There is very limited information available on this very important aspect. Typically, the ejector volume is 1–2% of the total system dispersion volume (Section 8.13). In the case of the conventional BLR, despite its relatively very small volume, the ejector contributes approximately 40% to the total mass transfer in the whole system (Dirix and van der Wiele 1990). Considering this partial volume of the ejector, for a given power input P_{Noz}, $[P_{Noz}/V_{Ej}]$ is higher by a factor of 50–100 than $[P_{Noz}/V_{HV}]$. This is the defining status of the ejector in the whole system. Dutta and Raghavan (1987) varied V_{HV}/V_{Ej} by sixfold and found that for the larger holding vessel, the gas holdup and $k_L \underline{a}$ values were lower. This was attributed to the inferior turbulence and circulation intensity in the larger holding vessel as compared to the smaller vessel. It is obvious that for the larger vessel, P_{Noz}/V_{HV} was lower, and hence, the mass transfer contribution of the holding vessel was lower. Evidently, the ratio V_{Ej}/V_{HV} is an important design parameter. It will be seen later that one of the scale-up criteria for the venturi loop system is constancy of V_{Ej}/V_{HV}. Understandably, Dirix and van der Wiele (1990) and Cramers et al. (2001) treated mass transfer in the ejector and holding vessel separately. However, only Dirix and van der Wiele (1990) have given separate correlations for the two zones. Cramers et al. (2001) have investigated the effects of important geometric parameters of the ejector on $k_L \underline{a}$. A summary of the important findings and correlations is given in Table 8.7. Data on $k_L \underline{a}$ using physical mass transfer should be used with caution. This is because the very high values of $k_L \underline{a}$ afforded by the venturi loop systems can lead to a pinch (zero or even negative driving force). Chemically reacting systems do not suffer from this drawback. However, most reacting systems contain inorganic electrolytes. These tend to prevent coalescence. Therefore, in general, they tend to yield higher values of $k_L \underline{a}$ as compared to coalescing systems. Because of the strong shear field leading to very small bubbles, in some studies, antifoaming agents were required to prevent foaming, particularly in the case of alkaline liquids commonly used in chemical methods for measuring $k_L \underline{a}$. These antifoaming agents lower the effective interfacial area (Betarbet and Pangarkar 1995). Therefore, caution needs to be exercised in the use of the correlations given in Table 8.7. Further, although the venturi loop system is widely used for reactions involving organic liquids (Table 8.1), there is no study on mass transfer in such systems. Therefore, as concluded in Section 8.7.2, there is a similar urgent need for studies on $k_L \underline{a}$ with organic liquids.

TABLE 8.7 Literature Correlations For Gas–liquid Mass Transfer Coefficient and Effective interfacial Area in Gas–liquid Ejectors in the Downflow Mode

Investigator (s)	System used	Range of physical properties covered	Geometrical parameters covered	Correlation proposed	Comments
Biswas et al. (1977)	CO_2–aqueous NaOH	Not appreciable.	D_N, 0.0033–0.0055 m; D_T, 0.01 m; A_R, 3.2–9.3; L_{TM}, 0.06 m; L_{Diff}, 0.12 m; D_S, 0.06 m; D_D, 0.025 m; Diffuser extended by 1 m. Bell-shaped entry.	$a = 60 \times \left(\dfrac{\varepsilon_{GTot}}{\varepsilon_{GNS}} \right)^{2.73} \times (A_R)^{2.33}$; correlation for $\left(\dfrac{\varepsilon_{GTot}}{\varepsilon_{GNS}} \right)$ correlated in terms of back pressure, nozzle Reynolds number, and Morton numbers for the primary and secondary fluids.	Horizontal flow. Geometry not the same as conventional venturi loop system. Noncoalescing system. Interfacial area per unit power input 1200–2500% higher than bubble column under similar conditions.
Dutta and Raghavan (1987)	Air + CO_2–aqueous sodium carbonate–bicarbonate buffer. Tricresyl phosphate used as antifoaming agent.	Not appreciable.	Two nozzles of 0.0045 and 0.0065 m. Two holding vessels (i) T_{HV} = 0.15 m and H_{HV}/T_{HV} = 5 and (ii) T_{HV} = 0.3 m and H_{HV}/T_{HV} = 4. Diffusers of design similar to C and E (Bhutada and Pangarkar, 1987).	C-type diffuser of Bhutada and Pangarkar (1987) Nozzle diameter is 0.0045: $$k_L\underline{a} = 0.0256 \left(\frac{P_{Noz}}{V_L} \right)^{0.76}$$ E-type diffuser of Bhutada and Pangarkar (1987) Nozzle diameter is 0.0065: $$k_L\underline{a} = 0.070 \left(\frac{P_{Noz}}{V_L} \right)^{0.70}$$	Overall $k_L\underline{a}$ value for the system reported $$k_L\underline{a} \propto \left(\frac{1}{V_{HV}} \right)$$

(Continued)

TABLE 8.7 (Continued)

Investigator (s)	System used	Range of physical properties covered	Geometrical parameters covered	Correlation proposed	Comments
Dierendonck et al. (1988)	Absorption of CO_2 in alkanolamines and oxidation of sodium sulfite.	No variation.	Buss AG design. $\left(\dfrac{H_{HV}}{T_{HV}}\right) = 8.$ Nozzle without swirl piece.	$k_L a \left(\dfrac{v}{g^2}\right)^{0.33} = 2.6 \times 10^{-4} \left(\dfrac{P_{Noz}}{V_L \rho (vg^4)^{0.33}}\right)^{0.9}$	Noncoalescing systems.
Dirix and van der Wiele (1990)	Stripping of oxygen from water by nitrogen.	No variation.	D_N, 0.004 and 0.006 m nozzles with swirl piece; D_{MT}, 0.012 m; $\left(\dfrac{L_{MT}}{D_T}\right) = 2$ and 10; $A_R = 9$; D_D, 0.04 m; diffuser length of 0.04 m, angle of divergence of $3°$.	Bubble flow regime: condition, $\left(\dfrac{Q_G}{Q_L}\right) < 1.3$ $\dfrac{k_L a V_{Ej}^{0.66}}{D} = 5.4 \times 10^{-4}$ $Re_{Noz}^2 \times \left(\dfrac{Q_G}{Q_G + Q_L}\right)$ Jet flow regime: condition: $1.3 < \left(\dfrac{Q_G}{Q_L}\right) < 3$ $\dfrac{k_L a V_{Ej}^{0.66}}{D} = 3.1 \times 10^{-4} Re_{Noz}^2$ Holding vessel: $\left(\dfrac{Q_G}{Q_L}\right) > 1.3$	$k_L a$ in jet flow regime $< k_L a$ in bubble flow regime. This is compensated by an increase in the $k_L a$ in the holding vessel

Reference	System	Dimensions	Variation	Correlation	Remarks
Changfeng et al. (1991) cf. Cramers et al. (2001)	No details given.	No details given.	No details given.	$\left(\dfrac{k_L a\,V_{HV}^{0.66}}{D_A}\right) = 3.5\times10^{-12}\,Re_T^{3.3}$, where Re_T is given by $Re_T = \left(\dfrac{4\rho_L(Q_G+Q_L)}{\mu_L\pi D_T}\right)$ $k_L a = 0.72\left(\dfrac{P}{V_L\times\rho_L}\right)^{0.492}(\varepsilon_G)^{0.88}$	No clarity about dependence of $k_L a$ on $\left(\dfrac{P}{M}\right)$ since ε_G also depends on $\left(\dfrac{P}{V_L\times\rho_L}\right)$. However, qualitatively, $k_L a \approx \left(\dfrac{P}{V_L\times\rho_L}\right)^n$ with n close to unity.
Cramers et al. (1993)	N_2–water.	D_N 0.019 m. Mixing tube diameter and length, 0.029 and 0.7 m, respectively.	No variation.	Literature correlations discussed. Qualitative discussion on the relative effects of the mixing tube and extension piece.	Nonstandard geometry. Ejector contributes 40% of the mass transfer. Use of a swirl device yields $(k_L a)_{Ej}$ in mixing tube zone lower by 50%.

(Continued)

TABLE 8.7 (Continued)

Investigator (s)	System used	Range of physical properties covered	Geometrical parameters covered	Correlation proposed	Comments
Betarbet and Pangarkar (1995)	Dilute carbon dioxide (7–$0 v/v\%$ in air– aqueous Na_2CO_3– $NaHCO_3$ (tricresyl phosphate used as antifoam)	No variation.	Diffusers E and C of Bhutada and Pangarkar (1987) with single- and multiorifice nozzles of total area equal to that $0.01\,m$ dia. nozzle. Four orifices of $0.005\,m$ and $0.006\,m$ diameter on 0.01 and $0.12\,m$ pitch, respectively. $T = 0.2\,m$; $\left(\dfrac{H_{HV}}{T_{HV}}\right) = 7.2$	$0.01\,m$ diameter single- and multiorifice nozzles with diffusers C and $12\,mm$ nozzle for diffuser $E^*: k_L\underline{a} = 0.0128 \left(\dfrac{P_{Noz}}{V_L}\right)^{0.88}$	Overall $k_L\underline{a}$ value for the system. Values reported are relatively low due to use of antifoaming agent. $k_L\underline{a}$ independent of nozzle type (single/ multiorifice) or diffuser. $k_L\underline{a}$ found to increase with nozzle diameter for diffuser E.
Cramers et al. (2001)	Oxygen desorption from water by N_2/He/Ar.		D_N, 0.004, 0.0047 and $0.0053\,m$; D_{MT} $0.012\,m$; A_R, 5.12–9; L_{MT} 0.024– $0.12\,m$; D_D, $0.035\,m$; angle of $3°$; D_{DT} and L_{DT}, 0.035 and $0.62\,m$, respectively. Ejector scale-up studied by varying the dimensions by a factor of 2. Exact nozzle and swirl piece details confidential.	For nozzle without a swirl piece: $k_L\underline{a} = \text{Constant}\,(\varepsilon_{Dis})^{0.65}$ $\times \varepsilon_G \left(\dfrac{1 + 0.2\varepsilon_G}{1 + \varepsilon_G}\right)^{1.2}$ $\left(\dfrac{\rho_L^2 \rho_G}{\sigma_L^3}\right)\left(\dfrac{L_{MT}}{D_{MT}}\right)^{0.42}$ $\left[1 - 0.55\left(0.38 - \left(\dfrac{D_N}{D_{MT}}\right)^2\right)\right]$	$k_L\underline{a}$ without swirl piece higher than with swirl piece. Optimum throat to nozzle area ratio, A_R: 6.25. $k_L\underline{a}$ is proportional to mixing tube length. $k_L\underline{a} \propto (\rho_G)^{0.2}$

For nozzle with a swirl piece and bubble flow regime:

$$k_L\underline{a} = \text{Constant}\left(\varepsilon_{\text{Dis}}\right)^{0.65}$$

$$\times\varepsilon_G\left(\frac{1+0.2\varepsilon_G}{1+\varepsilon_G}\right)^{1.2}\left(\frac{\rho_L^2\rho_G}{\sigma_L^3}\right)$$

$$\left(\frac{L_{\text{MT}}}{D_{\text{MT}}}\right)^{0.42}\left(\frac{D_N}{D_M}\right)^{0.65}$$

Jet-annular flow regime:

$$k_L\underline{a} = \text{constant}\left(\varepsilon_{\text{Dis}}\right)^{0.65}$$

$$\left(1-\varepsilon_G\right)\left(\frac{D_N}{D_M}\right)^{0.65}$$

$$\varepsilon_{\text{Dis}} = \frac{P_{\text{Noz}} - \left(\Delta P_G\right)_{\text{Ej}}Q_G}{\rho_L\left(1-\varepsilon_G\right)V_{\text{Ej}}}$$

(Continued)

TABLE 8.7 (Continued)

Investigator (s)	System used	Range of physical properties covered	Geometrical parameters covered	Correlation proposed	Comments
Gourich et al. (2008)	CO_2–aqueous monoethanolamine. Simultaneous estimation of k_L and a using Danckwerts' plot (Danckwerts plot method: Danckwerts et al. 1963; Linek et al. 2005).	No variation.	Four orifices of dia. 0.01 m; suction chamber length of 0.096 m, angle of convergence of 20°; $L_{MT} \sim 0$; diffuser length of 0.225 m, angle of divergence of 7.6°; $2 < H/T_{HV} < 4$.	Annular flow regime: $k_L = 0.0129 \times (Q_L)^{0.595}$; $$k_L a \propto \left(\frac{P_{Noz}}{V_L} \right)^{0.48}$$ Correlations obtained from regression of Figures 7a, 8a, and 6a of Gourich et al. (2008), respectively: $$a = 13.87 \left(\frac{P_{Noz}}{V_L} \right)^{0.456}$$ $$k_L = 4.21 \times 10^{-5} \left(\frac{P_{Noz}}{V_L} \right)^{0.28}$$ $$k_L a = 0.088 \left(\frac{P_{Noz}}{V_L} \right)^{0.643}$$	
Raghuram (2007)	Air–water/2 wt % aqueous sodium chloride; γ-ray attenuation technique used.		D_N, 0.003 m; D_T, 0.003 m; L_{MT}, 0.021 m; D_S, 0.44 m; L_{Diff}, 0.25 m; angle of divergence: 9°	Correlations obtained from regression of Figure 7 of Raghuram (2007) Air-aq. NaCl: $a = 9.02 \left(\dfrac{Q_L}{Q_G} \right)^{0.218}$ Air-water: $a = 8.75 \left(\dfrac{Q_L}{Q_G} \right)^{0.22}$	a for aq. NaCl, only 3–5% higher than for water.

8.7.3.1 Mass Transfer in the Ejector Zone: Effects of Various Geometric Parameters of the Ejector on $k_L a$ (Cramers et al. 2001)

Table 8.7 indicates that in correlations of the type $(k_L a)_{Overall} \propto (P_{Noz} / V_L)^n$, the exponent n varies from 0.65 to 0.88. These differences have been attributed to the different volume ratios V_{Ej}/V_{HV} used in the respective studies (Cramers et al. 1993). The effects of various components/parameters of the ejector on $k_L a$ are discussed in the following sections.

Effect of a Swirl Device in the Nozzle   As discussed earlier in Section 8.6.2.1, the presence of a swirl device in the primary nozzle creates a disturbance in the jet. This results in roughness on the nozzle surface and consequently higher rates of gas induction. It would therefore be expected that there may be concomitant changes in mass transfer parameters. Cramers et al. (2001) and Baier (2001) have shown that a swirl piece actually causes a decrease in $k_L a$. For an upflow system, Havelka et al. (2000) have also made similar observations. Cramers et al. (2001) found that for nozzles with a swirl piece, $k_L a \propto \varepsilon_{Dis} (D_N / D_{MT})^{0.65}$. On the other hand, for conventional nozzles (without a swirl piece), it was found that $k_L a \propto \varepsilon_{Dis} (1 - 0.55(0.38 - (D_N / D_{MT}))^2)$. It has been argued earlier that when a swirl piece is used, a certain part of the motive power is consumed in (i) conversion from axial to tangential velocity component and (ii) overcoming the friction in the swirl piece. This results in a weaker jet that tends to break prematurely. Thus, the level of turbulence in the mixing tube is lower. Cramers et al. (1993) have argued that the jet diameter D_J at the point of impact in the mixing tube is more relevant than the jet diameter at the exit of the nozzle. The presence of a swirl device produces a wider jet, and consequently, the actual D_J/D_{MT} ratio at the impact in the mixing tube is higher than that based on the nozzle diameter. These investigators have found that the interfacial area, a, in the mixing zone shows a maximum at $D_J/D_{MT} \sim 0.4$. Higher values of D_J/D_{MT} tend to cause a decrease in a. Further, if D_J is $\sim D_{MT}$ at the point of impact, there is a distinct possibility of choking of the mixing tube. Baier (2001) has presented a cogent discussion on the effect of the swirl piece for the case of the ABLR®. This discussion also supports the deleterious effect of a swirl piece on $k_L a$. In conclusion, it is recommended that a conventional nozzle such as that depicted in Figure 8.14 should be preferred.

Effect of A_R, the Throat to Nozzle Area Ratio: $(D_{MT}/D_N)^2$ For a nozzle containing a swirl piece, Dirix and van der Wiele (1990) found that $k_L a \propto (A_R)^{1.53}$. At equal P_{Noz}/V_{Ej}, Cramers et al. (2001) showed that $k_L a \approx (1 - 0.55(0.38 - (D_N / D_{MT}))^2)$. Further, for a conventional nozzle, Cramers et al. (2001) found that the best mass transfer performance is obtained at $A_R = 6.25$. This compares favorably with the optimum value of $A_R = 4$ for gas induction reported by Bhutada and Pangarkar (1987). The minor difference may be attributed to different ejector configurations used and also different mechanisms governing gas induction and mass transfer processes.

Effect of Mixing Tube Length: L_{MT}/D_T   As discussed in Section 8.7.2.5, the L_{MT}/D_T ratio has a positive effect on the gas holdup. Since the effective interfacial area component in $k_L a$ is related to the gas holdup, it is expected that the mixing tube length effect on $k_L a$ should also be positive. Dirix and van der Wiele (1990) studied the effect of L_{MT}/D_T on mass transfer at two levels: $L_{MT}/D_T = 2$ (defined as standard condition) and 10. Surprisingly, they did not observe any effect of the mixing tube length. However, the subsequent investigations of Baier (2001) and Cramers et al. (2001) yield proof of a distinct increase in $k_L a$ with increase in L_{MT}. This experimental evidence is in agreement with the logic outlined in Section 8.7.2.5 that a longer mixing tube results in smaller bubbles or higher a.

Effect of Pressure Recovery Diverging Section   Most of the investigations reported in Table 8.7 have used a pressure recovery diverging cone. This part is an extension of the mixing tube. It must be stressed that the level of turbulence in it is much lower than in the mixing tube because of considerable reduction in the velocity. There is no special mention of the effect of this part on $(k_L a)_{Ej}$. Its effect on the gas holdup has been discussed in Section 8.7.2.6, wherein it has been shown that inclusion of the pressure recovery section increases the gas holdup. In the absence of such specific information, an argument analogous to that in Section 8.7.3.1 would reveal that the pressure recovery diverging cone should help in increasing $(k_L a)_{Ej}$.

8.7.3.2   Mass Transfer in the Holding Vessel   With the exception of Dirix and van der Wiele (1990), all studies reported in Table 8.7 give overall $k_L a$ for ejector and holding vessel. The importance of mass transfer in the holding vessel should not be underrated since its contribution is 60% of the total mass transfer as reported by Dirix and van der Wiele (1990). The relative contribution will of course depend on the system configuration that includes the ejector geometry/size and the holding vessel volume and in particular V_{Ej}/V_{HV}. It will be shown later under the section on solid suspension that for actual industrial operating conditions mass transfer in the holding vessel is expected to be vigorous.

Effect of a Draft Tube   Bhutada and Pangarkar (1987) and Panchal et al. (1991) have shown that the presence of a draft tube improves the gas holdup above a power input, which allows the two-phase jet to reach the bottom of the holding vessel (Fig. 8.15). There is no information on $k_L a$ for the standard venturi loop system incorporating a draft tube. In this case also, the logic in Sections 8.7.2.8 and 10.10.2 indicates that a draft tube with proper configuration ($A_{DFT}/A_{HV} \sim 0.25$) may yield higher value of $k_L a$.

8.7.3.3   Effect of Physical Properties   There are no studies on the effect of liquid physical properties such as viscosity and surface tension on $k_L a$ in the vertical downflow mode. For the vertical upflow mode, Elgozali et al. (2002) have found that $k_L a \propto (\mu_L / \rho_L)^{-0.166}$. In the absence of information on the downflow mode, the above dependence as well that on gas holdup (Section 8.7.2.7) may be used as a guide. Thus, it may be assumed that the effect of liquid physical properties on $k_L a$ is insignificant. For the gas phase, its density is the sole important property. Cramers et al. (2001) have shown that $d_B \propto (\rho_G)^{-0.2}$. Thus, higher pressures/gas density yields a smaller average bubble size, which suggests that increase in $k_L a$ is mainly due to increase in a. The gas holdup is an indication of a. This is in agreement with the dependence of the gas holdup on ρ_G discussed in Section 8.7.2.7.

8.8 SOLID SUSPENSION IN VENTURI LOOP REACTOR

Solid suspension and uniformity in its concentration are as important in this case as for other gas dispersed-type of reactors. This matter has been highlighted in Chapter 7A, Section 7A.7.1, with respect to stirred tank reactors. In the venturi loop system, the catalyst/gas/liquid mixture issues out of the ejector and travels downward to the bottom of the reactor. The driving force for this flow is the momentum of the jet and the resistance to this flow is offered by the inertia of the dispersion in the holding vessel. The only reported investigation on solid suspension in venturi loop systems is that of Bhutada and Pangarkar (1989). These investigators used spherical glass beads of 150, 250, and 450 μm size. The solid loading was varied from 0.5 to 5 wt %. The solid concentration profiles were measured in both the radial and axial direction. The equipment used also allowed visual observation of the state of solid suspension. Thus, both objective (measurement of solid concentration profiles) and subjective (visual observation) techniques were used. It was found that at relatively low power inputs, the two-phase jet issuing out of the ejector is arrested well above the bottom of the holding vessel due to the resistance of the dispersion. After this, a flow reversal takes place and the gas portion rises vertically in the annulus. The turbulence below this location of reversal is very poor, and therefore, the solid phase tends to settle at the bottom of the holding vessel. Cramers et al. (1992b) have indicated that the flow regime in the ejector has a strong influence on the hydrodynamics in the holding vessel. Solid suspension can be visualized in the following steps: starting with a low power input, the two-phase jet is arrested before it reaches the bottom of the holding vessel. As the power input is increased, the two-phase jet penetrates further downward. Finally, at a certain power input $[P_{\text{Noz}}/V_{\text{L}}]_{\text{S}}$ termed critical power input, the two-phase jet reaches the bottom of the holding vessel. This phenomenon is similar to that explained in Chapter 7A, Section 7A.7.1, wherein the just-suspended condition in the stirred tank reactor is achieved at a power input that is sufficient to force the gas phase against its buoyancy to the bottom of the stirred tank. It is difficult to predict the depth of penetration of the two-phase jet in the present case. Information available on penetration of plunging jets can be used to obtain an idea of the penetration depth. Correlations dealing with plunging liquid jets used a criterion that the maximum penetration depth is defined by the point at which the local liquid velocity is equal to the free rise velocity of the bubbles. Two factors complicate this analysis: (i) the free rise velocity is related to the bubble size that itself is an unknown factor and (ii) liquid circulation caused by the buoyancy of the rising bubble swarm. Bin (1983), McKeogh and Ervine (1981), and other investigators (Bin, 1993) have suggested the following empirical correlation for the depth of penetration, L_{PT}:

$$L_{\text{PT}} = \text{Constant} \times (V_{\text{Noz}})^{n_1} (D_{\text{N}})^{n_2} \tag{8.21}$$

For $V_{\text{Noz}} \times D_{\text{N}} \geq 0.01 \, \text{m}^2/\text{s}$, constant $= 24$, $n_1 = n_2 = 0.66$, and for $V_{\text{Noz}} \times D_{\text{N}} < 0.01 \, \text{m}^2/\text{s}$, constant $= 61$, $n_1 = n_2 = 1.36$.

Bin (1993) suggested the following simplified form of Eq. 8.21:

$$L_{\text{PT}} = 2.1 \times (V_{\text{Jet}})^{0.775} (D_{\text{N}})^{0.67}, \quad \text{where} \quad V_{\text{Jet}} = (V_{\text{Noz}} + 2gL_{\text{Jet}})^{0.5} \tag{8.22}$$

The jet geometry has a complex dependence on the turbulence level in the jet itself. The latter is in turn decided by the nozzle configuration, which includes factors that cause jet roughness and instabilities. Because of these complicating factors, it is not possible to propose a unified and sufficiently accurate correlation for L_{PT} (Bin 1993). For a venturi loop system, owing to the relatively large gas induction rates, the situation is even more difficult to correlate/model. This underlines the importance of experimental data that need to be generated. Figure 8.16, which is similar to Figure 7A.8, shows the progressive penetration of the two-phase jet culminating in solid suspension at the critical power input $[P_{Noz}/V_L]_S$.

At and above $[P_{Noz}/V_L]_S$, Bhutada and Pangarkar (1989) found that the measured solid concentration profiles were independent of both the axial and radial positions in the holding vessel. This observation implies that the solid phase is completely mixed. This is an important advantage of the venturi loop reactor over stirred tank reactor. In the latter, such uniform solid concentration profile requires relatively very high (P/V_L). Therefore, stirred tank reactors generally operate at N_{SG} at which the solid concentration is far from uniform. At the top of the dispersion, solid concentration is relatively very low and it increases gradually toward the impeller and the region below it. Because of this variable catalyst concentration, the rate of the catalytic reaction can be significantly different across the stirred tank reactor height. Evidently, this problem does not arise in a venturi loop reactor operated above $[P_{Noz}/V_L]_S$. Bhutada and Pangarkar (1989) found that the induced gas flow rate was a major dependent deciding parameter. This in turn depends on the power input and the ejector configuration. This is illustrated by the following example. During an experiment after achieving uniform suspension of solids, the induction rate was reduced by increasing the "back pressure" on the ejector. There was no change in the power input. However, the higher "back pressure" resulted in a lower Q_G. Consequently, solid segregation from the gas–liquid jet was observed. The gas phase traveled downward and prematurely rose toward the top. The solid particles exhibited a distinct settling tendency. It was found that if the reduction in Q_G was sufficient, the solids settled down at the bottom. At this stage, if the power input was increased or "back pressure" decreased to restore the gas induction rate to its original value, the solids were resuspended. The effect of Q_G in the present case contrasts that in a stirred reactor, wherein increase in Q_G causes settling of solids. This must be seen in the light of the fact that for a venturi loop system, Q_G depends on the power input and is not an independent variable as is the case of a stirred reactor. The physical properties that affect solid suspension are (i) liquid-phase viscosity and density and (ii) solid density and particle size. The effect of these parameters can be combined into a single parameter, the terminal settling velocity of the solid, V_T. The range of V_T covered in the study of Bhutada and Pangarkar (1989) was 0.018–0.0778 m/s. This range is large enough to cover situations of industrial interest in which the catalyst particle size is ~25–50 μm and the catalyst density can range from 500 to 6500 kg/m^3 (activated carbon and nickel, respectively). The correlations proposed were as follows:

$$\text{Diffuser E}: 0.012 \text{ m diameter nozzle,} \quad \left(\frac{P_{Noz}}{V_L}\right)_S = 15.45(V_T)^{0.15}(X)^{0.334} \quad (8.23)$$

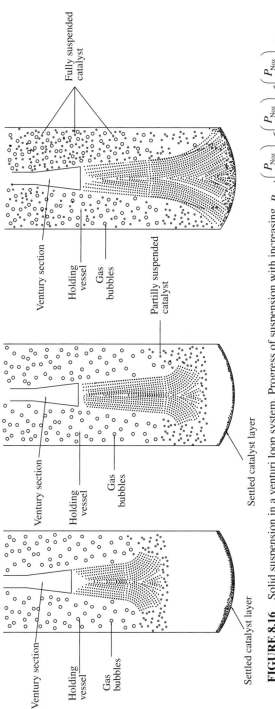

FIGURE 8.16 Solid suspension in a venturi loop system. Progress of suspension with increasing P_{Noz}: $\left(\dfrac{P_{Noz}}{V_L}\right)_a < \left(\dfrac{P_{Noz}}{V_L}\right)_b < \left(\dfrac{P_{Noz}}{V_L}\right)_s$.

Diffuser C : 0.01 m diameter nozzle, $\left(\dfrac{P_{Noz}}{V_L}\right)_s = 33.11\left(V_T\right)^{0.101}\left(X\right)^{0.376}$ (8.24)

On closer examination, it was found that the differences in the above correlations were mainly due to their different gas induction characteristics. These differences were eliminated when the solid suspension criterion was redefined in the form of Q_{GS}, the critical gas rate at which the two-phase jet reached the vessel bottom as shown in the last stage of Figure 8.16 denoted by the subscript "s." . This approach gave a unified correlation independent of the nozzle diameter, diffuser type, etc.

$$Q_{GS} = 5.61 \times 10^{-3}\left(V_T\right)^{0.05}\left(X\right)^{0.11}$$ (8.25)

The exponents on X in Eq. 8.23 and those in the correlations reported in Table 7A.3, Chapter 7A, are strikingly similar (0.11 and 0.13, respectively). This further supports the Q_{GS} criterion for solid suspension elaborated earlier.

Dutta and Raghavan (1987) have (Section 8.7.3) reported that for larger V_{HV}/V_{Ej} ratios, the circulation intensity in the holding vessel is relatively lower. With increasing V_{HV} at constant V_{Ej}, two phenomena cause a decrease in the circulation: (i) with increasing diameter of the vessel, there is a larger annular zone in which two-phase jet flow is weak/absent, and (ii) with increasing H_{HV}, the two-phase flow has to traverse a longer distance before it reaches the bottom to suspend the solids. In view of this, careful attention needs to be provided with respect to V_{HV}/V_{Ej} and H_{HV} to assure solid suspension.

8.9   SOLID–LIQUID MASS TRANSFER

There is no reported study on solid–liquid mass transfer in a venturi loop reactor. However, since the gas–liquid mass transfer rates are very high, it can be argued that the same very high levels of turbulence, which result in relatively high k_L values, should also yield correspondingly high K_{SL} values. Indeed, a major advantage of venturi loop reactor claimed by Buss AG pertains to the high hydrogen concentrations on the catalyst surface afforded by a venturi loop reactor. The latter prevents catalyst leaching commonly encountered when the catalyst surface is starved of hydrogen (Leuteritz et al. 1976). Continuing the argument, it is evident that because of the complete saturation of the liquid phase and very high levels of turbulence, the concentration of the gaseous species (e.g., H_2) on the solid surface should also correspond to this saturated liquid concentration. This proposition derives support from a number of examples (Table 8.1) in which a change over to a venturi loop system has allowed significant reduction in catalyst loading. Alternatively, the catalyst surface area per unit dispersion volume could be reduced without any deleterious effect. The volumetric solid–liquid mass transfer coefficient is: $K_{SL}A_P/V_D$. Venturi loop systems typically use ~1 wt % or lower catalyst loading, which is substantially lower than those in stirred reactors. Reduction in catalyst loading reduces A_p/V_D. If this reduction (in some cases by an order of magnitude) has no effect on the overall rate of reaction/ batch time, it implies that there is a concomitant order of magnitude increase in K_{SL}. Thus, the argument of relatively very high K_{SL} similar to high $k_L a$ values is justified. As a result, the solid–liquid mass transfer resistance is likely to be insignificant and can be ignored. It can therefore be concluded that the catalytic reaction can achieve its intrinsic (maximum) rate in the venturi section.

8.10   HOLDING VESSEL SIZE

The holding vessel acts as a bubble column that has a central core consisting of two-phase cocurrent downflow. In the annulus, the unreacted gas travels upward similar to the conventional bubble column. For the conventional venturi loop system, the combination of (i) ejector, (ii) holding vessel, and (iii) power input should be such that the two-phase jet is able to reach the bottom of the holding vessel and allow utilization of the entire volume of the holding vessel. Bhutada (1989) has found that at power inputs greater than $[P_{Noz}/V_L]_S$ (or the gas rate greater than Q_{GS}), the momentum of the two-phase jet may also force some part of the gas into external loop. This situation may create cavitation problems in the pump in the external loop if it is not designed to operate for a two-phase mixture (It should be noted that current pump designs can handle substantially high gas volume fractions). The H_{HV}/T_{HV} ratio of the holding vessel is an important parameter. A relatively high value of the same can cause problems with catalyst suspension as discussed earlier. In contrast, a low value can cause a two-phase jet to flow out of the bottom into the pump, thereby creating cavitation problems in the latter, unless it is designed to handle substantial gas volume fraction. The H_{HV}/T_{HV} ratio of the holding vessel is related to the design of the ejector configuration and the power input through the primary liquid phase. Dierendonck et al. (1988) have suggested an H_{HV}/T_{HV} ratio of 8. Panchal et al. (1991) have also reached a similar conclusion.

8.11   RECOMMENDED OVERALL CONFIGURATION

Dierendonck et al. (1998) have recommended the following ranges of geometric parameters for the ejector: $1.5 < D_T/D_N < 4.5$; $5 < L_{MT}/D_T < 8$; $8 < L_{Diff}/D_T < 12$. In Section 8.7.3, it has been indicated that typical V_{Ej}/V_{HV} ratios are 0.01–0.02. In the configuration used by Cramers et al. (1992b), the V_{Ej}/V_{HV} ratios were 0.014, 0.016, and 0.018. Dierendonck et al. (1998) suggested that the highest value of H_{HV} that ensures uniform gas distribution over the cross section should be used. This indicates a slender configuration with H_{HV}/T_{HV} in the range of 5 (Cramers et al., 1992b) to 8 (Betarbet and Pangarkar, 1995). For larger-diameter holding vessels ($T_{HV} > 1.5$–3 m and $H_{HV}/T_{HV} < 2$), Dierendonck et al. (1998) suggested either (i) a central draft tube or (ii) multiple venturi ejectors distributed over the holding vessel cross section.

8.12   SCALE-UP OF VENTURI LOOP REACTOR

Dierendonck et al. (1988) have suggested that during scale-up, besides the ejector configuration, V_{Ej}/V_{HV} should also be maintained constant. As discussed earlier, this ratio is in the range of 0.01–0.02. Further, these investigators also suggested that the $[P_{Noz}/V]_{Ej}$ ratio be kept constant during scale-up. As regards the holding vessel, the scale-up criterion suggested is in terms of $Q_L/V_{HV} = n$:

$$60 < Q_L / V_{HV} < 120 / h \qquad (8.26)$$

where n by definition is similar to "vvm" in Chapter 7A and represents the number of times the holding vessel volume is replaced per hour. The value of n may vary depending on requirements such as gas flow rate required by the stoichiometry of the reaction and suspension of solids. This is illustrated in the worked reactor design example.

8.13   WORKED EXAMPLES FOR DESIGN OF VENTURI LOOP REACTOR: HYDROGENATION OF ANILINE TO CYCLOHEXYLAMINE

Calculations for design of venturi loop reactor for manufacturing of 25,000 metric tonnes per year of cyclohexylamine by hydrogenation of aniline are presented in a separate Excel file. This file works in all Excel versions of Microsoft Office Excel 97–2003 and higher.

All calculation cells are linked by respective formulae with the capacity in cell D6. Any alteration done in the cell D6 yields new values of system size, energy required per unit mass, and other related parameters.

In the spreadsheet provided for a given set of conditions (capacity, conversion, ΔH_R, etc.), the main variable for optimization of energy required is ΔT_{HEX}. For calculation of the area of external heat exchanger, standard design procedures are available (Kern 1997). The external shell and tube heat exchanger should be 1-1 type with both fluids flowing vertically upward. The organic phase containing the catalyst *should always* be on the tube side. This is important since there is a distinct tendency of the catalyst to settle if the organic phase is on the shell side. In addition, the tube-side velocity should be sufficiently high (in the range of 1.5–2 m/s) to promote total transport of the catalyst with the organic phase back into the primary nozzle via the cross-flow filter. In the present case, the holding vessel is maintained at the reaction temperature of 130 K. The heat of reaction is removed in the external heat exchanger. The parameters for the same are shown in Figure 8.17.

The criteria suggested for satisfactory design are as follows: (i) P_{Noz}/V_{HV} of 3–10 kW/m³ and Q_L/V_{HV} of 60–120/h (Dierendonck et al. 1988) and (ii) H_{HV}/T_{HV} of 5–8. Zahradnik and Rylek (1991) suggested that pressure drop across the primary nozzle, ΔP_N, should be between 0.5 and 0.8 MPa, which corresponds to a nozzle liquid velocity range of 20–50 m/s. However, it has been found that ΔP_N as low as

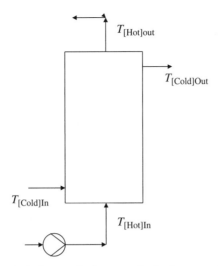

FIGURE 8.17 Block diagram for heat transfer in the external heat exchanger.

0.35 MPa yields smooth jets for properly designed primary nozzles (Bhutada 1989; Dirix and van der Wiele 1990). Lower nozzle velocity results in lower rate of gas induction. The combined operating conditions should be such that the rate of gas induction is sufficiently in excess of (i) the stoichiometric requirement and (ii) minimum flow rate Q_{GS} required to keep the catalyst in suspension (Eq. 8.25).

The liquid circulation rate is decided by the heat balance and the temperature drop for the organic phase across the external heat exchanger, ΔT_{HEX} (cell #D50). For a given heat removal duty, ΔT_{HEX} determines (i) Q_L or (ii) P_{Noz}/V_{HV} and thereby the gas induction rate Q_G. Q_G should be sufficiently in excess of (i) the stoichiometric requirement and (ii) Q_{GS}. Once ΔT_{HEX} (cell #D50) is provided, the energy balance calculates Q_L and other related parameters, for example, rate of gas induction, Re_{Noz}, Q_G/Q_L, Re_T, etc. Output of the spreadsheet for $\Delta T_{HEX} = 4\,K$ is given in the following and discussed in the light of available design information. Physicochemical data provided by Ramachandran and Chaudhari (1983) have been used. Heats of formation data are from Yaws et al. (1999) (Fig. 8.18).

Spreadsheet Output and Justification The value of $k_L a$ for the ejector section predicted from the correlation of Dirix and van der Wiele (1990) is 15.6/s and is of the same order as that obtained by Havelka (1997 *cf* Dierendonck et al. 1988). The calculated value of $k_L a$ for the holding vessel is, however, relatively very high. This may be because the operating Re_T in the present case is much higher than the maximum value used by Dirix and van der Wiele in their study (Figure 7 of Dirix and van der Wiele). Thus, the validity of their correlation for $Re_T > 2.8 \times 10^5$ may be uncertain. In any case, the overall $k_L a$ is significantly high to ensure that there is no mass transfer limitation.

The value of n (Q_L/V_{HV}) obtained using the spreadsheet is 42/h. This is lower than the lower limit of 60/h suggested by Dierendonck et al. (1988). This may be due to (i) the use of correlations for M_R (Bhutada and Pangarkar 1987) derived from a geometric configuration different from that for which the thumb rule of Dierendonck et al. (1988) applies or (ii) a higher and conservative value of n suggested by Dierendonck et al. (1988). It should be noted that most thumb rules are conservative, whereas in the aforementioned solved problem, the controlling regime, suspension of solids, etc. have been estimated from available literature correlations, and therefore, even if the

Design of ventury loop reactor for hydrogenation of aniline to cyclohexylamine. Physico-chemical data from Ramachandran and Chaudhari (1983), pp 300-307

Basis			25000	Metric tonnes per year of cyclohexilamine (CHA)
	Working days/year		300	days/year
		Aniline	$\rightarrow$	CHA
	Reaction	$C_6H_5NH_2$	$3H_2$	$C_6H_{13}N$
	Molecular weight	93	6	99
			25000	Metric tonne/year
			83.33	Metric tonne/day
	Design duty		842	kmol/day
			0.0097	kmol/s
			0.965	kg/s
Supporting data				
	Catalyst		Supported nickel catalyst	
	Particle density of catalyst		1750	kg/m^3
	Liquid density		900	kg/m^3
	Initial concentration of aniline (pure aniline, without solvent)		9.677	$kmol/m^3$
	Solubility of H2		0.0045	$kmol/m^3$
	Reactor pressure		1	MPa
	Reaction Temp		130	degree C
			403	degree K
	Catalyst loading (1 wt/vol %)		10	kg /m3
	Diameter of catalyst particle		10	microns
	Terminal settling velocity		0.0002	m/s
	Hydrogen diffusivity		1.16E-09	m^2/s
	Rate constant		0.05149	m^3/ kg catalyst / s
	With a catalyst loading of 10 kg/m^3, the first order rate constant		0.5149	1/s
Material Balance				
	Conversion of aniline desired		90%	
	Aniline required		0.0108	kmol/s
	CHA formed		0.0097	kmol/s
	H_2 required		0.0292	kmol/s
Energy Balance				
	Reaction temp		130	C
			403	K
	Heat of formation of CHA		-115.8	kJ/mol	Reference: Yaws et al (1999)
	Heat of formation of aniline		-75.39	kJ/mol
	Heat of reaction per mol of CHA		-40.41	kJ/mol
	Heat of reaction per kmol of CHA		-40410	kJ/kmol
	Heat evolved during reaction		-393.69	kJ/s
	Specific heat		2.23	kJ/kg,K
	$[\Delta T]_{Hex}$		4	K
	Mass of liq to circulate		44.14	kg/s
	Density		900	kg/m^3
	Vol rate of circulation		0.0490	m^3/s

Nozzle design

$$V_{PNT} = \left(\frac{\Delta P_G \times 2}{\zeta \rho_L} \right)^{0.5}$$

ΔP_N	0.4	MPa	
	400000	Pa	
ζ	1		Reference: Zahradnik and Rylek (1991)
Density of liquid	900	kg/m^3	
Velocity at nozzle exit, V_{PNT}	29.81	m/s	
Liq vol flowrate	0.049	m^3/s	
Dia of nozzle, D_N	0.046	m	
Viscosity of liq	0.0005	Poise	
Density of supported Ni	1750	kg/m^3	

FIGURE 8.18 Output of the spreadsheet for $\Delta T_{HEX} = 4°K$.

Ejector calculation

Throat dia, D_T	$D_{throat} = 2 \times D_N$	0.092	m
Diffuser entry diameter, D_E	$D_E = (40)^{0.5} D_N$	0.289	m
Length of conical contraction section, L_1		0.463	m
Length of conical expansion section, L_2		1.158	m

Vol of Sand clock (2 cone frustrum)

$$Vol_{ot} = \left(\frac{\pi}{12} [D_E^{\,2} + D_E \bullet D_{throat} + D_{throat}^{\,2}] \right)[L_1 + L_2]$$

Ejector volume	0.050	m^3

Mass ratio

$$\mathbf{Re}_{Noz} = \left(\frac{d_N V_{PNt} \rho_L}{\mu_L} \right)$$

Reynolds Number	2,455,922

$$M_R = 9.67 X 10^{-4} (AR)^{0.1} \left(\frac{\Delta P_G}{\rho_L V_{PNt}^{\,2}} \right)^{-0.135}$$

Ratio of nozzle area to diffuser throat area, AR	4	
Back pressure on diffuser, ΔP_G	20	kN/m^2
Mass ratio	4.64E-03	kg H$_2$/kg liquid to circulate
H$_2$ induced	0.205	kg/s
Q_G	0.102	kmol/s
H$_2$ induced	0.171	m^3/s
Hydrogen density	1.2	kg/m^3
kmol of CHA produced	0.00974	kmol/s
H$_2$ required per stoichiometry	0.0292	kmol/s

Therefore, H$_2$ supplied is greater than amount required which ensures that there is no short supply of H$_2$

Gas to liquid volumetric flow ratio G/L	3.483

Gas- liquid mass transfer in the ejector section
Dirix and van der Wiele (1990)

Regime	Condition	Corelation
Bubble flow regime	$(Q_G/Q_L) < 1.3$	$\left(\frac{k_L a V_{Ej}}{D} \right)^{0.66} = 5.4 \times 10^{-4} \, Re_{Noz}^{\,2} \times \left(\frac{Q_G}{Q_G + Q_L} \right)$
Jet flow regime	$1.3 < (Q_G/Q_L) < 3$	$\left(\frac{k_L a V_{Ej}}{D} \right)^{0.66} = 3.1 \times 10^{-4} \, Re_{Noz}^{\,2}$

FIGURE 8.18   (Continued)

Jet flow applies since G/L ~ 3

Reynolds number	2.46.E+06	
Diffusivity of H_2	1.16.E-09	m^2/s
Volume of ejector	0.0503	m^3
Ejector section mass transfer coefficient, $k_L a$	15.60	1/s

G/L >1.3

Gas-liquid mass transfer in the holding vessel   Dirix and van der Wiele (1990)

$$\left(\frac{k_L a V_L}{D} \right) = 3.5 \times 10^{-12} \, Re_T^{3.3} \qquad Re_T = \left(\frac{4 \rho_L (Q_G + Q_L)}{\mu_L \pi D_T} \right)$$

Density of liquid, ρ_L	900	kg/m^3
Volumetric flow rate of gas, Q_G	0.171	m^3/s
Volumetric flow rate of liquid, Q_L	0.049	m^3/s
Viscosity of liquid, μ_L	0.0005	Poise
Diameter of throat, D_T	0.092	m
Reynolds number, Re_T	5,507,935	
V_L, holding vessel volume	4.24	m^3
Diffusivity of H_2, D_{H2}	1.16.E-09	m^2/s
Holding vessel mass transfer coefficient, $k_L a$	27.51	1/s

Solid liquid mass transfer

As argued in section 8.9 in view of the very high levels of turbulence in the ejector and the holding vessel, the value of $K_{SL}(A_P/V_R)$ is expected to be substantially high to eliminate solid-liquid mass tranfer resistance.

Solid suspension

$$Q_{GS} = 5.61 \times 10^{-3} \left(V_T \right)^{0.05} \left(X \right)^{0.11}$$

V_T, terminal settling velocity	0.0002	m/s
X, solid loading	1.00	%
Q_{GS}	0.0037	m^3/s

Since actual gas flow rate >> Q_{GS} or D146 >> D166, solid suspension is assured

Holding Vessel Design

Reaction kinetics step/rate expression / controlling step

Rate constant	0.515	1/s

This value is much lower than $k_L a$ for both the ejector and the holding vessel. Thus, the gas liquid and solid liquid mass transfer resistances can be neglected.

Production rate required		0.00974	kmol/s
Hydrogen solubility		0.00446	$kmol/m^3$
Volume required		4.24	m^3
Ejector volume to holding vessel volume ratio (%)	1.1860		

Check on scale up

Volumetric liquid flow rate, Q_L	176.54	m^3/hr
Vessel volume, V_L	4.24	m^3
Allowing for 40 % disengagement zone, Total volume,	7.07	m^3
Circulation number, n= Q_L/V_L	41.61	hr^{-1}

The holding vessel diameter can be obtained by using $H_{HV}/T = 8$. Please refer Spreadsheet output and justification in the text.

Power required

Power required, P= $Q_L \times (\Delta P_N)$	19.62	kW
Assuming extra power	15%	more
Power required	22.56	kW
Power input per unit volume	5.32	kW/m^3
Energy input per unit mass of product, E/M	23.39	kJ/kg

FIGURE 8.18   (Continued)

value of n (42/h) is lower, it is likely to assure a satisfactory performance of the reactor. This argument is supported by the fact that the power input per unit volume from the spreadsheet (5.4 kW/m³) is within the range given in Table 8.2. The holding vessel dimensions can be obtained by using $H_{HV}/T=8$. The dispersion volume obtained from the spreadsheet is 4.25 m³. Accordingly, for $H_{HV}/T=8$, T_{HV} is 0.88 m. An additional disengagement zone height corresponding to 40% excess volume is required. Hence, the total height of the holding vessel, H_{HVT}, is 11.6 m out of which the working height is 7 m. For larger reactor volumes or T_{HV} greater than 1.5–3 m and $H_{HV}/T_{HV}<2$, the arrangement suggested in Section 8.11 may be used (Dierendonck et al. 1998).

The main contributions to the capital cost of the venturi loop system are (i) the special pump and (ii) the patented venturi ejector. Choice of the circulation pump is crucial to stable continuous operation (Dierendonck et al. 1998). The circulating pump cost varies with its rating (pressure developed and flow rate). The pressure rating is fixed by the process conditions. On the other hand, the pump flow rate that directly affects the power required ($P=M_L \times \Delta P_N$) as well as its cost can be varied by choosing appropriate ΔT_{HEX}. A simple analysis shows that a higher ΔT_{HEX} results in lower operating cost (lower P_{Noz}/V). Depending on the requirements of the reaction, suitable ΔT_{HEX} may be selected. However, as mentioned earlier, the combined operating conditions should be such that the rate of gas induction is sufficiently in excess of (i) the stoichiometric requirement and (ii) minimum flow rate Q_{GS} required to keep the catalyst in suspension (Eq. 8.25). Generally, $5<\Delta T_{HEX}<10\,°C$ is adequate.

Comparison with Stirred Reactor   Comparison of the previous results with those in Section 7A.10 for stirred reactor reveal the following:

1. The venturi loop system operates at the intrinsic productivity as defined by the kinetics, whereas the stirred reactor (Section 7A.10) operates under a serious mass transfer limitation. This should be a matter of great concern if a noble metal catalyst is used.

2. The reactor volume required is significantly lower as compared to the stirred reactor of Section 7A.10. This is due to the absence of mass transfer limitation explained in (1) previously.

NOMENCLATURE

$[A_B]$:	Dissolved gas concentration in the bulk liquid phase (mol/m³)
$[A^*]$:	Dissolved gas concentration at the gas–liquid interface (mol/m³)
$[A]$:	Dissolved gas concentration at a given location in the liquid phase (mol/m³)
$[A_S]$:	Dissolved gas concentration at the surface of the solid catalyst (mol/m³)
a:	Effective gas–liquid mass transfer area (m²/m³)
A_{1-n}, A_{2-n}:	Dimensionless area ratios defined in Eq. 8.9

A_{DT}:   Area of the draft tube (m²)

A_{EN}:   Ratio of diffuser entry area of to nozzle area (—)

A_{HV}:   Cross-sectional area of holding vessel (m²)

A_N: Cross-sectional area of nozzle (m²)

A_p: Total surface area of the catalyst (m²)

A_R: Ratio of throat area to primary fluid nozzle area (—)

A_{SN}:   Ratio of suction chamber area to nozzle area (—)

$[B]$:   Concentration of liquid-phase reactant (mol/m³)

C_t: Craya–Curtet number (dimensionless ratio of thrusts and jet momentum controlling flow in confined jets where the primary jet is much smaller than the dimensions of the confining enclosure)

D_A: Molecular diffusivity of solute A (m²/s)

D_D: Diameter of the diffuser at the exit (m)

T_{DFT}:   Diameter of draft tube (m)

D_H: Diameter of the orifice in a multiorifice nozzle (m)

D_J: Diameter of the primary fluid jet (m)

D_{MT}:   Diameter of the mixing tube (m)

D_N: Diameter of the nozzle (m)

D_p: Diameter of cylindrical section of nozzle (m)

D_S: Diameter of suction chamber (m)

D_T: Diameter of the throat of the ejector (m)

d_B: Bubble diameter (m)

E_{uS}:   Secondary fluid Euler number, $\left(\dfrac{\Delta P}{\rho_S V_{SO}^{\,2}}\right)$ (—)

Ha: Hatta number, $\left(\sqrt{\dfrac{2}{p+1} D_A k_{p-q} [B_B]^q [A^*]^{p-1}}\right)\Big/ k_L$ (—)

H_A: Henry's law coefficient for solute A (mol/m³ kN/m²)

H_{CL}:   Clear liquid height in the holding vessel (m)

H_{HV}:   Dispersion height of in the holding vessel (m)

H_{HVT}:   Total height of the holding vessel including 40% excess volume (m)

H_N: Height of tapered portion of the primary fluid nozzle (m)

k: Consistency index for pseudo-plastic non-Newtonian liquid (Pa sⁿ)

k_1:   First-order reaction rate constant (m³/kg catalyst s)

k_2:   Second-order reaction rate constant (m³/mol s)

$k_L a$:   Volumetric liquid side mass transfer coefficient (1/s)

k_R: Overall surface reaction rate constant (1/s)

k'_R: Surface reaction rate constant (1/kg catalyst s) or (m³/m² s)

L_D: Length of diverging section of the diffuser (m)

L_{DP}:	Length of straight portion following the diverging section of the diffuser (m)
L_{Jet}:	Length of a plunging jet (m)
L_{MT}:	Length of the mixing tube (m)
L_{NT}:	Distance between the tip of the nozzle and the throat (m)
L_p:	Length of cylindrical section of nozzle (m)
L_{PT}:	Length of penetration of a plunging jet (m)
M:	Mass of liquid in the reactor (kg)
M_G:	Mass flow rate of the gas phase (kg/s)
M_L:	Mass flow rate of liquid phase (kg/s)
M_{OP}:	Morton number for the primary fluid, $\left(\dfrac{g\mu_P^{\,4}}{\rho_P\sigma_P^{\,3}}\right)$ (—)
M_P:	Mass flow rate of primary fluid (kg/s)
M_R:	Ratio of mass flow rates of secondary and primary fluids (—)
M_{Sec}:	Mass flow rate of secondary fluid (kg/s)
n:	Flow behavior index for pseudo-plastic non-Newtonian liquid (—)
N_O:	Number of orifices in a given multiorifice nozzle (—)
N_{SG}:	Critical speed for solid suspension in a three-phase stirred reactor (rev/s)
p:	Order of reaction with respect to the gaseous solute (—)
P:	Total pressure (Pa or MPa)
p_A:	Partial pressure of solute A (Pa or MPa)
P_{Diff}:	Pressure at entry to the diffuser (Pa or MPa)
P_{MT}:	Pressure at entry to the mixing tube (Pa)
P_N:	Pressure at nozzle exit (Pa)
PR:	Projection ratio, L_{NT}/D_T (—)
$[P_{Noz}/V_L]_{Crit}$:	Power input per unit liquid volume at which the two-phase jet reaches the bottom of the holding vessel (kW/m³)
$[P_{Noz}/V_L]_S$:	Power input per unit liquid volume for just suspension of the solid (kW/m³)
ΔP_D:	Pressure recovery of the secondary fluid across the ejector (Pa)
ΔP_G:	Pressure drop across the diffuser or "back pressure" on the suction chamber (Pa)
ΔP_N:	Pressure drop of the primary fluid across the nozzle (Pa)
q:	Order of reaction with respect to the liquid-phase reactant (—)
Q_G, Q_S:	Flow rate of the induced gas/secondary fluid (m³/s)
Q_{GS}:	Gas induction rate at which solids get suspended in the holding vessel (m³/s)
Q_P, Q_L:	Flow rate of the primary fluid or liquid (m³/s)

Re_{Noz}:	Reynolds number of the primary fluid at nozzle exit $(D_N V_{Noz} \rho_P / \mu_P)$ (—)
Re_T:	Primary fluid Reynolds number based on the throat diameter $(D_T V_{Noz} \rho_P / \mu_P)$ (—)
R_V:	Volumetric rate of reaction (mol/m^3 s)
T_{HV}:	Diameter of the holding vessel (m)
ΔT_{HEX}:	Temperature drop of the reactant across the external heat exchanger (K)
U:	Overall heat transfer coefficient (W/m^2 K)
V_D:	Dispersion volume in the holding vessel (m^3)
V_{Ej}:	Volume of the ejector section (m^3)
V_F:	Flow ratio (Q_S/Q_P) (—)
V_G:	Superficial gas velocity based on empty cross section of the holding vessel (m/s)
V_{HV}:	Volume of the holding vessel (m^3)
V_{Jet}:	Velocity of plunging jet (m/s)
V_L:	Superficial liquid velocity in the holding vessel (m/s)
V_P:	Superficial velocity of primary fluid/motive fluid (m/s)
V_{PNT}, V_{Noz}:	Velocity of primary fluid/motive liquid at the nozzle tip (m/s)
V_{SI}:	Velocity of secondary fluid at point I (m/s)
V_T:	Terminal settling velocity of solid particles (m/s)
X:	Solid loading (wt %)

Greek letters

η':	Ejector efficiency, $\left(\dfrac{M_G}{M_L}\right)\left(\dfrac{\Delta P_G}{\Delta P_L}\right)$ (—)
μ:	Fluid viscosity (Pa s)
σ:	Surface tension (N/m)
$(\varepsilon_G)_{AN}$:	Fractional gas holdup in the annulus (—)
$(\varepsilon_G)_{AVG}$:	Average fractional gas holdup for the venturi loop system (—)
$(\varepsilon_G)_{DT}$:	Fractional gas holdup in the draft tube (—)
ε_L:	Fractional liquid holdup (—)
ε_{GNS}:	Gas holdup based on assumption of "no slip" (—)
ε_{GTot}:	Gas holdup based on the total ejector volume (—)
γ:	Shear rate (/s)
γ':	Pa (s)^n
ζ:	Nozzle discharge coefficient (—)
η_I:	Induction efficiency parameter, $\left(\dfrac{Q_G}{P}\right)$ (m^3/W)
μ_{Eff}:	Effective viscosity of non-Newtonian liquid (Pa s)

ν: Kinematic viscosity (m^2/s)

ρ: Density (kg/m^3)

ρ_{P-S}:   Ratio of the densities of primary and secondary fluids (—)

ψ: Stoichiometric factor in the reaction between the dissolved gas- and liquid-phase reactants (—)

Subscripts

A: Solute A

AN: Annulus

B: Bulk liquid-phase property

Diff: End of diffuser

DT: Draft tube

Ej: Ejector

HV: Holding vessel

J: Jet

Mix: Mixture property

MT: Mixing tube

P: Primary fluid

S: Secondary fluid

T: Throat

REFERENCES

Acharjee DK, Bhat PA, Mitra AK, Roy AN. (1975) Studies on momentum transfer in vertical liquid-jet ejector. Indian J. Technol., 13:205–210.

Aloqaily A, Kuhn DCS, Sullivan P, Tran A. (2007) Effect of burning non-condensable gases on limekiln flame patterns. http://www.tappi.org/Downloads/unsorted/UNTITLED---ICR0415pdf.aspx. Accessed Feb. 2012.

Baier FO. (2001) Mass transfer characteristics of a novel gas–liquid contactor, the Advanced Buss Loop Reactor. Doctoral thesis. Swiss Federal Institute of Technology Zurich.

BCT Loop® Reactor Technology (2009–02) http://www.buss-ct.com/e/company/publications/reaction_technology/BCT_Loop_Reactor_Technology_2009-02.pdf. Accessed Nov. 2011.

Behr A, Becker M, Dostal J. (2008) cf Behr et al., 2009. Hydrodynamic and lapsed time response of jet operation reactors and for the adoption in polyphase catalyst. Chem. Ing. Technol., 80(10):1501–1508.

Behr A, Becker M, Dostal J. (2009) Bubble-size distributions and interfacial areas in a jet loop reactor for multiphase catalysis. Chem. Eng. Sci., 64:2934–2940.

Ben Brahim A, Prevost M, Bugarel R. (1984) Momentum transfer in a vertical down flow liquid jet ejector: case of self-aspiration and emulsion flow. Int. J. Multiphase Flow, 10(1):79–94.

Berman LD, Efimochin GI. (1964) Design equations for water jet ejectors. Thermal Eng., 11 (7):57–62.

Betarbet HR (1991) Mass transfer in liquid jet loop reactors. M. Chem. Eng. Thesis, University of Mumbai, Mumbai, India.

Betarbet HR, Pangarkar VG. (1995) Gas–liquid mass transfer in liquid jet loop reactors. Ind. Chem. Eng., 37(1&2):68–71.

Bhat PA, Mitra AK, Roy AN. (1972) Momentum transfer in a horizontal liquid jet ejector. Can. J. Chem. Eng., 50:313–317.

Bhutada SR, Pangarkar VG. (1987) Gas induction and hold-up characteristics of liquid jet loop reactors. Chem. Eng. Commun., 61(1–6):239–258.

Bhutada SR, Pangarkar VG. (1988a) Solid suspension and mixing in liquid jet loop reactors. Chem. Eng. Sci., 44:2384–2387.

Bhutada SR, Pangarkar VG. (1989b) Gas induction and hold-up characteristics of liquid jet loop reactors: pseudo plastic non-Newtonian liquids. Chem. Eng. Sci., 43: 2904–2907.

Bhutada SR, Pangarkar VG. (1989) Solid suspension and mixing in liquid jet loop reactors. Chem. Eng. Sci., 44:2384–2387.

Bhutada SR. (1989) Design of liquid jet loop reactors. Ph.D. (Tech.) Thesis, University of Mumbai, Mumbai, India.

Bin AK. (1983) Oxygen transfer in jet mixers. Chem. Eng. J., 26:225–257.

Bin AK. (1993) Gas entrainment by plunging liquid jets. Chem. Eng. Sci., 48:3585–3630.

Biswas MN, Mitra AK. (1981) Momentum transfer in multi jet liquid–gas ejectors. Can. J. Chem. Eng., 59:634–637.

Bloor CJ, Anderson G, Willey AR. (1995) High rate aerobic treatment of brewery wastewater using the jet loop reactor. Wat. Res., 29:1217–1223.

Bonnington ST. (1956) Jet pumps. BHRA Publication. SP529, BHRA Fluid Engineering, Cranfield, Bedford, UK.

Bonnington ST. (1964) A guide to liquid jet pumps. Brit. Chem. Eng., 9(3):150–154.

Bonnington ST, King AL. (1976) Jet pumps and ejectors: a state of the art review and bibliography. 2nd ed. BHRA Fluid Engineering, Cranfield, Bedford, UK.

Burgess JM, Molloy NA. (1973) Gas absorption in the plunging jet reactor. Chem. Eng. Sci., 28:183–190.

Cairns JR, Na TY. (1969) Optimum design of water jet pump. Trans. ASME, 91(1):62–68.

Chaudhari RV, Mills PL. (2004) Multiphase catalysis and reaction engineering for emerging pharmaceutical processes. Chem. Eng. Sci., 59:5337–5344.

Concordia JJ. (1990) Batch catalytic gas/liquid reactors: types and performance characteristics. Chem. Eng. Prog., 86(3):50–54.

Cramers PHMR, van Dierendonck LL, Beenackers AACM. (1992a) Influence of the gas density on the gas entrainment rate and gas hold up in loop-venturi reactors. Chem. Eng. Sci., 47:2251–2256.

Cramers PHMR, Beenackers AACM, van Dierendonck LL. (1992b) Hydrodynamics and mass transfer characteristics of a loop-venturi reactor with a down flow liquid jet ejector. Chem. Eng. Sci., 47:3557–3564.

Cramers PHMR, Smit S, Leuteritz GM, van Dierendonck LL, Beenackers AACM. (1993) Hydrodynamics and local mass transfer characteristics of gas–liquid ejectors. The Chem. Eng. J., 53:67–73.

Cramers PHMR, Beenackers AACM. (2001) Influence of the ejector configuration, scale and the gas density on the mass transfer characteristics of gas–liquid ejectors. Chem. Eng. J., 82:131–141.

Cunningham RG. (1957) Jet pump theory and performance with fluids of high viscosity. Trans. ASME, 79:1807–1820.

Cunningham RG. (1974) Gas compression with the liquid jet pump. J. Fluid Eng., Series I, Vol. 96, No. 3 (Sept.):203–215.

Cunningham RG, Dopkin RJ. (1974) Jet breakup and mixing throat lengths for the liquid jet water pump. Trans ASME Series I, Vol. 96, No. 3 (Sept.):216–226.

Curtet R. (1958) Confined jets and recirculation phenomena with cold air. Combust. Flame, 2(4):383–411.

Danckwerts PV, Kennedy AM, Roberts D. (1963) Kinetics of CO_2 absorption in alkaline solutions—II absorption in a packed column and tests of surface renewal models. Chem. Eng. Sci., 18:63–72.

Davies GS, Mitra AK, Roy AN. (1967a) Momentum transfer studies in ejectors. Correlations for single-phase and two-phase systems, Ind. Eng. Chem. Proc. Des. Dev., 6(3):293–299.

Davies GS, Mitra AK, Roy AN. (1967b) Momentum transfer studies in ejectors. Correlation for three-phase (air–liquid–solid) system. Ind. Eng. Chem. Proc. Des. Dev., 6(3):299–302.

Davies TW, Jackson MK. (1985) A procedure for the design of nozzles used for the production of turbulent liquid jets. Int. J. Heat Fluid Flow, 6(4):289–305.

Deckwer W-D. (1992) Bubble column reactors (English translation) John Wiley & Sons, Chichester, UK.

Di Serio M, Tesser R, Santacesaria E. (2005) Comparison of different reactor types used in the manufacture of ethoxylated, propoxylated products. Ind. Eng. Chem. Res., 44: 9482–9489.

Dierendonck van LL, Meindersma GW, Leuteritz GM. (1988) Scale-up of G-L reactions made simple with loop reactors. In Proceedings of Sixth European Conference on Mixing, Pavia, Italy, 24–26 May, p 287–295.

Dierendonck van LL, Zahradnik J, Linek V. (1998) Loop venturi reactor—a feasible alternative to stirred tank reactors? Ind. Eng. Chem. Res., 37:734–738.

Dirix CAMC, van der Wiele K. (1990) Mass transfer in jet loop reactors. Chem. Eng. Sci., 45:2333–2340.

Donald MB, Singer H. (1959) Entrainment in turbulent jets. Trans. Inst. Chem. Eng. (UK), 37:255–267.

Duquenne AM, Guiraud P, Bertrand J. (1993) Swirl-induced improvement of turbulent mixing: laser study in a jet-stirred tubular reactor. Chem. Eng. Sci., 48:3805–3812.

Dutta NN, Raghavan KV. (1987) Mass transfer and hydrodynamic characteristics of loop reactors with down flow liquid jet ejector. The Chem. Eng. J., 36:111–121.

Duveen RF. (1998) High-performance gas–liquid reaction technology. In: Proceedings of the Symposium New Frontiers in Catalytic Reactor Design, The Royal Society of Chemistry, Applied Catalysis Group, Billingham, UK, October 1998.

Elgozali A, Linek, V, Fialova M, Wein O, Zahradnik J. (2002) Influence of viscosity and surface tension on performance of gas–liquid contactors with ejector type gas distributor. Chem. Eng. Sci., 57:2987–2994.

Engdahl RB, Holton WC. (1943) Overfire air jets. Trans. ASME., 65(10):741–754.

Engel MO. (1963) Some problems in the design and operation of jet ejectors. Proc. Inst. Mech. Eng., 177(3):347–362.

Engel OG. (1966) Crater depths in fluid impacts. J. Appl. Phys., 37(4):1798–1808.

Engel OG. (1967) Initial pressure, initial velocity and the time dependence of crater depths in fluid impacts. J. Appl. Phys., 38(10):3935–3940.

Evans GM, Jameson GJ. (1991) Prediction of the gas film entrainment rate for a plunging liquid jet reactor. In: Proceedings of AICHE Symposium on Multiphase Reactors, Houston, TX, USA.

Farizoglu B, Keskinler B. (2007) Influence of draft tube cross-sectional geometry on $K_L a$ and ε in jet loop bioreactors (JLB). Chem. Eng. J. 133:293–299.

FLUENT (2006) FLUENT User's Guide, Version 6.2, Fluent Inc.

Flugel G. (1939) The design of jet pumps. V. D. I. Forsch. 395, SS p 1–21. (March–April, 1939). (In German). Translation N.A.C.A. Tech. Mem. No. 982. (July 1941), cf: Engel, 1963.

Folsom RG. (1948) Jet pumps with liquid drive. Chem. Eng. Prog., 44(10):765–770.

Gamisans X, Sarrà M, Lafuente J. (2004) Fluid flow and pumping efficiency in an ejector-venturi scrubber. Chem. Eng. Process., 43:127–136.

Geser U, Malone RJ. (2000) New alkoxylation technologies. Paper presented at Alcoxilação-Il Seminário Interno, Sao Paulo, Brazil, May 2000. (http://hhtservices.com/hh-publications/New-Alkoxylation-Technologies-UGeser-RMalone.pdf) Accessed Sept., 2013.

Goff JA, Coogan CH. (1942) Some two-dimensional aspects of the ejector problem. J. Appl. Mech., 9(4):A-151–154.

Gosline JE, O'Brien MP. (1934) The water jet pump. Univ. of California Pubs., vol. III, 167–190.

Gourich B, Belhaj Soulami M, Zoulalian A, Ziyad M. (2005) Simultaneous measurement of gas hold-up and mass transfer coefficient by tracer dynamic technique in "Emulsair" reactor with an emulsion-venturi distributor. Chem. Eng. Sci., 60:6414–6421.

Gourich B, El Azher N, Vial Ch, Belhaj Soulami M, Ziyad M, Zoulalian A. (2007) Influence of operating conditions and design parameters on hydrodynamics and mass transfer in an emulsion loop-venturi reactor. Chem. Eng. and Process., 46:139–149.

Gourich B, Vial Ch, Belhaj Soulami M, Zoulalian A, Ziyad M. (2008) Comparison of hydrodynamic and mass transfer performances of an emulsion loop-venturi reactor in co current down flow and upflow configurations. Chem. Eng. J., 140:439–447.

Grandchamp X, Fujiso Y, Wu B, Van Hirtum A. (2012) Steady laminar axisymmetrical nozzle flow at moderate Reynolds numbers: modelling and experiment. J. Fluids Eng., 134(1):0112031–13.

Greenwood TS. (1986) Loop reactors for catalytic hydrogenations. Chem. & Ind., Feb 3:94–98.

Hammoud AH. (2006) Effect of design and operational parameters on jet pump performance. In: Proceedings of the Fourth WSEAS International Conference on Fluid Mechanics and Aerodynamics, Elounda, Greece, August 21–23, 2006, p 245–252. (http://www.wseas.us/e-library/conferences/2006elounda2/papers/538-161.pdf) Accessed Sept. 2013.

Hansen AG, Kinnavy R. (1965a) The design of water jet pumps. Part 1: Experimental determination of optimum design parameters. ASME Paper No.65-WA/FE-31. American Society of Mechanical Engineers, New York, USA.

Hansen AG, Kinnavy R. (1965b) The design of water jet pumps. Part 2: Jet pump system characteristics. ASME Paper No.65-WA/FE-32. American Society of Mechanical Engineers, New York, USA.

Hansen AG, Na TY. (1966) Optimization of jet pump systems. ASME-EIC Fluids Eng. Conf., Paper No. 66-FE-44. American Society of Mechanical Engineers, New York, USA.

Havelka P. (1997) Mass transfer capacity of gas–liquid reactors with ejector-type gas distributors. Ph.D. Thesis, Prague Institute of Chemical Technology, Czech Republic. *cf.* Dierendonck et al. (1998).

Havelka P, Linek V, Sinkule J, Zahradnik J, Fialova P. (1997) Effect of the ejector configuration on the gas suction rate and gas hold-up in ejector loop reactors. Chem. Eng. Sci., 52:1701–1713.

Havelka P, Linek V, Sinkule J, Zahradnik J, Fialova P. (2000) Hydrodynamic and mass transfer characteristics of ejector loop reactors. Chem. Eng. Sci., 55:535–549.

Hedges KR, Hill PG. (1972) A Finite-difference method for confined jet mixing. Thermal and Fluid Sciences Group, Department of Mechanical Engineering, Queen's University, Kingston, Ontario, Canada.

Helmbold H, Luessen G, Heinrich AM. (1954) An experimental comparison of constant pressure and constant diameter jet pumps. Engineering study No.147, School of Engineering, Univ. of Wichita, KS, USA.

Henzler HJ. (1981) Das sogverhalten von strahlsaugern fuer das stoffsystem: flussig-gasformig. Vt-Verfahrenstechnik, 15(10):738–749. *cf.* Baier (2001).

Henzler HJ. (1983) Design of ejectors for single-phase material systems. German Chem. Eng., 6(6):292–300.

Hill BJ. (1973) Two-dimensional analysis of flow in jet pumps. J. Hydraulics Div., ASCE, 99 (HY-7):1009–1026.

Huang BJ, Chang JM, Wang CP, Petrenko VA. (1999) A 1-D analysis of ejector performance. Int. J. Refrigeration 22:354–364.

Kandakure MT, Gaikar VG, Patwardhan AW. (2005) Hydrodynamic aspects of ejectors. Chem. Eng. Sci., 60:6391–6402.

Kandakure MT. (2006) Studies on gas–liquid contactor. Ph.D. (Tech.) Thesis, University of Mumbai, Mumbai, India.

Kar S, Reddy YS. (1968) Theory and performance of water jet pumps. J. Hydraulic Division ASCE, 94, HY:1261–1281.

Kastner LJ, Spooner JR. (1950) An investigation of the performance and design of the air ejector employing low-pressure air as the driving fluid. Proc. Inst. Mech. Eng., 162 (6):149–166.

Keenan JH, Neumann EP. (1950) An investigation of ejector design by analysis and experiment. J. Appl. Mech., Trans ASME, 72:299–309.

Kern DQ. (1997) Process heat transfer. Tata McGraw-Hill Publishing Company, New Delhi, India.

Kim MI, Kim OS, Lee DH, Kim SD. (2007) Numerical and experimental investigations of gas–liquid dispersion in an ejector. Chem. Eng. Sci. 62:7133–7139.

Kroll AE. (1947) The design of jet pumps. Chem. Eng. Prog., 1(2):21–24.

Lau R, Peng W, Velazquez-Vargas LG, Yang GQ, Fan LS. (2004) Gas–liquid mass transfer in high pressure bubble columns. Ind. Eng. Chem. Res., 43:1302–1311.

Leuteritz GM. (1973) Loop reactor gives fast, cool, liquid-phase hydrogenation reactions. Process Eng., (Dec. 1973):62–63.

Leuteritz GM, Reiman P, Vergeres P. (1976) Loop reactors: Better gas/liquid contact. Hydrocarb. Proc., 55(6):99–100.

Li C, Li YZ. (2011) Investigation of entrainment behavior and characteristics of gas–liquid ejectors based on CFD simulation. Chem. Eng. Sci., 66:405–416.

Li M, Niu F, Zuo X, Metelski PD, Busch DH, Subramaniam B. (2013) A spray reactor concept for catalytic oxidation of p-xylene to produce high-purity terephthalic acid. Chem. Eng. Sci., 104:93–102.

Liao C. (2008) Gas ejector modeling for design and analysis. Ph. D. Thesis, Texas A & M University, TX, USA (URL: http://hdl.handle.net/1969.1/ETD-TAMU-3206) Accessed Feb. 2012.

Linek V, Moucha T, Kordac M. (2005) Mechanism of mass transfer from bubbles in dispersions: Part I. Danckwerts' plot method with sulphite solutions in the presence of viscosity and surface tension changing agents. Chem. Eng. Process.: Proc. intensification, 44(3):353–361.

Lubbecke S, Vogelpohl A, Dewjanin W. (1995) Wastewater treatment in a biological high performance system with high biomass concentration. Wat. Res., 29:793–802.

Malone RJ. (1980) Loop reactor technology multiphase reactor improves catalytic hydrogenations. Chem. Eng. Prog., 33(6):53–59.

Malone RJ, Merten HL. (1992) A comparative mass transfer study in reductive N-alkylation of aromatic nitro compounds. In: Pascoe WL, editor. Catalysis of organic reactions. Marcel Dekker, Boca Raton, FL, USA.

Mandal A, Kundu G, Mukharjee D. (2005a) Comparative study of two-phase gas–liquid flow in the ejector induced upflow and down flow bubble column. Int. J. Chem. Reactor Eng., 3, Article A13:1–13.

Mandal A, Kundu G, Mukharjee D. (2005b) Energy analysis and air entrainment in an ejector induced down flow bubble column with non-Newtonian motive fluid. Chem. Eng. Tech., 28:210–218.

Manninen M, Taivassalo V, Kallio S. (1996) On the mixture model for multiphase flow. VTT Publications 288, Technical Research Centre of Finland. Valtion teknillinen tutkimuskeskus, Julkaisija-Utgivare Publishers, VTT, Espoo, Finland.

McKeogh MJ, Ervine DA. (1981) Air entrainment rate and diffusion pattern of plunging liquid jets. Chem. Eng. Sci., 36:1161–1172.

Mellanby AE (1928) Fluids jets and their practical applications. Trans. Inst. Chem. Eng. (UK), 6:66–84.

Mueller NHG. (1964) Water jet pump. J. Hyd. Div. ASCE, 90, No.HY3, Proc. 3908 (May): 83–113.

Nardin D, Cramers PHMR. (1996) Trends in Buss Loop reactor technology. Specialty Chem., 16(8):308–312.

Neve RS. (1988) The performance and modeling of jet pumps. Int. J. Heat Fluid flow, 9(2):156–164.

Neve RS. (1991) Diffuser performance in two-phase jet pumps. Int. J. Multiphase Flow, 17(2):267–272.

Neve RS. (1993) Computational fluid dynamics analysis of diffuser performance in gas powered jet pumps. Int. J. Heat Fluid flow, 14(4):401–407.

Nilavalagan S, Ravindran M, Radhakrishna HC. (1988) Analysis of mixing characteristics of flow in a jet pump using a finite-difference method. Chem. Eng. J., 39:97–1010.

Opre Z. (2009) Benefits of the advanced BUSS Loop reactor technology by removal of by-products through the gas circulation system: http://www.dechema.de/index.php?id=11 6220&tagung=9&file=7717&site=achema2009&lang=en&path=1%25252C79095. Accessed December, 2011.

Otake T, Tone S, Kuboi R, Takahashi Y, Nakao K. (1981) Dispersion of a gas by a liquid jet ejector. Int. Chem. Eng., 21(1):72–80.

Panchal NA. (1990) Design of liquid jet loop reactors. M. Chem. Eng. Thesis, University of Mumbai, Mumbai, India.

Panchal NA, Bhutada SR, Pangarkar VG. (1991) Gas induction and hold-up characteristics of liquid jet loop reactors using multi orifice nozzles. Chem. Eng. Commun., 102:59–68.

Park B, Hwang G, Haam S, Lee C, Ahn I-S, Lee K. (2008) Absorption of a volatile organic compound by a jet loop reactor with circulation of a surfactant solution: performance evaluation. J. Haz. Mat., 153:735–741.

Petruccioli M, Duarte JC, Eusebio A, Federici F. (2002) Aerobic treatment of winery wastewater using a jet loop activated sludge reactor. Process Biochem., 37(8):821–829.

Radhakrishnan VR, Sen-Chaudhury D and Mitra AK (1986) Momentum transfer in multi-jet liquid–gas ejectors. IX Australasian fluid mechanics conference, 8–12 Dec. 136–139, Auckland, New Zealand (http://www.mech.eng.unimelb.edu.au/people/staffresearch/AFMS%20site/9/RadhakrishnanEtAl.pdf). Accessed March 2012.

Raghuram PT. (2007) Interfacial area measurement in a gas liquid ejector. Ind. J. Chem. Tech., 16:278–282.

Rusly E, Aye L, Charters WWS, Ooi A. (2005) CFD analysis of ejector in a combined ejector cooling system. Int. J. of Refrigeration 28:1092–1101.

Sande van de E, Smith JM. (1973) Surface entrainment of air by high velocity water jets. Chem. Eng. Sci., 28:1161–1168.

Sande van de E, Smith JM. (1975) Mass transfer from plunging water jets. Chem. Eng. J., 10:225–233.

Sanger NL. (1969) Noncavitating and cavitating performance of several low area ratio water jet pumps having throat lengths of 3.54 diameter. NASA Report. National Aeronautics and Space Administration, Washington, DC, USA.

Sanger NL (1971) Fortran programs for the design of liquid-to-liquid jet pumps. TN D-6453. National Aeronautics and Space Administration, Washington, DC, USA.

Schugerl PK (1982) New bioreactors for aerobic processes. Int. Chem. Eng., 22:591–610.

Sokolov EY and Zinger NM. (1970) Jet Devices, Energia Publ., Moscow, Russia.

Stefoglo EF, Zhukova OP, Kuchin IV, Albrecht SN, Nagirnjak AT, Dierendonck LL. (1999) Reaction engineering of catalytic gas–liquid processes in loop-venturi reactors in comparison with stirred vessels operation. Chem. Eng. Sci., 54:5279–5284.

Sun DW, Eames IW. (1995) Recent developments in the design theories and applications of ejectors—a review. J. Inst Energy, 68:65–79.

Tinge JT, Rodriguez Casado AJ. (2002) Influence of pressure on the gas hold-up of aqueous activated carbon slurries in a down flow jet loop reactor. Chem. Eng. Sci., 57:3575–3580.

Utomo T, Jin Z, Rahman M, Jeong H, Chung H. (2008) Investigation on hydrodynamics and mass transfer characteristics of a gas–liquid ejector using three-dimensional CFD modeling. J. Mech. Sci. Tech., 22:1821–1829.

Vogel WJ, Knautz E, Kiefer PJ. (1956) Air lance for blowing out superheater elements and the like. U.S. Patent 2770830 A.

Vogel GH. (2005) Process development: from the initial idea to the chemical production plant. Wiley-VCH Verlag GmbH & Co. KGaA, Weinheim, Germany.

Von Pawell-Rammingen G. (1936) Dissertation, Braunschweig, Germany. *cf:* Witte (1965)

Witte JH. (1962) Mixing shocks and their effects on the design of liquid–gas ejectors. Ph. D. Thesis, University of Delft, The Netherlands.

Witte JH. (1965) Efficiency and design of liquid–gas ejectors. Brit. Chem. Eng., 10(9): 602–607.

Witte JH. (1969) Mixing shocks in two-phase flow. J. Fluid Mech., 36:639–655.

Yadav RL, Patwardhan AW. (2008) Design aspects of ejectors: Effects of suction chamber geometry. Chem. Eng. Sci., 63:3886–3897.

Yang Z, Fan C, Zhuang C, Chen G. (1996) The development of a new ethoxylation technology. Petrochem. Technol., 25 (10):700–705.

Yaws CL, Li B, Nijhawan S, Hooper JR, Pike RW. (1999) Chapter 12: Enthalpy of formation. In: Yaws CL, editor. Chemical properties handbook. McGraw-Hill, New York, USA.

Yuan G, Zhang L, Zhang H, Wang Z. (2011) Numerical and experimental investigation of performance of the liquid–gas and liquid jet pumps in desalination systems. Desalination 276:89–95.

Zahradnik J, Kratochvil J, Kaštánek F, Rylek M. (1982) Hydrodynamic characteristics of gas–liquid beds in contactors with ejector type gas distributors. Collect. Czech. Chem. Communications, 47:1939–1949.

Zahradnik J, Rylek M. (1991) Design and scale up of venturi tube gas distributors for bubble column reactors. Collect. Czech. Chem. Commun., 56:619–634.

Zehner P. (1975) Stoffaustauschflache und Gasverdichtung in einer neu entwickelten Ejektorstrahlduse. Wissenschaftliche Forschungsarbeit. 201/75, Synopse. Chem. Ing. Tech., 47(5):209–2114.

9

GAS-INDUCING REACTORS

9.1   INTRODUCTION AND APPLICATION AREAS
OF GAS-INDUCING REACTORS

Froth flotation cells, which are the forbearers of gas-inducing reactor, have been extensively used in the mineral processing industry. Although the basic principle is the same in froth flotation cells and gas-inducing reactors, these two have totally different objectives. Harris (1976) has presented a critical review of the equipment used for froth flotation. The mined ore is crushed to "release" the desired components of the ore from the undesired. In flotation cells, special types of surface-active agents are used to create uniform-sized bubbles with definite surface properties. These bubbles have specific attraction for a particular component in the crushed ore that attaches to the bubbles. Consequently, as the bubbles rise, they carry the attached component. This adhering material is lifted by the buoyancy of the bubbles and floated to the top surface from where it is withdrawn. The emphasis here is on bubble–solid interaction. The power input should be sufficient enough to suspend the solids and attach them to the bubbles. Too high a power input results in detachment of the solids from the bubble and the primary objective of floating the desired solid material is defeated (Evans et al. 2008). Yianatos and coworkers studied various aspects of froth flotation cells. These include mixing characteristics (Yianatos et al. 2005); time scale-up factor for laboratory to plant-scale flotation cells (Yianatos et al. 2006); fluid flow and kinetic modeling (Yianatos 2007); residence time distribution and gas holdup (Yianatos et al. 2010a); scale-up approach employing

Design of Multiphase Reactors, First Edition. Vishwas Govind Pangarkar.
© 2015 John Wiley & Sons, Inc. Published 2015 by John Wiley & Sons, Inc.

hydrodynamic regime represented by mixing, effective residence time in terms of solid segregation and froth recovery in flotation rates (Yianatos et al. 2010b); and modeling and simulation of flotation flow circuits (Yianatos et al. 2012). Gas-inducing reactors also require thorough mixing of the gas–liquid mixture in order to ensure uniformity in concentrations and enhance selectivity of the reaction to a particular product. The attachment/detachment of solids is not a factor in this application since no "collectors" are employed. The feed to the reactor is a liquid that contains solid catalyst. The reacting gas is induced by the special impellers used. Gas-inducing reactors are similar to venturi loop reactor in terms of some of the advantages. In particular, the self-aspirating nature of these reactors can provide a significant advantage over conventional stirred tank reactors.

9.1.1   Advantages

1. Self-aspirating nature makes gas-inducing reactors attractive for applications in dead-end situations involving pure/expensive gas phase (Table 3.3). As a result, recycle compressor for the gas is not required.

9.1.2   Drawbacks

1. On a stand-alone basis, the heat transfer area available is only through the external jacket on the vessel (Table 3.3). Further, as the reactor volume or diameter increases, the jacket area per unit volume $\{\propto(1/T)\}$ decreases. For fast, highly exothermic reactions ($\Delta H_R > 100\,\text{kcal/mol}$) and large-diameter reactors, the jacket may not be sufficient to allow reasonable temperature control (Section 7A.10). The alternative of internal cooling coil is counterproductive (Section 7A.8). Therefore, in such a situation, the only alternative is an external heat transfer loop (Section 7A.10). This entails additional capital cost of the pump around circuit comprising of the pump and heat exchanger besides the operating cost of the pump.

2. Gas induction rate is sensitive to the liquid level (submergence) above the main gas-inducing impeller (Conway et al. 2002).

3. In view of (2), the gas-inducing impeller has to be necessarily located closer to the top liquid level (Section 3.4.2.4.3 and Figure 3.3 Praxair liquid oxidation reactor). For three-phase reactions involving a solid catalyst, a multi-impeller system is required. This is because a single impeller used to induce the gas is not capable of suspending the solid settled at the bottom. Further, because of the poor gas dispersion by the top gas-inducing impeller, the second lower impeller has to also perform this duty of gas dispersion in the lower part of the reactor.

4. Total power required (sum of impeller power and external loop pump) for cases (1) and (3) above is relatively high, particularly on large scale (Conway et al. 2002; also Section 7A.10).

5. Requires careful maintenance of the sealing mechanism for the agitator shaft if the gas/liquid mixture is inflammable and reaction pressure is also high (>1 MPa).

The qualitative comparison in Table 3.3 and the earlier specific discussion indicates that gas-inducing systems have severe limitations. A survey of the relevant literature also does not reveal the use of gas-inducing systems in large-scale commercial applications for three-phase reactions.

Gas-inducing reactor can also be used for gas–liquid reactions employing a homogeneous catalyst and particularly when the gas phase is available at low pressure. Some industrial reactions for which gas-inducing reactor can be used are oxidations, ozonolysis, alkylation, low-pressure hydrogenations, etc. (Patil et al. 2005). This chapter gives a brief review of the literature on gas-inducing reactor and the corresponding design procedures. The discussion is limited because of the restrictions on its use in view of the drawbacks mentioned earlier (Conway et al. 2002). The literature references provided may be used for additional details on different aspects.

9.2 MECHANISM OF GAS INDUCTION

Consider a hollow shaft connected to a hollow impeller immersed in a liquid. The space above the liquid level is filled with the gas to be induced (Fig. 9.1a, b, c, d, e).

As the impeller rotates, according to Bernoulli's theorem, the increase in kinetic energy at the tip of the impeller causes a decrease in the local pressure. As a result, the liquid level in the hollow shaft decreases. With increasing speed of agitation, there is concomitant decrease in the local pressure. Thus, the liquid level in the hollow shaft continues to fall. At a certain speed, this level reaches the tip of the impeller. Any further increase in the speed of agitation will now cause the gas to be sparged into the liquid surrounding the impeller through orifices suitably located on the impeller. The speed of agitation at which the liquid level reaches the tip of the impeller (Fig. 9.1d) is denoted as the critical speed for gas induction, N_{CG}. This mechanism suggests a simple method to obtain the value of N_{CG}. Thus, under ideal conditions of no frictional losses, at N_{CG}, the kinetic head should equal the submergence of the impeller in the liquid. For real situations, frictional losses and a slip between the impeller and the surrounding liquid cause the actual N_{CG} to be somewhat higher than the theoretically calculated value.

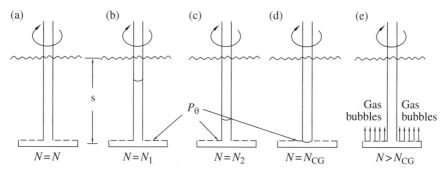

FIGURE 9.1 Principle of gas induction through a hollow rotating impeller.

9.3   CLASSIFICATION OF GAS-INDUCING IMPELLERS

The design of a gas-inducing impeller has a major bearing on its performance. Different types of gas-inducing impellers ranging from the simple hollow pipe type (Martin 1972; Fig. 9.1) to the more complex froth flotation or multi-impeller combinations (Patwardhan and Joshi 1999) can be used depending on the objective. Mundale (1993) has classified the different types available in the following three categories:

1. Single-phase flow (gas alone) at the inlet as well as at the outlet of the impeller (1–1 type system)
2. Single-phase flow (gas alone) at the inlet and two-phase flow (gas and liquid) at the outlet of the impeller (1–2 type system)
3. Two-phase flow (gas and liquid) at the inlet as well as at the outlet of the impeller (2–2 type system)

Type (1) involving single-phase flow is easier to analyze than (2) and (3). The three categories defined by Mundale (1993) encompass all types of impellers commercially available for different applications.

9.3.1   1–1 Type Impellers

Figure 9.2 shows a schematic of the arrangement used in a 1–1 type system. It consists of a hollow shaft, at the bottom of which a suitable impeller with orifices to discharge the induced gas is located.

Zlokarnik and Judat (1967) were the first to study the use of gas-inducing impellers for gas–liquid contacting using a simple system similar to that shown in Figure 9.2. Subsequently, a number of investigators have studied this system (Martin 1972; Topiwala and Hamer 1974; Joshi and Sharma 1977; Joshi 1980; Baczkiewicz and Michalski 1988; Evans et al. 1990, 1991; Rielly et al. 1992; Forrester and Rielly 1994; Heim et al. 1995; Forrester et al. 1994, 1998; Scargiali et al. 2007; Kasundra et al. 2008; Abdullah et al. 2011). Based on the results of the various investigations available in the literature, the following conclusions can be drawn:

1. Increasing the number of orifices for the gas in the pipes yields higher fractional gas holdup, effective gas–liquid interfacial area, and better mass transfer (Forrester and Rielly 1994).
2. Impellers shaped to avoid boundary layer separation on the downstream side yield better results than impellers with bluff shapes (Martin 1972; Evans et al. 1991; Rielly et al. 1992).
3. Standard and modified pipe/turbine impellers yield nearly the same value of $k_L a$ ($\sim 10^{-3}$ to $10^{-2}/\mathrm{s}$) as conventional stirred reactors employing similar impellers (Hsu and Huang 1996; Forrester et al. 1998; Kasundra et al. 2008).

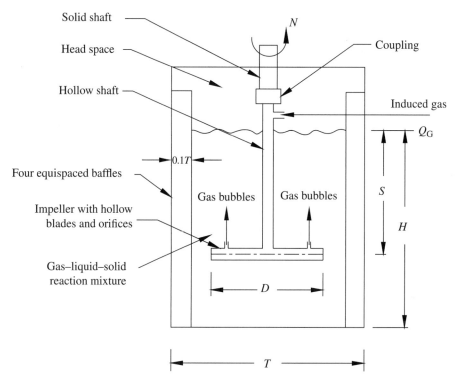

FIGURE 9.2 Schematic of 1–1 type gas-inducing impeller. (Reproduced from Forrester and Rielly, 1994, with permission from Elsevier Copyright © 1995 Published by Elsevier Ltd. http://www.journals.elsevier.com/chemical-engineering-science/)

The fact that most versions of the 1–1 impellers do not afford significant mass transfer advantage (relatively higher $k_L a$) over conventional stirred reactors severely limits their application as compared to 1–2 or 2–2 type. Therefore, the discussion on the 1–1 type will be limited only to the extent that it allows a theoretical analysis of the gas induction process.

9.3.1.1 Hydrodynamics As mentioned in Section 9.2, this type of impeller is the easiest to analyze due to lack of complications arising out of two-phase flow at any point in the impeller. The basic equation to predict the value of N_{CG} is Bernoulli's equation for an inviscid fluid. The difference form of Bernoulli's equation can be written as

$$\Delta\left(\frac{\rho_L U_T^2}{2}\right) + \Delta P_{Tot} = 0 \qquad (9.1)$$

In Equation 9.1, ΔP_{Tot} is the total pressure that is the sum of hydrostatic and pressure head that is acting on the orifice in the impeller from which gas bubbles come out. The tip velocity at the orifice, U_T, is generally used in estimating the kinetic head.

Martin's (1972) was the first investigation that attempted to understand the process of gas induction. This was based on Equation 9.1. For a 1–1 type impeller, the potential head of the gas in the hollow pipe can be neglected in comparison to the static liquid head on the orifice. Equation 9.1 further implies that the viscous dissipation in gas flow through the pipe is negligible (viscosity, $\mu_G \sim 0$). Considering flow past a cylindrical body (most first-generation impellers were cylindrical in shape), Martin (1972) derived an equation for the pressure on the surface of the cylinder at a point located at an angle θ from the axis of the cylinder:

$$\frac{h_S - h_L}{\left(\dfrac{U_T^2}{2g}\right)} = 1 - 4\sin^2\theta \tag{9.2}$$

Here, h_L is the liquid head above the orifice and h_S is that outside the orifice in the absence of gas flow.

In reality, with fluids of finite viscosity, the frictional losses occurring during flow cannot be ignored. Martin experimentally measured the local pressure on the cylinder surface. He defined a dimensionless parameter that is the ratio of local potential and kinetic energies:

$$\Phi_{P-K} = \frac{h_S - h_L}{\left(\dfrac{U_T^2}{2g}\right)} \tag{9.3}$$

Φ_{P-K} in Equation 9.1 is the value measured using fluids of finite viscosity. It incorporates the nonideal effects due to energy dissipation by frictional factors. Martin's analysis had a number of flaws. This analysis was subsequently improved by Evans et al. (1991), Rielly et al. (1992), Forrester et al. (1994), and finally Forrester et al. (1998), as will be discussed later.

9.3.1.2 Prediction of the Critical Speed of Agitation for Gas Induction
Subsequent to Martin's analysis, Evans et al. (1990, 1991) made attempts to remove the lacunae. This basic approach was further employed by Rielly et al. (1992). The first step in this direction was introduction of a slip factor, K, and a pressure coefficient, C_p. The slip factor was needed to be introduced because an impeller rotating in a baffled vessel causes a swirling of the liquid with a rotational velocity that is different from the local impeller velocity. The actual fluid velocity is thus given by $2\pi NK$, where K is the slip factor. The relative velocity between the fluid upstream of the orifice and that at the orifice located at a radial distance R is given by

$$U_T' = (1 - K) \times 2\pi NR \tag{9.4}$$

For relatively high Reynolds numbers, K is invariant with respect to the impeller speed, N (Rielly et al. 1992).

The concept of pressure coefficient, C_p, was borrowed from the conventional aerofoil theory (Abbot and von Doenhoff 1959). As defined later, it is related to the difference between the pressure at the stagnation point, P_S, and that at the orifice, P_θ. P_S as defined by Rielly et al. (1992) is given below:

$$P_S = P_0 + \rho_L gS + \frac{1}{2}\{2\pi NR(1-K)\}^2 \qquad (9.5)$$

where P_0 is the pressure in the headspace and S the submergence of the impeller in terms of the clear liquid height:

$$P_\theta = P_0 + \rho_L gS + \frac{1}{2}\rho_L \{2\pi NR(1-K)\}^2 (1-C_{P\theta}) \qquad (9.6)$$

The pressure coefficient $C_{P\theta}$ was defined in terms of the headspace pressure P_0 and the local pressure P_θ at the orifice surface as follows:

$$C_{P\theta} = \left(\frac{P_0 + \rho_L gS - P_\theta}{\frac{1}{2}\rho_L \times \{(1-K)2\pi NR\}^2} \right) \qquad (9.7)$$

For flow past a cylinder, $C_{P\theta}$ can be obtained by combining Equations 9.2 and 9.7.

For geometries significantly different than a cylinder such as a turbine blade, it is necessary to experimentally determine the value of $C_{P\theta}$ under actual conditions of operation.

According to the mechanism of gas induction discussed in Section 9.2, the critical speed, N_{CG}, is that at which $P_\theta = P_0$. This yields (Rielly et al. 1992)

$$N_{CG} = \sqrt{\frac{gS}{2\{(1-K)\pi R\}^2 (C_{P\theta} - 1)}} \qquad (9.8)$$

Equation 9.8 can yield positive values for finite S only when $C_{P\theta} > 1$. According to Rielly et al. (1992), Equation 9.8 allows independent evaluation of the effects of vessel and impeller blade geometry. The former decides K, whereas the latter has a dominant effect on $C_{P\theta}$. In the case of simple impeller blade shapes and high impeller Reynolds numbers, $C_{P\theta}$ can be estimated by employing the inviscid flow theory. However, for complex blade shapes, $C_{P\theta}$ needs to be experimentally determined. The local pressure coefficient is independent of the impeller speed at relatively high Reynolds numbers. Rielly et al. (1992) suggested that the blade slip factor, K, can be determined by measurements in a geometrically similar scaled-down vessel. However, in view of the complex scale effects associated with stirred vessels discussed in Chapters 6 and 7A, it is difficult to speculate about the validity of these small-scale measurements for use on large scale.

Rielly et al. (1992) measured N_{CG} over a wide range of vessel diameters ($0.3 < T < 0.6$ m) submergence of the orifice ($0.03 < S < 0.6$ m). Their data could be satisfactorily correlated by Equation 9.8. Forrester and Rielly (1994) extended this approach to impellers having multiple orifices at different radii on the blade. The correlation proposed was a minor modification of Equation 9.8:

$$N_{CGI} = \sqrt{\frac{gS}{2\left\{(1-K_I)\pi R_I\right\}^2 (C_{P\theta I} - 1)}} \tag{9.9}$$

where K_I and $C_{P\theta I}$ represent the values for the ith orifice located at a radial distance of R_I from the impeller axis.

9.3.1.3 Prediction of the Rate of Gas Induction
The approach used is a typical phenomenological approach involving the driving force and a resistance. The driving force for gas induction is the pressure differential between the orifice pressure and that in the headspace (Rielly et al. 1992):

$$\Delta P_D = P_\theta - P_0 \tag{9.10}$$

$$\Delta P_D = \rho_L gS + \frac{1}{2}\rho_L \left\{2\pi NR(1-K)\right\}^2 (1 - C_{P\theta}) \tag{9.11}$$

As defined, ΔP_D is negative because gas induction occurs in the direction of decreasing pressure.

At steady state, the driving force should be balanced by the resistance (pressure losses), which yields

$$\Delta P_D + \Delta P_F + \Delta P_0 + \Delta P_{KE} + \Delta P_\sigma = 0 \tag{9.12}$$

In Equation 9.11, ΔP_F, ΔP_0, ΔP_{KE}, and ΔP_σ are the pressure losses due to (i) friction during flow through the hollow pipe and blades, (ii) flow through the orifice, (iii) kinetic energy imparted to the fluid during formation of the bubble at the orifice, and (iv) work required to overcome the surface tension forces during bubble formation, respectively. Rielly et al. (1992) suggested that ΔP_F and ΔP_0 should be determined experimentally. ΔP_σ is given by

$$\Delta P_\sigma = 2\left(\frac{\sigma}{R_{DB}}\right) \tag{9.13}$$

R_{DB} is the radius of the detached bubble and σ is the surface tension of the liquid used. During bubble formation, the bubble is moving along the impeller at a velocity of $2\pi NR(1-K)$. The physical process of bubble detachment has been described by Rielly et al. (1992). Accordingly, the detached bubble radius was calculated from

$$R_{DB} = \sqrt{\frac{3Q_{AV}}{4\pi^2 N \times R(1-K)}} \tag{9.14}$$

where Q_{AV} is the time-averaged rate of gas induction. ΔP_{KE} was obtained using the model proposed by Davidson and Schueler (1960):

$$\Delta P_{KE} = \frac{3 Q_{AV} \times \rho_L}{32 \pi^2 R_{DB}^4} \tag{9.15}$$

Knowing ΔP_F and ΔP_0, Equations 9.11–9.15 can be used to calculate the rate of gas induction. Rielly et al. (1992) found that the predicted rates were higher than the experimental rates. This was attributed to Equation 9.7 that was employed to calculate C_{p0}. The pressure driving force thus calculated is applicable only at the orifice. In practice, C_{p0} and ΔP_D decrease as the evolving bubble expands and departs from the orifice. Information on the variation of bubble radius with time along with the fluid velocity in the vicinity of the orifice is required to predict the variation in the pressure driving force.

Forrester et al. (1998) improved the original analysis of Evans et al. (1991) and Rielly et al. (1992) and extended it to the case of an impeller having multiple orifices. The model of Evans et al. (1991) gave an unrealistic prediction of ΔP_σ in the vicinity of N_{CG}. In this region, the gas induction rate tends to zero implying that $R_{DB} \to 0$. Since $\Delta P_\sigma \propto (1/R_{DB})$ with $R_{DB} \to 0$, $\Delta P_\sigma \to \infty$, which is unacceptable. Forrester et al. (1998) rightly argued that R_{DB} cannot be lesser than the orifice radius and therefore resolved this anomaly by assuming that the original bubble formed has a radius the same as that of the orifice and, in the event that additional gas is supplied ($N > N_{CG}$), the bubble grows from this initial orifice radius. The second improvement was in terms of ΔP_{KE}. In the original model, ΔP_{KE} was estimated using the model of Davidson and Schueler (1960) based on growth of a bubble in an infinite stagnant liquid. In the case of a gas-inducing impeller, there is a relative cross-flow of liquid at the orifice. Therefore, the ΔP_{KE} term was modified using the model of Witze et al. (1968) for growth of a bubble at a point on a planar surface under the influence of potential flow. The third improvement was in terms of use of a time-averaged total pressure driving force. These modifications yielded a better fit of the data than the original model of Evans et al. (1991). Further, it was found that increasing the number of orifices from one to two doubled the rate of gas induction. However, further increase in the number of orifices did not yield proportionate increase in the gas induction rate. The model predictions were reasonably satisfactory for the multiorifice blades.

Ju et al. (2009) investigated the performance of modified versions of 1–1 type gas-inducing impellers. These consisted of modifications of disc turbine (DT), downflow marine propeller (PD), and down flow pitched blade turbine (PTD) coupled with the use of a PTD impeller as gas dispersion device below the inducing impeller. They concluded that PD and PTD were more energy efficient as compared to DT. Gas induction in the presence of gas sparging was also studied using a ring sparger located midway between the lower PTD and vessel bottom (i.e., at $0.5 \times C_{2-B}$ from bottom; Figure 9.10). They found that the gas induction rate was unaffected for the range $0.5 \times 10^{-3} < V_G < 7 \times 10^{-3}$ m/s, whereas there was a marginal decrease in Q_G over the range $7 \times 10^{-3} < V_G < 15 \times 10^{-3}$ m/s.

9.3.1.4   Gas–Liquid Mass Transfer   It has been mentioned earlier that the 1–1 type of gas-inducing impellers does not yield substantially higher values of $k_L a$ as compared to conventional stirred reactor with gas sparging (particularly for $P/V < 1.5\,kW/m^3$). Forrester et al.'s (1998) data bear testimony to this statement. In view of this, it is felt that further discussion on this aspect is not warranted. For more recent details, the reader is referred to Poncin et al. (2002), Sardeing et al. (2006), Kasundra et al. (2008), and Achouri et al. (2013). Achouri et al. used a hollow disc mounted directly on the hub. The ports for gas sparging were provided on the edge of the disc. This geometry also obeyed Equation 9.8 for N_{CG}, indicating that the mechanism of gas induction remains unaltered irrespective of the geometry of the rotor employed. Further, the values of $k_L a$ were in the same range as those obtained by other investigators for 1–1 type systems.

9.3.2   1–2 and 2–2 Type Impellers

Before discussing the hydrodynamics and mass transfer aspects of gas-inducing impellers in this category, it is desirable to briefly describe these two variants.

1–2 type of impellers has been widely used in the mineral dressing process. Representative commercial designs in this class are Outokumpu (Fallenius 1981), Aker (Barbery 1984), and BCS (Barbery 1984). The AIRE-O$_2$ system of Aeration Industries International, Inc., United States, and Acqua Co. Italy's EOLO$_2$ system also belong to this category.

9.3.2.1   1–2 Impellers   The important components of 1–2 type gas-inducing impeller are (i) hollow shaft, (ii) impeller, (iii) orifices for gas entry, and (iv) a stator/diffuser (Fig. 9.3). The impeller is mounted on the hollow shaft.

The principle of gas induction is the same as that described in Section 9.2. Impeller rotation results in reduction of pressure in the hollow shaft till a stage is reached when the reduction in pressure equals the static liquid head on the gas port in the hollow shaft. At an impeller speed slightly higher than this critical speed, N_{CG}, gas enters the impeller region. As shown in Figure 9.3, simultaneously liquid also enters the disperser as a result of the suction generated by the impeller rotation. The radial flow generated by the impeller mixes the gas/liquid mixture and propels it through the stator/diffuser and then into the surrounding contents of the vessel. This two-phase mixture contains relatively small ($\leq 1\,mm$) bubbles. In view of this, the mass transfer efficiency of 1–2 devices is expected to be better than 1–1 devices. However, as observed by Zundelvich (1979), there is rapid coalescence of the bubbles once they are outside the shear zone of impeller. This may be because the turbulence is strong only in the disperser region and dies out quickly as the dispersion exits the disperser. This results in relatively coarse bubbles having diameter in the range of 2–3 mm.

The rate of gas induction is decided by the type of impeller used, the design of the stator, the speed of the impeller, and the static head above the gas ports. Analogous to the venturi loop reactor (Sections 8.6 and 8.7), the rate of gas induction cannot be varied independent of the aforementioned design and operating features. However, in the present case of gas-inducing impeller, in the event that the induced

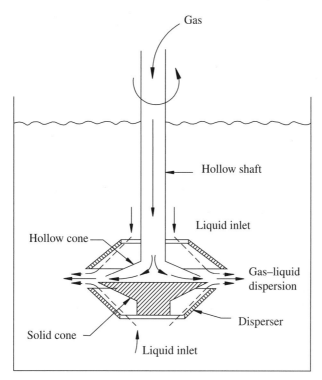

FIGURE 9.3 Schematic of 1–2 type gas-inducing impeller. (Adapted from Koen and Pingaud (1977) with permission from BHRA, The Fluid Engineering Centre, Cranefield, England, Copyright © BHRA 1977 and reproduced from Patwardhan and Joshi (1999) with permission from American Chemical Society, Copyright © 1999, American Chemical Society.)

gas flow is insufficient to meet the requirements of the process, an additional gas quantity can be introduced through a conventional sparger designed in such a manner as not to disturb the natural flow generated by the gas-inducing system. Wang et al. (2013) modified the 1–1 type gas-inducing impellers employed by Ju et al. (2009). The modification consisted of drilling holes upstream of the gas outlet orifices. This resulted in part of the impeller acting as 1–2 type. Wang et al. studied the effect of diameter of liquid inlet orifices and found that an optimum value (3 mm in their case) existed such that for smaller- and larger-diameter holes, (i) the critical speed for gas induction increased and (ii) the rate of gas induction decreased. It was argued that ingress of the liquid resulted in an increase of the velocity head that more than compensates for the loss in potential energy consumed for the additional liquid flow.

9.3.2.2   2–2 Impellers   The major difference between 1–1 and 2–2 devices is that in the 2–2 devices a solid shaft (as against a hollow shaft) is surrounded by an annular standpipe (Fig. 9.4). The annular standpipe is open to the gas space above the liquid surface. Thus, the standpipe plays the role of conveying the gas to the impeller

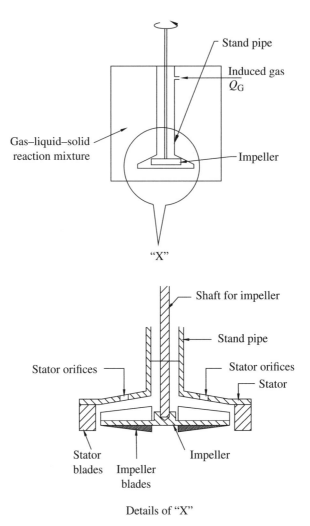

FIGURE 9.4 Schematic of 2–2 type gas-inducing impeller with impeller details. (Reproduced from Sawant et al. (1981) with permission from Elsevier. Copyright © 1981. Published by Elsevier B.V. http://www.journals.elsevier.com/chemical-engineering-journal/.)

region. This standpipe replaces the hollow shaft in 1–1 devices. The other parts are (i) the impeller and (ii) the stator/diffuser. These could be designed on similar lines although different brands have their own special designs.

Some of the prominent commercial designs are Denver (Taggart 1960; Evans et al. 2008), Wedag (Arbiter and Steininger 1965), Agitair (Arbiter and Steininger 1965), Wemco (Degner and Treweek 1976; Kind 1976; Evans et al. 2008), Sala (Harris 1976), Turboaerator (Zundelevich 1979), Booth using a second impeller for adequate gas dispersion (Barbery 1984), and Bateman/Batequip (Nessett et al. 2006; van der Westhuizen and Deglon 2007, 2008). Besides the patented impeller designs, conventional impellers such as downflow pitched turbine have also been used in single- (Raidoo et al. 1987; Mundale 1993; Saravanan et al. 1994) and multiple-impeller systems (Barbery 1984; Saravanan and Joshi 1995; Yu and Tan 2012). The mechanism of gas induction is similar to that in the first two cases. The standpipe is filled with the working liquid when the impeller is stationary. As the impeller starts rotating, a vortex is formed in the standpipe. With increasing speed, the vortex moves downward in a manner similar to that shown in Figure 9.1, till it reaches the impeller level. This is followed by formation and entrapment of the bubbles. The surrounding liquid is also induced into the suction zone created by the impeller through the ports in the stator. The induced gas–liquid flow is forced radially into the diffuser. However, unlike the 1–2 type system, both the gas and liquid phases enter and exit the impeller region together. The dispersion is generated under the action of the impeller. The diffuser of a 2–2 system is akin to the diffuser in a centrifugal pump. The flow area increases as the two-phase flow traverses the diffuser. This results in a pressure recovery (at the expense of kinetic energy) similar to that in the pressure recovery diverging cone of the ejector in a venturi loop system (Fig. 8.3a and b). Arbiter et al. (1976) have presented an exhaustive review of the geometrical aspects of conventional froth flotation cells that belong to this category. The general range of geometrical parameters used in industrial flotation cells is:

1. D/T: The overall range reported is 0.25–0.64 with most systems employing D/T between 0.4 and 0.5.
2. C/T: Specific values are not available. However, analysis of the diagrams/pictures given in the literature of the manufacturers indicated that relatively low C/T values (~0.2) were used in these machines. These low values are probably a result of the high solid loadings that prevail in the ore dressing industry. In the application of gas-inducing systems as reactors for solid-catalyzed gas–liquid reactions, the solid loading is relatively low, typically less than 5 wt% when precious metal catalysts are used. Further, low C/T values imply high submergence and consequently lower gas induction rates (White and de Villiers 1977). The negative effect of submergence in this situation is analogous to that of the "back pressure" on the diffuser in a venturi loop reactor discussed in Section 8.6.1 and Table 8.3. Indeed, this is an important drawback since the gas-inducing impeller has to be located close to the liquid surface (high C/T) for obtaining reasonably high gas induction rates. A major part of the vessel from the gas-inducing impeller to the bottom of the vessel is consequently deprived of turbulence. This drawback makes it mandatory to have an additional impeller(s) for solid suspension/gas dispersion.
3. The effective volume, V_{eff}, of the cell defined as the volume after deducting the volume occupied by the impeller system comprising of standpipe and shroud is related to the diameter of the tank by $V_{Eff} = k(T)^n$.

   Arbiter et al. (1976) reviewed the proportions of 20 industrial flotation cells and found that the range of k and n are $0.6 < k < 1.3$ and $2 < n < 3$, respectively. The average values were $k = 0.9$ and $n = 2.6$. This review indicates that cells with n values of 2 and above are outdated designs.

4. The power number reported for Wemco cells varied from 2.9 for laboratory scale (2 lit) to 1.45 for 1.75 m³ (Arbiter et al. 1976). Kind (1976) has reported the power number for vessel diameters from ~1 to 3.5 m. These varied over a relatively narrow range from 3.3 to 4.4. Since the larger sizes have better geometrical uniformity, the range given by Kind (1976) is more reliable.

9.3.2.3 Prediction of Critical Speed for Gas Induction

As mentioned earlier, because of two-phase flow, this situation is not amenable to a theoretical analysis similar to that discussed in Section 9.3.1.2. White and de Villiers (1977) used the conventional approach based on dimensional analysis as discussed in Section 6.3.1a (Empirical Correlations Using Conventional Dimensionless Groups). This approach based on dynamic similarity has a number of drawbacks. White and de Villiers identified the following independent variables that affect the gas induction rate: speed of agitation, N; gravitational constant, g; submergence of the impeller, S; and diameter of the impeller, D. It should be pointed out that the type of impeller itself is an important parameter. The gas induction rate, Q_G, was included through the dimensionless flow number F_{LG}:

$$F_{LG} = \frac{Q_G}{ND^3}$$
(9.16)

Replacing N by N_{CG} defines the flow number at the critical speed for gas induction, F_{LGC}:

$$F_{LGC} = \frac{Q_G}{N_{CG}D^3}$$
(9.17)

An analysis based on the Buckingham π theorem yielded the following relationship:

$$F_{LGC} = \frac{Q_G}{N_{CG}D^3} = f\left(Fr, \frac{S}{D}\right)$$
(9.18)

White and de Villiers (1977) defined a modified Froude number. For $N=N_{CG}$,

$$Fr_C = \left(\frac{N_{CG}^2 D^2}{S \times g}\right)$$
(9.19)

At this stage, the basic principle of gas induction (Bernoulli's theorem) was invoked. White and de Villiers used dimensional considerations to define the critical suction due to impeller rotation by

$$\Delta P_D = k(\rho_L N_{CG}^2 D^2)$$
(9.20)

The hydrostatic head on the orifice is

$$\rho_L g S$$
(9.21)

Noting the equality of Equations 9.20 and 9.21 at $N=N_{CG}$ and simplifying yield

$$\text{Fr}_C = \left(\frac{N_{CG}^2 D^2}{S \times g}\right) = \frac{1}{k} \tag{9.22}$$

k is a constant for a given rotor/stator geometric configuration. For the system employed by White and de Villiers and $1 < S/D < 2.2$, $\text{Re}(ND^2\rho_L/\mu_L)$ in the turbulent region, mean value of Fr_C or $\{1/k\}$ was found to be 0.23 with a standard deviation of 0.022. This value is close to 0.21 obtained by Sawant et al. (1980) for Wemco cell and Denver cells of four different sizes (Sawant et al. 1981). Using a different definition of the critical Froude number (N^2D/g), Heim et al. (1995), on the other hand, obtained somewhat different values for pipe (0.155 for four-pipe impeller, 0.162 for six-pipe impeller) and disc (0.23) impeller. Fr_C is based on a simple pressure balance. It does not involve any factor defining the shape of the impeller, and therefore, it is logical that different impellers afford nearly the same value of Fr_C. Using the value obtained by Sawant et al. (1980, 1981), the expression for N_{CG} is

$$N_{CG} = \sqrt{\frac{0.21gS}{D^2}} \quad \text{or} \quad 0.45 \times \sqrt{\frac{gS}{D^2}} \tag{9.23}$$

The effect of viscosity was incorporated by the following modification of Equation 9.22 (Sawant and Joshi 1979):

$$\text{Fr}_C \left(\frac{\mu_W}{\mu}\right)^{0.11} = 0.21 \tag{9.24}$$

Evidently, liquid-phase viscosity has a marginal effect on Fr_C. This is expected since both Equations 9.20 and 9.21 from which Fr_C is derived do not contain liquid-phase viscosity. The small effect observed nonetheless is likely to be due to finite nonidealities and viscosity effects such as the blade slip factor introduced by Rielly et al. (1992). A major application of gas-inducing systems is in solid-catalyzed gas–liquid reactions. Aldrich and Deventer (1994) studied the effect of presence of solid particles on N_{CG}. It was found that N_{CG} was unaffected up to solid loading of 15 wt%. Above this loading, there was an increase in N_{CG} presumably due to increase in the effective viscosity of the suspension. This study, however, used a relatively very small-size system ($T=0.2$ m and $D=0.065$ m), and therefore, it is difficult to derive quantitative conclusions.

Zundelevich (1979) reviewed the literature emanating from Eastern Europe (then Soviet Bloc) on gas-inducing impellers. He extended the approach of Braginsky (1964). In this approach, it was assumed that the liquid velocity in the stator is relatively small as compared to the velocity of the tip of the impeller. The head generated by the impeller was given by

$$\Delta P_{KE} = K_1 \rho_L \left(\frac{\pi DN^2}{2}\right) \tag{9.25}$$

At $N = N_{CG}$,

$$\Delta P_{KE} = \rho_L g S \tag{9.26}$$

and

$$N_{CG} = \sqrt{\frac{2gS}{K_1 \pi^2 D^2}} \tag{9.27}$$

K_1 is the coefficient of head losses due to finite viscosity. Zundelevich measured the critical speed and gas induction rate in a pilot scale system. This system was similar to that shown in Figure 9.4. Two types of impeller–stator systems were used. In the first, the liquid entered from the top cover of the stator, while in the second the liquid entry was from bottom cover of the stator. The induced gas (from the top port) and sucked liquid were mixed by the impeller action and ejected radially out of the stator. The critical speed and gas induction rates were measured for both types of devices. The range of parameters covered was not very wide (for instance, for device 2, $0.2 < D/T < 0.3$). The value of K_1 obtained from the experimental data was $= 1.15$. Using this value in Equation 9.27,

$$N_{CG} = 0.42 \sqrt{\frac{gS}{D^2}} \tag{9.28}$$

The values of N_{CG} predicted from Equations 9.23 and 9.28 are within 6.6%, indicating that the basic criterion for critical speed is relatively simple and the difference between different impeller designs is mainly due to the energy/pressure losses suffered in the impeller/stator region.

9.3.2.4 Prediction of the Rate of Gas Induction
In this case also, White and de Villiers (1977) used an empirical approach based on dimensional analysis. The flow number, F_{LG}, was given as

$$F_{LG} = \left(\frac{Q_G}{ND^3} \right) = f \left(\frac{ND^2}{gS}, \frac{S}{D} \right) \tag{9.29}$$

Equation 9.29 is applicable for $N > N_{CG}$. White and de Villiers defined a modified flow number as follows:

$$F'_{LG} = \left(\frac{Q_G}{ND^3} \right) \left(\frac{D}{S} \right)^{0.5} \tag{9.30}$$

A log-log plot of F'_{LG} and $[\text{Fr} - \text{Fr}_C]$ gave a satisfactory fit described by

$$F'_{LG} = 0.023(\text{Fr} - \text{Fr}_C)^{1.84} \tag{9.31}$$

Equation 9.31 allows the calculation of Q_G for a given set of parameters. White and de Villiers made a curious observation about the effect of increased gas induction rate on gas holdup. In the case of water as the liquid (coalescing system), they found that despite an increase in the induction rate, the dispersion height or gas holdup was constant. This observation indicates that at higher induction rates the action of the disperser is inadequate. The bubble size generated probably increases in such a manner that the higher rise velocity of these larger bubbles nullifies the increase in induction rate. On the other hand, the use of 0.1 wt% Teepol ($\sigma = 3\,N/m$) resulted in significant foaming. Gorain et al. (1995a) also found that higher gas rates caused increase in bubble size.

Sawant et al. (1980, 1981) also used White and de Villiers' empirical approach and correlated their data for Denver and Wemco cells, respectively. Their correlation for Denver cell was explicit in terms of Q_G unlike that of White and de Villiers. The correlations obtained were

$$\text{Denver cell: } Q_G = 2.1 \times 10^{-3} \left\{ N^2 - N_{CG}^2 \right\}^{0.75} D^3 \tag{9.32}$$

For Wemco cell however, a correlation similar to White and de Villiers was suggested:

$$\text{Wemco cell: } Q_G = 51.2 \times \left\{ \text{Fr} - \text{Fr}_{CG} \right\}^{0.83} \times \left(\frac{D}{S} \right)^{0.5} \tag{9.33}$$

In an altogether different approach, Zundelevich (1979) attempted to extend the theory of water jet ejectors to gas-inducing impellers. Admittedly, the geometries of the two systems are significantly different. However, the high-velocity rotational liquid motion causing gas induction is analogous to the suction-creating high-kinetic-energy jet in water jet ejectors. Thus, the liquid sucked in by the impeller is the primary fluid, and it is mixed with the gas (secondary fluid) entering the diffuser through the standpipe surrounding the impeller shaft. The analogy is depicted in Figure 9.5.

The volume ratio of the secondary (gas) and primary (liquid) fluids also known as injection coefficient, ψ is

$$\psi = \frac{Q_G}{Q_L} \tag{9.34}$$

Zundelevich used the expression of Sokolov and Zinger (1960) for ψ:

$$\psi = \sqrt{\frac{K_1 - (\Delta P_G / \Delta P_L) \times (A_{MT} / A_N) \times (\varphi_D / 2\varphi_N)}{(1 - (\varphi_D / \varphi \zeta_N)) \times (A_N / A_{MT})}} - 1 \tag{9.35}$$

$\Delta P_G = P_N - P_S$ and $\Delta P_L = P_N - P_O$ (notation as per Figure 9.5).

The parameters K (=0.834) and $\dfrac{\varphi_D}{2\varphi_N}$ (=0.47) were obtained by Sokolov and Zinger, fitting Equation 9.35 to their experimental data for a typical water jet ejector

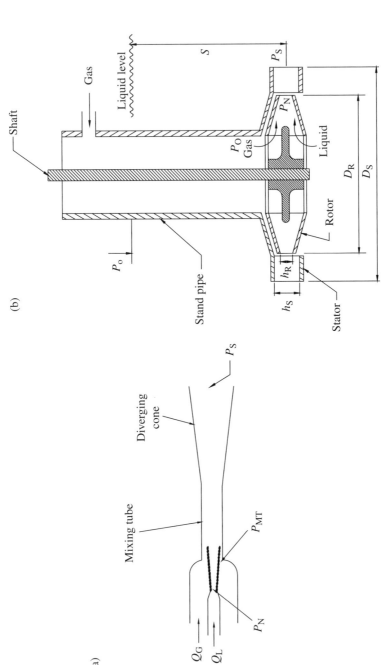

FIGURE 9.5 Analogy between water jet ejector and 2–2 type gas-inducing impeller. (a) Water jet ejector. (b) 2–2 type gas-inducing impeller. (Reproduced from Zundelevich, 1979 with permission from John Wiley and Sons. Copyright © 1979 American Institute of Chemical Engineers.)

shown in Figure 9.5a. Considering that the rotor in the impeller is analogous to the nozzle and stator to the mixing tube followed by the diverging section (Fig. 9.5b), respectively, Zundelevich defined the nozzle area and mixing tube areas, respectively, as follows:

$$A_N = \pi D_{RT} \times \frac{h_{RT}}{2}; \quad A_{MT} = \pi D_{ST} h_{ST} \tag{9.36}$$

Zundelevich neglected the single-phase pressure drop in the standpipe. Though not mentioned specifically, the potential energy difference in the standpipe was also neglected. Sokolov and Zinger (1960) suggested the following optimum proportions:

$$D_{ST} = 1.4 D_{RT} \tag{9.37}$$

$$h_{ST} = 2 h_{RT} \tag{9.38}$$

In the experimental part of Zundelevich's work, an optimized impeller system as per Sokolov and Zinger was compared with another having diverse proportions. The impeller configurations used were similar to that of a Denver cell. A comparison of the performances discussed later confirms the superiority of the optimized impeller system.

The pressure at the outlet of the stator, P_S, that now includes the kinetic head generated by the rotor is given by

$$P_{ST} = \rho_L g S + P_O + \rho_L K_2 \frac{\left(\pi D_{RT} N\right)^2}{2} \tag{9.39}$$

K_2 in Equation 9.39 is the head loss coefficient that is analogous to the slip factor, K, in Equation 9.4. Using the definitions of A_N and A_{MT} (Eq. 9.36), optimized geometric parameters (Eqs. 9.37 and 9.38), and Equation 9.39 in Equation 9.35, the following expression was obtained:

$$\psi = 3 \times \left[\sqrt{1 - 3\frac{2Fr'}{2Fr' + K_2 \pi^2}} - 1 \right] \tag{9.40}$$

where

$$Fr' = Fr^{-1}. \tag{9.41}$$

In Equation 9.40, the only operating variable is Fr', which is the ratio of the hydrostatic to kinetic head. According to this equation, the induction rate is solely decided by this ratio (Fr') that is independent of the liquid-phase density. Continuing with this expression for ψ, Zundelevich introduced the impeller performance parameter in terms of the pumping capacity ($= K_p N D^3$) and power input through the impeller ($P = N_p \rho_D N^3 D^5$). The dispersion density, ρ_D, for use in estimation of power input was defined as $\rho_D = \dfrac{\rho_L}{\psi + 1}$. The following expression for ψ was obtained by combining the definition of ψ with the pumping capacity $(Q_L = K_P N D^3)$ and Equation 9.40 along with $Fr' = Fr^{-1}$:

$$\psi = \frac{1}{K_P}\sqrt{\frac{Fr'}{Eu_G}} \tag{9.42}$$

where Eu_G is the Euler number based on gas flow rate $[Eu_G = gS/(Q_G/D_{RT}^2)^2]$. Zundelevich's investigation was aimed at optimizing the performance of the gas induction system in terms of the volume of gas induced per unit power input for a given submergence, S. For this purpose, the equations derived earlier were rearranged to yield

$$\frac{Q_G S}{P} = \frac{Fr'}{N_P \rho_L g}\left(\frac{Fr'}{K_P Eu_G} + \sqrt{\frac{Fr'}{Eu_G}}\right) \tag{9.43}$$

The term $Q_G S/P$ is the impeller performance index and is synonymous with the induction efficiency parameter of a venturi loop system in Section 8.7.1 and Table 8.4. Submergence, S, in this case is the equivalent of the "dip" or back pressure in the case of the venturi loop reactor. Equation 9.43 can predict the performance of a gas-inducing impeller of the type shown in Figure 9.5b if K_P, Fr', and Eu_G are known. Values of K_P for conventional impellers are available in the literature (Gray 1966). Alternatively, for a substantially different geometry, K_P can be obtained from experimentally measured values of Q_G and power input over a range of speeds of agitation, preferably over the range in which Q_G/P exhibits a maximum. Substituting Q_G/P (for a given value of S) in Equation 9.43 yields the value of K_P. The power number N_P may not be available for different designs of self-inducing impellers. Independent measurement of power can be used for estimation of the same. Zundelevich (1979) measured gas induction rates for two different impeller designs (i) optimized as per Sokolov and Zinger (1960) and (ii) a diverse nonoptimum geometry. The submergence was varied by a factor of 5. It was found that the plots of Eu_G against Fr' were unique for geometrically similar aerator designs and did not depend on the submergence. Thus, Eu_G and Fr' are the key defining parameters of a 2–2 type gas-inducing impeller device. This was further buttressed by agreement between the model predictions and experimental data for Eu_G versus Fr' plots for the two geometric configurations used. Similarly, data on $Q_G S/P$ were generated using Equation 9.43 for both the impellers. The model predictions again agreed satisfactorily with the experimental data as shown in Figure 9.6.

Such a plot allows comparison of the efficacy of different designs of 2–2 type devices. For a given impeller system, it was found that $Q_G S/P$ is independent of the submergence as evinced from plots of Eu_G against Fr' (Zundelevich 1979). It was also found that the device based on the optimized configuration of Sokolov and Zinger (Fig. 9.6) yielded a better performance than the nonoptimum geometry. Most importantly, from a design/scale-up point of view, plots similar to Figure 9.6 yield the optimum operating condition with respect to the speed of agitation since $Q_G S/P$ exhibits a maximum with respect to Fr', Fr'_{Opt}. For the optimized aerator in

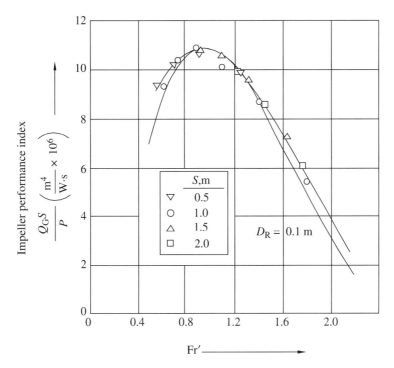

FIGURE 9.6 Impeller performance index as a function of modified Froude number, Fr. (Reproduced from Zundelevich, 1979 with permission from John Wiley and Sons. Copyright © 1979 American Institute of Chemical Engineers.)

Zundelevich's study, Fr'_{Opt}, Eu_{GOpt}, and $[Q_G S/P]_{max}$ were found to be 0.9, 53, and 1.08×10^{-5} m^4/W s, respectively. Thus, initially with increasing value of N, $Q_G S/P$ increases reaching a maximum and then decreases with further increase in N. The optimum value of N is given by

$$N_{Opt} = \sqrt{\frac{gS}{Fr'_{Opt} D_{RT}^2}} \qquad (9.44)$$

Zundelevich further analyzed the implications of complete and partial geometric similarities. In the case of complete geometric similarity, $(Q_G/P) \propto (T)^{-1}$. He suggested that for scale-up, the vessel diameter should be some multiple of the rotor diameter (keeping submergence and superficial gas velocity, V_G constant) such that the scaled-up version yields similar Q_G/P. Raidoo et al. (1987) also found a unique single-valued dependence of Eu_G and Fr' for a given D_{RT}/T. However, when D_{RT}/T was varied at constant T, this observation was not sustained. This is likely to be due to the observation of Zundelevich that single-valued dependence between Eu_G and Fr' holds only for the similarity criteria suggested by Zundelevich.

Optimal Design Procedure for Estimating Dimensions for Three-Phase System   On the basis of the model developed by Zundelevich the following optimal design procedure can be adopted. This procedure is for estimating the dimensions of a reactor using the Denver-type of impeller system used by Zundelevich:

1. In the case of a reaction, estimate stoichiometric requirement of Q_G. This can be done from a material balance based on the stoichiometry of the reaction and rate of consumption of the gas similar to that in Section 7A.10. Actual Q_G may be fixed at 20% higher than the theoretically required.

    Assume a realistic value of S. This should be related to the total reactor (dispersion) volume required for the given duty as explained in Section 7A.10. Due to the significant negative impact of S on Q_G regardless of the vessel diameter, the former should be in the range of 0.2 m. (Further, as will be shown later, the optimum value of S is related to the rotor blade height.) Using these values of Q_G and S, the power required for optimized operating conditions is $\left(\dfrac{Q_G S}{1.08 \times 10^{-5}}\right)$.

2. The rotor diameter is obtained using $Eu_{GOpt} = 53$:

$$D_{RT} = \sqrt[4]{\frac{53 \times Q_G^2}{gS}}$$

3. The optimum rotor speed is obtained at $Fr' = 0.9$ from

$$N_{Opt} = \sqrt{\frac{gS}{0.9 \times D_{RT}^2}}$$

Zundelevich (1979) suggested that the best scale-up criterion in terms of energy efficiency should be on the basis of constant V_G and S. However, when no bench-scale data are available, V_G may be selected in the range of 1×10^{-4} to 1×10^{-3} m/s, which is typical of the range used in stirred reactors. Using this value of V_G and the operating Q_G, the vessel diameter is $T = \sqrt{4 \times Q_{GOperating} / \pi V_G}$. The dispersion volume is $(\pi T^2 S/4)$. This should be checked with the value of dispersion volume required for obtaining the specified conversion as outlined in Section 7A.10.

Comments   This model is aimed solely at obtaining optimum configuration for maximum gas induction efficiency, $Q_G S/P$. When the aerator is to be used as a self-inducing system for heterogeneous reactions, the $Q_G S/P$ criterion or N_{Opt} is unlikely to be the major decisive factor. Other factors, for instance, gas dispersion/solid suspension, gas–liquid, and solid–liquid mass transfer. are also equally important in this latter application. The problem of gas dispersion and solid suspension may be accentuated particularly for large reactors where S is low or C/T is high (Conway et al. 2002). The optimum agitation speed obtained by the procedure given earlier may not be able to create enough pumping action to propel the gas–liquid dispersion up to the relatively distant bottom of the vessel in order to yield both desired quality of gas dispersion and solid suspension. Consequently, as proposed in the Booth design (Barbery 1984) [and earlier for a fermenter by Matsumura 1982a], a second impeller located at a lower level than the gas-inducing impeller is essential to provide dispersion of the gas and suspension of the solid phase in the lower part of the reactor, as discussed in the next section.

9.3.2.5   Gas–Liquid Mass Transfer in 2–2 Type Systems   Based on the results of the various available investigations in the literature, the following conclusions can be drawn:

1. True k_L is a very weak function of power input (Sawant et al. 1980, 1981).
2. The effective interfacial area, a, is a much stronger function of the power input per unit volume. The dependence of a ranged from $[P/V]^{0.86}$ for Wemco cells (Sawant et al. 1980) to $[P/V]^{0.5}$ for Denver cells (Sawant et al. 1981).

The $k_L a$ values are typically of the order of $2–6 \times 10^{-2}$/s for both Wemco (Sawant et al. 1980) and Denver (Sawant et al. 1981) systems in the range of P/V commonly employed: $0.5–1 \, kW/m^3$.

9.4   MULTIPLE-IMPELLER SYSTEMS USING 2–2 TYPE IMPELLER FOR GAS INDUCTION

The need for an additional impeller in a gas induction system has been outlined in Section 9.3.2.4 (Matsumura 1982a; Barbery 1984). Figure 9.7 shows the variation of the rate of induction with the submergence.

The data in Figure 9.7 were extracted from Figure 12A of Patwardhan and Joshi (1997). It is evident that there is an optimum value of S. Too low or high levels of submergence are detrimental. This behavior can be explained as follows: when the impeller rotates, it creates a vortex that lowers the liquid level in the hood. Consequently, at the lowest submergence, the liquid level is likely to fall much lower. This exposes part of the rotor blades to the gas phase resulting in significant loss of power transmitted to the impeller. With increasing submergence, the rotor is progressively submerged in the liquid, which improves the power transmitted. When the submergence reaches a certain critical value, which is probably such that the rotor blade is completely immersed in the liquid ($S=H_S$), the power transmitted is likely to attain a maximum value. Above this value of S, its negative impact on Q_G detailed in Section 9.3.2.4 starts playing an important role and Q_G decreases.

According to Zundelevich (1979) "although extremely fine gas bubble breaking takes place inside the aerator, the bubbles instantly coalesce and leave the stator with a diameter of 2 to 3 mm." This observation is corroborated by several investigators who found that the gas holdup in the vicinity of the impeller is higher than at the wall (Grau and Heiskanen 2003; Nesset et al. 2006). Consequently, the high rates of mass

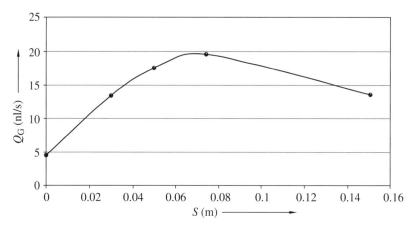

FIGURE 9.7   Variation of gas induction rate with submergence (analysis of data from Fig. 12A of Patwardhan and Joshi, 1997: $P = 500$ W; V_G: 0 (no gas sparging); Impeller combination: Upper PTD–lower PTU).

transfer arising out of smaller bubbles are limited to the radial zone in the proximity of the disperser. This problem is further aggravated by another constraint that the gas dispersion due to the top impeller (coarse as it may be) is also limited to a relatively small zone below the impeller (Patwardhan and Joshi 1997). This drawback can be illustrated using a simple example: we consider a reactor with $T = 2$ m and $H_D/T = 1$. Allowing a liberal set of values for the submergence of the top impeller $= 0.2$ m and a depth of its influence $= 0.5$ m, the fractional volume under the influence of the top impeller $= 0.5/2 = 0.25$. Since the depth of this zone of influence of the gas-inducing impeller is constant irrespective of the vessel diameter for $T = H$, the fraction of reactor volume under the influence of the gas-inducing impeller $\propto (T)^{-1}$. The aforementioned semiquantitative analysis is a supporting evidence for the observation of Conway et al. (2002) that gas-inducing systems are not suited for applications involving large industrial reactors. This predicament was first addressed by Matsumura et al. (1982a) who employed a small, high-speed impeller near the surface for gas induction. A larger-diameter, high-pumping-capacity impeller located closer to the bottom of the fermenter served the purpose of dispersion of the gas induced by the top impeller. This latter impeller was bottom mounted on a separate shaft. This arrangement could resolve the problem of gas dispersion in the lower part of the fermenter that was relatively distant from the gas-inducing impeller near the gas–liquid interface at the top. This system is cumbersome since it requires separate shafts, separate seals, and drives. In addition, when the reaction stoichiometry demands supply of large volume of gas, the limitations of the gas-inducing impeller outweigh its self-inducing advantage. Praxair liquid oxidation reactor (Roby and Kingsley 1996) uses a two-impeller arrangement similar to Matsumura et al. (1982a) along with gas sparging. However, the Praxair liquid oxidation reactor design employs a common shaft for both the upper gas-inducing and lower impeller (for gas dispersion/solid suspension) along with a draft tube. Additional gas is sparged in the annulus outside the draft tube. This is probably because the gas-inducing impeller may not be able to supply the amount required by reaction stoichiometry. Patwardhan and Joshi (1997) also employed a similar system but without a draft tube (Section 9.4.5).

9.4.1 Critical Speed for Gas Induction

The critical speed for gas induction is solely decided by the ability of the rotor to generate a suction higher than the sum of static head and other pressure losses. The rotor design obviously plays an important role. Even when a second impeller is employed for gas dispersion/solid suspension, its action is limited to the role assigned to it in a region substantially away from the gas-inducing device. Therefore, the presence of the second impeller should not have any effect on the critical speed for gas induction in a multiple-impeller system. This obvious fact was experimentally proved by Saravanan and Joshi (1995). Consequently, Equations 9.23 or 9.28, which yield similar predictions, can be used to predict N_{CG}. The effect of liquid viscosity can be accounted through Equation 9.24.

9.4.2 Rate of Gas Induction (Q_G)

In multiple-impeller systems, the interaction between the flows generated by the two impellers has a bearing on the rate of gas induction. Saravanan and Joshi (1995) investigated a variety of bottom impellers using PTD as the rotor in the top gas-inducing system. Their results indicate that at low power inputs [$P/V \leq 0.7\,kW/m^3$], there is very little difference in Q_G afforded by different designs of bottom impellers. However, for relatively higher power inputs [$P/V \geq 0.9\,kW/m^3$], they found that the use of an upflow impeller at the bottom afforded substantially higher (33–100%) Q_G as compared to a single PTD impeller at the bottom or a combination of two PTDs. Upflow propeller in particular gave better results as compared to 6 PTU. Figure 9.8 forms the basis of the explanation provided by Patwardhan and Joshi (1999).

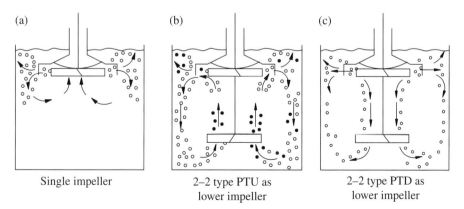

(a)	(b)	(c)
Single impeller	2–2 type PTU as lower impeller	2–2 type PTD as lower impeller

FIGURE 9.8   Flow patterns of (a) single PTD, (b) PTD–PTU, and (c) PTD–PTD multiple-impeller systems. (Reproduced from Patwardhan and Joshi, 1999 with permission from American Chemical Society. Copyright © 1999 American Chemical Society.)

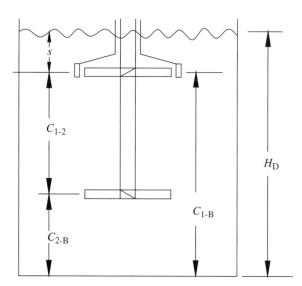

FIGURE 9.9   Various geometric parameters relevant to multi-impeller gas-inducing system as used by Saravanan and Joshi (1995). (Reproduced from Saravanan and Joshi, 1995 with permission from American Chemical Society. Copyright © 1995 American Chemical Society.)

It was argued that the liquid flow from an upflow impeller is directed such that it supplies liquid to the gas-inducing impeller through the bottom port of the diffuser. This supply augments the ability of the top impeller for carrying the induced gas bubbles. Therefore, upflow impellers have a positive effect on Q_G. When PTD is used as the lower impeller, it was suggested that the PTD withdraws liquid from the gas-inducing impeller, thereby hampering its bubble carrying capability. Consequently, Q_G for PTD–PTD is lower as compared to PTD–PTU or PTD–PU combinations. There are a number of other geometric factors that affect Q_G in multiple-impeller systems. Saravanan and Joshi (1995, 1996) and Patwardhan and Joshi (1997, 1999) have discussed these effects. The conclusions drawn by Saravanan and Joshi (1995) with reference to Figure 9.9 are:

(i) Increasing the submergence of the top impeller has a detrimental effect on Q_G. The top impeller used was the same as in Figure 9.7. However, the minimum submergence employed was 0.15 m, which is higher than the optimum value. Data extracted from Figure 12B of Saravanan and Joshi (1995) yield results similar to Figure 9.7 for $S > 0.15$ m. However, from the trend of these extracted data, it is evident that a maximum in Q_G is likely for $S < 0.15$ m and is probably the same as in Figure 9.7 and is related to the height of the rotor blades as discussed later. The relationship obtained for $S > 0.3$ m was $Q_G \propto (S)^{-1.75}$ indicating the very strong negative effect of S on Q_G. (ii) Q_G increases with a decrease in the interimpeller clearance, C_{1-2} in the range $0.68 \times D < C_{1-2} < 2 \times D$. Data extracted from Figure 14B of Saravanan and Joshi (1995) are plotted in Figure 9.10.

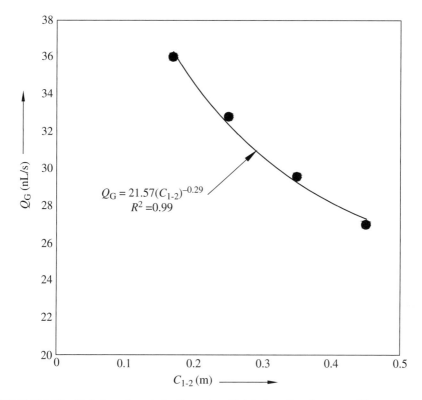

FIGURE 9.10   Variation of gas induction rate with interimpeller clearance. (Data generated from Fig. 14B of Saravanan and Joshi 1995.)

A regression yielded the following dependence: $Q_G \propto (C_{1\text{-}2})^{-0.29}$. Evidently, $C_{1\text{-}2}$ has a much weaker negative effect on Q_G as compared to that of S. (iii) It has been suggested that Q_G increases with decreasing clearance of the bottom impeller, $C_{2\text{-}B}$ defined in Figure 9.9. However, Figure 17 of Saravanan and Joshi (1995) (PU as the lower impeller) does not show appreciable changes in Q_G with varying $C_{2\text{-}B}$ in the relatively wide range $(0.19 < (C_{2\text{-}B}/T) < 0.5)$. It is understood that too low a value of $C_{2\text{-}B}$ is detrimental since the impeller action is likely to be hampered by close proximity to the bottom of the vessel.

9.4.2.1   Recommended Configuration for Maximum Q_G   The overall picture that emerges from the aforementioned analysis is that for realizing high Q_G, $C_{2\text{-}B}$ should be at least equal to $T/3$ (Nienow and Bujalski 2004). Therefore, with $C_{2\text{-}B} = T/3$, for a given value of $H_D = [S + C_{1\text{-}2} + C_{2\text{-}B}]$, the choice is between a low value of S or $C_{1\text{-}2}$. From the analysis of Saravanan and Joshi's (1995) and Patwardhan and Joshi's (1997) data discussed earlier, $Q_G \propto (C_{1\text{-}2})^{-0.29}$ and $Q_G \propto (S)^{-1.75}$, respectively. It is evident that the choice is clearly in favor of low S. From a comparison of the optimum value of $S \approx 0.07$ (Fig. 9.7) and the geometry of the gas-inducing assembly used by

Saravanan and Joshi, it transpires that the optimum value of S is very close to the height of the stator used. This agrees with the logic pertaining to power transmitted by the rotor outlined in the discussion pertaining to Figure 9.7. Figure 9.7 also indicates that the decrease in Q_G is not very steep in the vicinity of $S = H_S$. It may thus be argued that $S = 2H_S$ may not lead to a severe drop in Q_G. Thus, with $S \sim 2H_S$ and $C_{2-B} = T/3$, the recommended value of $C_{1-2} = [H_D - 2H_S - T/3]$. This should give the necessary specifications for a 6PTD–6 PTU system with respect to maximum Q_G. It may be noted that for larger reactors, such that absolute H_D is relatively high, two impellers may not be sufficient to cater to the desired gas dispersion/solid suspension duties.

9.4.3 Critical Speed for Gas Dispersion

9.4.3.1 2–2 Type Single Impeller   Several studies pertaining to gas dispersion in 2–2 type froth flotation cells are available in the literature (Gorain et al. 1995b; Grau and Heiskanen 2003; Nesset et al. 2006; Evans et al. 2008; Yianatos et al. 2010; Miscovic 2011; Shabalala et al. 2011; Vinnett et al. 2012). In most cases, the focus was on measurement of gas holdup/bubble size and its distribution. Miscovic (2011) has investigated loading/flooding and complete dispersion regimes besides gas holdup. In order to simulate the actual working of flotation cells frothers/collectors were used. The presence of surface-active agents has a significant effect on gas holdup and its structure. The data obtained are specific to the type and amount of the surface-active agent used. Therefore, these data are not relevant to the use of a gas-inducing system in three-phase systems devoid of surface-active agents.

9.4.3.2 2–2 Type Multiple Impellers   As repeatedly mentioned earlier, a single gas-inducing impeller located very close to the top ($S \approx 0.2$ m) provides a limited dispersing action. A second (possibly third also, depending on the absolute value of H) impeller is required to give the desired dispersion in the lower zone far away from the top impeller. The type, size, and speed of the bottom impeller are the governing factor in this latter phenomenon. In the case of a conventional stirred reactor (Section 7A.5.4), the gas to be dispersed is generally supplied through a sparger located below the impeller used. In this case, flow from the impeller has to counteract the buoyancy of the gas as the first step in dispersion. N_{CD} increases with increase in the volume of gas sparged, which is generally represented by V_G. The extent of increase depends on the type of impeller and is least for upflow impellers wherein the discharge stream from the impeller and gas buoyancy are in the same direction. The situation is reversed in the case of a gas-inducing system. The gas is supplied to the bottom impeller by the gas induction system located well above the former. As discussed in Section 9.4.2, the rate of gas induction is a dependent variable that is itself a complex function of various geometric factors (Fig. 9.9). The extent of dispersion is related to the volume of gas, Q_G (or V_G) that is handled by the lower impeller. The induced gas flows out of the diffuser in a radial direction (Figure 9.8 from Patwardhan and Joshi 1997). At the vessel wall, this flow divides such that two distinct streams, (1) traveling up and (2) traveling down, are formed. The upward component rapidly exits the dispersion because of proximity to the top of the dispersion and buoyancy of the bubbles. The part going down penetrates into the zone $C_{1\text{-}2}$. The penetration depth is expected to increase with increase in the discharge of the top impeller, which is related to Q_G. The latter increases with decrease in S. The fate of this penetrating component is decided by the pumping action of the bottom impeller. At this stage, the ability of the bottom impeller to pull this penetrating component further down is expected to lead to better dispersion. In the case of PTU, the flow direction is upward near the central axis and downward at the wall. The directions of the two flows at the wall are same as shown in Figure 9.8b. Therefore, PTU is expected to yield better gas dispersion and also lower values of N_{CD}. Due to the improved gas dispersion, the gas holdup is also expected to be higher in this case. On the other hand, for PTD as the bottom impeller, the directions are exactly reversed (Figs. 9.8c and 7A.13a). The upward wall region flow of the lower PTD opposes the downward flow of gas induced from above. N_{CD} values for PTD–PTD combination have not been reported. However, Saravanan and Joshi (1996) observed that ε_G for PTD–PTU combination is higher than PTD–PTD by 100% at $P/V \sim 0.3\,\text{kW/m}^3$ and this difference increases with increasing P/V (Figure 3 of Saravanan and Joshi 1996). The increasing difference in ε_G with increasing P/V clearly indicates that the PTU in a PTD–PTU combination pulls increasing quantities of the induced gas in the lower zone as P/V increases. Another interesting difference between conventional stirred vessels and gas-inducing systems is the effect of V_G on N_{CD}. In the former, N_{CD} increases with increase in V_G or Q_G. Data from Table 4 of Saravanan and Joshi (1996) can be used to discern the effect of V_G on N_{CD}. The last data set gives the variation of N_{CD} with increasing S at constant $C_{1\text{-}B}$ and $C_{1\text{-}2}$. This data set clearly shows that with decreasing S (increasing Q_G), there is clear and substantial decrease in N_{CD} ($N_{CD} = 4.68/s$ for $S = 0.7\,\text{m}$ decreasing to $1.71/s$ for $S = 0.15\,\text{m}$). Higher Q_G (or V_G) results in a greater momentum and penetration of the induced stream that improves its carriage to the lower zone. Subsequently, this induced stream is assisted in the downward direction by the cooperative flow generated by the bottom PTU impeller. This "direction"-based logic explains why N_{CD} actually decreases with increase in V_G in a gas-inducing system. Table 4 of Saravanan and Joshi (1996) also shows that for PTD–PTU, increase in $C_{1\text{-}2}$ results in $\sim 25\%$ increase in N_{CD} for $0.16 < (C_{1\text{-}2}/T) < 0.45$ for $(C_{2\text{-}B}/T) = 0.16$ and $T = 1\,\text{m}$. However, there is a marginal ($\sim 6\%$) increase in N_{CD} over the range $0.16 < (C_{2\text{-}B}/T) < 0.5$ for $(C_{1\text{-}2}/T) = 0.17$ and $T = 1\,\text{m}$.

The following logical conclusions can be drawn for the case when PTU is used as the bottom impeller (data of Saravanan and Joshi 1996): (i) The strength of the flow generated by the lower impeller is $\propto D^3$. Therefore, N_{CD} should decrease with increase in diameter of the lower impeller. (ii) N_{CD} increases with an increase in the interimpeller distance, $C_{1\text{-}2}$, and (iii) $C_{2\text{-}B}$ does not have a significant effect on N_{CD} for $0.16 < (C_{2\text{-}B}/T) < 0.5$ (for $C_{1\text{-}2}/T = 0.17$ and $T = 1\,\text{m}$).

These conclusions must be seen in the light of higher Q_G and optimum configuration recommended in Section 9.4.2.

9.4.4 Critical Speed for Solid Suspension

As discussed in Chapters 7A and 8, solid suspension is critical when the solid phase acts as a catalyst for a reaction between a dissolved gas and the liquid phase. Most solids used as catalysts are heavier than common organic liquids. Therefore, off-bottom suspension is the criterion that needs to be applied.

9.4.4.1 2–2 Type Single Impeller The widespread use of froth flotation machines based on gas-inducing principle has attracted substantial studies on solid suspension. Solid suspension is as important as gas dispersion in the overall beneficiation process in the ore dressing industry. Several investigators studied solid suspension in flotation cells (Cliek and Yilmazer 2003; Zheng et al. 2005; van der Westhuizen and Deglon 2007, 2008; Lima et al. 2009; Miscovic 2011). In the case of froth flotation, suspended solids are of two types: (1) attached to rising bubbles because of the action of the collector and (2) suspended due to mean flow and turbulence. Johnson (1972 c.f. Zheng et al. 2005) suggested that in order to study truly hydrodynamic suspension, the focus should be on "nonfloating" particles in the ore. However, such tracers are usually not present in real ores. Therefore, Zheng et al. (2005) suggested that "tailings" may be considered as representative of that part of the feed that is not attached to bubbles ("nonfloating"). A classification function, C_{FI}, was defined for such particles as follows:

$$C_{FI} = \left(\frac{W_I^P}{W_I^T} \right) \tag{9.45}$$

Their results on a $150\,m^3$ Outokumpu cell showed that C_{FI} was correlated by an exponential type expression:

$$C_{FI} = \alpha' \exp - (\lambda' d_{PI}) \tag{9.46}$$

The parameter λ was related to cell hydrodynamics represented by the cell size, design, and operating conditions (impeller speed, air rate, etc.). The proportionality constant α' was found to be close to unity in most of the cases studied. Analogous to the case of conventional stirred reactors, Zheng et al. (2005) found that the degree of suspension decreased with increase in air rate. Curiously, at relatively higher air rates, C_{FI} for coarser particles increased.

van der Westhuizen and Deglon (2007, 2008) studied solid suspension in a Batequip cell. The 1 s criterion of Zwietering (1958) was used to determine the critical speed for solid suspension. In addition, solid concentration was measured to determine the cloud height. The latter has also been used as a criterion for solid suspension (Section 7A.7.1). van der Westhuizen and Deglon (2007) used three parameters to characterize solid suspension: (i) degree of off-bottom suspension, (ii) axial solid concentration profile, and (iii) suspension zone or cloud height. Results obtained and conclusions in respect of these three parameters were as follows.

Off-bottom suspension was quantified in terms of a ratio X/X_S as a function of relative speed of suspension N/N_S for two-phase and N/N_{SG} for three-phase systems, respectively. Plots of X/X_S versus N/N_S showed that settling of solids became significant at $N/N_S \sim 0.6$. Therefore, for relatively better suspension, N/N_S should be above 0.6.

Plots of dimensionless cloud height, H_C/H versus N/N_S obtained from measured solid concentrations also exhibited similar behavior. For $N/N_S = 1$, the cloud height was consistently 90% of the dispersion height. Further, the 90% cloud height criterion showed good agreement with Zwietering's criterion. Significant decrease in H_C/H was observed at and below $N/N_S \sim 0.6$ in a manner similar to that for the X/X_S versus N/N_S plots. In conclusion, speeds of agitation below 60% of the critical speed for suspension should be avoided.

This investigation also used the Relative Standard Distribution (RSD) as a measure of homogeneity in suspension. A low value of RSD reflects better homogeneity. Introduction of gas was found to result in inferior solid suspension in a manner similar to that in stirred reactors.

In a subsequent investigation, van der Westhuizen and Deglon (2008) studied suspension of silica, rutile, and zircon in a $0.125\,m^3$ Batequip cell. The data obtained were used to develop a Zwietering-type correlation for the critical speed of suspension. The major conclusion derived was that the critical speed for solid suspension is proportional to $(d_p)^{0.33}; (X)^{0.17}; (\Delta\rho/\rho_L)^{0.7}$ and varies linearly with the gas rate. The exponents on X and slurry viscosity were not significantly different, whereas those on d_p and ρ_S were substantially higher than in Zwietering's correlation. The authors argued that the stator reduces the mean flow and favors increase in turbulence. The relatively higher exponent on d_p was attributed to this reduced mean flow. By corollary, they concluded that the mean flow plays a dominant role in solid suspension. This conclusion is in agreement with the importance of mean flow elaborated in Sections 6.3.1b, 7A.5.2, 7A.5.3, 7A.6, and 7A.7.1. The effect of gas rate was incorporated using two variants:

(i) $\Delta N_{SG} = \text{constant} \times (\text{vvm})$ [similar to Chapman et al. (1983: Table 7A.4)]   (9.47)

(ii) $N_{SG} = N_S[1 + \beta' \times vvm]$ similar to Nienow (2000)   (9.48)

Equation 9.48 was found to give better representation of the data. Accordingly, the following expression for N_{SG} was suggested:

$$N_{SG} = S_G d_P^{0.33} X^{0.17} \left(\frac{\rho_S - \rho_L}{\rho_L} \right)^{0.7} \left(\frac{v}{v_W} \right)^{0.05} (1 + \beta' \times vvm) \qquad (9.49)$$

S_G and β' are constants that are specific to the geometric configuration and impeller design $\left\{ f\left[T; \dfrac{D}{T}; \dfrac{C}{T} \right] \right\}$. van der Westhuizen and Deglon (2008) also gave a correlation for the power input per unit volume required to achieve the just-suspended condition as follows:

$$\left(\frac{P}{V} \right)_S = \left[\frac{N_{PG} \rho_{SL} N_{SG}^3 D^5}{V} \right] \qquad (9.50)$$

For the system employed, the following equation was suggested:

$$N_{PG} = 6.5 \exp(-0.68 \times \text{vvm}) \tag{9.51}$$

Lima et al. (2009) investigated solid suspension in laboratory-scale Wemco and Denver cells and compared their results with those of van der Westhuizen and Deglon (2008). These investigators could also correlate N_{SG} using Zwietering-type correlation (Eq. 9.49). Expectedly, the parameters S_G and β, which are dependent on the type and size of the impeller/cell design, were different for the three types of cells.

9.4.4.2 2–2 Type Multiple Impeller

The bottom impeller should simultaneously play the role of gas dispersion and solid suspension. Solid suspension in this category has received very limited attention. The only study available is that of Saravanan et al. (1997). These investigators studied solid suspension characteristics of 2–2 type multiple-impeller systems. The top gas-inducing impeller used was 6 PTD, while a variety of impellers that generated different flow patterns (Fig. 9.8) were used as the bottom impeller. The difference between multiple-impeller gas-inducing systems and stirred reactors *vis-à-vis* the location of the gas sparger has already been discussed in Section 9.4.3. Saravanan et al. (1997) found that the value of N_{SG} obeys the following order for different impeller combinations:

$$\text{PTD–PTD} < \text{PTD–SBT} < \text{PTD–DT} < \text{PTD–PTU} < \text{PTD–PU}$$

In Section 7A.7.1c, it has been shown that downflow and radial flow impellers are inferior to upflow impellers for suspension of solids in conventional three-phase stirred reactors (Nienow and Bujalski 2004). It must be pointed out that the choice of the bottom impeller should involve three considerations: (1) high Q_G and ε_G, (2) low N_{CD}, and finally (3) low N_{SG}. PTU as the lower impeller satisfies (i) and (ii) but evidently not (iii) as per the results of Saravanan and Joshi (1997). The latter indicate that N_{SG} for PTU is approximately twice that for PTD. This should result in a correspondingly higher power (approximately eight times higher) if the power numbers for PTU and PTD are assumed to be approximately equal. However, according to Figure 3B of Saravanan and Joshi (1997), PTU requires approximately three times higher power than PTD as bottom impeller. In general, $N_{CD} < N_{SG}$ (Section 7A.4). Further, as discussed in Section 9.4.3, N_{CD} for PTU is lower than that for PTD. Therefore, unless $(N_{SG} - N_{CD})_{PTU} >> (N_{SG} - N_{CD})_{PTD}$, it is expected that N_{SG} for PTU should also be approximately same as that for PTD. These arguments indicate that for design purposes it is not incorrect to assume that N_{SG} for PTU and PTD are substantially similar. Therefore, PTU should be preferred as the lower impeller on both counts (N_{CD} and N_{SG}). Saravanan and Joshi (1997) have correlated their data for N_{SG} in the form of Equation 9.52:

$$N_{SG} = 0.031 \left(\frac{\rho_s - \rho_L}{\rho_L} \right) A^{-0.83} \left(\frac{W}{D} \right)^{-0.314} d_P^{0.17} X^{0.15} \left(\frac{C_{1-2}}{D} \right)^{-0.446} D^{-2.42} \left(\frac{C_{2-B}}{T} \right)^{0.127} T^{1.35} \tag{9.52}$$

where A is the blade angle in radian and W blade width of the lower impeller in meters.

The constant 0.031 has the following units ($m^{1.072} rad^{0.83} \mu m^{-0.17}/s$). The data used for the correlation given by Equation 9.52 pertain to vessel sizes ranging from 0.57 to 1.5 m. Equation 9.52 has a number of geometry-related parameters, and therefore, its extrapolation to conditions outside the data limits as well as for different geometric configurations can pose problems.

9.4.5   Operation of Gas-Inducing Reactor with Gas Sparging

Litz (1985), Litz et al. (1990), Kingsley et al. (1994), and Roby and Kingsley (1996) suggested a design that combines sparging and gas induction to increase the mass transfer rate and energy efficiency of the Praxair liquid oxidation reactor. In this mode, pure gas is sparged in the annulus as shown in Figure 3.3. The sparged gas is entrained in the swirling action of the impeller in the draft tube. The gas liquid mixture exits at the bottom of the draft tube and is recirculated. The details and advantages of such a combination are discussed in Section 3.4.2d. Patwardhan and Joshi (1997) studied a modification of the original concept of the aforementioned investigators for a 2–2 type multi-impeller gas-inducing system. Evans et al. (1991), Forrester and Rielly (1994), and Ju et al. (2009) have also studied a similar combination. However, the impeller used was a pipe impeller (1–1 type), and therefore, there was either no effect or a marginal decrease in the rate of gas induction in the presence of the sparged gas. Evans et al. (1991) and Forrester and Rielly (1994) further did not notice any change in N_{CG} due to sparging.

For the 2–2 type system (Fig. 9.4), the bottom part of the stator is open and allows fluid exchange with the vessel. When gas is sparged from a lower location, it rises and can enter the diffuser of the gas-inducing impeller. At low speeds of the top impeller, the sparged gas can flood the diffuser because of insufficient opposing flow that can prevent its entry into the stator zone. As the speed increases, there is increasing opposing flow/resistance to gas entry. With further increase in N, the sparged gas is prevented from entering the stator zone. Finally, at still higher speeds, gas inductions start in a manner similar to that in the conventional gas-inducing systems without gas sparging. It is evident that N_{CG} in the presence of gas sparging, N_{CG+S}, is higher than in its absence and the extent of increase in N_{CG} for a given gas-inducing impeller is related to the rate of gas sparging. The difference $\Delta N_{CG+S} = [N_{CG+S} - N_{CG}]$ is an indication of the additional power $\{\propto [\Delta N_{CG+S}]^3\}$ that is required.

In applications of 2–2 type systems as three-phase reactor, a relatively large stoichiometric amount of gas may have to be supplied. The 2–2 type system on its own may not be energy efficient for such large requirement. Gas sparging coupled with a 2–2 type multiple-impeller system can be considered as an alternative (Litz 1985; Litz et al. 1990; Kingsley et al. 1994; Roby and Kingsley 1996; Patwardhan and Joshi 1997). The required quantity (including about 20% excess above the stoichiometric requirement) of pure gas is sparged in a manner similar to the conventional stirred reactor (Patwardhan and Joshi 1997). An impeller located above the sparger disperses the gas that travels along the dispersion height. Unreacted gas escapes into the gas space. The gas-inducing impeller now plays the role of recirculating this relatively small amount of unreacted gas albeit, over a small zone (Section 9.4). The power requirement for the top impeller is thus substantially reduced. Such a system can operate in a dead-end mode although bulk of the gas is supplied through sparging. It can ensure a balance between sparging, consumption in reaction and recirculation of the pure gas. Patwardhan and Joshi (1997) have discussed various aspects of this type of arrangement. The following correlation was suggested to calculate ΔN_{CG+S} or N_{CG+S} for a given sparging rate represented by the superficial gas velocity, V_G:

$$[\Delta N_{CG+S}] = \sqrt{\frac{2 \times S}{(\pi D)^2 \varphi}} \times \alpha_1 \times \left(\frac{C_{1-2}}{D}\right)^{\alpha_2} \left(\frac{S}{T}\right)^{\alpha_3} (V_G)^{\alpha_4} \qquad (9.53)$$

Saravanan and Joshi (1995) and Patwardhan and Joshi (1997) have given empirical correlations for Q_G in the absence and presence of gas sparging. Patwardhan and Joshi (1997) have suggested the following correlation for estimating the reduction in gas induction rate due to sparging:

$$\left(1 - \frac{Q_{G+S}}{Q_G}\right) = \left[\beta_1 \left(\frac{Q_S}{(N - N_{CG})D^3}\right)^{\beta_2} \left(\frac{C_{1-2}}{D}\right)^{\beta_3}\right] \qquad (9.54)$$

Patwardhan and Joshi (1999) subsequently commented that Equation 9.54 is completely empirical and caution must be exercised in its use for other systems and importantly also for scales larger than those used in obtaining the empirical constants in Equation 9.54. Zundelevich's procedure discussed in Section 9.3.2.4 is relatively simple in so far as a quick estimate of power required to achieve a given Q_G in the absence of sparging is concerned. Alternatively, knowing the amount of gas to be recirculated, the power input required can be estimated as per the stepwise procedure for optimal design outlined in section Optimal Design Procedure for Estimating Dimensions for Three-Phase System. Unfortunately, this procedure cannot be directly used in the presence of gas sparging. However, it can yield the lower bound of N_{CG} and power required for comparison with other alternatives (Section 9.5). This is illustrated for the design of a multiple-impeller 2–2 type gas-inducing system for hydrogenation of aniline (Section 9.5).

9.4.6 Solid–Liquid Mass Transfer Coefficient (K_{SL})

There is not a single investigation on this important aspect for any type of gas-inducing impeller discussed in this chapter. Information on K_{SL} is particularly important for three-phase systems involving a solid phase as a catalyst. In the case of multiple-impeller system, the lower impeller is accorded the duty of solid suspension. As discussed in Sections 7A.5.4 and 7A.7, 6 PTU affords superior performance in three-phase systems. Therefore, the same is advisable in the present case. In the absence of such information, K_{SL} afforded by 6 PTU in a geometrically similar three-phase stirred tank (Section 7A.7) configuration may be used.

9.4.6.1   Summary on 2–2 Type Systems   The earlier discussion indicates that design of a 2–2 type system for three-phase reactions needs to serve the following minimum objectives: (i) high Q_G and ε_G for a given power input, (ii) adequate gas dispersion and solid suspension, and (iii) high $k_L a$ and K_{SL}. It is not possible to satisfy all these requirements simultaneously. The process designer therefore must take a holistic view and make compromises in such a manner that none of the basic requirement is substantially transgressed.

9.5   WORKED EXAMPLE: DESIGN OF GAS-INDUCING SYSTEM WITH MULTIPLE IMPELLERS FOR HYDROGENATION OF ANILINE TO CYCLOHEXYLAMINE (CAPACITY: 25,000 METRIC TONNES PER YEAR)

This system has been used in Chapters 7A and 8 to illustrate design of the respective reactors. The configuration employed in this chapter is a Turboaerator combined with a lower 6 PTU. The total amount of hydrogen required is sparged below the lower PTU. There is considerable similarity between this gas-inducing system and conventional stirred reactor. Further, in both the cases, the mass transfer coefficients ($k_L a$ and K_{SL}) are governed by the lower impeller. Consequently, the controlling regime is unlikely to be different than in the conventional stirred reactor (Section 7A.10). The gas–liquid mass transfer step contributes more $\sim 99\%$ of the resistance to absorption of hydrogen. The solid–liquid mass transfer and reaction steps that contribute $< 1\%$ can therefore be ignored. The inadequacy of a stirred reactor for this reaction has been brought out in Section 7A.10. The gas-inducing system does not offer any significant advantage over the stirred reactor in terms of the gas–liquid and solid–liquid mass transfer coefficients. The volume of stirred reactor obtained in Section 7A.10 for 20% excess hydrogen is daunting. A dual impeller gas-inducing system is unlikely to yield any different result in view of similar $k_L a$. However, it can be advantageous in terms of recycling the unreacted hydrogen internally. Therefore, as a matter of academic interest, this aspect is examined in the following.

9.5.1   Geometrical Features of the Reactor/Impeller (Dimensions and Geometric Configuration as per Section 7A.10 and Figure 9.9, Respectively)

Reactor diameter, $T = 12.75\,m$

Impeller = multiple impeller and PTD–PTU. Lower PTU diameter = $(T/3) = 4.25\,m$.

Submergence of the top impeller = $0.2\,m$.

Dispersion height, $H = T = 12.75\,m$.

Clearance of bottom PTU impeller, $C_{2\text{-}B} = D = (T/3) = 4.25\,m$.

Distance between top and bottom impeller, $C_{1\text{-}2} = H - S - C_{1\text{-}2} = 8.3\,m$.

The large absolute value of $C_{1\text{-}2}$ reflects the difficulty that will be encountered in dispersion of the gas induced by the top impeller.

9.5.2 Basic Parameters

Since the reactor production capacity is same as in Section 7A.10, the flow rates of the reactants remain unaltered. Further, the other major operating parameters for the lower impeller (N_{CD}, N_{SG}, etc.) remain the same. For the lower 6 PTU with $D = 4.25$ m, the operating speed is increased by 10% to ~0.25 rev/s to provide for any deviation in this nonstandard combination. The power required at 0.25 rev/s is 34.4 kW.

Amount of excess hydrogen (20% of the theoretical requirement) that is supplied 70.5 m³/h or 2×10^{-2} m³/s. A Turboaerator is to be used for recycling this unreacted hydrogen. The simplified stepwise procedure outlined in section Optimal Design Procedure for Estimating Dimensions for Three-Phase System (Zundelevich 1979) can be used for estimation of the dimensions and other operating parameters of the Turboaerator. It must be noted that actual parameters (N_{CG+S} and P_{G+S}) will be higher than those estimated by this method:

1. Power required = ~ 370 W
2. Rotor diameter, $D_{RT} = 0.32$ m
3. $N_{Opt} = 4.6$ rev/s

The optimum agitation speed obtained from Zundelevich's method is significantly different than the operating speed for the lower impeller (0.25 rev/s). If two impellers are to operate at these substantially (1800%) different speeds, separate shafts/ seals and gearboxes need to be used. It must be stressed that for a hydrogenation reactor operating at 1 MPa, this is not a comfortable situation. Safety/maintenance and other precautions dictate that it is preferable to mount both the impellers on a single shaft running at a suitable speed. The options available are (i) the use of a smaller bottom impeller operating at 4.6 rev/s and (ii) larger gas-inducing impeller operating at 0.25 rev/s.

Option (i): In this option, the diameter of a suitable 6 PTU that yields solid suspension (as well as gas dispersion since $N_{SG} > N_{CD}$ at 4.6 rev/s; Section 7A.10) is very low [D ~ 0.06 m and $D/T = 0.025$]. D/T is particularly very low as compared to the optimum value of $D/T = 0.33$ (Fig. 7A.1; Section 7A.7; McCabe et al. 1986). Therefore, this option is ruled out.

Option (ii): The gas-inducing agitator has to operate at 0.25 rev/s and yield an induction rate of 2×10^{-2} m³/s. This will be a nonoptimized configuration. The corresponding rotor diameter can be obtained using the following procedure. Zundelevich's (1979) Figure 7, design "a," which affords better performance than design "b," is used for this purpose. The data for Figure 7a can be correlated by

$$Eu_G = 0.4912e^{3.886 \times Fr'} \tag{9.55}$$

Solving Equation 9.55 for Q_G at $N = 0.25$ rev/s yields $D_{RT} = 3$ m (a spreadsheet with "goal seek" can be used). For the Turboaerator used by Zundelevich, $N_P = 3.1$ and pumping capacity $K_P = 0.071$. Using these values of Eu_G, Fr', and N_P in Equation (19) to (21) of Zundelevich (1979) yields the power required as 10.6 kW. This is substantially higher than the power required for the recycle compressor in Section 7A.10 for the base case of 20% excess hydrogen. This could be attributed to the nonoptimum geometry in which the rotor diameter is very high and the fact that $P \propto \left(D_{RT}^5 \right)$. The 9.4-fold increase in D_{RT} over the optimum value of 0.32 m obtained using Zundelevich's procedure is the main source of this substantial increase in power required for gas induction.

In conclusion, the gas-inducing impeller, which affords identical values of $k_L a$, has the same limitation as that of the stirred reactor for this relatively fast reaction (Section 7A.10). The venturi loop reactor appears to be the best option both in terms of capital and in operating cost.

NOMENCLATURE

A_{MT}:   Cross-sectional area of mixing section (m²)

A_N: Cross-sectional area of nozzle (m²)

C: Clearance of impeller from bottom (m)

C_{1-2}:   Interimpeller distance (m)

C_{2-B}:   Clearance of lower impeller (m)

C_{FI}:   Classification function defined by Equation 9.45 (—)

C_P: Pressure coefficient (—)

$C_{P\theta}$:   Pressure coefficient defined by Equation 9.7 (—)

D: Impeller diameter (m)

d_P: Particle diameter (m or μm)

d_{PI}:   Particle diameter of species I (m)

D_{RT}:   Diameter of rotor in gas-inducing impeller (m)

D_{ST}:   Diameter of stator in gas-inducing impeller (m)

Eu_G: Euler number for gas phase, $\dfrac{gS}{\left(Q_G / D_{RT}^2 \right)^2}$ (—)

F_{LG}:   Flow number, $\dfrac{Q_G}{ND^3}$ (—)

F_{LG}':   Modified flow number, $\left(\dfrac{Q_G}{ND^3} \right) \left(\dfrac{D}{S} \right)^{0.5}$ (—)

F_{LGC}:   Flow number at critical speed for gas induction, $\dfrac{Q_G}{N_{CG} D^3}$ (—)

Fr: Froude number $\left(\dfrac{N^2 D^2}{S \times g} \right)$ (—)

Fr':   Reciprocal of the Froude number, $\left(\dfrac{S \times g}{N^2 D_{RT}^2} \right)$ (—)

Fr_C: Froude number at critical speed for gas induction, $\left(\dfrac{N_{CG}^2 D^2}{S \times g} \right)$ (—)

g: Gravitational constant, 9.81 m/s²

H_C: Cloud height (m)

H_D: Dispersion height (m)

h_L: Liquid head above the orifice (m)

h_{RT}:   Height of rotor (m)

h_S: Liquid head outside the orifice in the absence of gas flow (m)

h_{ST}:   Height of stator (m)

K: Slip factor (—)

K_1:   Head loss coefficient due to nonideal effects defined in Equation 9.25 (—)

K_2:   Head loss coefficient in Equation 9.36 (—)

$k_L \underline{a}$:   Volumetric gas–liquid mass transfer coefficient (1/s)

K_P: Impeller pumping coefficient, $\left(\dfrac{Q_L}{ND^3}\right)$ (—)

N: Impeller speed (rev/s)

N_{CG}:   Critical speed for gas induction (rev/s)

N_{CG+S}:   Critical speed for gas induction in the presence of gas sparging (rev/s)

N_P: Impeller power number (—)

N_{PG}:   Impeller power number in the presence of gas (—)

N_S: Critical speed for suspension in solid–liquid (two-phase) system (rev/s)

N_{SG}:   Critical speed for "just-suspended" condition in gas–liquid–solid system (rev/s)

P: Power input (W)

P_0:   Pressure in the headspace (N/m², Pa)

P_S: Pressure at the stagnation point (N/m², Pa)

P_{ST}:   Pressure at the outlet of stator (N/m², Pa)

P_{Tot}:   Total pressure on the orifice, sum of hydrostatic and pressure head (m)

Q_{AV}:   Time-averaged rate of gas induction (m³/s)

Q_G: Gas induction rate (m³/s)

R: Radius of pipe impeller (m)

R_{DB}:   Radius of detached bubble (m)

Re: Impeller Reynolds number $(ND^2 \rho_L / \mu_L)$ (—)

R_i: Radius of the ith orifice on the pipe impeller (m)

S: Submergence of the impeller (m)

S_G: Zwietering-type constant in Equation 9.49 (rev/min)

T: Vessel or tank diameter (m)

U_T: Tip velocity at the orifice (m/s)

V_{eff}:   Effective volume of the froth flotation cell (kT^n) (m³)

W_I: Weight fraction of component I (—)

X: Solid concentration (wt%)

X_S: Solid concentration at the "just-suspended" condition (wt.%)

ΔH_R: Heat of reaction (kcal/mol or kJ/mol)

ΔN_{CG+S}: Difference in the critical speed for gas induction with and without gas sparging $[N_{CG+S} - N_{CG}]$ (rev/s)

ΔN_{SG}: Difference between critical speed for suspension in solid–liquid (two-phase) and gas–liquid–solid (three-phase) system (rev/s)

ΔP_{D}: Driving force for gas induction (pressure differential between the orifice pressure and that at the headspace) (N/m²)

ΔP_{F}: Frictional pressure loss due to flow through the hollow pipe and blades (N/m²)

ΔP_{G}: Pressure drop suffered by gas phase defined in Equation 9.35 (N/m²)

ΔP_{G+S}: Difference in the power for gas induction with and without gas sparging $[P_{CG+S} - P_{CG}]$ (W)

ΔP_{KE}: Frictional pressure loss due to the kinetic energy imparted to the fluid (N/m²)

ΔP_{L}: Pressure drop suffered by liquid phase defined in Equation 9.35 (N/m²)

ΔP_{O}: Frictional pressure loss due to flow through the orifice in hollow pipes (N/m²)

$\Delta P\sigma$: Pressure loss due to work required to be done to overcome the surface tension forces during bubble formation at the orifice in hollow pipes (N/m²)

Greek Letters

α': Constant in Equation 9.46 (—)

β': Constant in Equation 9.48 (min)

$\beta_1, \beta_2, \beta_3$: Constants in Equation 9.54

θ: Angle measured from the axis of the cylindrical pipe impeller (degree)

λ': Constant in Equation 9.46 (—)

μ: Fluid viscosity (Pa, s)

ρ_D: Dispersion density (kg/m³)

ρ_L: Liquid density (kg/m³)

σ: Liquid surface tension (N/m)

Φ_{P-K}: Ratio of local potential and kinetic energies defined by Equation 9.3 (—)

ϕ: Vortexing constant in Equation 9.53

φ_D: Diffuser velocity coefficient defined in Equation 9.35 (—)

φ_N: Nozzle velocity coefficient defined in Equation 9.35 (—)

ψ: Injection coefficient (Q_G/Q_L) (—)

Subscripts

G: Gas phase

L: Liquid phase

S: Solid phase

W: Water

Superscripts

P: Product
T: Tailing

REFERENCES

Abbott IH, von Doenhoff AE. (1959) Theory of wing sections. Dover Publications, New York, USA.

Abdullah B, Dave C, Nguyen TH, Cooper CG, Adesina AA. (2011) Electrical resistance tomography-assisted analysis of dispersed phase hold-up in a gas-inducing mechanically stirred vessel. Chem. Eng. Sci., 66:5648–5662.

Achouri R, Hamza SB, Dhaouadi H, Mhiri H, Bournot P. (2013) Volumetric mass transfer coefficient and hydrodynamic study of a new self-inducing turbine. Energy Conv. Management, 71:69–75.

Aldrich C, van Deventer JSJ. (1994) Observations on induced aeration in agitated slurries. Chem. Eng. J., 54:199–205.

Arbiter N, Harris CC. (1962) Flotation machines. In: Fuerstenau DW, editor. Froth flotation. AIME, New York, USA. p 347–364.

Arbiter N, Steininger J. (1965) Hydrodynamics of flotation machines. In Roberts A, editor. Mineral processing. Proceedings of Sixth International Mineral Processing Congress. Pergammon Press, London, UK. p 595–608.

Arbiter N, Harris C, Yap RF. (1976) The air flow number in flotation machine scale up. Int. J. Miner. Process., 3:257–280.

Baczkiewicz J, Michalski M. (1988) Oxygen transfer during mixing of acetic acid fermentation medium with self aspirating tube agitator. In: Proceedings of Sixth European Conference on Mixing, Pavia, Italy, May 24–26, BHRA, Cranfield, UK.

Barbery G. (1984) Engineering aspects of flotation in minerals industry: flotation machines, circuits and their simulation, in the scientific basis of flotation. In: Ives KJ, editor. NATO-ASI series, Martinus Nijhoff Publishers, The Hague. p 289–348.

Braginsky LN (1964) The study of aeration characteristics of impellers for autoclave leaching processes. Scientific Report on Research Project, Research and Design Institute NIIKhIMMASH, Leningrad, Russia. *cf*: Zundelevich (1979).

Chapman CM, Nienow AW, Cooke M, Middleton JC. (1983) Particle–gas–liquid mixing in stirred vessels. Part III. Three phase mixing. Chem. Eng. Res. Des., 61(3):167–181.

Cliek EC, Yilmazer BZ. (2003) Effects of hydrodynamic parameters on entrainment and flotation performance. Miner. Eng., 16:745–756.

Conway K, Kyle A, Rielly CD. (2002) Gas–liquid–solid operation of a vortex-ingesting stirred tank reactor. Chem. Eng. Res. Des., 80:839–845.

Davidson JF, Schueler BOG. (1960) Bubble formation at an orifice in a viscous liquid. Trans. Inst. Chem. Eng., 38:144–154.

Degner VR, Treweek HB. (1976) Large flotation cell designs and development. In flotation: A.M. Gaudin Memorial edition. AIME, New York, USA. p 824–831.

Dietrich E, Mathieu C, Delmas D, Jenck J. (1992) Raney-nickel catalyzed hydrogenations: Gas-liquid mass transfer in gas-induced stirred slurry reactors. Chem. Eng. Sci., 47: 3597–3604.

Evans GM, Rielly CD, Davidson JF, Carpenter KJ. (1990) A fundamental study of gas inducing impeller design. In: Benkreira H, editor. Fluid mixing IV. Institution of Chemical Engineers Symposium Series 121; Institution of Chemical Engineers, Rugby, UK. p 137–152.

Evans GM, Rielly CD, Davidson JF, Carpenter KJ. (1991) Hydrodynamic characteristics of a gas inducing impeller. In: Proceedings of Seventh European Conference on Mixing, Kiav, Brugge, September 18–20, Belgium, p 515–523.

Evans GM, Doroodchi E, Lane GL, Koh PTL, Schwarz MP. (2008) Mixing and gas dispersion in mineral flotation cells. Chem. Eng. Res. Des., 86:1350–1362.

Fallenius K. (1976) A new set of equations for the scale-up of flotation cells. In: Proceedings of Mineral Processing 13th Congress, Dev. Mineral Process., p 1353–1373.

Forrester SF, Rielly CD, Carpenter KJ. (1994) Bubble formation from the moving blades of a gas-inducing impeller. In: Proceedings of Eighth European Conference on Mixing, BHRA, Cambridge, England, p 333–340.

Forrester SF, Rielly CD. (1994) Modelling the increased gas capacity of self-inducing impellers. Chem. Eng. Sci., 49:5709–5718.

Forrester SF, Rielly CD, Carpenter KJ. (1998) Gas-inducing impeller design and performance characteristics. Chem. Eng. Sci., 53:603–615.

Gorain BK, Franzidis JP, Manlapig EV. (1995a) Studies on impeller type, impeller speed and air flow rate in an industrial flotation cell. Part 1. Effect on bubble-size distribution. Miner. Eng., 8:615–635.

Gorain BK, Franzidis JP, Manlapig EV (1995b) Studies on impeller type, impeller speed and air flow rate in an industrial scale flotation cell. Part 2. Effect on gas holdup. Miner. Eng., 8:1557–1570.

Gray JB (1966) Flow patterns, fluid velocities and mixing in agitated vessels. In: Uhl VW, Gray JB, editors. Mixing theory and practice. Vol. I. Academic Press, New York, USA.

Grau RA, Heiskanen K. (2002) Visual technique for measuring bubble size in flotation machines. Miner. Eng., 15:507–513.

Harris CC. (1976) Flotation machines. In: Flotation: A.M. Gaudin memorial edition, AIME, New York, p 753–815.

Harris CC, Mensah-Biney RK. (1977) Aeration characteristics of laboratory flotation machine impellers. Int. J. Mineral Processing, 4:51–67.

Heim A, Kraslawski A, Rzyski E, Stelmach J. (1995) Aeration of bioreactors by self aspirating impellers. Chem. Eng. J., 58:59–63.

Hichri H, Accary A, Puaux JP, Andrieu J. (1992) Gas-liquid mass transfer coefficients in a slurry batch reactor equipped with a self-gas-inducing agitator. Ind. Eng. Chem. Res., 31:1864–1867.

Hsu Y, Huang C. (1996) Characteristics of a new gas-induced reactor. AICHEJ, 42:3146.

Joshi JB, Sharma MM. (1977) Mass transfer and hydrodynamic characteristics of gas inducing type of agitated contactors. Can. J. Chem. Eng., 65:683–695.

Joshi JB. (1980) Modifications in the design of gas inducing impellers. Chem. Eng. Commun., 5(1–4):109–114.

Ju FJ, Cheng Z-M, Chen J-H, Chu X-H, Zhou Z-M, Yuan PQ. (2009) A novel design for a gas-inducing impeller at the lowest critical speed. Chem. Eng. Res. Des. 87:1069–1074.

Kasundra RB, Kulkarni AV, Joshi JB. (2008) Hydrodynamic and mass transfer characteristics of single and multiple impeller hollow self-inducing reactors. Ind. Eng. Chem. Res., 47: 2829–2841.

Kind P. (1976) Design criteria and recent developments in large-capacity Wemco flotation cells. J. South African Inst. Mining and Metallurgy, 76(3):345–362.

Kingsley JP, Roby AK, Litz LM. (1994) Terephthalic acid production. U.S. Patent 5,371,283.

Koen C, Pingaud B. (1977) Development of a self-inducing disperser for gas/liquid and liquid/liquid systems. In: Proceedings of Second European Conference on Mixing, March 30–April 1, 1977, BHRA, Cambridge, England, Paper F5, p 67–81.

Lima OA, Deglon DA, Leal Filho LS. (2009) A comparison of the critical impeller speed for solids suspension in a bench-scale and a pilot-scale mechanical flotation cell. Miner. Eng., 22:1147–1153.

Litz LM. (1985) A novel gas–liquid stirred reactor. Chem. Eng. Prog., 81(11):36–39.

Litz LM, Weise MK, Adis M. (1990) Gas-liquid mixing. U.S. Patent 4,900,480.

Matsumura M, Sakuma H, Yamagata T, Kobayashi J. (1982a) Gas entrainment in a new gas entraining fermentor. J. Fermentation Tech., 60:457–462.

Matsumura M, Sakuma H, Yamagata T, Kobayashi J. (1982b) Performance of oxygen transfer in a new gas entraining fermentor. J. Fermentation Tech., 60:551–563.

Martin GQ. (1972) Gas-Inducing Agitator. Ind. Eng. Chem. Process Des. Dev. 11:397–404.

McCabe WL, Smith JC, Harriott P. (1986) Unit operations of chemical engineering. IV International Student Edition, McGraw-Hill Chemical Engineering Series, McGraw-Hill Book Co., New York, USA.

Miskovic S. (2011) An investigation of the gas dispersion properties of mechanical flotation cells: an in-situ approach. Ph.D. Thesis, Virginia Polytechnic Institute and State University, USA.

Mundale VD. (1993) Design of gas inducing type agitated reactors. Ph.D. (Tech) Thesis, Bombay University, India.

Nesset JE, Hernandez-Aguilar JR, Acuna C, Gomez CO, Finch JA. (2006) Some gas dispersion characteristics of mechanical flotation machines. Miner. Eng., 19:807–815.

Nienow AW. (2000) The suspension of solid particles. In: Harnby N, Edwards MF, Nienow AW, editors. Mixing in the process industries. 2nd ed. Butterworth-Heinemann, Oxford, UK. p 364–393.

Nienow AW, Bujalski W. (2004) The versatility of up-pumping hydrofoil agitators. Trans. IChemE, Part A, Chem. Eng. Res. Des., 82(A9):1073–1081.

Patil SS, Mundale VD, Joshi JB. (2005) Mechanism of gas induction in self-inducing impellers. Ind. Eng. Chem. Res., 44:1322–1328.

Patwardhan AW, Joshi JB. (1997) Hydrodynamics of stirred vessel equipped with a gas inducing impeller. Ind. Eng. Chem. Res., 36:3942–3958.

Patwardhan AW, Joshi JB. (1999) Design of gas-inducing impellers. Ind. Eng. Chem. Res., 38:49–80.

Poncin S, Nguyen C, Midoux N, Breysse J. (2002) Hydrodynamics and volumetric gas–liquid mass transfer coefficient of a stirred vessel equipped with a gas-inducing impeller. Chem. Eng. Sci., 57:3299–3306.

Raidoo AD, Raghav Rao KSMS, Sawant SB, Joshi JB. (1987) Improvements in gas inducing impeller design. Chem. Eng. Commun., 54:241–264.

Rielly CD, Evans GM, Davidson JF, Carpenter KJ. (1992) Effect of vessel scale-up on the hydrodynamics of a self aerating concave blade impeller. Chem. Eng. Sci., 47:3395–3402.

Roby AK, Kingsley JP. (1996) Oxide safely with pure oxygen. Chemtech, February:39–46.

Saravanan K, Mundale VD, Joshi JB. (1994) Gas-inducing type mechanically agitated contactors. Ind. Eng. Chem. Res., 33:2226–2241.

Saravanan K, Joshi JB. (1995) Gas inducing type mechanically agitated contactors: hydrodynamic characteristics of multiple impellers. Ind. Eng. Chem. Res. 34:2499–2514.

Saravanan K, Joshi JB. (1996) Fractional gas hold-up in gas inducing type of mechanically agitated contactors. Can. J. Chem. Eng. 74:16–30.

Saravanan K, Patwardhan AW, Mundale VD, Joshi JB. (1996) Power consumption in gas inducing type mechanically agitated contactors. Ind. Eng. Chem. Res., 35:1583–1602.

Saravanan K, Patwardhan AW, Joshi JB. (1997) Critical impeller speed for solid suspension in gas inducing type mechanically agitated contactors. Can. J. Chem. Eng., 75:664–676.

Sardeing R, Xuereb C, Poux M. (2006) Improvement of the performances of a gas-inducing system for application in waste water treatment. Int. J. Chem. Reaction Eng., 4, A30:1–21.

Sawant SB, Joshi JB. (1979) Critical impeller speed for onset of gas inducing in gas inducing types of agitated contactors. Chem. Eng. J., 18:87-91.

Sawant SB, Joshi JB, Pangarkar VG. (1980) Mass transfer and hydrodynamic characteristics of the Wemco type of flotation cell. Ind. Chem. Eng., 22(4):ChE 89-96.

Sawant SB, Joshi JB, Pangarkar VG, Mhaskar RD. (1981) Mass transfer and hydrodynamic characteristics of the Denver type of flotation cells. Chem. Eng. J., 21:11–19.

Scargiali F, Russo R, Grisafi F, Brucato A. (2007) Mass transfer and hydrodynamic characteristics of a high aspect ratio self-ingesting reactor for gas-liquid operations. Chem. Eng. Sci., 62:1376–1387.

Shabalala NZP, Harris M, Leal Filho LS, Deglon DA. (2011) Effect of slurry rheology on gas dispersion in a pilot-scale mechanical flotation cell. Miner. Eng., 24:1448–1453.

Sokolov EYa, Zinger NM. (1960) The Jet Devices, Moscow, Gosenergoizdat. *cf* Zundelevich (1979).

Taggart AF (1960) Handbook of mineral dressing, Wiley Interscience, New York, USA.

Topiwala HH, Hamer G. (1974) Mass transfer and dispersion properties in a fermentor with gas inducing impeller, Trans. Inst. Chem. Eng., 52:113–120.

Van der Westhuizen AP, Deglon DA. (2007) Evaluation of solids suspension in a pilot-scale mechanical flotation cell: the critical impeller speed. Miner. Eng., 20:233–240.

Van der Westhuizen AP, Deglon DA. (2008) Solids suspension in a pilot-scale mechanical flotation cell: a critical impeller speed correlation. Miner. Eng., 21:621–629.

Vinnett L, Contreras F, Yianatos J. (2012) Gas dispersion pattern in mechanical flotation cells. Miner. Eng., 26:80–85.

Wang Z, Peng X, Li X, Wang S, Cheng Z, Ju F. (2013) Impact of liquid driving flow on the performance of a gas-inducing impeller. Chem. Eng. Process., 69:63–69.

White DA, de Villiers JU. (1977) Rates of induced aeration in agitated vessels. Chem. Eng. J., 14:113–118.

Witze CP, Schrock VE, Chambre PL. (1968) Flow about a growing sphere in contact with a plane wall. Int. J. Heat Mass Transfer, 11:1637–1652.

Yianatos J, Bergh LG, Díaz F, Rodríguez R. (2005) Mixing characteristics of industrial flotation equipment. Chem. Eng. Sci., 60:2273–2282.

Yianatos J, Henrıquez FH, Oroz AG. (2006) Characterization of large size flotation cells. Miner. Eng., 19:531–538.

Yianatos JB. (2007) Fluid flow and kinetic modeling in flotation related processes columns and mechanically agitated cells—a review. Chem. Eng. Res. Des., Trans. IChemE, Part A, 85(A12):1591–1603.

Yianatos J, Contreras F, Díaz F. (2010a) Gas holdup and RTD measurement in an industrial flotation cell. Miner. Eng., 23:125–130.

Yianatos J, Contreras F, Morales P, Coddou F, Elgueta H, Ortíz J. (2010b) A novel scale-up approach for mechanical flotation cells. Miner. Eng., 23:877–884.

Yianatos J, Carrasco C, Bergh L, Vinnett L, Torres C. (2012) Modelling and simulation of rougher flotation circuits. Int. J. Miner. Process., 112–113:63–70.

Yu H-S, Tan Z-C. (2012) New correlations of volumetric liquid-phase mass transfer coefficients in gas-inducing agitated tank reactors. Int. J. Chem. Reaction Eng., 10(Article A-50):1–18.

Zheng X, Franzidis J-P, Johnson NW, Manlapig EV. (2005) Modelling of entrainment in industrial flotation cells: the effect of solids suspension. Miner. Eng., 18:51–58.

Zlokarnik M, Judat H. (1967) Tube and disk stirrers—two efficient stirrers for the gassing of liquids. Chem. Ing. Tech., 39:1163–1168.

Zundelevich Y. (1979) Power consumption and gas capacity of self- inducing turbo aerators. AICHEJ, 25:763–773.

Zwietering ThN. (1958) Suspending of solid particles in liquids by agitators. Chem. Eng. Sci., 8:244–253.

10

TWO- AND THREE-PHASE SPARGED REACTORS

10.1   INTRODUCTION

The various advantages of two-/three-phase sparged reactors (TPSRs) over other types of multiphase reactors have been briefly discussed in Chapter 3. These pertain mainly to the simplicity of construction and operation, facile heat and mass transfer, etc. In view of these advantages, these sparged reactors are used for a variety of applications ranging from aerobic fermentations to Fischer–Tropsch (F–T) synthesis. Sparged reactors, which are also referred to as bubble columns, have therefore received attention worldwide. Every text on multiphase reactors includes a separate chapter on these reactors. Their importance can also be judged from the fact that Deckwer's (1992) book is solely devoted to bubble column reactors. These reactors provide unmatched flexibility in the residence time of the liquid phase. The high level of turbulence prevailing in the liquid phase results in higher transport rates common to reactors in which gas is dispersed. A major drawback of sparged reactors is extensive axial mixing in both the liquid and gas phases. This can, however, be reduced by introducing partitioning internals such as sieve plates at specified intervals (Zahradnik et al. 1992; Saxena and Chen 1994). Such sectionalized columns can also yield certain flexibility with respect to the residence time of the gas phase. Their utility in three-phase systems is, however, severely limited (Section 10.6).

   In a conventional sparged reactor, the gas phase is sparged at the bottom of a cylindrical column. Generally, low pressure drop spargers, such as those used in packed distillation columns albeit with smaller orifices, are used. The bubbling gas

Design of Multiphase Reactors, First Edition. Vishwas Govind Pangarkar.
© 2015 John Wiley & Sons, Inc. Published 2015 by John Wiley & Sons, Inc.

phase flows upward due to the natural buoyancy. The energy required for creating the high level of turbulence is supplied by the isoentropic expansion of the gas phase and can be precisely calculated as will be seen later. The behavior of this dispersed gas phase is quite complicated. The size of the discrete bubble phase is a very important parameter that is decided not only by the physical properties of the system but also by the operating conditions (particularly, total pressure besides gas velocity), the diameter of the column, the height to diameter ratio, and, in some cases, the type of sparger used.

10.2 HYDRODYNAMIC REGIMES IN TPSR

For a given set of liquid-phase properties, the hydrodynamic regime in a TPSR mainly depends on (i) the superficial gas velocity, V_G, and (ii) the column diameter, T.

Figure 10.1 shows variation of the hydrodynamic regime with respect to the two main parameters: superficial gas velocity and column diameter.

Three main regimes that can be identified are as follows.

10.2.1 Slug Flow Regime

This regime is relevant to small ($T<0.1$ m) column diameter and high ($V_G>0.1$ m/s) superficial gas velocities. In this regime, large slugs of gas of practically the same dimensions as the diameter of the column traverse the height of the column. The liquid is simply pushed up by the slugs. There is no liquid circulation in the column. These large slugs have relatively high rise velocity and, therefore, spend very less time in the dispersion. As a result, the gas holdup, ε_G, for this regime is comparatively low. The overall result is manifest in a low level of turbulence in the liquid phase and low rates of heat and mass transfer. This regime has very little industrial relevance since most industrial columns are much larger than 0.1 m in diameter.

10.2.2 Homogeneous Bubble Flow Regime

This regime prevails at low superficial gas velocity ($V_G<0.05$ m/s) and particularly in noncoalescing systems for relatively small-diameter columns. Discrete bubbles are formed and released at the sparger. These bubbles have practically the same size as the orifices in the sparger ($d_B \sim 1\text{--}2 \times 10^{-3}$ m). Although the bubbles oscillate in a direction transverse to the mean flow, the amplitude is small. For all practical purposes, their rise is unidirectional. Thus, there is very little interaction between neighboring bubbles. The bubble diameter is determined by the type of sparger and physical properties of the liquid.

In view of the fact that chains of bubbles are formed, this regime is sometimes referred to as chain bubbling regime. The terminal rise velocity of a small discrete bubble is relatively very low. However, the drag on an individual bubble is affected by the presence of bubbles in its vicinity. The resulting hindrance increases with increasing number of bubbles or gas holdup. The drag and consequent turbulence is

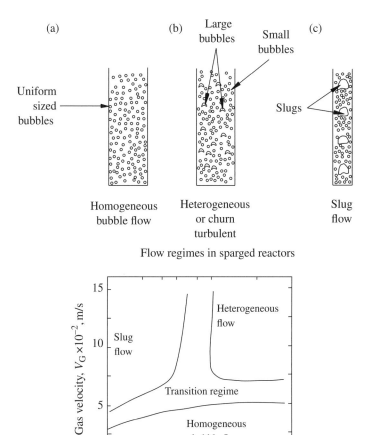

FIGURE 10.1 Approximate dependence of flow regime on column diameter and superficial gas velocity for water and dilute aqueous solutions. (Reproduced from Shah et al. 1982 with permission from John Wiley & Sons, Inc. Copyright © 1982 American Institute of Chemical Engineers.)

therefore a function of the gas holdup. The rise velocity in such cases is lower than the terminal rise velocity and is referred to as hindered rise velocity. The uniformity in bubble size and absence of coalescence/breakup of bubbles result in radially uniform gas holdup. The number of bubbles increases as the gas velocity increases. A combination of low rise velocity and increase in the number of bubbles with increasing gas throughput results in a strong dependence of the gas holdup on the superficial gas velocity. The sparger design and type of liquid (coalescing/noncoalescing) have important effects on the bubble size and gas holdup (Ruzicka et al. 2001a; Ajbar et al. 2009). For relatively large column diameters (>0.2 m), the wall effect is substantially reduced, and therefore, gas holdup is independent of the column diameter.

Typical industrial applications that conform to the homogeneous regime are (Shaikh and Al-Dahhan 2007) cultivation of bacteria and molds/fungi, production of single-cell protein, sewage treatment, and some specific cases of animal cell culture. Hydrotreatment of heavy fractions and coal to hydrocarbon conversion are some application relevant to the chemical industry.

10.2.3   Heterogeneous Churn-Turbulent Regime

As the gas velocity increases further ($V_G > 0.05$ m/s), there is excessive crowding of bubbles, and interaction between different bubbles cannot be prevented. This results in coalescence/breakup of bubbles, particularly for liquids that afford coalescence. This phenomenon is particularly prominent for larger (>0.2 m) column diameters simultaneously with relatively high ($V_G > 0.1$ m/s) superficial gas velocity. Such combinations are typical of industrial reactors, and hence, this regime is of great industrial relevance. Because of the high frequency of bubble coalescence and breakup, this regime exhibits a wide bubble size distribution. The rise velocity of a bubble depends on its size. The wide variation in bubble size results in wide variation in their rise velocity. This feature generates nonuniformity in radial gas holdup. Consequently, liquid velocities also exhibit radial variation, both in terms of direction and value. The liquid-phase flow is mainly upward in the central core and downward at the column wall. This liquid circulation is the main source of the high rates of heat and mass transfer, afforded by this regime. It is also the cause of the high level of axial mixing in the liquid phase. Ruzicka et al. (2001a) have defined two types of heterogeneous regimes: (1) type I is produced at high gas velocities by spargers that have small orifices and small pitch of the orifices. The high gas velocity initiates instability at the gas distributor level. The instability propagates to finally yield the heterogeneous regime (2) type II that is produced by spargers with relatively large orifices and larger orifice pitch. Nonhomogeneous gas flow at the sparger is the source of the instabilities that lead to heterogeneities. This type II was termed "pure" heterogeneous regime by Ruzicka et al. (2001a). The two types have essentially similar hydrodynamic features, identified by wide bubble size distribution that causes the strong liquid circulation. The only difference is in the type of perturbation that leads to the transition from homogeneous to heterogeneous regime as indicated earlier. Gas holdup in this regime is independent of column diameter for $T > 0.15$ m and also of H/T for >5 (Deckwer 1992).

The concept of hydrodynamic regime similarity has been discussed in Section 5.2.2.2.1. As pointed out by Shaikh and Al-Dahhan (2010), hydrodynamic regime similarity is a necessary condition while scaling up. It is futile to generate data on small scale if they do not conform to the hydrodynamic regime that prevails on the large scale. Therefore, for this regime, data on laboratory scale are meaningful only when the column diameter >0.2 m. The heterogeneous regime is of interest in reactions that are relatively exothermic. Some examples are F–T synthesis and liquid-phase synthesis of methanol (Shaikh and Al-Dahhan 2007).

10.2.4 Transition from Homogeneous to Heterogeneous Regimes

This subject has attracted great interest in view of the distinctly different behavioral pattern of the two regimes. Deckwer (1992) has discussed the literature for the relevant period. In the recent past, several investigators have made significant contributions to this specific aspect. These include, among others, Zahradnik et al. (1997), Ruzicka et al. (2001a, b, 2003, 2008, 2013), Mena et al. (2005), Shaikh and Al-Dahhan (2007), Su et al. (2008), Ribeiro (2008), Ajbar et al. (2009), Hur et al. (2013), and Shiea et al. (2013).

A major distinguishing feature for the two regimes is the dependence of the gas-phase holdup on the superficial gas velocity. The general dependence is $\varepsilon_G \propto (V_G)^n$. For homogeneous bubble flow regime, the value of "n" ranges from 0.8 to 1, whereas in the heterogeneous regime, "n" is much lower in the range of $0.4 < n < 0.7$ (Deckwer et al. 1980; Deckwer 1992). This distinct difference has been used by many investigators to arrive at the broad range of the transition region. Since transition is a physical phenomenon, there is no sharp change over. Indeed, Mudde and van den Akker (1999) have shown that there is a transition zone over which the change from one regime to another occurs. The small-scale structures typical of the homogeneous regime are not significant contributors to the ultimate longtime average. In the event that small-scale structures amplify to the scale of the column diameter and get oriented, coherence is produced on the long scale. This coherence is the well-known circulation pattern typical of the heterogeneous regime. This amplification is intermittent, and therefore, both the homogeneous and heterogeneous regimes can coexist in space and time (Ruzicka et al. 2001a). Figure 10.2 shows typical gas holdup variation with V_G.

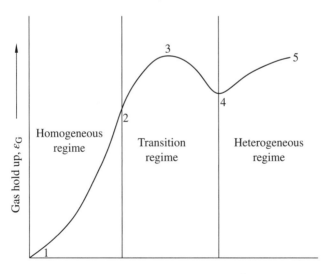

FIGURE 10.2 Typical variation of gas holdup with superficial gas velocity in a bubble column. (Reproduced from Ruzicka et al. 2003 with permission from Elsevier B.V. Copyright © 2003 Elsevier B.V. http://www.journals.elsevier.com/chemical-engineering-journal/)

In their review, Shaikh and Al-Dahhan (2007) have listed the following different techniques used in the literature for studying the transition from homogeneous to heterogeneous regime:

1. **Visual observation:** This is the simplest method to demarcate the regimes. Vertical chains of small spherical bubbles are the distinguishing feature of the homogeneous regime, whereas an intense circulatory pattern with wide bubble size distribution identifies the heterogeneous regime. In the latter, the larger bubbles are typical spherical cap type rather than spherical. There is a significant element of subjectivity in this technique. Therefore, it is very difficult to precisely demarcate the beginning or conclusion of the transition.

2. **Measurement of gas holdup as a function of the gas velocity:** Gas holdup can be measured by simply noting the static liquid and dispersion height, H_S and H_D, respectively. Gas holdup is then estimated from $\varepsilon_G = H_D - H_S/H_S$. Shaikh and Al-Dahhan (2005) suggested that the point at which a sharp change in the slope of ε_G versus V_G plot is observed should be considered as the transition point. However, as shown by Ruzicka et al. (2001a) and also evident from Figure 10.2, the transition occurs over a broad range of V_G, and the sharp change in slope suggested is elusive. In cases where such a diffuse change in the slope is observed (Rados et al. 2003), the gas velocity at which a maximum in gas holdup occurs is taken as the transition velocity. However, in many cases, a well-defined maximum in holdup versus V_G plot is also intangible. In such cases, two alternative methods are available. Wallis' (1969) model defines a drift velocity flux: $J_{DFV} = V_G(1 - \varepsilon_G) \pm V_L \times \varepsilon_G$. A plot of the drift flux velocity against ε_G such as that given by Deckwer et al. (1981) yields a much better transition state through a sharper change in the slope of the plot (Shaikh and Al-Dahhan 2007). Another approach is that of Zuber and Findlay (1965). Gourich et al. (2006) analyzed their data for air–water and air–aqueous propanol (0.05 vol. %) using (i) the simple ε_G versus V_G plot, (ii) Wallis' model, and (iii) Zuber and Findlay's model. For air–water system, it was found that the Wallis model gives a satisfactory prediction of the point at which the homogeneous regime ends or the transition to heterogeneous regime starts. However, it cannot predict the condition for transition to the fully developed heterogeneous regime. For noncoalescing systems (alcohol–water), it was found that the Wallis model-based plot does not reveal the beginning of the transition to heterogeneous regime. On the other hand, for this system, a simple ε_G versus V_G plot clearly indicates the beginning of the transition through a sharp decrease in ε_G. Gourich et al. (2006) found that Zuber and Findlay's model could predict the beginning of the transition for both air–water (coalescing) and air–aqueous propanol (noncoalescing) systems rightly and comparable to the simple ε_G versus V_G plot. However, this model also was not able to predict the end of the transition zone or beginning of the fully established heterogeneous regime.

3. **Temporal signatures of properties associated with the hydrodynamic state:** A number of techniques have been used to measure temporal signatures

to discern the changes in regimes. The techniques used most widely are (Shaikh and Al-Dahhan, 2007) (i) wall pressure fluctuations (Vial et al. 2001; Gourich et al. 2006; Vazquez et al. 2008; Fraguio et al. 2009; Shiea et al. 2013), (ii) optical probes (Bakshi et al. 1995; Briens et al. 1997; Mena et al. 2008), and (iii) acoustic probes (Mannasseh et al. 2001; Holler et al. 2003; Al-Masry et al. 2005, 2006, 2007). Optical and other similar intrusive probes have the drawback that they interfere with the flow (Ellis et al. 2004). Further, they yield point values of the data. Pressure transducers have a number of advantages: they are robust, inexpensive, nonintrusive, and unaffected by any corrosive liquids inside the column. Their major drawbacks are as follows: they reflect conditions at the wall and not inside the column, assumption of incompressible fluids inside the column, and high complexity of data interpretation (Drahoš et al. 1991; Letzel et al. 1997). In view of this, analysis of the voluminous data generated for extracting meaningful information requires relatively high skill and care.

4. **Advanced measuring techniques:** With the advent of sophisticated imaging and velocimetry techniques, better understanding of the flow behavior and consequently a better prediction of regime transition is possible. The techniques used in the literature are listed below with the corresponding references. Readers interested in a specific technique may refer to these references:

Electrical capacitance tomography (ECT) (Bennett et al. 1999; Tapp et al. 2003)

Electrical impedance tomography (EIT) (Tapp et al. 2003)

Electrical resistance tomography (ERT) (Dong et al. 2001; Tapp et al. 2003; Murugaian et al. 2005; Ismail et al. 2011)

Particle image velocimetry (Chen et al. 1994; Lin et al. 1996; Thet et al. 2006)

Laser Doppler anemometry (LDA) (Olmos et al. 2003)

Computer-automated radioactive particle tracking (CARPT) (Cassanello et al. 2001; Nedeltchev et al. 2003; Shaikh and Al-Dahhan 2010)

γ-ray computed tomography (CT) (Shaikh and Al-Dahhan 2005)

Nuclear gauge densitometry (Nedeltchev et al. 2011; Shaikh and Al-Dahhan 2013)

10.3 GAS HOLDUP

The most important factor that determines the gas holdup in sparged reactors is the mean bubble size. The number of bubbles is decided by gas flow. Together, these two parameters govern the total gas volume in a given volume of the gas–liquid dispersion or the fractional gas holdup. The mean bubble size is in turn related to the initial bubble size at the sparger (type of sparger: high or low pressure drop sparger) and the liquid-phase properties, namely, viscosity, surface tension, and presence of surface-active materials. The effects of different parameters are briefly discussed in the following sections.

10.3.1 Effect of Sparger

Spargers can be broadly classified into two classes: (1) high pressure drop type represented by sintered discs with very small openings and (2) the perforated plates or ring spargers that belong to the low pressure drop class. The average bubble size decreases with decrease in the opening from which the gas ensues. Consequently, sintered discs yield smaller primary bubbles than orifice plates. In the case of the homogeneous regime, due to lack of interaction between neighboring bubbles, the average bubble size is relatively close to the primary bubble size. In the heterogeneous regime, coalescence and breakup of bubbles cause significant changes, and hence, the average bubble size is insensitive to the sparger type.

10.3.2 Effect of Liquid Properties

10.3.2.1 Effect of Surface-Active Substances   Most pure liquids promote coalescence. However, the addition of small amounts of another liquid (lower alcohols in water, Krishna et al. 2000a) or the presence of impurities/surface-active substances/salts causes significant hindrance to coalescence. Surface-active materials can be broadly divided in two classes: organic and inorganic. Organic surface-active substances typically cause a decrease in the surface tension of the liquid. The nonpolar (hydrophobic) part of the organic type is attracted and strongly adsorbed at the gas–liquid interface. Organic surface-active substances can be further subdivided into two categories: surfactants and antifoam agents as discussed in Section 7B.13. Both types reduce the equilibrium surface tension but have different effects. The former hinders coalescence, thereby producing a lower average bubble size. Antifoams, on the other hand, promote coalescence that disrupts a stable foam. Antifoams are specific substances/compositions that are generally added to prevent foaming, whereas foaming agents can be intrinsic to the system or can enter inadvertently.

Inorganic substances that affect bubble size/gas holdup are typically salts of mineral acids/ionic substances. Yawalkar et al. (2002a) have discussed the theoretical basis that governs salt effects on bubble size/gas holdup. The ionic species cause a small increase in the surface tension and, hence, are referred to as negative surfactants. These are repelled from the interface (negative adsorption). Behavior of adjacent bubbles in such systems is dictated by a very large number of interconnected phenomena as pointed out by Marrucci (1969). The inhibition of coalescence is caused by hindrance to the thinning of the liquid film that separates two bubbles (Prince and Blanch 1990). As the film becomes thinner, its surface area increases. This results in an increase in the surface excess of the salt in the film as compared to that on the bubble surface. Consequently, the surface tension of the film increases. This gives rise to a force that opposes the direction of flow at the gas–liquid interface. Accordingly, there is a delay in the thinning of the intervening liquid film that otherwise would have resulted in coalescence (Marrucci 1969; Prince and Blanch 1990). Viewed in their entirety, these phenomena caused by the presence of salts decrease the bubble diameter. With increasing electrolyte concentration in the solution, the liquid film between adjacent bubbles is stabilized due to increase in its surface tension. This further retards bubble coalescence and retains the initial bubble diameter. This behavior continues up to a certain concentration of the salt at which the augmented surface tension immobilizes the film. At this concentration, the bubble size achieves a certain minimum value. Further addition of salt does not show any marked effect. This concentration is termed "transition electrolyte concentration," C_T (Marrucci 1969; Prince and Blanch 1990). Prince and Blanch (1990) gave the following modified version of the original expression for C_T proposed by Marrucci (1969):

$$C_T = 1.18 \times z \left(\frac{B_W \sigma}{R_B} \right)^{0.5} \times R_G \times \Theta \left(\frac{d\sigma}{dC_E} \right)^{-2} \tag{10.1}$$

Prince and Blanch used Equation 10.1 to estimate the values of C_T for different electrolytes. In deriving Equation 10.1, it was assumed that the interaction between molecules in the liquid film separating adjacent bubbles is governed by van der Waal's forces of attraction. According to Prince and Blanch, predictions based on Equation 10.1 exhibited fairly good agreement with experimentally measured values of C_T. The concentration of the electrolyte varies across the film thickness. It is lower at the surface of the film and increases toward the center of the film. This dependence is given by (Marrucci 1969; Prince and Blanch 1990)

$$\Delta\sigma = -\left(\frac{1}{z \times \delta_F} \right) \left(\frac{2C_E}{R_G \times \Theta} \right) \left(\frac{d\sigma}{dC_E} \right)^2 \tag{10.2}$$

Equation 10.2 yields $\Delta\sigma$ that is a characteristic of the electrolyte–liquid system and defines the effect of the concentration of a given electrolyte on the bubble size. Equation 10.2 can be split into two parts: (1) surface tension and (2) turbulence parameters. The film thickness, δ_F, is governed by turbulence parameters. Yawalkar et al. (2002a) defined a surface tension factor (STF) that represents the effect of surface tension in Equation 10.2 as $(C_E/z) (d\sigma/dC_E)^2$. Yawalkar et al. found that at $C_E = C_T$, the values of d_B and STF were approximately the same irrespective of the type of electrolyte. Since the bubble diameter does not vary for $C_E > C_T$, the effect of surface tension on gas holdup should be absent for higher concentrations. Yawalkar et al. (2002a) confirmed this proposition in the case of stirred vessels. In all probabilities, bubble columns should also exhibit similar behavior.

10.3.2.2 *Effect of Liquid-Phase Viscosity*

Liquid-phase viscosity has a negative effect on the gas holdup due to increased coalescence that generates large bubbles. This feature is common to both Newtonian and non-Newtonian liquids (Deckwer 1992; Kantak et al. 1995; Kuncová et al. 1995). Bucholz et al. (1978) and Bach and Pilhofer (1978) found that for viscosities in excess of 40 mPa s, gas holdup attained a constant value. In the case of non-Newtonian solutions, for a fixed gas velocity, the effective viscosity did not have any effect on the gas holdup. Ruzicka et al. (2003) found that for air–aqueous glycerol system, in the lower viscosity range (1–3 mPa s), the homogeneous regime was stabilized. For the relatively viscous liquids

(3–22 mPa s), the transition to heterogeneous regime began at lower gas velocities as determined by the drift flux model of Wallis (1969). The critical values of gas velocity, V_{GCrit}, and holdup, ε_{GCrit}, at which the transition began could be correlated as follows ($\mu > 3$ mPa s) (Ruzicka et al. 2003):

$$V_{GCrit} = 4.8 \times 10^{-2} (\mu_L)^{-0.24} \tag{10.3}$$

$$\varepsilon_{GCrit} = 0.28(\mu_L)^{-0.37} \tag{10.4}$$

Similar to air–water system, in relatively high-viscosity systems, column height had a destabilizing effect on the homogenous regime (Ruzicka et al. 2001b).

10.3.3 Effect of Operating Pressure

TPSRs are used in industrially important high-pressure reactions such as liquid-phase methanol synthesis and F–T synthesis. The effect of pressure on the fractional gas holdup is an important parameter that has attracted the attention of many investigators (Wilkinson and van Dierendonck 1990; Krishna et al. 1991; Wilkinson et al. 1992; Krishna et al. 1994; Reilly et al. 1994; Luo et al. 1999; Krishna et al. 2000b; Jordan et al. 2002; Urseanu et al. 2003; Lau et al. 2004; Behkish et al. 2007; Han and Al-Dahhan 2007; Chilekar et al. 2010; Rollbusch et al. 2013; Jin et al. 2013). Most investigations have found that the gas holdup increases with increase in pressure. It has been argued that higher pressures have a stabilizing effect on bubbles leading to smaller bubbles and also prevention of coalescence. Krishna et al. (1994), Letzel et al. (1999), and Krishna et al. (2000b), among others, found that increasing pressure delayed the transition from homogeneous to heterogeneous regime. Wilkinson et al. (1992), Reilly et al. (1994), and Letzel et al. (1999) proposed correlations for the gas holdup at which transition to heterogeneous regime occurs. Reilly et al. (1994) and Letzel et al.'s (1999) correlations are relatively simple to use:

$$(\varepsilon_G)_{Crit} = 4.45 \sqrt{\frac{\rho_G^{0.96} \sigma^{0.12}}{\rho_L}} \quad \text{(Reilly et al. 1994)} \tag{10.5}$$

$$(\varepsilon_G)_{Crit} = 0.17 \rho_G^{0.22} \quad \text{(Letzel et al. 1999)} \tag{10.6}$$

Reilly et al. (1994) found that V_{GCrit} increases with increase in pressure/gas density. Letzel et al., on the other hand, found no such effect. Letzel et al. (1999) explained the lower rise velocity of large bubbles on the basis of the Kelvin–Helmholtz stability analysis. They concluded that $V_B \propto (\rho_G)^{-0.5}$. Urseanu et al. (2003) carried out simultaneous investigation of the effect of operating pressure and liquid viscosity on gas holdup. These investigators found that the effect of pressure on the gas holdup varied with viscosity. For the low viscosity, N_2–water ($\mu_L = 1 \times 10^{-3}$ Pa s) data, the gas holdup is a strong function of pressure. For the higher-viscosity (N_2–Tellus oil; $\mu_L = 7 \times 10^{-2}$ Pa s) system, the pressure effect was not as marked. Finally, for the highest viscosity used ($\mu_L > 0.1$ Pa s), increasing pressure had a negligible effect on the gas holdup. Therefore, it appears that the strong coalescing effect of higher viscosity is dominant in comparison to the stabilizing effect of higher pressure.

10.3.4 Effect of Presence of Solids

Solid catalysts are widely used in three-phase bubble columns, and therefore, their effect on gas holdup is an important aspect. It is therefore not surprising that a large number of investigations have been aimed at elucidating the effect of the presence of solids on the gas holdup. Some of these are Kojima et al. (1987), Sada et al. (1986a, b), Saxena et al. (1992), Saxena and Rao (1993), Saxena and Chen (1994), Krishna et al. (1997), Garcia-Ochoa et al. (1997), Inga and Morsi (1999), Luo et al. (1999), Gandhi et al. (1999), Li and Prakash (2000), Mena et al. (2005), Moshtari et al. (2007), Behkish et al. (2006, 2007), van der Schaaf et al. (2007), and Mena et al. (2011).

Introduction of solid particles has complex effects on the gas holdup. These range from initial increase in the gas holdup, then decrease, and again increase (Gandhi et al. 1999) to modification of the beginning of transition region (Mena et al. 2005). The results obtained depend on various factors such as size, density and loading of the solid used, the gas velocity range (homogeneous/heterogeneous), operating pressure, viscosity, coalescing nature of the liquid, and affinity of the solid to the liquid phase. It is, therefore, not surprising that there is considerable discrepancy in the observations by different investigators. Most investigators found a maximum in the gas holdup with respect to the solid loading, although the maximum occurred at different solid loading depending on the conditions of the studies. There are differences in the effect of solid loading observed by different investigators. In the following list, some of the factors that cause the differences in the observations are discussed:

1. **Effect of size and density of the particles used, solid loading, and effective slurry viscosity:** Garcia-Ochoa et al. (1997) measured axial concentration profile of three different glass bead sizes. They found that for the smallest particles ($38\,\mu m$), the axial concentration profile was relatively flat ($Z/H=0$, $X \sim 20\%$; $Z/H=1$, $X \sim 14\%$) for $V_G = 0.01$ and $0.06\,m/s$. The latter is most likely in the transition region. There was a sharp reduction in the concentration of the solids along the height as the particle size increased from 38 to $85\,\mu m$ and then to $160\,\mu m$. In the last case, the solid concentration at the base was as high as 40%, reducing to practically no solids at the top of the liquid surface. This finding is readily explained on the basis of the higher settling velocity of the larger particles (Zhang 2002). Thus, for a given gas velocity, the fraction of the solid suspended increases with decrease in the settling velocity or particle size. The effect of particle density can be similarly explained: high-density particles have a higher settling velocity and are therefore relatively difficult to suspend. The solid loading effect, however, cannot be explained in a similar straightforward manner. Gandhi et al. (1999) measured the gas holdup as a function of the vertical distance from the sparger at high gas velocities (0.2 and $0.25\,m/s$). The glass beads used had the same density and particle size ($d_p = 35\,\mu m$) as in the study of Garcia-Ochoa et al. (1997). The low d_p combined with high V_G is

likely to lead to relatively better solid suspension and hence a flat axial solid concentration profile similar to the study of Garcia-Ochoa et al. (1997). The high V_G used most likely reflect conditions in the heterogeneous regime. The column height could be broadly segregated into three zones: (1) distributor zone, (2) bulk zone in the middle, and (3) foam zone at the top. In all the cases, the gas holdup started with a low value in the sparger zone and then increased sharply up to a height of approximately $Z/H = 0.16$ from the sparger. Above this height, for 10 vol.% solid loading, a flat gas holdup profile resembling the case with no solids was observed. Further increase in solid loading resulted in another point of inflection, giving a minimum gas holdup at $Z/H = 0.24$. At positions higher than this, there was a relatively small increase in the gas holdup. The points of inflection became prominent with increasing solid loading. Gandhi et al. explained this behavior through changes in the effective viscosity of the slurry. It was found that in the higher solid loading range from 30 to 40 vol.%, the gas holdup in the sparger region was substantially low. The apparent viscosity of the slurry was estimated using the equation proposed by Barnea and Mizrahi (1973). This apparent viscosity exhibited a sharp, almost exponential increase in the 30–40 vol.% range. The formation of large bubbles was attributed to the high apparent viscosity. It was argued that the high effective viscosity causes a sharp increase in viscous dissipation in the sparger region, which is directly reflected in lower energy available for bubble breakup at the sparger. At higher axial positions corresponding to the bulk zone, the curious decrease–increase in gas holdup was attributed to the existence of the heterogeneous regime. This conclusion was based on the observation that this type of behavior was not observed at low V_G (≤ 0.1 m/s) that are likely to correspond to transition or homogeneous regime.

2. **Gas velocity**: In the low gas velocity range, two effects are important: the high solid concentration near the sparger and prevalence of the homogenous bubble flow regime. In the higher V_G range, solid suspension may improve (depending on d_p, ρ_p, and solid loading), but the strong coalescence/breakup typical of the heterogeneous regime may lead to the type of behavior described earlier.

3. **Regime transition**: Mena et al. (2005) have presented an exhaustive discussion concerning the influence of solids on transition from homogeneous to heterogeneous regime. Their experimental investigation used large, neutrally buoyant calcium alginate beads ($d_p = 2.1$ mm and $\rho_p = 1023$ kg/m³). Although most slurry reactor systems use much higher-density catalysts, the relatively low density used by Mena et al. eliminated doubts about incomplete particle suspension. The results showed that with increasing solid loading up to C_V of ~3 vol.%, there is an initial increase in gas holdup, implying that the homogeneous regime is stabilized (transition delayed). In contrast to this, higher solid loadings promoted an early transition. For example, at a solid loading of $C_V = 3$ vol.%, the transition began at a critical gas holdup, $(\varepsilon_G)_{Crit}$, as high as 20% and $(V_G)_{Crit}$ of ~0.04 m/s. $(\varepsilon_G)_{Crit}$ decreased steadily as the solid loading increased, which could be described by $(\varepsilon_G)_{Crit} = 0.21 - 2.5 \times 10^{-3} C_V$. A number of possibilities for the stabilization below 3% were examined. These included the following:

1. Lower fraction of the reactor volume available to gas because of the additional solid phase ($\varepsilon_G = 1 - \varepsilon_L - \varepsilon_S$ as against $\varepsilon_G = 1 - \varepsilon_L$ in the absence of solids). This argument was not tenable because of the relatively low solid holdup (3–5 vol.%) at which the transition began.

2. The solid–liquid density difference: this possibility was also eliminated since $\rho_P - \rho_L$ was very small.

3. Viscosity effect: in this case, also because of the low solid loading at the beginning of transition (and corresponding low effective viscosity), the viscosity effect was not pronounced.

4. Liquid–solid affinity: this refers to lyophobicity/lyophillicity of the solid. van der Schaaf et al. (2007) have shown that lyophobic solids delay the transition, whereas the lyophilic type causes earlier transition. Solids having affinity to bubbles are likely to adsorb and prevent coalescence, leading to smaller bubbles and hence homogeneous regime. *Vice versa*, lyophilic particles do not interfere with bubble coalescence, thereby encouraging larger bubbles. This leads to an earlier start of transition. This effect is pronounced when $d_p \ll$ bubble size. Mena et al. (2005) used relatively very large particles (d_p nearly same as d_B), and hence, adsorption of solid particles having a size nearly the same as bubble diameter can be precluded.

5. For heavier particles that tend to settle at the sparger, there is a significantly higher solid concentration at the sparger. This downward force has a bearing on the bubble size formed at the sparger. In regard to this factor also, since $\rho_P - \rho_L$ was small, extensive settling resulting in a strong downward force was disregarded.

6. The effect of solids on bubble rise velocity was examined. The solid particles are a distinct separate phase. However, since $\rho_P \sim \rho_L$, they follow the liquid flow patterns. It is evident that they are an impediment to the rising bubbles and reduce the bubble rise velocity. This effect was substantiated with photographs of solid–bubble interactions and the resulting deflection of bubbles from their normal path. The rise velocity was reduced by 5–15%, resulting in correspondingly higher gas holdups.

7. The last effect related to the effect of coalescence in slurries, which is similar to point (4) discussed earlier. This effect is quite complex and depends on particle size, density, their lyophobic/lyophilic nature, etc. These phenomena are similar to other surface phenomena and have not been completely understood. Mena et al.'s (2005) photographic visualization of bubble rise gave evidence of hindered bubble rise that stabilizes the homogeneous regime. At higher solid content, however, sufficiently long periods of contact are likely to allow drainage of the intervening film and consequent bubble coalescence.

The foregoing discussion clearly reflects the complex behavior of gas holdup in three-phase bubble columns. Correlations proposed by different investigators would therefore be of limited value while extrapolating to significantly different d_p, ρ_p, solid loading, solid surface characteristics, etc, although the range of other parameters may

TABLE 10.1 Correlations for Fractional Gas Holdup in Sparged Reactors Available in the Literature

Reference	Column diameter, system used, range of variables	Correlation proposed	Comments
Hikita et al. (1980)	T, 0.1 m; gas phase, air/O_2/CH_4/C_3H_8/ H_2/H_2+N_2/CO_2; liquid phase, water, aqueous methanol/butanol/ sucrose; and methanol, butanol, aniline. Range of dimensionless numbers covered: $1.1 \times 10^{-3} < \left(\dfrac{V_G \mu_L}{\sigma} \right) < 8.9 \times 10^{-3}$ $2.5 \times 10^{-11} < \left(\dfrac{\mu_L^4 g}{\rho_L \sigma^3} \right) < 1.9 \times 10^{-6}$ $8.4 \times 10^{-5} < \left(\dfrac{\rho_G}{\rho_L} \right) < 1.9 \times 10^{-3}$ $1 \times 10^{-3} < \dfrac{\mu_G}{\mu_L} < 1.8 \times 10^{-2}$	$\varepsilon_G = 0.672 \left(\dfrac{V_G \mu_L}{\sigma} \right)^{0.578} \times \left(\dfrac{\mu_L^4 g}{\rho_L \sigma^3} \right)^{-0.131}$ $\times \left(\dfrac{\mu_G}{\mu_L} \right)^{0.107} \times \left(\dfrac{\rho_G}{\rho_L} \right)^{0.062}$	Experimental conditions covered a wide range of organic chemicals. Column diameter too small to allow reliable correlation for churn-turbulent regime
Akita and Yoshida (1973)	$0.15 < T < 0.6$ m; gas–liquid systems used: air, oxygen, CO_2, helium/ water, air/methanol/glycol/CCl_4, air/aqueous electrolyte solutions	$\left(\dfrac{\varepsilon_G}{(1-\varepsilon_G)^4} \right) = 0.2 \left(\dfrac{gT^2 \rho_L}{\sigma} \right)^{0.12}$ $\times \left(\dfrac{gT^3}{v_L^2} \right)^{0.081} \times \left(\dfrac{V_G}{\sqrt{gT}} \right)$	Experimental conditions covered a range of organic liquids as well as aqueous electrolytes

Schumpe and Deckwer (1987)	$0.06 < T < 0.3$ m. Various Newtonian and non-Newtonian liquids studied. Literature data also included	$$\varepsilon_G = 0.2\left(\frac{gT^2\rho_L}{\sigma}\right)^{-0.13}\times\left(\frac{gT^3\rho_L^2}{\mu_{\text{eff}}^2}\right)^{0.11}$$ $$\times\left(\frac{V_G}{(gT)^{0.5}}\right)^{0.54}$$ $$\gamma_{\text{Eff}} = 2800V_G$$	Single correlation satisfying Newtonian and non-Newtonian liquids
Inga and Morsi (1999)	Column diameter, 0.31 m; V_G, 0.05–0.32 m/s; system used, H_2, N_2, CO, CH_4/mixture of hexanes; solid phase, iron-based F–T catalyst (0–50 wt.%)	$$\varepsilon_G = \left(\frac{17}{6}\right)(M_G)^{2.7}\exp(-3C_v)$$ $$M_G = \left(\frac{\rho_G V_G}{\rho_L(1-\varepsilon_G)}\right)$$	Wide variation in gas-phase species
Urseanu et al. (2003)	T, 0.15 and 0.23 m; V_G, up to 0.3 m/s; operating pressure, 0.1–1 MPa, temperature, ambient; $5.3 < (H/T) < 8.1$. System used: N_2/Tellus oil, glucose solutions	$$\varepsilon_G = 0.21\times(V_G)^{0.58}(T)^{-0.18}(\mu_L)^{-0.12}$$ $$\rho_G^{[0.3\exp(-9\mu_L)]}$$	Tenfold variation in system pressure. Organic as well as aqueous liquids with good variation of viscosity
Behkish et al. (2006)	Literature data covering a very wide range of system properties and column sizes used to derive correlations. Separate correlations given for holdup due to large and small bubbles	$$\varepsilon_G = 4.94\times10^{-3}\times\left(\frac{\rho_L^{0.415}\rho_G^{0.177}}{\mu_L^{0.174}\sigma_L^{0.27}}\right)V_G^{0.553}$$ $$\left(\frac{P_T}{P_T-P_S}\right)^{0.203}\left(\frac{T}{T+1}\right)^{-0.117}\Gamma^{0.053}$$ $$\times\exp\{-2.231C_V-0.157(\rho_P d_P)$$ $$-0.242X\}$$ X weight fraction of the primary liquid in the mixture. $(1 \geq X \geq 0.5)$ (w/w); Γ: represents effect of gas sparger, correlations given for various sparger designs	System and correlation relevant to F–T synthesis

be same. Table 10.1 gives representative correlations covering a wide range of system properties ranging from coalescing to noncoalescing and also organic liquids. For design purposes, the particular correlation obtained for systems with similar physical properties and also similar range of operating conditions should be used. Since small diameter columns (<0.1 m diameter) tend to develop slug flow at higher gas velocities (>0.15 m/s), correlations based on such small-diameter column data should not be used for design of large industrial columns.

10.4 SOLID–LIQUID MASS TRANSFER COEFFICIENT (K_{SL})

Unlike previous chapters, the discussion on K_{SL} precedes that on $k_L a$ in this chapter. As discussed in Chapter 6, it is relatively simple to develop a rational correlation for K_{SL} based on the turbulence intensity. The linear dimension in the Reynolds number to be used is the catalyst linear dimension, d_p, which is usually known.

Investigations on K_{SL} in TPSRs date back to the 1960s. Jadhav and Pangarkar (1991) presented an exhaustive review of the literature on K_{SL} covering the earliest investigations to the 1990s and critically examined the same. It was pointed out in Sections 6.2 and 6.3 that in view of the lack of understanding of the turbulence structure, the initial investigators resorted to empirical correlations such as Equation 6.1 or 6.2. Subsequently, correlations based on dynamic similarity such as the Froessling and Ranz and Marshall (Eqs. 6.3 and 6.4) involving dimensionless groups were used. Another approach based on dimensionless groups that found favor in the studies was the use of Kolmogorov's theory of isotropic turbulence. Reynolds number based on power dissipation was used to correlate data on K_{SL} with operating variables and system properties (Sano et al. 1974; Sanger and Deckwer 1980). As discussed in Sections 6.2 and 6.3, the earlier approaches had a number of basic flaws. Consequently, these correlations were valid for the specific range of operating variables and system properties as pointed out by Jadhav and Pangarkar (1991). In the following section, a brief analysis of literature results pertaining to effects of different operating parameters and system properties is presented.

10.4.1 Effect of Gas Velocity on K_{SL}

Jadhav and Pangarkar (1988, 1991) carried out extensive measurements of K_{SL} over a wide range superficial gas velocities ($0.08 < V_G < 0.4$ m/s) and column diameters ($0.1 < T < 0.4$ m). Evidently, all the regimes (slug flow, homogeneous, and heterogeneous) were covered. Literature data were also included in their analysis. It was clearly shown that for smaller column diameters ($T < 0.1$ m), K_{SL} becomes practically independent of superficial gas velocity for $V_G > 0.2$ m/s. On the other hand, for larger column diameters ($T > 0.2$ m), such a plateau was not observed. The constancy in transport coefficient in the smaller diameter was attributed to the change over from homogeneous to slug flow regime at $V_G > 0.2$ m/s. The large slugs do not contribute effectively to the liquid-phase turbulence, and hence, no increase in K_{SL} is observed with increasing gas velocity. Similar constancy in K_{SL} was reported by Jadhav and Pangarkar (1991) for high-viscosity systems, which also have a pronounced slugging tendency. As discussed in Section 6.4, the main contributor to the transport characteristics in any multiphase reactor is the turbulence intensity, u'. In slug flow, there is no liquid circulation, and hence, the turbulence level is relatively low. Thus, having reached a plateau value, K_{SL} remains practically constant beyond $V_G > 0.2$ m/s for smaller column diameters/high-viscosity liquids.

10.4.2 Effect of Particle Diameter d_p on K_{SL}

For most catalytic applications, the particle size is less than $100 \, \mu m$. Jadhav and Pangarkar (1991) examined the extensive information available in the literature and showed that for the low size range ($d_p < 1000 \, \mu m$), $K_{SL} \propto (d_p)^{-0.5}$ under otherwise similar conditions. For $d_p > 1000 \, \mu m$, K_{SL} increases up to $1800 \, \mu m$ and then remains practically constant. The region of higher d_p ($>2000 \, \mu m$) generally corresponds to three-phase fluidized beds or beds supported, mainly by net liquid flow rather than gas flow. Such particle sizes are not relevant to applications involving catalytic reactions but may be encountered in cell culture applications. In this latter case, the cells are supported on larger supports as discussed in Section 7B.8.2.2. However, sparged reactors suffer from the serious drawback of cell damage during bubble bursting (Section 7B.8).

10.4.3 Effect of Column Diameter on K_{SL}

Jadhav and Pangarkar (1991) extended the original approach of Whalley and Davidson (1974) and its subsequent modifications to obtain the following expression for the turbulence intensity, u', in TPSRs:

$$u' = 0.43 \left\{ gT \left(V_G - \varepsilon_G V_{B\infty} - \varepsilon_S V_T \left(\frac{\rho_S - \rho_L}{\rho_L} \right) \right) \right\}^{0.33} \qquad (10.7)$$

Equation 10.7 has been derived for low solid loadings (~1 wt.%) and no net liquid flow ($V_L = 0$). It is evident that u' increases with increase in T under otherwise identical conditions. This may be the reason for higher values of K_{SL} in larger-diameter columns. Jadhav and Pangarkar (1991) analyzed data for different column diameters and obtained the following relationship between K_{SL} and T:

$$K_{SL} \propto (T)^{0.29} \qquad (10.8)$$

Based on their analysis of the extensive data on K_{SL}, Jadhav and Pangarkar (1991) reached the following conclusions:

1. Tower diameter has no effect on the K_{SL}–d_p relationship.
2. For low- to moderate-viscosity liquids, K_{SL} increases steadily with increasing T for $T > 0.2$ m. On the other hand, for $T < 0.2$ m, K_{SL} approaches a constant value at $V_G > 0.2$ m/s. For high-viscosity liquids and $V_G > 0.2$ m/s, K_{SL} exhibits constancy even in larger ($T > 0.2$ m) column diameters.

3. K_{SL} increases with increasing T. The dependence is given by $K_{SL} \propto (T)^{0.29}$. This dependence is nearly the same as that for u' predicted from the energy balance. This observation further strengthens the premise that the turbulence intensity is the main causative factor.

10.4.4   Correlation for K_{SL}

For the catalyst particle size of interest in multiphase reactor ($d_p < 100\,\mu m$), Jadhav and Pangarkar (1991) regressed more than 150 data points from their own work as well as those of other workers spread globally and covering a wide range of T, superficial gas velocity, physical properties, etc. The following correlation was proposed:

$$\text{Sh}_p = 2 + 0.091 \left(\frac{d_p u' \rho_L}{\mu_L} \right)^{0.617} \left(\frac{\mu_L}{\rho_L D_A} \right)^{0.45} \tag{10.9}$$

(standard deviation = 15%)

This correlation requires information on u', which can be estimated using Equation 10.7. This latter equation requires data on ε_G as a function of superficial gas velocity to evaluate the terminal rise velocity of the bubble, $V_{B\infty}$. These data can be obtained through simple gas holdup measurements. The drift flux model of Zuber and Findlay (1965) can be used to obtain $V_{B\infty}$ as per Equation 10.10:

$$\varepsilon_G = \frac{V_G}{V_{B\infty} + b V_G} \tag{10.10}$$

Thus, a plot of $1/\varepsilon_G$ versus $1/V_G$ should yield a straight line with $V_{B\infty}$ as the slope. This value of $V_{B\infty}$ can be used in Equation 10.7 to obtain u'.

As pointed out by Jadhav and Pangarkar (1991), Equation 10.9 could satisfactorily correlate data from various groups covering a wide range of operating conditions and physical properties. The success of this correlation using u' as the characteristic velocity is most likely because the expression for u' includes all the pertinent parameters that characterize the dispersion. These are mainly ε_G and $V_{B\infty}$ besides V_G, which is representative of the operating conditions. In particular, $V_{B\infty}$ characterizes the ε_G–V_G relationship that is specific to a given dispersion

10.5   GAS–LIQUID MASS TRANSFER COEFFICIENT ($k_L a$)

Similar to gas holdup, a very large number of empirical correlations have been reported for estimating $k_L a$ in TPSRs. The correlations relevant to the heterogeneous regime are listed in Table 10.2. It is evident that different investigators have obtained different dependences of $k_L a$ on the operating conditions and system physical properties. Ideally, it would have been desirable to derive a correlation of the type presented earlier for K_{SL}:

$$k_L a = f(\text{Re}, \text{Sc}) \tag{10.11}$$

TABLE 10.2 Correlations for Liquid-Side Mass Transfer Coefficient Available in the Literature

Reference	Correlation suggested	Comments
Akita and Yoshida (1973)	$$\left(\frac{k_L a T^2}{D_A}\right) = 0.6 (\mathrm{Sc})^{0.5} \left(\frac{g T^2 \rho_L}{\sigma}\right)^{0.62} \left(\frac{g T^3}{\nu_L^2}\right)^{0.31} \varepsilon_G^{1.1}$$ Column diameter, 0.152 m; dispersion height, 2–3 m; gas phase, air, O_2, CO_2, and helium; liquid phase, water, aqueous glycol, NaCl, Na_2SO_3, and CCl_4; $4 \times 10^{-3} < V_G < 0.33$ m/s	$k_L a$ based on completely mixed liquid phase. $k_L \underline{a}$ is a weak function of T ($\propto T^{-0.17}$)
Ozturk et al. (1987)	$$\left(\frac{k_L \underline{a} d_B^2}{D_A}\right) = 0.62 \left(\frac{\mu_L}{\rho_L D_A}\right)^{0.5} \left(\frac{g \rho_L d_B^2}{\sigma}\right)^{0.33}$$ $$\left(\frac{g \rho_L^2 d_B^3}{\mu_L^2}\right)^{0.29} \left(\frac{V_G}{\sqrt{g d_B}}\right)^{0.68} \left(\frac{\rho_G}{\rho_L}\right)^{0.04}$$ Large variation in physical properties and gas–liquid pairs. $0.008 < V_G < 0.1$ m/s	d_B assumed $= 0.003$ m. Correlation found reliable for many systems including organic liquids and over a wide range of dimensionless numbers. Column diameter used ($T = 0.095$ m) is below the minimum suggested
Kawase and Moo-Young (1990)	$$k_L a \underline{a} = 0.452 \left(\frac{\rho_L^{0.6} \nu^{0.25}}{\sigma^{0.6}}\right) \varepsilon_M^{59/60} \left(\frac{1}{T^{0.166} g^{0.5}}\right)^{0.5} \left(\frac{\mu_G}{\mu_L}\right)^{-0.25} \mathrm{Sc}^{-0.5}$$ Literature data used to obtain empirical constants	Correlation based on Kolmogorov's theory of isotropic turbulence. Drawbacks discussed in Section 6.3.2. Does not include ε_G term
Seno et al. (1990)	$$\frac{k_L \underline{a} T^2}{D_{AL}} = 0.6 \left(\frac{V_G}{V_G + V_L}\right)^{-0.39} \left(\frac{\mu_L}{\rho_L D_{AL}}\right)^{0.5} \left(\frac{g T^2 \rho_L}{\sigma}\right)^{0.62} \left(\frac{g T^3 \rho_L^2}{\mu_L^2}\right)^{0.31} \varepsilon_G^{1.1}$$ Cocurrent mode Column diameter, 0.046 m; dispersion height, 1.2 m; gas phase, O_2; liquid phase, distilled water, aqueous butanol/glycerol/surface-active agent T-20 + silicone oil, CMC, and poly(acrylamide); $0.01 < V_G < 0.06$ m/s	Column diameter well below the minimum suggested. Akita and Yoshida's (1974) correlation has been modified for cocurrent and countercurrent mode of operation

(Continued)

TABLE 10.2 (Continued)

Reference	Correlation suggested	Comments
Cockx et al. (1995)	$k_L = (0.1 \pm 0.02) \dfrac{u^*}{Sc^{0.5}}$ $T = 0.2$ m; dispersion height, ~3 m; V_G, 0.002–0.007 m/s. V_L, 0.004–0.02 m/s. Cobalt-catalyzed absorption of oxygen in aqueous sodium sulfite	The mean shape factor of the bubbles considered in the calculation of a used to account for the regime of operation. Correlation based on analogy between momentum and mass transfer. Found to fit a wide range of physical properties of the gas–liquid system and mode of operation
Letzel et al. (1999)	$k_L a = 0.5 \times \varepsilon_G$ $T = 0.15$ m. Air (oxygen)–water. Step change from N_2 to O_2 or pressure of air. V_G up to 0.4 m/s. Operating pressure: 0.1–1 MPa	Requires information on ε_G for the system. ε_G ↑ with increase in total pressure. Hence, $k_L a$ also ↑ with increase in total pressure
Behkish et al. (2002)	$k_L a = 0.18(Sc)^{-0.6} \left(\dfrac{\rho_L \nu_A}{M_B} \right)^{-2.84} (\rho_G \times V_G)^{0.49} \exp(-2.66 C_V)$ T, 0.31 m; dispersion height, ~2.5 m; gas phase, H_2, CO_2, N_2, and CH_4; liquid phase, Isopar M (C10–C16—molecular weight, 192) and mixed hexanes (molecular weight, 86); solid phase, iron catalyst (d_p: 1.5 and 40 μm, density: 4000 kg/m³) and glass beads (d_p 11.4 μm; density, 2500 kg/m³; vol.%, 0–36%); 0.08 < V_G < 0.25 m/s; operating pressure: 0.17–0.79 MPa)	$k_L a$ ↑ with ↑ in V_G and pressure but ↓ dramatically with ↑ in solid loading due to probably a decrease in the effective interfacial area, a, caused by bubble coalescence. Provision of sparger to create small bubbles/sieve plates (with very low free area) suggested to break the large bubbles formed
Lau et al. (2004)	$k_L a = 1.77\sigma^{-0.22} \exp(1.65 V_L - 65.3\mu_L) \varepsilon_G^{1.2}$ $T = 0.05$ and 0.1 m; 8 × 10−4 < V_L < 8.9 × 10−3 m/s; V_G up to 0.4 m/s. Gas phase, air/N_2; liquid phase, water and Paratherm NF heat transfer fluid. Operating pressure and temperature: up to 4.24 MPa and 92°C, respectively	Data obtained cover a wide range of pressure and temperature for both aqueous and organic liquids. Column diameters below the minimum suggested
Chaumat et al. (2005)	$0.04 < V_G < 0.14$ m/s; $0.035 < V_L < 0.08$ m/s. $T = 0.2$ m; dispersion height, 1.6 m; 4 sieve plates each having 31.5% free area; gas phase, N_2 and CO_2; liquid phase, cyclohexane and water; operating pressure, 0.1 MPa. $(k_L a / \varepsilon_G)$ is not constant but varies from 0.1 to 0.3/s. $k_L a = (\varepsilon_G / 6)$ can be used as an approximation for the range of variables covered	No significant effect of sectionalization/liquid physical properties found. No effect of sparger design in the heterogeneous regime. Increasing liquid velocity decreases ε_G and hence $k_L a$

The representative characteristic velocity to be used in Re can be obtained from Equation 10.7. Unfortunately, unlike the linear dimension d_p, which is well defined in the case of K_{SL}, it is not possible to define a representative linear dimension for the present case. For instance, the bubble diameter, d_B, which should be the preferred choice for representative linear dimension, exhibits a wide variation, particularly in the heterogeneous regime. Thus, in the absence of a representative value for d_B, an approach similar to Equation 10.9 cannot be adopted. In view of this difficulty, most investigators resorted to correlations based on dynamic similarity involving dimensionless groups (Section 6.3.1) as is evident from Table 10.2.

Observations made in a few important studies available in the literature are discussed in the following sections. Some of these point out the key role of gas holdup in gas–liquid mass transfer.

Vandu and Krishna (2004) determined $k_L a$ using the dynamic oxygen absorption technique over a wide range of column diameters ($0.1 < T < 0.63$ m), liquid viscosity ($1 < \mu_L < 75$ mPa s), and surface tension ($23 < \sigma < 72$ mN/m). The liquid density range was relatively limited ($800 < \rho_L < 1000$ kg/m³). Porous silica particles representative of heterogeneous catalyst were used ($d_p \sim 38$ μm; solid loading $0 < C_V < 25$ vol.%). The following major conclusions were derived:

1. For both two- and three-phase systems and for $V_G > 0.08$ m/s (implying heterogeneous regime, particularly for $T > 0.2$ m), the quotient $\left(k_L a/\varepsilon_G\right)$ is invariant with respect to the column diameter.

2. For pure liquids, $\left(k_L a / \varepsilon_G\right) \propto \left(Sc\right)^{-0.33}$. This is suggestive of boundary layer-type mechanism, which is surprising for a relatively turbulent liquid phase and mobile bubble–liquid interface. It may be mentioned that several investigators (Ozturk et al. 1987; Kawase and Moo-Young 1990; Seno et al. 1990; Cockx et al. 1995) have obtained $k_L a \propto \left(D_A\right)^{0.55}$ (also Jadhav and Pangarkar (1991) for K_{SL}: Equation 10.9 for a particulate phase that was a solid) that is closer to that described by the unsteady-state convective mass transfer mechanism.

3. Vandu and Krishna (2004) found a significant reduction in $\left(k_L a / \varepsilon_G\right)$ in the presence of solid particles. This suggests strong coalescing tendency due to the solid phase. Vandu and Krishna's (2004) study buttressed the importance of gas holdup in scale-up of TPSRs. In Section 10.3.4, the important role of surface interactions in deciding gas holdup has been discussed. Literature studies give contradictory results. Such contradictions can be possibly explained through peculiar gas holdup variations due to surface effects. Vandu et al. (2004) found that $k_L a$ in an alumina–paraffin oil slurry was unaffected by an increase in solid loading as against the increase or decrease observed in the literature. This apparent anomaly is probably due to the relative affinities of solid surface and the liquid (Section 10.3.4.3) that can affect gas holdup through peculiar coalescence behavior.

4. The operating pressure also has an important effect on gas holdup and in turn on $k_L a$. Letzel et al. (1999) reported a strong positive dependence of $k_L a$ on operating pressure, which was attributed to an increase in ε_G (or a). $\left(k_L a / \varepsilon_G\right)$ was found to be constant and approximately 0.5. This study also emphasized the importance of ε_G. Thus, according to Letzel et al., if ε_G is known, $k_L a = 0.5 \times \varepsilon_G$. However, Chaumat et al.'s (2005) data showed that $\left(k_L a / \varepsilon_G\right)$

is not constant (=0.5) but varies from 0.1 to 0.3/s from V_G=0.04 to 0.14 m/s. Han and Al-Dahhan (2007) observed variations in k_L and $k_L a$, which could be explained through the different effects of operating pressure on bubble size/gas holdup (Section 10.3.3).

In conclusion, most investigators have found that ε_G plays a significant role in deciding $k_L a$. Therefore, it is important that for design purposes, the estimation of $k_L a$ should rely on correlations based on systems with similar physical properties/range of operating conditions such that ε_G has a similar value. Alternatively, correlations that give due importance to ε_G, such as Equation 10.9 for K_{SL}, should be preferred.

10.6 AXIAL DISPERSION

In a sparged reactor, the behavior of the liquid and gas phase deviates significantly from plug flow, particularly at high gas and low liquid velocities. This deviation is generally accounted for by the axial dispersion coefficient, D_{Ax}. Deckwer (1992) has discussed this matter in detail outlining the various approaches used to quantify axial dispersion. Tables 10.3 and 10.4 list some of the correlations available in the literature for estimating the liquid- and gas-phase axial dispersion coefficient, D_L and D_G, respectively.

Liquid-phase back mixing is a serious issue for reactions that have nonzero-order kinetics with respect to the liquid-phase reactant. Sectionalization of bubble column using sieve plates of relatively low free area is an attractive choice in such a case. Although this choice has been mentioned in the literature, its application in solid-catalyzed reactions has not attracted any attention. The sieve plate design must be such that it prevents weeping (Prince 1960). The free area in such sieve plates is

TABLE 10.3   Correlations for D_L

Reference	Correlation	Comments
Towell and Ackerman (1972)	$D_L = 1.23 \times T^{1.5} V_G^{0.5}$ (CGS units)	Empirical correlation based on $0.40 < T < 1.0$ m
Deckwer et al. (1973, 1974)	$D_L = 2.7 \times T^{1.4} V_G^{0.3}$ (CGS units)	T=0.2 m, H/T=36; T=0.15 m, H/T=29; V_G relatively low (0.05 m/s). Liquids used: tap water, aqueous NaCl, Na_2SO_4, molasses. Empirical equation based on investigator's own and literature data (CGS units)
Joshi and Sharma (1979)	$D_L = 0.31 \times T V_C$ $V_C = 3[T(V_g - \varepsilon_G V_{B\infty})]^{0.33}$	Equation derived from liquid circulation velocity based on energy balance approach of Whalley and Davidson (1974)
Garcia-Calvo and Leton (1994)	$D_L = T \left(\dfrac{V_C \times V_{B\infty}}{2} \right)^{0.5}$	Correlation valid for Newtonian as well as non-Newtonian liquids over a wide range of conditions

TABLE 10.4   Correlations for D_G

Reference	Correlation	Comments
Towell and Ackerman (1972)	$D_G = 0.2\ T^2 V_G$	Empirical correlation based on authors' own and literature data (CGS units)
Field and Davidson (1980)	$D_G = 9.36 \times 10^{-5}\ \{T^{1.33}\ V_G^{*3.56}\}$	Empirical correlation based on authors' own and literature data (CGS units)
Mangartz and Pilhofer (1980, 1981)	$D_G = 5 \times 10^{-5}\ \{T^{1.5}\ V_G^{*3}\}$	Empirical correlation based on authors' own and literature data (CGS units)
Wachi and Nojima (1990)	$D_G = \left(\dfrac{180}{\alpha}\right) V_G \times T^{1.5}$ $\alpha = 9$. $T = 0.2$ and 0.5 m; dispersion height: 4.5 m; $T = 0.5$ m with/without three sieve plates spaced at 1 m interval (free area, 44%); $0.029 < V_G < 0.45$ m/s	Introduction of sieve plates (free area: 44%) in the 0.5 m dia. column gave lower dispersion for $V_G > 0.1$ m/s (heterogeneous regime)

CGS units: D_L, cm²/s; T, cm; V_G, V_C, $V_{B\infty}$, cm/s; $V_G^* = V_G/\varepsilon_G$.

relatively low (free area $< 1.5\%$ with small holes, as outlined in Section 10.8.9). Therefore, the major problem in this application is the obvious apprehension of plugging of the sieve plates. Further, it is known that bubble columns exhibit marked variation in axial solid concentration profile even without sieve plates (Gandhi et al. 1999). Introduction of sieve plates/internals is likely to aggravate this problem to the extent that the solid phase may be nonexistent in the top of dispersion. However, for gas–liquid reactions involving a homogeneous liquid phase (Section 10.8) or involving very fine neutrally buoyant solids (Prokop et al. 1969; Zahradnik et al. 1992) such that the settling velocity is relatively low and also $d_P \ll d_H$, sectionalization can be beneficial. An example that belongs to this category is of cell culture systems. The cell sizes range from 1 to 40 μm (Table 7B.6), and the liquid phase can be considered as pseudohomogeneous. However, curtailing axial mixing implies higher mixing times. As elaborated in Section 7B.9.1, poor mixing is detrimental in cell culture systems, and hence, such internals have no relevance. Zahradnik et al. (1992) have investigated gas holdup, mass transfer, and axial mixing in a sectionalized bubble column containing zinc oxide particles ($d_P \sim 3$ μm; solid loading up to 5 wt%) in a 0.29 m diameter bubble column having total height $= 2$ m. Sieve plates with 0.25% free area and hole diameter $= 1.8$ mm fitted with pipe downcomers were used. The column could be divided into three or six equal sections using corresponding number of sieve plates. Sectionalization was found to improve gas holdup ($\sim 100\%$ increase) and $k_L a$ ($>200\%$ increase). Liquid back mixing was satisfactorily described by completely mixed tanks in series model with back flow between stages. The number of sections was considered as the number of completely mixed tanks in series. The coefficient of back flow stages was the major parameter that reflected the extent of back mixing. The data obtained reflected the beneficial effect of sectionalization.

10.7 COMMENTS ON SCALE-UP OF TPSR/BUBBLE COLUMNS

Several state-of-the-art reviews on scale-up of bubble columns are available (Kantarci et al. 2005; Shaikh and Al-Dahhan 2013b). These have discussed the relative merits/demerits of various approaches used in the literature. With the advent of various advanced measurement techniques, our understanding of the phenomena involved has improved. The scale-up procedures used have been based mainly on global parameters such as overall gas holdup, mass transfer coefficients, etc. It has been suggested that industrial-scale bubble columns should be designed using sound computational fluid dynamics (CFD) models. Rafique and Dudukovic (2004) had pointed out the uncertainties involved in developing such fundamentally sound CFD models. As regards the approach based on global parameters (dynamic similarity), Shaikh and Al-Dahhan (2010) showed that it does not necessarily yield hydrodynamic regime similarity/turbulence similarity. Therefore, this approach may not yield reliable scale-up rules. In view of the earlier discussion, Shaikh et al. (2013b) concluded that the complexity of the hydrodynamics involved precludes an *a priori* design procedure. In Sections 2.2, 2.3, and 2.5, the scale-up conundrum has been discussed. From this discussion, it is evident that intrinsic kinetics being scale invariant, the only problem in reliable scale-up is prediction of the transport coefficients. In the absence of sound science-based models, an approach based on reliable correlations for the respective mass transfer coefficients (Jadhav and Pangarkar 1991; Pangarkar et al. 2002; Yawalkar and Pangarkar 2002a, b; Nikhade and Pangarkar 2007; Pangarkar 2011) can be resorted to. This will be illustrated in Section 10.8.

10.8 REACTOR DESIGN EXAMPLE FOR FISCHER–TROPSCH SYNTHESIS REACTOR

10.8.1 Introduction

Section 3.4.1 provides detailed discussion on F–T synthesis using coal as the feedstock. This route can be described by the following overall sequence:

$$\text{Coal} + \text{steam} \quad \xrightarrow{\text{Gasification}} \quad \text{CO} + \text{H}_2(\text{syngas}) \quad \xrightarrow{\text{F – T synthesis}} \quad \text{hydrocarbons}$$

F–T synthesis is known to yield a very high-quality diesel in comparison to the diesel obtained from petroleum crude (Dry 1990, 2001, 2002). This F–T diesel has great relevance in view of the stringent specifications imposed in almost all countries in the world. Sasol has been successfully using F–T synthesis for the production of a wide variety of hydrocarbons since 1954.

The actual F–T reactions involved are complex. A simplified representation is

$$CO + 2H_2 \rightarrow -CH_2- + H_2O \quad \Delta H_R = -0.172\,\text{MJ/mol}_{CO} \qquad (10.12)$$

The following discussion illustrates the design of a TPSR (slurry) for F–T synthesis using cobalt on silica catalyst. The slurry phase reactor for F–T synthesis has a number of advantages over the fixed bed reactor. Woo et al. (2010) have indicated that the capital cost for a commercial TPSR is less than 40% of that for a multitubular fixed bed reactor designed for the same production duty. The catalyst loading for the TPSR is approximately one-third of that for the packed bed tubular reactor. The small size of the catalyst particles used in the TPSR curtails internal diffusion resistance and allows operation of the catalyst at its intrinsic rate. In addition, both heat and mass transfer rates are significantly higher for slurry reactors. Krishna and Sie (2000) have made a comparison of the various reactor options for the F–T synthesis process. Their study also indicates that the TPSR is the best alternative for large-scale application of the F–T synthesis. As explained in Section 3.4.1, the major hindrance to the commercial application of slurry reactors was nonavailability of a suitable separation system for the rather fragile, fine catalyst used. Once this problem was solved, the slurry reactor succeeded the fixed bed shell-and-tube-type reactors for low-temperature F–T synthesis (Espinoza et al. 2000).

10.8.2   Physicochemical Properties

The kinetics of this system are reported by Yates and Satterfield (1991).

The following rate expression for the F–T reaction using a Co on silica catalyst was proposed:

$$R_{syngas} = \frac{a p_{H_2} p_{CO}}{\left(1 + b p_{CO}\right)^2} \qquad (10.13)$$

$$a = 8.8533 \times 10^{-3} \exp\left[4494.41\left(\frac{1}{493.15} - \frac{1}{T}\right)\right] \quad (\text{gmol/s,[kg]}_{cat}\,\text{bar}^2) \quad (10.14)$$

$$b = 2.226\ \exp\left[-8236\right]\left(\frac{1}{493.15} - \frac{1}{T}\right)\ \text{bar}^{-1} \qquad (10.15)$$

The rates of synthesis gas consumption at different temperatures based on the kinetics represented by Equations (10.13) through (10.15) are given in Table 10.5.

TABLE 10.5   Rate of Synthesis Gas Consumption at Different Temperatures

Temperature (°K)	Rate of syngas consumption (gmol/kg catalyst, s)
503	7.222×10^{-3}
513	1.57×10^{-2}
523	3.22×10^{-2}

The reactor operating pressure is fixed at 3 MPa. The feed to the reactor is a 2:1 H$_2$: CO mixture. The liquid used as solvent is a typical C28 hydrocarbon. Marano and Holder (1997) have given vapor–liquid equilibria and the physical properties of this solvent–synthesis gas mixture system: liquid density, 680 kg/m^3; viscosity, 7×10^{-4} Pa s; surface tension, 0.01 N/m; average molecular weight of the gas, 10.5; gas density, 7.15 kg/m^3; gas molar volume, 1.4 m^3/kmol; and diffusivity of CO, 1.7×10^{-8} m^2/s. These values give a Schmidt number for CO (slower diffusing component), $[Sc]_{CO} = 60.5$. The catalyst particle density (Co supported on silica) is 1200 kg/m^3, and the catalyst particle size, 60×10^{-6} m.

10.8.3 Basis for Reactor Design, Material Balance, and Reactor Dimensions

The basis for the reactor design: 1000 kg mol/h of feed gas or 0.4 m^3/s, superficial gas velocity $V_G = 0.2$ m/s, and L/D of the reactor $= 6$.

Using the basis given above, the dimensions of the sparged reactor are as follows: tank diameter $= 1.6$ m and the corresponding dispersion height ($H/T = 6$) equal to 9.6 m.

With the above data and the range of catalyst loadings suggested by Maretto and Krishna (1999), the catalyst weight/m^3, total catalyst weight, and catalyst area, m^2/m^3 of dispersion volume (A_p/V_L), are as follows.

Catalyst loading (wt.%)	Catalyst weight (kg/m^3)	Catalyst area, m^2/m^3 of dispersion volume (A_p/V_L)
4	27.2	2,600
12	81.6	7,800
20	136	13,000
28	190.4	18,200
35	238	22,750

10.8.4 Calculation of Mass Transfer Parameters

For designing the large-scale reactor, it is required to establish the controlling step. As outlined in Chapter 1, for this solid-catalyzed gas–liquid reaction, there are three resistances in series: (1) gas–liquid mass transfer, (2) solid–liquid mass transfer, and (3) reaction at the catalyst surface. The detailed calculation procedure and equations used are given in the following section. The catalyst loading to be used is 20 wt.%.

Gas–Liquid Mass Transfer Coefficient ($k_L a$)
STEP I: GAS HOLDUP, ε_G, AT THE OPERATING CONDITIONS   The correlation proposed by Inga and Morsi (1999) is appropriate for the present case since it has been derived using conditions representative of F–T synthesis:

$$\varepsilon_G = \left(\frac{17}{6}\right)(M_G)^{2/7} \exp(-3C_v) \tag{10.16}$$

where M_G is given by

$$M_G = \left(\frac{\rho_G V_G}{\rho_L (1-\varepsilon_G)} \right) \tag{10.17}$$

Table 10.6 gives the values of ε_G obtained from Equation 10.17.
  Figure 10.3 gives a plot of ε_G versus V_G.

STEP 2: CALCULATION OF $k_L \underline{a}$ A recent correlation provided by Behkish et al.
(2002) (Table 10.2) is appropriate for the conditions representative of the present
case: at $V_G = 0.2$ m/s, Sc = 60.5 and $C_V = 0.12$. The calculated value of $k_L \underline{a}$ is 0.25/s.
This value is obtained for ambient conditions with diffusivity of $\sim 1.5 \times 10^{-9}$ m^2/s.
According to the penetration model, $k_L \propto \sqrt{D_A}$. Therefore, k_L under actual conditions ($D_{Actual} = 1.7 \times 10^{-8}$ m^2/s):

$$k_{L\ actual} = \sqrt{\frac{1.7 \times 10^{-8}}{1.5 \times 10^{-9}}} = 0.84 / s \tag{10.18}$$

Solid–Liquid Mass Transfer Coefficient, K_{SL}
STEP 3: ESTIMATION OF K_{SL} The correlation proposed by Jadhav and Pangarkar
(1991) is used to calculate K_{SL}:

$$Sh_P = 2 + 0.091 \left(\frac{d_p u' \rho_L}{\mu_L} \right)^{0.617} \left(\frac{\mu_L}{\rho_L D_A} \right)^{0.45} \tag{10.9}$$

TABLE 10.6   Gas Holdup Values at Different Gas Velocities

V_G (m/s)	ε_G (—)
0.10	0.2080
0.15	0.2360
0.20	0.2580
0.25	0.2770

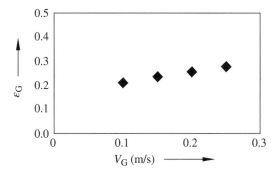

FIGURE 10.3   Variation of ε_G with V_G for TPSR for F–T synthesis.

TABLE 10.7   Data for $1/V_G$ Versus $1/\varepsilon_G$ Plot to Obtain $V_{B\infty}$

$1/V_G$ (m/s)$^{-1}$	$1/\varepsilon_G$ (—)
10.00	4.807692
6.67	4.237288
5.00	3.875969
4.00	3.610108

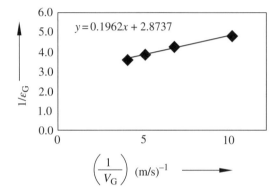

FIGURE 10.4   Plot of $1/\varepsilon_G$ versus $1/V_G$ for TPSR for F–T synthesis.

The turbulence intensity, u', required in Equation 10.9 is obtained by (i) evaluating $V_{B\infty}$ from a plot $1/V_G$ versus $1/\varepsilon_G$ as shown in Figure 10.4. The calculated values are given in Table 10.7. The value of $V_{B\infty}$ obtained is 0.19 m/s. (ii) Using this value in the equation for u' (Eq. 10.7) and $T=1.6$ m, the value u' is 0.45 m/s. Substituting this value of u', $d_p=60\times10^{-6}$ m, $[Sc]_{CO}=60.5$, and other physical properties given in Section 10.8.2 in Equation 10.9, the value of K_{SL} is 1.79×10^{-3} m/s.

STEP 4: *Estimation of Relative Rates of the Various Steps* CO, which has a lower diffusivity than H_2, diffuses slower, and hence, it is the rate-limiting component in the diffusion process across the gas–liquid and solid–liquid films. The solubility of CO, $[A^*]_{CO}$, in the C28 hydrocarbon liquid is 55 gmol/m^3 (Marano and Holder 1997).

10.8.5   Estimation of Rates of Individual Steps and Determination of the Rate Controlling Step

The estimated rates of the various steps and their reciprocals are as follows:

1. Rate of gas–liquid mass transfer step, $\left(k_L\underline{a}\right)_{actual}\times\left[A^*\right]_{CO}$: 46.2 (gmol / m^3 s).
   Resistance due to gas–liquid mass transfer step, $(1/k_LaA^*)$: 0.021
2. Rate of solid–liquid mass transfer step, $K_{SL}\times(A_p/V_L)\times[A^*]_{CO}$: 1280 (gmol/m^3 s).
   Resistance due to solid–liquid mass transfer step, $(1/K_{SL}(A_p/V_L)A^*)$: 7.8×10^{-4}

3. Rate of reaction at the catalyst surface: $R_{syngas} \times kg_{Cat} = 3.22 \times 10^{-2} \times 136 \approx 4.37$ (gmol/m^3 s). Resistance due to reaction at the catalyst surface, $(1/R_{syngas})$: 0.228.

The sum of individual resistances is ~0.25.
The contribution of each step to the total resistance is as follows:

1. Gas–liquid: $\left(\dfrac{0.024}{0.25} \right) \times 100 = 9.6\%$

2. Solid–liquid: $\left(\dfrac{0.00078}{0.25} \right) \times 100 = 0.31\%$

3. Reaction at the catalyst surface: $\left(\dfrac{0.228}{0.25} \right) \times 100 = 91.2\%$

The above analysis indicates that the rates of the mass transfer steps are significantly higher than the rate of the reaction on the catalyst surface. This is anticipated since (i) turbulence in the liquid phase is much higher than that in the fixed beds used earlier (Krishna and Sie 2000) and (ii) the small particle size used ensures relatively very high catalyst surface area. Inga and Morsi (1997) had suggested an approach similar to that outlined earlier in step 4. They defined a dimensionless parameter, β, which is a ratio of the mass transfer resistance to the total resistance:

$$\beta = \left(\frac{1/k_L a}{\left(1/k_L a + 1/k_1 \right)} \right) \tag{10.19}$$

The reaction resistance was represented by a rate constant, k_1, similar to that given in Table 10.5. In Equation 10.19, the solid–liquid mass transfer resistance was ignored. This is probably because the small size of the catalyst combined with high catalyst surface area results in a very high rate of solid–liquid mass transfer (or low solid–liquid mass transfer resistance). Similar to the arguments made in Sections 7A.10, 8.13, and 7B.15 for the F–T bubble column also, it has been possible to discern the rate controlling step in the overall sequence that can yield a rate expression. The volume of the reactor required can be subsequently obtained using the rate expression for the rate controlling step so determined. In addition, this exercise also gives a direction for future course of action: The solid–liquid and gas–liquid mass transfer resistances together constitute less than 10% of the total resistance. Thus, the mass transfer rates are more than adequate to meet the needs of the reaction on the solid catalyst surface. Therefore, attention should be focused on developing catalysts with higher activity rather than increasing the rates of mass transfer steps by providing higher energy inputs. The assessment presented here is a straightforward order of magnitude analysis based on simple correlations for the mass transfer. Maretto and Krishna (1999) simulated the performance of a TPSR for a typical CO + H$_2$ conversion to liquid hydrocarbons. A detailed sensitivity analysis using a relatively elaborate model for the F–T reaction in a bubble column was performed. These authors also concluded that interphase mass transfer does not play an important role in deciding the reactor productivity. This conclusion is exactly similar to that based on the simple order of magnitude analysis presented earlier.

The total syngas consumed per second can be obtained from the dispersion volume and $R_{syngas} \times kg_{Cat}$. To calculate R_{syngas}, the partial pressures of H_2 and CO are needed. The feed is at 3 MPa. Since the reactants are fed in stoichiometric proportion, the mole fractions are assumed to remain constant. The only change is in the pressure at the top of the dispersion volume. This pressure = feed pressure − pressure drop due to static head of the liquid. The dispersion height is 9.6 m and the gas holdup is 0.25. Hence, the actual static head is $9.6 \times (1 - \varepsilon_G) = 7.2$ m of C28 liquid having a density of 680 kg/m^3. This pressure drop is not significant as compared to the feed pressure of 3 MPa. Therefore, it is assumed that the volumetric rate of reaction is the same as that calculated using Equation (10.13) (and the assumption that the pressure does not change substantially across the column height) for estimating the resistances: $R_{syngas} \times kg_{Cat}$. $= 4.37$ (gmol/m^3 s). The total rate of syngas consumption per second is $4.37 \times [(\pi/4) \times 1.6^2 \times 9.6] = 84.3$ gmol/s $= 302$ kmol/h. This implies a syngas conversion of ~30% per pass.

10.8.6 Sparger Design

The sparger should be designed such that there is no weeping of the liquid through the orifices in the sparger. Considerable information is available on weeping in sieve plates used in absorption/distillation duties. In these applications, the maximum submergence is limited to 0.15 m (Treybal 1981), and the free area in the sieve plate is also relatively high (~10%). In the case of a bubble column, the liquid height (equivalent of submergence) can be much higher, in the range of several meters. For sieve plates to be used in distillation, Prince (1960) suggested the following criterion using the F factor based on the gas velocity through the hole: $(F_H)_{min} = [V_H]_{Min} (\rho_G)^{0.5} > 10$ (SI units).

Mersmann (1978) has suggested the following criteria for uniform gassing of all the orifices:

$$\left(We_{Crit} \right)_H = \left(\frac{V_H^2 d_H \rho_G}{\sigma} \right)_{Crit} \geq 2 \tag{10.20}$$

$$\left(d_H \left[\frac{\rho_G \times g}{\sigma} \right]^{0.5} \left[\frac{\rho_L - \rho_G}{\rho_G} \right]^{0.625} \right) \leq 2.32 \tag{10.21}$$

Thorat et al. (2001) have given Equation 10.22 for the critical Froude number for bubble columns when the dispersion height is much larger than the maximum encountered in sieve trays:

$$Fr_C = 0.37 + 140 H_L \left(\frac{P}{d_H} \right)^{-1.6} \left(\frac{t_T}{d_H} \right)^{0.75} \tag{10.22}$$

Equation 10.22 is valid for relatively large submergences (of the order of several meters) that are common to bubble columns and is therefore recommended. Treybal (1981) has given the (t_T/d_H) values for stainless steel and carbon steel. The holes in the plates are punched. Smaller holes cannot be punched in thicker plates. For instance, for a 3 mm hole, Treybal (1981) recommends (t_T/d_H) of 0.65 for stainless steel plates. Thus, t_T is 2 mm, implying that plates of greater thickness cannot be used. On the other hand, mechanical integrity considerations may dictate larger plate thickness.

10.9 TPSR (LOOP) WITH INTERNAL DRAFT TUBE (BCDT)

10.9.1 Introduction

An important adaptation of TPSRs that may be useful in some special industrial applications is a bubble column with a draft tube. This variant is also known as internal airlift reactor. It has a draft tube that is positioned coaxially inside the main bubble column. Two further variations of this bubble column with draft tube (BCDT) are possible. The gas sparging may be in the draft tube or in the annulus between the draft tube and the main column. In this section, the focus will be on BCDT with gas sparging in the draft tube. The objective of the draft tube is to provide well-directed bulk circulation of gas, liquid, and solid phases in the reactor (Fan et al. 1987; Chisti et al. 1988). Chisti and Moo-Young (1987) and Fan et al.'s (1987) exhaustive reviews should be referred for more details. Similar to a bubble column, BCDT is simple, easy to operate and maintain since it also has no moving parts. It has been widely used in continuous production of beer, vinegar, citric acid, etc., for a wide range of capacities (Chisti and Moo-Young 1987). In these fermentations, BCDT has provided distinctly improved productivities in comparison to the conventional stirred tank fermenters. BCDT is also used in municipal and industrial wastewater treatment. However, as discussed in Section 3.4.2.4, other hitherto unexplored applications are also good candidates where a retrofit of a bubble column in the form of BCDT can yield significant benefits. The improvements afforded by BCDT have been ascribed to the highly ordered circulation of the gas, liquid, and solid phases; higher rates of gas–liquid mass transfer at lower power inputs (higher kg O_2/kW), etc. Facile solid suspension in BCDT can open totally new applications discussed later in Sections 10.9.7 and 10.9.8.

10.9.2 Hydrodynamic Regimes in TPSRs with Internal Draft Tube

The version with internal draft tube exhibits a well-directed and upward flow of both the gas and liquid phase in the riser portion. At low gas velocities, most of the gas coming out of the riser escapes into the disengaging zone, whereas the liquid over-flows into the annulus. With increasing gas velocity, the liquid circulation rate increases rapidly. The high liquid velocity directed upward in the draft tube sustains the homogeneous bubble flow regime at gas velocities much higher than those at which the transition to heterogeneous regime begins in a conventional bubble column

(Section 10.2.4). At theses high circulation rates, the momentum of the downflowing liquid draws gas bubbles along with it into the annulus. In particular, smaller bubbles having lower rise velocities are preferentially propelled into the annulus and then to the bottom for recirculation through the riser. Increasing V_G leads to concomitant increase in liquid circulation. A stage is now reached where the gas recirculation is extensive and the entire volume of the contactor is effective for gas–liquid contacting (Fig. 3.3). This is an important operating regime for the BCDT, which has the advantage of the homogeneous bubble flow regime (high gas holdup or effective interfacial area).

10.9.2.1 *Liquid Circulation Velocity* (V_{Circ}) Kushalkar and Pangarkar (1994) measured V_{Circ} in 0.2 and 0.4 m dia. for $1.33 < T/T_{DFT} < 2.63$ and $H_D/T_{DFT} > 4$. They found that V_{Circ} attains a maximum at $T/T_{DFT} = 2$. The mixing time for a BCDT was found to be higher than that for a bubble column. Both of these observations are in agreement with Jones (1985) and Chisti and Moo-Young (1987). Mixing in bubble columns is caused solely by dispersion. On the other hand, in a BCDT, a strong liquid circulation is superimposed on the dispersion phenomenon. This results in a typical tracer response, exhibiting a maximum in tracer concentration each time the tracer passes the probe. The amplitude of the maximum decays till it reaches the "mixing cup" concentration. The time interval between two successive peaks is the circulation time from which V_{Circ} can be calculated. Kushalkar's (1993) measurements showed that V_{Circ} increased with V_G, while mixing time decreased.

10.9.3 Gas–Liquid Mass Transfer

10.9.3.1 *Gas Holdup* Chisti and Moo-Young (1987) have examined the extensive literature data on gas holdup in BCDT. They found that the riser to downcomer area ratio had no effect on the gas holdup. For the case of BCDT, these authors rightly argued that comparison with bubble column should be on the superficial gas velocity based on total cross-sectional area. Most investigators have used this norm while reporting data/comparing performance of BCDT with conventional bubble column. This ensures comparison at equal power input per unit volume since $P/V_L = V_G \times \rho_L \times g$. These studies found that the total gas holdup for BCDT and bubble column were the same when compared at equal V_G. Chisti and Moo-Young (1987) argued that the decrease in gas holdup in the draft tube due to high upward-directed liquid circulation velocity is compensated by the gas holdup in the annulus due to downward-directed liquid velocity. Departures from this conclusion were attributed to differences in the extent of separation of the gas–liquid mixture in the headspace above the draft tube.

The driving force for the circulation between the riser and the downcomer is the difference in gas holdups in these sections. Bello et al. (1985) gave Equation 10.23 relating the holdups in the case of a BCDT for air–water system with air sparging in the annulus:

$$[\varepsilon_G]_{Downcomer} = 0.89 \times [\varepsilon_G]_{Riser} \qquad (10.23)$$

The BCDT system gave higher gas holdups than a corresponding external loop (EL) system. This was attributed to a longer connector between the riser and downcomer in the EL system, which allowed escape of substantial amount of gas. For rheologically complex liquids, Chisti et al. (1986) obtained Equation 10.24:

$$[\varepsilon_G]_{\text{Downcomer}} = 0.46 \times [\varepsilon_G]_{\text{Riser}} - 0.024 \qquad (10.24)$$

Effect of the Presence of Solids on Gas Holdup   Several investigators have studied the effect of the presence of solids on the gas holdup. Using activated carbon particles [$0.25 \times 10^{-3} < d_p < 2.2 \times 10^{-3}$ m; $\rho_p = 1600-1900$ kg/m^3; solid loading up to 10 vol.%], Muroyama et al. (1985) found no effect of the presence of solids on the gas holdup. On the other hand, Koide et al. (1985) [for glass spheres, $\rho_p = 2500$ kg/ m^3, $79 < d_p < 198$ μm, and $\rho_p = 4680$ kg/m^3, $d_p = 79$ μm, and for bronze spheres, $\rho_p = 9770$ kg/m^3, and $d_p = 89$ μm] found that the gas holdup decreased by 10–30% in the presence of a solid phase. The decrease was generally found to be correlated with the terminal settling velocity of the solid, V_T. Evidently, high-density, larger-size particles had a negative effect on ε_G as compared to low-density, smaller particles at similar loading. Miyahara and Kawate's (1993) study involved much lower density particles [$1046 < \rho_p < 1135$ kg/m^3 and $2.6 \times 10^{-3} < d_p < 12.68 \times 10^{-3}$ m]. They observed no effect of solids on ε_G for solid holdup < 20 %. For higher solid hold loading, the results were similar to those of Koide et al. (1985). However, no effect of particle size on ε_G was observed. These literature data indicate that conditions that lead to higher solid settling tendency reduce the gas holdup. It is interesting to note that Muroyama et al.'s (1985) measurements of ε_G were carried out under conditions such that solid settling did not occur. Their data showed no effect of the presence of solids. Thus, it may be argued that the reduction in gas holdup observed in some studies could be due to settling of solid particles (Freitas and Teixeira 2001). The settled solid layer partially blocks the clearance between the annulus and riser at the bottom of the reactor. This gives rise to a higher resistance for the circulating liquid. Consequently, the circulation velocity decreases. This decrease is manifest in lower carryover of gas bubbles into the annulus/lower gas holdup in the annulus/lower average gas holdup.

10.9.3.2 Gas–Liquid Mass Transfer

Table 10.8 gives a summary of the literature investigations on gas–liquid mass transfer in BCDT.

It is evident that there are considerable discrepancies in the observations of various investigators. It is also apparent that except for Koide et al. (1983a, b) and Kawase and Moo-Young (1986), all other investigators have used relatively very small column and draft tube diameters. Particularly, the small column diameters data are likely to be suspect. It is interesting to note that these data show a negative dependence of $k_L a$ on T. For these data, the occurrence of slug flow cannot be ruled out even with large recirculation velocities. For instance, Bello et al. (1985) observed that there is no contribution of the downcomer to mass transfer. In their study, the maximum draft tube diameter was ~0.09 m. Additionally, physical

TABLE 10.8 Summary of Literature Investigations on Gas–liquid Mass Transfer in BCDT

Reference	T/T_{DFT} and T	H_D/T_{DFT}	System used	Correlation for $k_L a$	Comments
Koide et al. (1983a). Gas sparged in annulus	1.5–2.12. $T=0.1$, 0.14, 0.22, and 0.3 m.	8.5–25.6	Air–water, aqueous $BaCl_2$, Na_2SO_4, glycerol (35–65 vol.%), and glycol (70 vol.%). Dynamic absorption of oxygen. $0.02 < V_G < 0.2$ m/s (V_G based on total column area)	$$\left(\frac{k_L a \times \sigma_L}{D_A \times g \times \rho_L}\right)^{0.5} = 2.25 \times \left(\frac{\mu_L}{\rho_L D_A}\right)^{0.13} \left(\frac{\rho_L \sigma_L^{\ 3}}{g \mu_L^{\ 4}}\right)^{-0.09} \left(\frac{d_N}{T_{DFT}}\right) (\varepsilon_G)_{Avg}^{1.26}$$	Minor effect of T on $k_L a$ and ε_G. At equal V_G, $k_L a$ and ε_G with draft tube > without draft tube
Koide et al. (1983b). Gas sparged in draft tube	1.5–2.12. $T=0.1$, 0.14, 0.22, and 0.3 m	8.5–25.6	Air–water, aqueous $BaCl_2$, Na_2SO_4, glycerol (35–65 vol.%), and glycol (70 vol.%). Dynamic absorption of oxygen. $0.02 < V_G < 0.2$ m/s (V_G based on total column area)	$$\left(\frac{k_L a T^2}{D_A}\right) = 0.477 \times \left(\frac{\mu_L}{\rho_L D_A}\right)^{0.5} \left(\frac{g \times T^2 \rho_L}{\sigma_L}\right)^{0.84} \left(\frac{g \times T^3 \rho_L^{\ 2}}{\mu_L^{\ 2}}\right)^{0.26} \left(\frac{T_{DFT}}{T}\right)^{-0.54} (\varepsilon_G)_{Avg}^{1.36}$$	$(\varepsilon_G)_{Avg}$ unaffected by T_{DFT} at constant T/T_{DFT}. $k_L a$ increases with increase in T. $[k_L a]_{BCDT}$ > bubble column but < for BCDT with gas sparging in annulus

Muroyama et al. (1985) Gas introduced in draft tube	1.42 and 1.56 for $T=0.1$ m; 1.41 and 1.5 for $T=0.148$ m	5–12.62	Air–water. Dynamic absorption of oxygen. Solid used: activated carbon particles, $0.25 \times 10^{-3} < d_p < 2.2 \times 10^{-3}$ m. $[\rho_p]_{Dry}$: 1680–1933 kg/m³. Solid vol.%: 1.48–5.9. $0.003 < (V_G)_{Riser} < 0.2$ m/s	$\left(\dfrac{k_L a T^2 \rho_L}{\mu_L}\right) = 3.87$ $\left(\dfrac{T V_G \rho_L}{\mu_L}\right)(T)^{1.8}$ $k_L a \propto T^{0.8}$	Correlations for critical gas velocity for solid suspension given. No effect of solids on $k_L a$. $k_L a$ with draft tube > without draft tube for $V_G > 0.03$ m/s
Bello et al. (1985)	1.7–3. $T=0.15$ m	17.4–30.4	Air–tap water, aqueous NaCl. Dynamic absorption/desorption of oxygen	$\left[\dfrac{[k_L a]_{Riser} \times H_D}{V_L}\right] = 2.28 \times \left(\dfrac{V_G}{V_L}\right)^{0.9}$ $\left(1+\dfrac{A_{DFT}}{A_R}\right)^{-1}$	Sparging in annulus. $[k_L a]_{Bubble\ column} >$ $[k_L a]_{BCDT}$ $[k_L a]_{Downcomer}$ found to be negligible
Koide et al. (1985)	1.5–2.12. $T=0.1, 0.14, 0.22,$ and 0.3 m	8.5–25.6	Air–water, aqueous BaCl₂, Na₂SO₄, glycerol (35–65 vol.%), and glycol (70 vol.%). Dynamic absorption of oxygen. $0.02 < V_G < 0.2$ m/s (V_G based on total column area). Glass spheres: $\rho_p = 2500$ kg/m³, $79 < d_p < 198$ μm and 4680 kg/m³, $d_p = 79$ μm. Bronze spheres: $\rho_p = 9770$ kg/m³, $d_p = 89$ μm	$\left(\dfrac{k_L a T^2}{D_A}\right) = 2.66$ $\times \left(\dfrac{\mu_L}{\rho_L D_A}\right)^{0.5} \left(\dfrac{g \times T^2 \rho_L}{\sigma_L}\right)^{0.715}$ $\left(\dfrac{g \times T^3 \rho_L^2}{\mu_L^2}\right)^{0.25}$ $\left(\dfrac{T_{DFT}}{T}\right)^{-0.43} (\varepsilon_G)_{Avg}^{1.34}$	$k_L a$ decreases with increase in solid loading, d_p and ρ_p

(Continued)

TABLE 10.8 (Continued)

Reference	T/T_{DFT} and T	H_D/T_{DFT}	System used	Correlation for $k_L \underline{a}$	Comments
Kawase and Moo-Young (1986b)	1.74. $T = 0.23$ m.	6.36	CO_2 diluted with air–water, aqueous glycerin, CMC of different concentrations. Dynamic absorption of CO_2 following a step change in the inlet gas mixture	$$\left(\frac{k_L \underline{a} \times T^2}{D_A}\right) = 6.8 \times n^{-6.72} \times \left(\frac{TV_G \rho_L}{\mu_L}\right)^{0.38n+0.52} \left(\frac{\mu_L}{\rho_L D_A}\right)^{0.38n+0.14}$$	$k_L \underline{a}$ decreases with increase in T
Goto et al. (1989)	1.3–3.6. $T = 0.042$ and 0.097 m	Not specified	Air–water. Dynamic desorption of oxygen	$k_L \underline{a} = 0.107 \times T^{-1.1} \times V_G^{0.9}$	$k_L \underline{a}$ decreases with increase in T. T lower than the minimum suggested
Wei et al. (2000)	2.66 $T = 0.08$ m	28.6	Air–water, aqueous CMC (1–2 wt.%) for two phase. Air–water–resin for three-phase system. Resin: $300 < d_P < 1200$ μm. $\rho_P \sim 1250$ kg/m^3. Solid loading: 0–9 wt.%	Complex correlation for $$Sh = \left(\frac{k_L \underline{a} T_{DFT}^2}{D_A}\right)$$	$k_L \underline{a}$ decreases with increase in T. T lower than the minimum suggested

Freitas and Teixeira (2001)

12.7–19

1.6–3.18. $T = 0.07$ m

Air–water, aqueous ethanol (1 wt.%). Solid used: calcium alginate, $\rho_P \sim 1023$ and $1048\,\text{kg/m}^3$. $d_P \sim 2.1 \times 10^{-3}$ m

Air–water–low density solids, $\rho_P = 1023\,\text{kg/m}^3$:

$$k_L \underline{a} = \times \left(\begin{array}{c} \left[-0.93 \times (V_G)_{\text{Riser}}^{\,2} \right. \\ + \\ 0.43(V_G)_{\text{Riser}} \\ \left. -6.4 \times 10^{-3}\right] \\ \left(8 \times 10^{-5} \varepsilon_S^{\,2} \right. \\ + \\ 7.9 \times 10^{-4} \\ \left. \times \varepsilon_S + 0.075 \right) \end{array} \right)$$

Air–water–high-density solid, $\rho_P = 1048\,\text{kg/m}^3$:

$$k_L \underline{a} = \left[\begin{array}{c} -0.33 \times (V_G)_{\text{Riser}}^{\,2} \\ +0.43 \times (V_G)_{\text{Riser}} \\ -6.4 \times 10^{-3} \end{array} \right] \times \left(\begin{array}{c} 7.2 \times 10^{-5} \varepsilon_S^{\,2} + \\ 7.9 \times 10^{-4} \varepsilon_S + \\ 7.5 \times 10^{-2} \end{array} \right)$$

$k_L \underline{a}$ decreases with increase in T. T lower than the minimum suggested absorption/desorption was used for measuring $k_L a$. In such measurements, there is a distinct possibility of equilibration of the liquid or a relatively very low driving force for mass transfer. In such cases, serious errors can arise in estimation of $k_L a$.

Koide et al.'s (1983b) results for relatively larger column diameters (particularly for $T=0.22$ and 0.3 m) may be considered more reliable. The following conclusions can be derived:

1. There is no effect of clearance of the draft tube for the range $0.125 < C/T_{DFT} < 1$.
2. For $1.4 < T/T_{DFT} < 2.12$, $k_L a$ decreases marginally with increasing T_{DFT} at constant T.
3. At constant T_{DFT}, $k_L a$ increases with increase in T. The relationship can be approximately described by $k_L a \propto T^{0.5}$. This is in contrast to the small column diameter data, which show a negative dependence on T.
4. For coalescing system (pure water, for instance), the mass transfer performance of BCDT and conventional bubble column is not substantially different.

For noncoalescing systems, BCDT yields better performance. Most liquids contain contaminants that may lead to a noncoalescing behavior. This probably explains the general observation of most investigators that BCDT performs better than conventional bubble columns. (v) The presence of solid has a deleterious effect on $k_L a$. Koide et al. (1985) quantified this negative effect through the terminal settling velocity, V_T, of the solid particles used. They found that $k_L a$ decreased with increasing V_T. The major factors that determine V_T are d_p and ρ_p. An explanation to this observation has been given in Section 10.9.3.1 under Effect of Presence of Solids. The lower $k_L a$ values obtained by various investigators (Table 10.8) could also be due to this effect. However, this must be unambiguously established by careful measurements of $k_L a$ under conditions of complete suspension similar to those used by Muroyama et al. (1985).

10.9.4 Solid Suspension

Effective utilization of a solid catalyst in a three-phase reactor is achieved when all catalyst particles are suspended and actively exchange solutes with the surrounding phase. It has been mentioned earlier in Section 10.9.1 that BCDT provides facile solid suspension at relatively low gas velocities.

Muroyama et al. (1985) (Section 10.9.3.1, Effect of Presence of Solids) reported solid suspension data in the form of critical volumetric solid loading, ε_{SCV} (solid loading at which all the solids are in suspension at a given V_G based on total column cross-sectional area). Plots of ε_{SCV} versus V_G have been given from which the critical gas velocity for complete suspension for a given solid volume fraction can be obtained. Kushalkar (1993) measured axial solid concentration profiles in relatively large column sizes ($T=0.2$ and 0.4 m). Figure.10.5 shows the measured solid holdup profiles for 0.2 m dia. BCDT and corresponding bubble column without the draft tube.

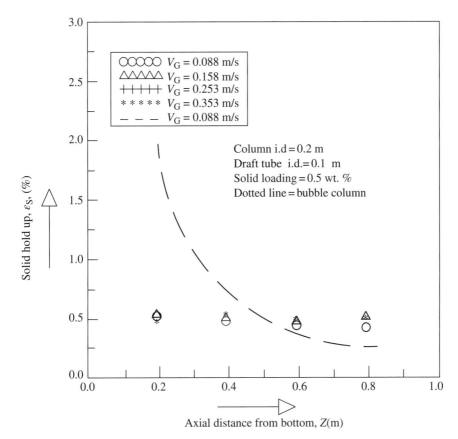

FIGURE 10.5 Comparison of axial solid concentration profiles for BCDT and conventional bubble columns. (Reproduced from Kushalkar 1993 with permission from the author.)

It is evident that BCDT shows no variation in the solid concentration along the column height. The conventional bubble column, on the other hand, shows approximately eightfold variation in the solid holdup over the same height. This is a noteworthy benefit of converting a bubble column into BCDT.

Muroyama et al. (1985) have shown that for a given V_G, the BCDT can sustain three to four times higher solid loadings without settling. This is an important observation, particularly for reactions that have a finite order kinetics with respect to the catalyst (Sections 7A.10, 8.13, and 10.8). It is also important when high-density solid catalysts (e.g., nickel) need to be suspended in a hydrogenation reactor at relatively low hydrogen input. Muroyama et al.'s findings also show that the aforementioned benefits are realized without decrease in gas holdup or gas–liquid mass transfer as long as the solid phase is in a state of suspension. It should be noted here that the maximum solid loading in Muroyama et al.'s study was ~6 vol.%. Systematic studies over a wider range, particularly for high solid loadings (20–40%), are required to confirm this advantage.

10.9.5   Solid–Liquid Mass Transfer Coefficient (K_{SL})

Goto et al. (1989) and Kushalkar and Pangarkar (1994) measured solid–liquid mass transfer coefficient in BCDT. Whereas Goto et al. used relatively very small columns ($T=0.042$ and 0.097 m), Kushalkar and Pangarkar's study used relatively large columns and hence should be considered representative of large-scale applications. A major part of the following discussion is derived from the study of Kushalkar and Pangarkar.

10.9.5.1   Effect of Particle Size, d_p Kushalkar and Pangarkar studied the effect of d_p over a very wide range ($550 < d_p < 2234$ μm). All the data collected related to complete suspension of the particles analogous to Muroyama et al. (1985). Similar to Jadhav and Pangarkar (1991), the overall range of d_p covered could be broadly split into three zones: (1) for $d_p < 1100$ μm, K_{SL} decreased with increasing d_p; (2) for $1100 < d_p < 1500$ μm, K_{SL} increased with d_p; and (3) finally, for $d_p > 1500$ μm, K_{SL} became independent of d_p. The first zone is relevant to three-phase catalytic reactions. For this range, the relationship could be described by $K_{SL} \propto [d_p]^{-n}$ with an average value of $n=0.5$.

10.9.5.2   Effect of T/T_{DFT} Kushalkar and Pangarkar found that K_{SL} attains a maximum at $T/T_{DFT}=2$. This is supported by the measurements of V_{Circ} in their and Koide et al.'s (1988) study. Both found that V_{Circ} also exhibits a maximum in the vicinity of $T/T_{DFT}=2$.

10.9.5.3   Effect of C/T   The C/T ratio was varied from 0.25 to 1. It was found that K_{SL} attained a maximum value at $C/T=0.5$. Kushalkar and Pangarkar argued that for $C/T<0.25$, the entrance area available for the recirculating liquid being too low, a greater resistance to flow is exerted. In addition, the turbulence in the sparger zone is also an impediment to the recirculating flow from the annulus. As a result, V_{Circ} decreases, which is manifest in lower K_{SL}. On the other hand, for $C/T>0.5$, for the configuration used by them, the gas partially escaped into the annular region. The buoyancy of this escaping gas opposes downward liquid circulation in the annulus. The overall effect is lowering of $\Delta\varepsilon_G$ $\{[\varepsilon]_{Riser}$ and $[\varepsilon]_{Downcomer}\}$, which causes lower K_{SL}.

10.9.5.4   Effect of Sparger Design   Two types of spargers were studied: (1) single-point and (2) multipoint sparger. The latter gave higher $\Delta\varepsilon_G$, V_{Circ}, and K_{SL}.

10.9.5.5   Effect of Column Diameter, T Kushalkar and Pangarkar (1994) observed that a twofold increase in column diameter (from $T=0.2$ to 0.4 m) resulted in 10–15% higher K_{SL} values, which is insignificant in relation to the increase in T. Similar observations have been made with respect to $k_L a$ (Table 10.8), particularly for the larger column diameters (Koide et al. 1983b). This finding will be substantiated in the light of the correlation proposed for K_{SL} in the following section.

10.9.6   Correlation for K_{SL}

The effects of various geometric parameters discussed in Sections 10.9.5.1 to 10.9.5.5 clearly point to the important role of $\Delta\varepsilon_G$, V_{Circ}, and, in turn, the turbulence intensity, u', in determining K_{SL}. In view of certain similarities between BCDT and bubble column, Kushalkar and Pangarkar (1994) adopted the approach of Jadhav and Pangarkar (1991), which was found to yield a unique correlation for K_{SL}. This approach needs the value of the turbulence intensity, u', to be used in a correlation of the type given by Equation 10.9. Equation 10.9 has been derived for the case when the net superficial liquid velocity is negligible or zero. As the net transverse liquid velocity increases, the internal liquid circulation in the riser decreases (Jadhav and Pangarkar 1991; Eq. 10.7). At very high net liquid rates, the internal circulation ceases altogether. This net liquid flow must be reckoned in the case of internal (BCDT) or EL reactors. The circulation velocity is decided by the net difference in static heads (or $\Delta\varepsilon_G$) between the two sections. The latter is now decided by the liquid head (H) and not the column diameter, as is the case in bubble columns. Kushalkar and Pangarkar gave the following modified expression (Eq. 10.25) for the turbulence intensity in a BCDT:

$$u' = 0.25 \left\{ g \times H_{DFT} \left[V_G - \varepsilon_G V_{B\infty} - \varepsilon_S V_T \left[\frac{\rho_S - \rho_D}{\rho_D} \right] \right] \right\}^{0.33} \qquad (10.25)$$

The particle Sherwood and Reynolds numbers were defined exactly in the same manner as by Jadhav and Pangarkar (1991). Mass transfer in BCDT also occurs by the same turbulent convective mode as in a bubble column. Therefore, the dependence of K_{SL} on the Schmidt number was assumed to be the same as obtained by Jadhav and Pangarkar. Equation 10.26 was proposed:

$$\left(\frac{K_{SL} d_P}{D_A} \right) = 2 + 0.156 \left(\frac{d_P u' \rho_L}{\mu_L} \right)^{0.48} \left(\frac{\mu_L}{\rho_L D_A} \right)^{0.45} \qquad (10.26)$$

Equation 10.26 incorporates all important parameters that characterize a gas–liquid dispersion such as variation of ε_G with V_G and, in particular, $V_{B\infty}$. It should therefore be applicable to other gas–liquid dispersions, such as for coalescing, noncoalescing, Newtonian/non-Newtonian media, etc. In view of this, Equation 10.26 is recommended for estimation of K_{SL}. The insignificant effect of T on K_{SL} recorded in Section 10.9.5.5 is also obvious from Equations 10.25 and 10.26, which do not contain any term related to the column diameter.

10.9.7   Application of BCDT to Fischer–Tropsch Synthesis

In this section, a configuration consisting of BCDT for the case of the bubble column design detailed in Section 10.8 will be considered. It has been shown that a BCDT gives superior performance with respect to (i) gas holdup, (ii) $k_L a$, (iii) solid concentration profile, (iv) K_{SL}, and (v) less back mixing. As a safe design practice, it is assumed that the mass transfer coefficients ($k_L a$ and K_{SL}) are the same as in Section 10.8. The discussion presented earlier suggests the following optimum configuration: (i) $T/T_{DFT} = 2$ and (ii) $C/T = 0.5$. Koide et al. (1985) have shown that the clear liquid height/dispersion height or height of the draft tube has no effect on ε_G and

$k_L a$. However, K_{SL} depends on H_{DFT}. Fortunately, the solid–liquid mass transfer resistance is a meager 0.31% of the total resistance at 20 wt.% catalyst loading. As discussed later, the catalyst loading can be increased to 30–40 wt.% in BCDT operating at the same V_G as the conventional bubble column. This will provide a relatively higher catalyst surface area. The advantage is apparent, for instance, in the F–T reaction as explained later in this section.

The basis for the calculations is the same as in Section 10.8: 1000 kg mol/h of feed gas or 0.4 m³/s; $H_D/T = 6$; $T = 1.6$ m. The corresponding dispersion height ($H_D/T = 6$) is equal to 9.6 m. The resultant optimum dimensions of the BCDT are $T_{DFT} = 0.8$ m and $C = 0.8$ m. Using the above values, $(V_G)_{Riser} \sim 0.8$ m/s. In Section 10.8, it has been shown that the overall synthesis process is kinetic controlled in conventional bubble column. In the present case of BCDT, it is anticipated that there will be improvements in the rates of the mass transfer steps. Evidently, for the BCDT, the synthesis process will be unambiguously kinetic controlled. In addition, advantages arising from a uniform solid concentration will be achieved. Further, for the aforementioned operating conditions and reactor dimensions, a BCDT can readily sustain a 150–200% higher catalyst loading at the same V_G as used in a conventional bubble column (Section 10.8). This should offer an enhanced (150–200% higher) reactor productivity since the rate of the F–T reaction is directly proportional to the solid loading. For example, if the catalyst loading is increased from 20 to 30 wt.%, $R_{syngas} \times kg_{Cat} = 4.37 \times 1.5 = 6.55$ (gmol/m³ s). The total rate of syngas consumption per second $= 6.55 \times [(\pi/4) \times 1.6^2 \times 9.6] = 126.5$ gmol/s $= 453$ kmol/h. This implies a syngas conversion of $\sim 45\%$ per pass, which is 50% higher than that in Section 10.8. This higher conversion has been achieved by a simple, inexpensive retrofit of the original conventional bubble column. Augmentation of heat transfer can also be achieved by circulating a cooling medium in a hollow draft tube as shown in Figure 3.2b. In conclusion, simple retrofitting of a slurry bubble column by adding a draft tube can yield substantial benefits.

10.9.8   Application of BCDT to Oxidation of *p*-Xylene to Terephthalic Acid

The use of sparged reactors for this application has been discussed at length in Section 3.4.2.4. A major drawback of the conventional sparged reactor is poor suspension and variable axial concentration profile of the solid phase. A special feature of the direct oxidation process is the presence of substantial amount of suspended crude terephthalic acid (varying from 25 to 35 wt.% depending on the feed *p*-xylene concentration, reaction temperature/pressure/type and concentration of the catalyst used, residence time, etc.). It is imperative that this solid phase be kept in suspension to prevent fouling of the internal reactor surface. A conventional bubble column is likely to require relatively very high air velocities for suspending such a large solid population (Section 3.4.2.4). This entails a higher compression power, possibility of oxygen concentration in the exhaust gas exceeding the permissible 5 vol.% safety limit, large tail gas treatment unit, etc. BCDT, on the other hand, affords facile solid suspension along with a uniform axial solid concentration profile (Section 10.9.4). This must be considered as a major benefit accruing from use of BCDT. With this modification, a sparged reactor can be a good alternative to the popular stirred reactors since a BCDT does not have the expensive titanium shaft, impeller, and its sealing and maintenance problems. The importance of uniformity in gas holdup/ $k_L a$ on the purity of terephthalic acid has been discussed in Section 3.4.2.3. Li et al. (2013) have concluded that low dissolved oxygen concentrations are mainly responsible for the formation of undesired products. In this regard, also BCDT appears to be attractive since the gas holdup in the riser and downcomer are not significantly different (Eq. 10.23). However, in view of the relatively high solid loading, systematic studies are required to understand its impact on (i) gas holdup/uniformity of gas holdup (particularly in the annulus), (ii) gas–liquid mass transfer, and (iii) solid suspension under conditions relevant to the Amoco Mid-Century technology. These efforts should be rewarding in the light of advantages described earlier.

NOMENCLATURE

$[A*]$: Interfacial concentration of the solute A (mol/m^3)

$[B_O]$: Concentration of liquid-phase reactant in the bulk liquid phase (mol/m^3)

a: Effective gas–liquid interfacial area (m^2/m^3)

B_W Retarded van der Waals coefficient in Equation 10.1 $(J m)$

B: Liquid-phase reactant

C: Concentration (mol/m^3)

C_T: Transition electrolyte concentration defined by Equation 10.1 (mol/m^3)

C_V: Solid loading (vol.%)

D_A: Diffusivity of solute A in the liquid phase (m^2/s)

D_{Ax}: Axial dispersion coefficient (m^2/s)

d_B: Bubble diameter (m)

D_G: Gas-phase axial dispersion coefficient (m^2/s)

D_L: Axial dispersion coefficient for liquid phase (m^2/s)

d_p: Particle diameter (m or µm)

d_H: Diameter of hole in sparger (m)

F_H: F factor based on gas velocity through the holes in the sparger $(V_H \rho_G^{0.5})$ $[(m/s \, (kg/m^3)^{0.5}]$

Fr_C: Critical Froude number for weeping in sparger $\left(\dfrac{V_H^2 \rho_G}{g d_H (\rho_L - \rho_G)} \right)$ $(—)$

g: Acceleration due to gravity $(9.81 \, m/s^2)$

H: Total column height (m)

Ha: Hatta number, $Ha = \left(\dfrac{\sqrt{D_{AL} k_{p-q} B_O}}{k_L} \right)$ $(—)$

H_D: Dispersion height (m)

H_{DFT}:   Height of the draft tube (m)

H_S: Clear liquid height or static head (m)

J_{DFV}:   Drift flux velocity (m/s)

k_1:   First-order rate constant in Equation 10.19 (1/s)

k_2:   Second-order rate constant (m³/mol s)

k_L: True gas–liquid mass transfer coefficient (m/s)

$k_L a$:   Volumetric gas–liquid mass transfer coefficient (1/s)

K_{SL}:   Particle liquid mass transfer coefficient (m/s)

L_C: Characteristic linear dimension (m)

l: Fractional liquid holdup (—)

M_B: Molecular weight of liquid (kg/kmol)

M_G: Gas-phase momentum based on unit mass of liquid defined in Equation 10.17, $\left(\dfrac{\rho_G V_G}{\rho_L (1 - \varepsilon_G)} \right)$ (m/s)

n: Flow behavior index for power law fluids (—)

P: Pitch of holes in sparger (m)

p: Order of reaction with respect to the gas-phase reactant (—)

Pe_G: Peclet number for the gas phase, $(V_G T/D_G)$ (—)

P_S: Vapor pressure of liquid (Pa or MPa)

P_T: Total system pressure (Pa or MPa)

Q_G: Gas flow rate (m³/s)

q: Order of reaction with respect to the liquid-phase reactant (—)

R: Total rate of absorption rate (mol/s)

R_A: Rate of absorption based on unit gas–liquid interfacial area (mol/m² s)

R_B: Radius of bubble (m)

Re: Reynolds number, $(L_C u_C \rho / \mu)$ (—)

$(Re_G)_H$:   Gas-phase Reynolds number based on velocity through the hole in the sparger and hole diameter (—)

R_G: Ideal gas law constant (8.3145 J/mol K)

R_{Syngas}:   Rate of syngas consumption (gmol/kg catalyst s)

Sc: Schmidt number, $[\mu/\rho D_A]$ (—)

Sh_p: Particle Sherwood number, $(K_{SL} d_p/D_A)$ (—)

STF: Surface tension factor, $(C/z)\,(d\sigma/dC)^2$ (kg² m³/mol s⁴)

T: Tower/column diameter (m)

T_{DFT}:   Diameter of draft tube (m)

u^*:   Friction velocity (m/s)

u':   Turbulence intensity (m/s)

u_C: Characteristic velocity (m/s)

$V_{B\infty}$:   Terminal rise velocity of bubble (m/s)

V_C: Liquid circulation velocity in conventional bubble column (m/s)

V_{Circ}:   Circulation velocity in BCDT (m/s)

V_G^*: $\left(\dfrac{V_G}{\varepsilon_G}\right)$ in Mangartz and Pilhofer's correlation in Table 10.4 (m/s)

V_G: Superficial gas velocity (m/s)

V_{GCrit}: Superficial gas velocity at which transition to heterogeneous regime begins (m/s)

V_H: Gas velocity through the holes in the sparger (m/s)

V_L: Superficial liquid velocity (m/s)

$\Delta(V_G)_{lm}$:   Log mean gas velocity, $\left(\dfrac{(V_G)_{In}-(V_G)_{Out}}{\ln((V_G)_{In}/(V_G)_{Out})}\right)$ (m/s)

z: Number of ions formed by dissociation of an electrolyte molecule (—)

Z: Vertical height from the bottom of the column (m)

Greek letters

ε_M: Power input per unit mass, P/m (W/kg)

ε_{SCV}: Critical volume fraction of solids at suspended condition (—)

Θ: Temperature (°K)

γ_{Eff}:   Effective shear rate in non-Newtonian systems (1/s)

δ_F: Liquid film thickness between coalescing bubbles (m)

ε_G: Fractional gas holdup (—)

ε_{GCrit}:   Gas holdup at which transition to heterogeneous regime begins (—)

ε_L: Fractional liquid holdup (—)

ε_S: Fractional holdup of solid phase (—)

μ_{eff}: Effective liquid-phase viscosity for non-Newtonian liquid (Pa s)

μ, μ_L: Liquid viscosity (Pa s) or (mPa s)

μ_W: Viscosity of water (Pa s)

ν: Kinematic viscosity (m²/s)

ρ_G: Gas density (kg/m³)

ρ_L: Liquid density (kg/m³)

ρ_P, ρ_S: Particle or solid density (kg/m³)

σ, σ_L: Surface tension of liquid phase (N/m)

υ_A: Molar volume of solute (m³/mol)

Subscripts

A: Solute A

Avg: Average property

B: Reactant B
G: Gas phase
L: Liquid phase
S, P: Solid or particle phase

Abbreviations

BCDT: Bubble column with draft tube
DFT: Draft tube
TPSR: Three phase sparged reactor

REFERENCES

Ajbar A, Al-Masry W, Ali E. (2009) Prediction of flow regimes transitions in bubble columns using passive acoustic measurements. Chem. Eng. Process., 48:101–110.

Akita K, Yoshida F. (1973) Gas hold-up and volumetric mass transfer coefficient in bubble columns. Ind. Eng. Chem. Proc. Dev. Des., 12:76–80.

Al-Masry W, Ali E, Aqeel Y. (2005) Determination of bubble characteristics in bubble columns using statistical analysis of acoustic sound measurements. Chem. Eng. Res. Des., 83:1196–1207.

Al-Masry W, Ali E, Aqeel Y. (2006) Effect of surfactant solutions on bubble characteristics in bubble columns based on acoustic sound measurement. Chem. Eng. Sci., 61:3610–3622.

Al-Masry W, Ali E. (2007) Identification of hydrodynamics characteristics in bubble columns through analysis of acoustic sound measurements—influence of the coalescence behavior of the liquid phase. Chem. Eng. Process., 46:127–138.

Bach HF, Pilhofer Th. (1978) Variation of gas hold-up in bubble columns with physical properties of liquids and operating parameters of columns. Ger. Chem. Eng., 1:270–275.

Bakshi BR, Zhong H, Jiang P, Fan L-S. (1995) Analysis of flow in gas–liquid bubble columns using multiresolution methods. Chem. Eng. Res. Des., 73(A6):608–614.

Barnea E, Mizrahi J. (1973) A generalized approach to the fluid dynamics of particulate systems: Part 1. General correlation for fluidization and sedimentation in solid multiparticle systems. Chem. Eng. J., 5(2):171–189.

Behkish A, Men Z, Inga JR, Morsi BI. (2002) Mass transfer characteristics in a large-scale slurry bubble column reactor with organic liquid mixtures. Chem. Eng. Sci., 57:3307–3324.

Behkish A, Lemoine R, Oukaci R, Morsi BI. (2006) Novel correlations for gas holdup in large-scale slurry bubble column reactors operating under elevated pressures and temperatures. Chem. Eng. J., 115:157–171.

Behkish A, Lemoine R, Sehabiague L, Oukaci R, Morsi BI. (2007) Gas holdup and bubble size behavior in a large-scale slurry bubble column reactor operating with an organic liquid under elevated pressures and temperatures. Chem. Eng. J., 128:69–84.

Bello RA, Robinson CW, Moo-Young M. (1985) Gas holdup and overall volumetric oxygen transfer coefficient in airlift contactors. Biotechnol. Bioeng., 27:369–381.

Briens LA, Briens CL, Margaritis A, Hay J. (1997) Minimum liquid fluidization velocity in gas–liquid–solid fluidized bed of low-density particles. Chem. Eng. Sci., 52:4231–4238.

Bucholz H, Bucholz R, Niebeschultz H, Schugerl K. (1978) Absorption of oxygen in highly viscous Newtonian and non-Newtonian fermentation model media in bubble column reactors. Europ. J. Appl. Microbiol. Biotechnol., 6:115–126.

Bennett MA, West RM, Luke SP, Jia X, Williams RA. (1999) Measurement and analysis of flows in gas–liquid column reactor. Chem. Eng. Sci., 54:5003–5012.

Cassanello M, Larachi F, Kemoun A, Al-Dahhan MH, Dudukovic MP. (2001) Inferring liquid chaotic dynamics in bubble columns. Chem. Eng. Sci., 56:6125–6134.

Chaumat H, Billet-Duquenne AM, Augier F, Mathieu C, Delmas H. (2005) Mass transfer in bubble column for industrial conditions—effects of organic medium, gas and liquid flow rates and column design. Chem. Eng. Sci., 60:5930–5936.

Chen RC, Reese J, Fan L-S. (1994) Flow structure in a three-dimensional bubble column and three-phase fluidized bed. AICHEJ, 40:1093–1104.

Chilekar VP, Singh C, van der Schaaf J, Kuster BFM, Schouten JC. (2007) A gas hold-up model for slurry bubble columns. AICHEJ, 53:1687–1702.

Chilekar VP, Singh C, van der Schaaf J, Kuster BFM, Schouten JC. (2010) Influence of elevated pressure and particle lyophobicity on hydrodynamics and gas–liquid mass transfer in slurry bubble columns. AICHEJ, 56:584–596.

Chisti MY, Moo-Young M. (1987) Airlift reactors: characteristics, applications and design considerations. Chem. Eng. Commun., 60:195–242.

Chisti MY, Halard B, Moo-Young M. (1988) Liquid circulation in airlift reactors. Chem. Eng. Sci., 43:451–457.

Cockx A, Rouston M, Line A, Herbrard G. (1995) Modeling of mass transfer coefficient kL in bubble columns. Trans. Ins. Chem. Eng. UK, 73(A):627–631.

Deckwer W-D, Graeser U, Langemann H, Serpemen Y. (1973a) Zones of different mixing in the liquid phase of bubble columns. Chem. Eng. Sci., 28:1223–1225.

Deckwer W-D, Burckhart R, Zoll G. (1974) Mixing and mass transfer in tall bubble columns. Chem. Eng. Sci., 29:2177–2188.

Deckwer W-D, Louisi Y, Zaidi A, Ralek M. (1980) Hydrodynamics of the Fischer–Tropsch slurry process. Ind. Eng. Chem. Process. Des. Dev., 19:699–708.

Deckwer W-D, Serpemen Y, Ralek M, Schmidt B. (1981) On the relevance of mass transfer limitations in the Fischer–Tropsch slurry process. Chem. Eng. Sci., 36:765–771.

Deckwer W-D. (1992) Bubble column reactors (English translation). John Wiley & Sons, Chichester, UK.

Deckwer W-D, Schumpe A. (1993) Improved tools for bubble column reactor design and scale-up. Chem. Eng. Sci., 48:889–911.

Dong F, Liu X, Deng X, Xu L, Xu L-A. (2001) Identification of two-phase flow regime using electrical resistance tomography. In: Proceedings of Second World Congress on Industrial Process Tomography, Hannover, Germany. cf: Shaikh and Al-Dahhan, 2005.

Drahoš J, Zahradnik J, Puncochar M, Fialova M, Bradka F. (1991) Effect of operating conditions on the characteristics of pressure fluctuations in a bubble column. Chem. Eng. Process., 29:107–115.

Dry ME. (1990) The Fischer–Tropsch process—commercial aspects. Cat. Today, 6:183–206.

Dry ME. (2001) High quality diesel via the Fischer–Tropsch process—a review. J. Chem. Technol. Biotechnol, 77:43–50.

Dry ME. (2002) The Fischer–Tropsch process: 1950–2000. Cat. Today, 71:227–241.

Ellis N, Bi HT, Lim CJ, Grace JR. (2004) Influence of probe scale and analysis method on measured hydrodynamic properties of gas-fluidized beds. Chem. Eng. Sci., 59:1841–1851.

Fan LS, Ramesh TS, Tang WT, Long TR. (1987) Gas–liquid mass transfer in a two-stage draft tube gas–liquid–solid fluidized bed. Chem. Eng. Sci., 42:543–553.

Freitas C, Teixeira JA. (2001) Oxygen mass transfer in a high solids loading three-phase internal-loop airlift reactor. Chem. Eng. J., 84:57–61.

Espinoza R, Du Toit E, Santamaria J, Menendez M, Coronas J, Irusta S. (2000) Use of membranes in Fischer–Tropsch reactors. Stud. Surf. Sci. Catal. 130A (International Congress on Catalysis, 2000, Pt. A):389–394.

Field RW, Davidson JF. (1980) Axial dispersion in bubble columns. Trans. Ins. Chem. Eng. UK, 58:228–236.

Fraguio MS, Casanello MC, Degaleesan S, Dudukovic MP. (2009) Flow regime diagnosis in bubble columns via pressure fluctuations and computer-assisted radioactive particle tracking measurements. Ind. Eng. Chem. Res., 48:1072–1080.

Gandhi B, Prakash A, Bergougnou MA. (1999). Hydrodynamic behaviour of slurry bubble column at high solids concentrations. Powder Tech., 103:80–94.

Garcia-Calvo E, Leton P. (1994) Prediction of fluid dynamics and liquid mixing in bubble columns. Chem. Eng. Sci., 49:3643–3649.

Garcia-Ochoa J, Khalfet R, Poncin S, Wild G. (1997) Hydrodynamics and mass transfer in a suspended solid bubble column with polydispersed high density particles. Chem. Eng. Sci., 52:3827–3834.

Goto S, Matsumoto Y, Gaspillo P. (1989) Mass transfer and reaction in bubble column slurry reactor with draft tube. Chem. Eng. Commun., B5:181–191.

Gourich B, Vial C, Essadki AH, Allam F, Soulami MB, Ziyad M. (2006) Identification of flow regimes and transition points in a bubble column through analysis of differential pressure signal—influence of the coalescence behavior of the liquid phase. Chem. Eng. Process., 45:214–223.

Han L, Al-Dahhan MH. (2007) Gas–liquid mass transfer in a high pressure bubble column reactor with different sparger designs. Chem. Eng. Sci., 62:131–139.

Hikita H, Asai S, Tanigawa K, Segawa K, Kitao M. (1980) Gas hold-up in bubble columns. Chem. Eng. J., 20:59–67.

Hikita H, Asai S, Tanigawa K, Segawa K, Kitao M. (1981) The volumetric liquid phase mass transfer coefficient in bubble column. Chem. Eng. J., 22:61–69.

Holler V, Ruzicka M, Drahoš J, Kiwi-Minsker L, Renken A. (2003) Acoustic and visual study of bubble formation processes in bubble columns staged with fibrous catalytic layers. Cat. Today, 79–80:151–157.

Hur YG, Yang JH, Jung H, Park SB. (2013) Origin of regime transition to turbulent flow in bubble column: orifice and column-induced transitions. Int. J. Multiphase Flow, 50:89–97.

Inga JR, Morsi BI. (1997) Effect of catalyst loading on gas/liquid mass transfer in a slurry reactor: a statistical experimental approach. Can. J. Chem. Eng., 75:872–881.

Inga JR, Morsi BI. (1999) Effect of operating variables on the gas holdup in a large-scale slurry bubble column reactor operating with an organic liquid mixture. Ind. Eng. Chem. Res., 38:928–937.

Ismail I, Shafquet A, Karsiti MN. (2011) Application of electrical capacitance tomography and differential pressure measurement in an air–water bubble column for online analysis of void fraction. In: Proceedings of Fourth International Conference on Modeling, Simulation and Applied Optimization, ICMSAO 2011, art. no. 5775613.

Jadhav SV, Pangarkar VG. (1988) Particle–liquid mass transfer in three phase sparged reactors. Can. J. Chem. Eng., 66:572–578.

Jadhav SV, Pangarkar VG. (1991a) Particle–liquid mass transfer in three phase sparged reactors: scale up effects. Chem. Eng. Sci., 46:919–927.

Jadhav SV, Pangarkar VG. (1991b) Particle–liquid mass transfer in mechanically agitated contactors. Ind. Eng. Chem. Res., 30:2496–2503.

Jin H, Qin Y-J, Yang S-H, He G-X, Guo Z-W. (2013) Radial profiles of gas bubble behavior in a gas–liquid bubble column reactor under elevated pressures. Chem. Eng. Technol., 36:1721–1728.

Jones AG. (1985) Liquid circulation in a draft-tube bubble column. Chem. Eng. Sci., 40:449–462.

Jordan U, Terasaka K, Kundu G, Schumpe A. (2002) Mass transfer in high-pressure bubble columns with organic liquids. Chem. Eng. Technol., 25:262–265.

Joshi JB, Sharma MM. (1979) A circulation cell model for bubble columns. Trans. Insn. Chem. Eng., 57:244–251.

Kantak MV, Hesketh RP, Kelkar BG. (1995) Effect of gas and liquid properties on gas phase dispersion in bubble columns, Chem. Eng. J., 59:91–100.

Kantarci N, Borak F, Ulgon KO. (2005) Bubble column reactors. Process Biochem., 40:2263–2283.

Kawase Y, Moo-Young M. (1986a) Liquid phase mixing in bubble columns with Newtonian and non-Newtonian fluids. Chem. Eng. Sci., 41:1966–1977.

Kawase Y, Moo-Young M. (1986b) Influence of non-Newtonian flow behavior on mass transfer in bubble columns with and without draft tubes. Chem. Eng. Commun., 40(1–6):67–83.

Kawase Y, Moo-Young M. (1990) Mathematical models for design of bioreactors: applications of Kolmogorov's theory of isotropic turbulence. Chem. Eng. J., 45:B19–B41.

Koide K, Sato H, Iwamoto S. (1983a) Gas holdup and volumetric liquid-phase mass transfer coefficient in bubble column with draught tube and with gas dispersion into annulus. J. Chem. Eng. Jpn., 16:407–413.

Koide K, Kurematsu K, Iwamoto S, Iwata Y, Horibe K. (1983b) Gas holdup and volumetric liquid-phase mass transfer coefficient in bubble column with draught tube and with gas dispersion into tube. J. Chem. Eng. Jpn., 16:413–419.

Koide K, Horribe K, Kawabata H, Ito S. (1985) Gas holdup and volumetric liquid-phase mass transfer coefficient in solid-suspended bubble column with draught tube. J. Chem. Eng. Jpn., 18:248–254.

Koide K, Kimura M, Nitta H, Kawabata H. (1988) Liquid circulation in bubble column with draught tube. J. Chem. Eng. Jpn., 21:393–399.

Kojima H, Uchida Y, Ohsawa T, Iguchi A. (1987) Volumetric liquid phase mass transfer coefficient in gas sparged three phase stirred vessel. J. Chem. Eng. Jpn., 20:104–106.

Krishna R, Wilkinson PM, van Dierendonck LL. (1991) A model for gas holdup in bubble columns incorporating the influence of gas density on flow regime transitions. Chem. Eng. Sci., 46:2491–2496.

Krishna R, de Swart JWA, Hennephof DE, Ellenberger J, Hoefsloot HCJ. (1994). Influence of increased gas density on the hydrodynamics of bubble column reactors. AICHEJ, 40:112–119.

Krishna R, de Swart JWA, Ellenberger J, Martina GB, Maretto C. (1997) Gas holdup in slurry bubble columns: effect of column diameter and slurry concentrations. AICHEJ, 43:311–316.

Krishna R, Urseanu MI, Dreher AJ. (2000a). Gas hold-up in bubble columns: influence of alcohol addition versus operation at elevated pressures. Chem. Eng. Process., 39:371–378.

Krishna R, Sie ST. (2000b) Design and scale-up of the Fischer–Tropsch bubble column slurry reactor. Fuel Process. Technol., 64:73–105.

Kuncová G, Zahradnik J. (1995) Gas holdup and bubble frequency in a bubble column reactor containing viscous saccharose solutions. Chem. Eng. Process., 34:25–34.

Kushalkar KB. (1993) Mass transfer in multiphase contactors. Ph.D. (Tech.) Thesis, University of Bombay, India.

Kushalkar KB, Pangarkar VG. (1994) Particle–liquid mass transfer in a bubble column with a draft tube. Chem. Eng. Sci., 49:139–144.

Lau R, Peng W, Velazquez-Vargas LG, Yang GQ, Fan LS. (2004) Gas–liquid mass transfer in high pressure bubble columns. Ind. Eng. Chem. Res., 43:1302–1311.

Letzel HM, Schouten JC, Krishna R, van den Bleek CM. (1997) Characterization of regimes and regime transitions in bubble columns by chaos analysis of pressure signals. Chem. Eng. Sci., 52:4447–4459.

Letzel HM, Schouten JC, van den Bleek CM, Krishna R. (1998) Influence of gas density on the large-bubble holdup in bubble column reactors. AICHEJ, 44:2333–2336.

Letzel HM, Schouten JC, Krishna R, van den Bleek CM. (1999) Gas holdup and mass transfer in bubble column reactors operated at elevated pressure. Chem. Eng. Sci., 54:2237–2246.

Li H, Prakash A. (2000) Influence of slurry concentrations on bubble population and their rise velocities in three-phase slurry bubble column. Powder Tech., 113:158–167.

Lin T-J, Reese J, Hong T, Fan L-S. (1996) Quantitative analysis and computation of two-dimensional bubble columns. AICHEJ, 42:301–318.

Luo X, Lee DJ, Lau R, Yang G, Fan L-S. (1999) Maximum stable bubble size and gas hold up in high pressure slurry bubble columns. AICHEJ, 45:665–680.

Mangartz KH, Pilhofer Th. (1980) Untersuchungen zur Gasphasendispersion in Blasensaulenreaktoren. Verfahrenstechnik (Mainz) 14, 40–46. *cf*: Deckwer W-D, Schumpe A. (1993).

Mangartz KH, Pilhofer Th. (1981) Interpretation of mass transfer measurements in bubble columns considering dispersion of both phases. Chem. Eng. Sci., 36:1069–1077.

Manasseh R, Lafontaine RF, Davy J, Shepherd IC, Zhu Y. (2001) Passive acoustic bubble sizing in sparged systems. Exp. Fluids, 30:672–682.

Marano JJ, Holder GD. (1997) Characterization of Fischer–Tropsch liquids for vapor–liquid equilibria calculations. Fluid Phase Equilibria, 138:1–21.

Maretto C, Krishna R. (1999) Modeling of a bubble column slurry reactor for Fischer–Tropsch synthesis, Cat. Today, 52:279–289.

Marrucci GA. (1969) A theory of coalescence. Chem. Eng. Sci., 24:975–985.

Mena PC, Ruzicka MC, Rocha FA, Teixeira JA, Drahoš J. (2005) Effect of solids on homogeneous–heterogeneous flow regime transition in bubble columns. Chem. Eng. Sci., 60:6013–6026.

Mena PC, Rocha FA, Teixeira JA, Sechet P, Cartellier A. (2008) Measurement of gas phase characteristics using a monofibre optical probe in a three-phase flow. Chem. Eng. Sci., 63:4100–4115.

Mena A, Ferreira A, Teixeira JA, Rocha F. (2011) Effect of some solid properties on gas–liquid mass transfer in a bubble column. Chem. Eng. Process., 50:181–188.

Mersmann A. (1978) Design an scale-up of bubble and spray column, Ger. Chem. Eng. 1:1–11.

Miyahara T, Kawate O. (1993) Hydrodynamics of a solid-suspended bubble column with a draught tube containing low-density particles. Chem. Eng. Sci., 48:127–133.

Miyauchi T, Furusaka S, Mooroka S, Ikeda Y. (1981) In: Advances in Chem. Eng., Drew TB, editor. Transport phenomena and reaction in fluidized beds. Vol. 11. Academic Press, New York, USA. p 276–448.

Moshtari B, Moghaddas JS, Gangi JS. (2007) A hydrodynamic experimental study of bubble columns. Stud. Surf. Sci. Catal., 167:67–72.

Mudde RF, van den Akker HEA. (1999). Dynamic behaviour of the flow field of a bubble column at low to moderate gas fractions. Chem. Eng. Sci., 54:4921–4927.

Muroyama K, Mitani Y, Yasunishi A. (1985) Hydrodynamic characteristics and gas–liquid mass transfer in a draft tube slurry reactor. Chem. Eng. Commun., 34:87–98.

Murugaian V, Schaberg HI, Wang M. (2005) Effect of sparger geometry on the mechanism of flow pattern transition in a bubble column. In: Proceedings of FourthWorld Congress on Industrial Process Tomography, Aizu, Japan.

Nedeltchev S, Kumar SB, Dudukovic MP. (2003) Flow regime identification in a bubble column based on both Kolmogorov entropy and quality of mixedness derived from CARPT data. Can. J. Chem. Eng., 81:367–374.

Nedeltchev S, Shaikh A, Al-Dahhan M. (2011) Flow regime identification in a bubble column via nuclear gauge densitometry and chaos analysis. Chem. Eng. Technol., 34:225–233.

Nikhade BP, Pangarkar VG. (2007) A theorem of corresponding hydrodynamic states for estimation of transport properties: case study of mass transfer coefficient in stirred tank fitted with helical coil. Ind. Eng. Chem. Res., 46:3095–3100.

Olmos E, Gentric C, Poncin S, Midoux N. (2003) Description of flow regime transitions in bubble columns via laser Doppler anemometry signals processing. Chem. Eng. Sci., 58:1731–1742.

Ozturk SS, Schumpe A, Deckwer W-D. (1987) Organic liquids in a bubble column: holdups and mass transfer coefficients. AICHEJ, 33:1473–1480.

Prince RGH. (1960) Characteristics and design of perforated plate columns, In: Proceedings. of International Symposium on Distillation. Institution of Chemical Engineers, London., UK, Brighton, 4–6 May 1960, p 177–184.

Pangarkar VG, Yawalkar AA, Sharma MM, Beenackers AACM. (2002) Particle–liquid mass transfer coefficient in two-/three-phase stirred tank reactors. Ind. Eng. Chem. Res., 41:4141–4167.

Prince MJ, Blanch HW. (1990) Transition electrolyte concentrations for bubble coalescence. AICHEJ, 36:1425–1429.

Prokop A, Eriksson LE, Fernandez J, Humphrey AE. (1969) Design and physical characteristics of a multistage, continuous tower fermentor. Biotechnol. Bioeng., 2:945–966.

Rados N, Al-Dahhan MH, Dudukovic MP. (2003) Modeling of the Fischer–Tropsch synthesis in slurry bubble column reactors. Cat. Today, 79–80:211–218.

Rafique PC, Dudukovic M. (2004) Computational modeling of gas–liquid flow in bubble columns. Revs. Chem. Eng., 20(3–4):225–375.

Reilly IG, Scott DS, De Bruijn TJW, MacIntyre D. (1994) The role of gas phase momentum in determining gas hold-up and hydrodynamic flow regimes in bubble column operations. Can. J. Chem. Eng., 72:3–12.

Ribeiro CP Jr. (2008) On the estimation of the regime transition point in bubble columns. Chem. Eng. J., 140:473–482.

Rollbusch P, Tuinier M, Becker M, Ludwig M, Grünewald M, Franke R. (2013) Hydrodynamics of high-pressure bubble columns. Chem. Eng. Technol., 36:1603–1607.

Ruzicka MC, Zahradnik J, Drahoš, J, Thomas NH (2001a) Homogenous–heterogeneous regime transition in bubble columns. Chem. Eng. Sci., 56:4609–4626.

Ruzicka MC, Drahoš, J, Fialova M, Thomas NH. (2001b) Effect of bubble column dimensions on flow regime transition. Chem. Eng. Sci., 56:6117–6124.

Ruzicka MC, Drahoš, J, Mena PC, Teixeira JA. (2003). Effect of viscosity on homogeneous–heterogeneous flow regime transition in bubble columns. Chem. Eng. J., 96:15–22.

Ruzicka MC, Vecer MM, Orvalho S, Drahoš J. (2008) Effect of surfactant on homogeneous regime stability in bubble column. Chem. Eng. Sci., 63:951–967.

Ruzicka MC. (2013) On stability of a bubble column. Chem. Eng. Res. Des., 91:191–203.

Seno T. (1990) Mass transfer in co-current and counter current bubble columns. Chem. Eng. Technol.,13:133–118.

Thorat BN, Kulkarni AV, Joshi JB (2001) Design of sieve plate spargers for bubble columns: role of weeping. Chem. Eng. Technol., 24:841–828.

Treybal RE. (1981) Mass transfer operations. 3rd ed., McGraw-Hill International editions, McGraw-Hill Book Co., Singapore.

Sada E, Kumazawa H, Lee C, Iguchi T. (1986a) Gas holdup and mass transfer characteristics in a three-phase bubble column. Ind. Eng. Chem. Process Des. Develop., 25:472–476.

Sada E, Kumazawa H, Lee C. (1986b) Influence of suspended fine particles on gas holdup and mass-transfer characteristics in a slurry bubble column. AICHEJ, 32:853–856.

Sanger P, Deckwer W-D. (1980) Liquid–solid mass transfer in aerated suspensions. Chem. Eng. J., 22:179–186.

Sano Y, Yamaguchi N, Adachi T. (1974) Mass transfer coefficients for suspended particles in agitated vessels and bubble columns. J. Chem. Eng. Jpn. 7:255–261.

Saxena SC, Rao NS, Thimmapuram PR. (1992) Gas phase holdup in slurry bubble columns for two- and three-phase systems. Chem. Eng. J., 49:151–159.

Saxena SC, Rao NS. (1993) Estimation of gas holdup in a slurry bubble column with internals: nitrogen–Therminol–magnetite system. Powder Tech., 75:153–158.

Saxena SC, Chen ZD. (1994) Hydrodynamics and heat transfer of baffled and unbaffled slurry bubble columns. Revs. Chem. Eng., 10:193–400.

Schumpe A, Deckwer W-D. (1987) Viscous media in tower bioreactors: hydrodynamic characteristics and mass transfer properties. Bioprocess Eng., 2:79–94.

Shaikh A, Al-Dahhan M. (2005) Characterization of the hydrodynamic flow regime in bubble columns via computed tomography. Flow Measurement and Instrumentation, 16:91–98.

Shaikh A, Al-Dahhan M. (2007) A review on flow regime transition in bubble columns. Int. J. Chem. React. Eng., 5(Review R1):1–68.

Shaikh A, Al-Dahhan M. (2010) A new methodology for hydrodynamic similarity in bubble columns. Can. J. Chem. Eng., 88:503–517.

Shaikh A, Al-Dahhan M. (2013a) A new method for online flow regime monitoring in bubble column reactors via nuclear gauge densitometry. Chem. Eng. Sci., 89:120–132.

Shaikh A, Al-Dahhan M. (2013b) Scale-up of bubble column reactors: a review of current state-of-the-art. Ind. Eng. Chem. Res. DOI: 10.1021/ie302080m, Publication Date (Web): 28 Mar 2013.

Shah YT, Kelkar BG, Godbole SP, Deckwer W-D. (1982) Design parameters estimations for bubble column reactors, AICHEJ, 28:353–379.

Sharma MM, Chandrasekaran K. (1978) Absorption of oxygen in aqueous solutions of sodium sulphide. Chem. Eng. Sci., 33:1294–1295.

Sharma MM, Chandrasekaran K. (1978) On gas absorption with autocatalytic reaction. Chem. Eng. Sci., 33:1294–1295.

Shiea M, Mostoufi N, Sotudeh-Gharebagh R. (2013) Comprehensive study of regime transitions throughout a bubble column using resistivity probe. Chem. Eng. Sci., 100:15–22.

Su Y, Wang Y, Zeng Q, Li J-H, Yu G, Gong X, Yu Z. (2008) Influence of liquid properties on flow regime and back mixing in a special bubble column. Chem. Eng. Process., 47:2296–2302.

Tapp HS, Peyton AJ, Kemsley EK, Wilson RH. (2003) Chemical engineering applications of electrical process tomography. Sensors and Actuators B 92:17–24.

Thet MK, Wang C-H, Tan RBH. (2006) Experimental studies of hydrodynamics and regime transition in bubble columns. Can. J. Chem. Eng., 84:63–72.

Thorat BN, Shevade AV, Bhilegaonkar KN, Aglave AV, Parasuveera U, Thakre SS, Pandit AB, Sawant SB, Joshi JB. (1998) Effect of sparger design and height to diameter ratio on fractional gas hold-up in bubble columns. Chem. Eng. Res. Des., 76(A):823–833.

Towell GD, Ackermann GH. (1972) Axial mixing of liquid and gas in large bubble reactors. In: Proceedings of the Fifth European—Second International Symposium on Chemical Reaction Engineering. Vol. B3-1. Elsevier, Amsterdam, the Netherlands.

Urseanu MI, Guit RPM, Stankiewicz A, van Kranenburg G, Lommen JHGM. (2003) Influence of operating pressure on the gas hold-up in bubble columns for high viscous media. Chem. Eng. Sci., 58:697–704.

van der Schaaf J, Chilekar VP, van Ommen JR, Kuster BFM, Tinge JT, Schouten JC. (2007) Effect of particle lyophobicity in slurry bubble columns at elevated pressures. Chem. Eng. Sci., 62:5533–5537.

Vandu CO, Krishna R. (2004) Volumetric mass transfer coefficients in slurry bubble columns operating in the churn-turbulent flow regime. Chem. Eng. Process., 43:987–995.

Vazquez A, Manasseh R, Sánchez RM, Metcalfe G. (2008) Experimental comparison between acoustic and pressure signals from a bubbling flow. Chem. Eng. Sci., 63:5860–5869.

Vial C, Poncin S, Wild G, Midoux N. (2001) A simple method for regime identification and flow characterization in bubble columns and airlift reactors. Chem. Eng. Process., 40:135–151.

Wachi S, Nojima Y. (1990) Gas-phase dispersion in bubble columns. Chem. Eng. Sci., 45:901–905.

Wallis GB. (1962) A simplified one-dimensional representation of two-component vertical flow and its application to batch sedimentation. In: Proceedings of the Symposium on the Interaction between Fluids and Particles, London, June 20–22, 1962. Institute of Chemical Engineers, London. p 9–16.

Wallis GB. (1969) One dimensional two phase flow. McGraw Hill, New York.

Whalley PB, Davidson JF. (1974) Liquid circulation in bubble columns. In: Proceedings of the Symposium on Two Phase Flow Systems, Inst. Chem. Eng. Symp. Series, vol. 38, Paper J5, p 1–29.

Wilkinson PM, van Dierendonck LL. (1990) Pressure and gas density effects on bubble breakup and gas holdup in bubble columns. Chem. Eng. Sci., 45:2309–2315.

Wilkinson PM, Spek AP, van Dierendonck LL. (1992). Design parameters estimation for scale up of high-pressure bubble columns. AICHEJ, 38:544–554.

Woo K-J, Kang S-H, Kim S-M, Bae J-W, Jun K-W. (2010) Performance of a slurry bubble column reactor for Fischer–Tropsch synthesis: determination of optimum condition. Fuel Proc. Tech., 91:434–439.

Yates IC, Satterfield CN. (1991) Intrinsic kinetics of the Fischer–Tropsch synthesis on a cobalt catalyst. Energy Fuels, 5:168–173.

Yawalkar AA, Heesink ABM, Versteeg GF, Pangarkar VG. (2002a) Gas holdup in stirred tank reactors in the presence of inorganic electrolytes. Can. J. Chem. Eng., 80:791–799.

Yawalkar AA, Heesink ABM, Versteeg GF, Pangarkar VG. (2002b) Gas–liquid mass transfer coefficient in stirred tank reactor. Can. J. Chem. Eng., 80:840–848.

Zahradnik J, Drapal L, Kaštánek F, Reznikova J. (1992) Hydrodynamic and mass transfer characteristics of sectionalized aerated slurry reactors. Chem. Eng. Process., 31:263–272.

Zahradnik J, Fialová M, Ruzicka M, Drahoš J, Kaštánek F, Thomas NH. (1997) Duality of the gas–liquid flow regimes in bubble column reactors. Chem. Eng. Sci., 52:3811–3826.

Zehner P, Kraume M. (2000) Bubble columns. Ullmann's Encyclopedia of Industrial Chemistry.

Zhang K. (2002) Axial solid concentration distribution in tapered and cylindrical bubble columns. Chem. Eng. J., 86:299–307.

Zuber N, Findlay JA. (1965) Average volume concentration in two phase flow systems. J. Heat Transfer, 87:453–468.

INDEX

Design of Multiphase Reactors, First Edition. Vishwas Govind Pangarkar.
© 2015 John Wiley & Sons, Inc. Published 2015 by John Wiley & Sons, Inc.